LANGUAGE, PROOF AND LOGIC

LANGUAGE, PROOF AND LOGIC

2nd Edition

DAVE BARKER-PLUMMER

JON BARWISE & JOHN ETCHEMENDY

In collaboration with

Albert Liu

Michael Murray

Emma Pease

CSLI Publications
Center for the Study of
Language and Information
Stanford, California

Copyright © 1999, 2000, 2002, 2003, 2007, 2008, 2011
CSLI Publications
Center for the Study of Language and Information
Leland Stanford Junior University
First Edition 1999
Second Edition 2011
Printed in the United States
21 20 19 18 17   6 7 8 9

Library of Congress Cataloging-in-Publication Data

Barker-Plummer, Dave.
   Language, proof, and logic. – 2nd ed. / Dave Barker-Plummer, Jon
   Barwise, and John Etchemendy in collaboration with Albert Liu, Michael
   Murray, and Emma Pease.
      p. cm. –
   Rev. ed. of: Language, proof, and logic / Jon Barwise & John Etchemendy.
   Includes index.
   ISBN 978-1-57586-632-1 (pbk. : alk. paper)
   1. Logic. I. Barwise, Jon. II. Etchemendy, John, 1952- III. Barwise, Jon.
   Language, proof, and logic. IV. Title.

   BC71.B25 2011
   160–dc23   2011019703

CIP

   ISBN 978-1-57586-736-6 (electronic version)

∞ The acid-free paper used in this book meets the minimum requirements of the
American National Standard for Information Sciences—Permanence of Paper for Printed
Library Materials, ANSI Z39.48-1984.

NOTICE: The authors make no representations or warranties with respect to
the contents hereof and specifically disclaim any implied warranties of fitness
for any particular purpose.

Please send comments and suggestions to one or both of us at:

Dave Barker-Plummer / John Etchemendy
Center for the Study of Language and Information
Stanford University
Stanford, CA 94305-4115

Previous printings of this book contained a CD-ROM. For the current version of this
package, the software and all the other files accompanying the textbook can be
downloaded by using the Registration/Book ID# printed on the reverse side of the card in
the plastic packet.

WARNING: This book may not be returned once the plastic packet on the
inside back cover has been unsealed or removed.

Acknowledgements

Our primary debt of gratitude goes to our main collaborators on this project: Gerry Allwein and Albert Liu. They have worked with us in designing the entire package, developing and implementing the software, and teaching from and refining the text. Without their intelligence, dedication, and hard work, LPL would neither exist nor have most of its other good properties.

In addition to the five of us, many people have contributed directly and indirectly to the creation of the package. First, over two dozen programmers have worked on predecessors of the software included with the package, both earlier versions of Tarski's World and the program Hyperproof, some of whose code has been incorporated into Fitch. We want especially to mention Christopher Fuselier, Mark Greaves, Mike Lenz, Eric Ly, and Rick Wong, whose outstanding contributions to the earlier programs provided the foundation of the new software. Second, we thank several people who have helped with the development of the new software in essential ways: Rick Sanders, Rachel Farber, Jon Russell Barwise, Alex Lau, Brad Dolin, Thomas Robertson, Larry Lemmon, and Daniel Chai. Their contributions have improved the package in a host of ways.

Prerelease versions of LPL have been tested at several colleges and universities. In addition, other colleagues have provided excellent advice that we have tried to incorporate into the final package. We thank Selmer Bringsjord, Rensselaer Polytechnic Institute; Tom Burke, University of South Carolina; Robin Cooper, Gothenburg University; James Derden, Humboldt State University; Josh Dever, SUNY Albany; Avrom Faderman, University of Rochester; James Garson, University of Houston; Christopher Gauker, University of Cincinnati; Ted Hodgson, Montana State University; John Justice, Randolph-Macon Women's College; Ralph Kennedy, Wake Forest University; Michael O'Rourke, University of Idaho; Greg Ray, University of Florida; Cindy Stern, California State University, Northridge; Richard Tieszen, San Jose State University; Saul Traiger, Occidental College; and Lyle Zynda, Indiana University at South Bend. We are particularly grateful to John Justice, Ralph Kennedy, and their students (as well as the students at Stanford and Indiana University), for their patience with early versions of the software and for their extensive comments and suggestions. We would also like to thank the many instructors and students who have offered useful feedback since the initial publication of LPL.

We would also like to thank Stanford's Center for the Study of Language and Information and Indiana University's College of Arts and Sciences for their financial support of the project. Finally, we are grateful to our publisher, Dikran Karagueuzian and his team at CSLI Publications, for their skill and enthusiasm about LPL, and to Lauri Kanerva for his dedication and skill in the preparation of the final manuscript.

Acknowledgements for the Second Edition

One part of developing courseware packages for publication is the continual challenge of maintaining software in the face of developments in the commercial computer market. Since the initial publication of LPL, many operating system versions have come and gone, each requiring modifications, small and large, to the applications that are part of the package. With the publication of the second edition of LPL we are releasing the 3.x series of the Fitch, Boole and Submit applications, and the 7.x series of the Tarski's World application. While retaining the same functionality, these are essentially complete rewrites of the applications that appeared with the initial publication of LPL. The Grade Grinder too has undergone many updates and changes — it is now running in its third major incarnation, having been ported from the Solaris operating system to Mac OS X along the way. Three generations of the LPL web site have also come and gone.

All of this maintenance and development requires the talent of skilled programmers, and we have been fortunate in the people that have contributed to the project. Albert Liu contributed to many aspects of the software over a long period. Michael Murray's work is evident in all aspects of the desktop applications, and is primarily responsible for the web site as it currently appears. Emma Pease has served as the project's system administrator for the past two years. We have modernized the appearance of the LPL applications with this new release, and the graphical design work of Aaron Kalb is evident in every aspect of the software including the web site design, which was realized by Deonne Castaneda. Nik Swoboda of the Universidad Politécnica de Madrid created and maintains the Linux ports of the LPL software. Leslie Rogers served as the lead QA engineer for the new applications. The LPL package would not be what it is without the dedication and hard work of all of these people, and many others, and we thank them all.

Dave Barker-Plummer frequently teaches using LPL in Stanford's Philosophy 150 class. He uses this as an excuse to try out new teaching material and sometimes beta versions of the software. Dave would like to thank all of the students who have enrolled in those classes over the years for their patience and good humor.

We have benefitted greatly from the feedback of the many instructors who have adopted the LPL package in their teaching. We would particularly like to thank Richard Zach, University of Calgary; S. Marc Cohen, University of Washington and Bram van Heuveln, Rensselaer Polytechnic Institute for much appreciated comments on the package. Bram suggested to us the addition of the "Add Support Steps" feature of the new Fitch program. Richard Johns of the University of British Columbia suggested the new "goggles" features which are also included in that program.

The Openproof project continues to benefit from generous funding from Stanford University and from its home in the intellectually stimulating environment of Stanford's Center for the Study of Language and Information (CSLI). As always, we are grateful to our publisher, Dikran Karagueuzian, and his team at CSLI Publications for their continued enthusiasm for LPL.

Contents

II   Quantifiers

To Sol Feferman and Pat Suppes,
teachers, colleagues, and friends.

Introduction

The special role of logic in rational inquiry

What do the fields of astronomy, economics, finance, law, mathematics, medicine, physics, and sociology have in common? Not much in the way of subject matter, that's for sure. And notcomputer: all that much in the way of methodology. What they do have in common, with each other and with many other fields, is their dependence on a certain standard of rationality. In each of these fields, it is assumed that the participants can differentiate between rational argumentation based on assumed principles or evidence, and wild speculation or nonsequiturs, claims that in no way follow from the assumptions. In other words, these fields all presuppose an underlying acceptance of basic principles of logic.

For that matter, *all* rational inquiry depends on logic, on the ability of people to reason correctly most of the time, and, when they fail to reason correctly, on the ability of others to point out the gaps in their reasoning. While people may not all agree on a whole lot, they do seem to be able to agree on what can legitimately be concluded from given information. Acceptance of these commonly held principles of rationality is what differentiates rational inquiry from other forms of human activity.

logic and rational inquiry

Just what are the principles of rationality presupposed by these disciplines? And what are the techniques by which we can distinguish correct or "valid" reasoning from incorrect or "invalid" reasoning? More basically, what is it that *makes* one claim "follow logically" from some given information, while some other claim does not?

Many answers to these questions have been explored. Some people have claimed that the laws of logic are simply a matter of convention. If this is so, we could presumably decide to change the conventions, and so adopt different principles of logic, the way we can decide which side of the road we drive on. But there is an overwhelming intuition that the laws of logic are somehow more fundamental, less subject to repeal, than the laws of the land, or even the laws of physics. We can imagine a country in which a red traffic light means *go*, and a world on which water flows up hill. But we can't even imagine a world in which there both are and are not nine planets.

logic and convention

The importance of logic has been recognized since antiquity. After all, no

1

laws of logic

science can be any more certain than its weakest link. If there is something arbitrary about logic, then the same must hold of all rational inquiry. Thus it becomes crucial to understand just what the laws of logic are, and even more important, *why* they are laws of logic. These are the questions that one takes up when one studies logic itself. To study logic is to use the methods of rational inquiry on rationality itself.

Since the beginning of the twentieth century the study of logic has undergone rapid and important advances. Spurred on by logical problems in that most deductive of disciplines, mathematics, it developed into a discipline in its own right, with its own concepts, methods, techniques, and language. The *Encyclopedia Brittanica* lists logic as one of the seven main branches of knowledge. More recently, the study of logic has played a major role in the development of modern day computers and programming languages. Logic continues to play an important part in computer science; indeed, it has been said that computer science is just logic implemented in electrical engineering.

goals of the book

This book is intended to introduce you to some of the most important concepts and tools of logic. Our goal is to provide detailed and systematic answers to the questions raised above. We want you to understand just how the laws of logic follow inevitably from the meanings of the expressions we use to make claims. Convention is crucial in giving meaning to a language, but once the meaning is established, the laws of logic follow inevitably.

More particularly, we have two main aims. The first is to help you learn a new language, the language of first-order logic. The second is to help you learn about the notion of logical consequence, and about how one goes about establishing whether some claim is or is not a logical consequence of other accepted claims. While there is much more to logic than we can even hint at in this book, or than any one person could learn in a lifetime, we can at least cover these most basic of issues.

Why learn an artificial language?

This language of first-order logic is very important. Like Latin, the language is not spoken, but unlike Latin, it is used every day by mathematicians, philosophers, computer scientists, linguists, and practitioners of artificial intelligence. Indeed, in some ways it is the universal language, the *lingua franca*, of the symbolic sciences. Although it is not so frequently used in other forms of rational inquiry, like medicine and finance, it is also a valuable tool for understanding the principles of rationality underlying these disciplines as well.

FOL

The language goes by various names: the lower predicate calculus, the functional calculus, the language of first-order logic, and FOL. The last of these is pronounced *ef–oh–el*, not *fall*, and is the name we will use.

Certain elements of FOL go back to Aristotle, but the language as we know it today emerged during the twentieth century. The names chiefly associated with its development are those of Gottlob Frege, Giuseppe Peano, and Charles Sanders Peirce. In the late nineteenth century, these three logicians independently came up with the most important elements of the language, known as the *quantifiers*. Since then, there has been a process of standardization and simplification, resulting in the language in its present form. Even so, there remain certain dialects of FOL, differing mainly in the choice of the particular symbols used to express the basic notions of the language. We will use the dialect most common in mathematics, though we will also tell you about several other dialects along the way. FOL is used in different ways in different fields. In mathematics, it is used in an informal way quite extensively. The various connectives and quantifiers find their way into a great deal of mathematical discourse, both formal and informal, as in a classroom setting. Here you will often find elements of FOL interspersed with English or the mathematician's native language. If you've ever taken calculus you have probably seen such formulas as:

logic and mathematics

$$\forall \epsilon > 0 \ \exists \delta > 0 \ldots$$

Here, the unusual, rotated letters are taken directly from the language FOL.

In philosophy, FOL and enrichments of it are used in two different ways. As in mathematics, the notation of FOL is used when absolute clarity, rigor, and lack of ambiguity are essential. But it is also used as a case study of making informal notions (like grammaticality, meaning, truth, and proof) precise and rigorous. The applications in linguistics stem from this use, since linguistics is concerned, in large part, with understanding some of these same informal notions.

logic and philosophy

In artificial intelligence, FOL is also used in two ways. Some researchers take advantage of the simple structure of FOL sentences to use it as a way to encode knowledge to be stored and used by a computer. Thinking is modeled by manipulations involving sentences of FOL. The other use is as a precise specification language for stating axioms and proving results about artificial agents.

logic and artificial intelligence

In computer science, FOL has had an even more profound influence. The very idea of an artificial language that is precise yet rich enough to program computers was inspired by this language. In addition, all extant programming languages borrow some notions from one or another dialect of FOL. Finally, there are so-called logic programming languages, like Prolog, whose programs are sequences of sentences in a certain dialect of FOL. We will discuss the logical basis of Prolog a bit in Part III of this book.

logic and computer science

artificial languages

FOL serves as the prototypical example of what is known as an artificial language. These are languages that were designed for special purposes, and are contrasted with so-called natural languages, languages like English and Greek that people actually speak. The design of artificial languages within the symbolic sciences is an important activity, one that is based on the success of FOL and its descendants.

logic and ordinary language

Even if you are not going to pursue logic or any of the symbolic sciences, the study of FOL can be of real benefit. That is why it is so widely taught. For one thing, learning FOL is an easy way to demystify a lot of formal work. It will also teach you a great deal about your own language, and the laws of logic it supports. First, FOL, while very simple, incorporates in a clean way some of the important features of human languages. This helps make these features much more transparent. Chief among these is the relationship between language and the world. But, second, as you learn to translate English sentences into FOL you will also gain an appreciation of the great subtlety that resides in English, subtlety that cannot be captured in FOL or similar languages, at least not yet. Finally, you will gain an awareness of the enormous ambiguity present in almost every English sentence, ambiguity which somehow does not prevent us from understanding each other in most situations.

Consequence and proof

logical consequence

Earlier, we asked what makes one claim follow from others: convention, or something else? Giving an answer to this question for FOL takes up a significant part of this book. But a short answer can be given here. Modern logic teaches us that one claim is a logical consequence of another if there is no way the latter could be true without the former also being true.

This is the notion of logical consequence implicit in all rational inquiry. All the rational disciplines presuppose that this notion makes sense, and that we can use it to extract consequences of what we know to be so, or what we think might be so. It is also used in disconfirming a theory. For if a particular claim is a logical consequence of a theory, and we discover that the claim is false, then we know the theory itself must be incorrect in some way or other. If our physical theory has as a consequence that the planetary orbits are circular when in fact they are elliptical, then there is something wrong with our physics. If our economic theory says that inflation is a necessary consequence of low unemployment, but today's low unemployment has not caused inflation, then our economic theory needs reassessment.

Rational inquiry, in our sense, is not limited to academic disciplines, and so neither are the principles of logic. If your beliefs about a close friend logically imply that he would never spread rumors behind your back, but you find that he has, then your beliefs need revision. Logical consequence is central, not only to the sciences, but to virtually every aspect of everyday life.

One of our major concerns in this book is to examine this notion of logical consequence as it applies specifically to the language FOL. But in so doing, we will also learn a great deal about the relation of logical consequence in natural languages. Our main concern will be to learn how to recognize when a specific claim follows logically from others, and conversely, when it does not. This is an extremely valuable skill, even if you never have occasion to use FOL again after taking this course. Much of our lives are spent trying to convince other people of things, or being convinced of things by other people, whether the issue is inflation and unemployment, the kind of car to buy, or how to spend the evening. The ability to distinguish good reasoning from bad will help you recognize when your own reasoning could be strengthened, or when that of others should be rejected, despite superficial plausibility.

It is not always obvious when one claim is a logical consequence of others, but powerful methods have been developed to address this problem, at least for FOL. In this book, we will explore methods of proof—how we can *prove* that one claim is a logical consequence of another—and also methods for showing that a claim is *not* a consequence of others. In addition to the language FOL itself, these two methods, the method of proof and the method of counterexample, form the principal subject matter of this book.

proof and counterexample

Obtaining the Software

Language, Proof and Logic is a courseware package which includes this textbook, four applications—Tarski's World, Fitch, Boole and Submit—and a manual that explains how to use them. With the purchase of this package, you are given access to the Grade Grinder, an Internet grading service that can check whether your homework is correct and the ability to download the latest version of the software. Without registration, you will not be able to do many of the exercises or follow many of the examples used in the book. Registration also gives you access to a PDF version of the textbook.

Tarski's World, Fitch, Boole and Submit the Grade Grinder

With the purchase of the paperless version you are automatically registered; with the physical package you must register. To register, first locate the plastic packet attached to the inside back cover of this book. Once this packet is opened, you will be unable to return the book. Remove the card from the packet and read the registration information. You will see the URL of the registration page on our web site, a registration id beginning with the letter L, and also a Quick Response (QR) code. If you have a QR code reader (software that is available, free for most computers, tablets and phones) you can simply scan this code. Alternatively, you can enter the URL manually into your web browser. In either case, you will see a form which requires you to enter your name, email address and registration code, which you should fill out to complete your registration. Note registration is not transferable.

Once your registration is complete, you will be able to log into the web site, and download a software installer appropriate for your computer. Finally, launch the installer and follow the prompts to install the software and documents. The Read Me file which is installed with the installation will describe what is installed in detail.

Essential instructions about homework exercises

About half of the exercises in the first two parts of the book will be completed using the software. These exercises typically require that you create a file or files using Tarski's World, Fitch or Boole, and then submit these solution files using the program Submit. When you do this, your solutions are not submitted directly to your instructor, but rather to our grading server, the Grade Grinder, which assesses your files and sends a report to both you and your instructor. (If you are not using this book as a part of a formal class, you can have the reports sent just to you.)

Exercises in the book are numbered $n.m$, where n is the number of the chapter and m is the number of the exercise in that chapter. Exercises whose solutions consist of one or more files created with the LPL applications that you are to submit to the Grade Grinder are indicated with an arrow (✔), so that you know the solutions are to be sent off into the Internet ether. Exercises that are not completed using the applications are indicated with a pencil (✎). For example, Exercises 36 and 37 in Chapter 6 might look like this:

✔ vs. ✎

6.36 Use Tarski's World to build a world in which the following sentences
✔ are all true. . . .

6.37 Turn in an informal proof that the following argument is logically
✎ valid. . . .

The arrow on Exercise 6.36 tells you that the world you create using Tarski's World is to be submitted electronically, and that there is nothing else to turn in. The pencil on Exercise 6.37 tells you that you do not complete the exercise using one of the applications. Your instructor will tell you how to turn in the solution. A solution to this exercise can be submitted as a text file using Submit, or your instructor might prefer to collect solutions to this exercise on paper.

Some exercises ask you to turn in something to your instructor in addition to submitting a file electronically. These are indicated with both an arrow and a pencil (✐|✎). This is also used when the exercise *may* require a file to be submitted, but may not, depending on the solution. For example, the next problem in Chapter 6 might ask:

6.38
✐|✎
Is the following argument valid? If so, use Fitch to construct a formal proof of its validity. If not, explain why it is invalid and turn in your explanation to your instructor.

Here, we can't tell you definitely whether you'll be submitting a file or turning something in without giving away an important part of the exercise, so we mark the exercise with both symbols.

By the way, in giving instructions in the exercises, we will reserve the word "submit" for electronic submission, using the Submit program. We use "turn in" when you are to turn in the solution to your instructor either as a text file via Submit, or on paper. The Grade Grinder is unable to assess the solutions to exercises submitted as text files. You will however receive an acknowledgement that the file has been received, and your instructor will be able to retrieve a copy of your work from the Grade Grinder web site.

submitting vs. turning in exercises

When you create files to be submitted to the Grade Grinder, it is important that you name them correctly. Sometimes we will tell you what to name the files, but more often we expect you to follow a few standard conventions. Our naming conventions are simple. If you are creating a proof using Fitch, then you should name the file Proof n.m, where $n.m$ is the number of the exercise. If you are creating a world or sentence file in Tarski's World, then you should call it either World n.m or Sentences n.m, where $n.m$ is the number of the exercise. If you are creating a truth table using Boole, you should name it Table n.m, and finally if you are submitting a text file it must be named Solution n.m. The key thing is to get the right exercise number in the name, since otherwise your solution will be graded incorrectly. We'll remind you of these naming conventions a few times, but after that you're on your own.

naming solution files

When an exercise asks you to construct a formal proof using Fitch, you will find a file on your disk called Exercise n.m. This file contains the proof set up, so you should open it and construct your solution in this file. This is a lot easier for you and also guarantees that the Grade Grinder will know which exercise you are solving. So make sure you always start with the packaged Exercise file when you create your solution.

starting proofs

Exercises may also have from one to three stars (⋆, ⋆⋆, ⋆⋆⋆), as a rough indication of the difficulty of the problem. For example, this would be an

⋆ stars

exercise that is a little more difficult than average (and whose solution you turn in to your instructor):

6.39 Design a first-order language that allows you to express the following
✎★ English sentences....

Remember

1. The arrow (✔) means that you submit your solution electronically.

2. The pencil (✎) means that you turn in your solution to your instructor.

3. The combination (✔|✎) means that your solution may be either a submitted file or something to turn in, or possibly both.

4. Stars (★, ★★, ★★★) indicate exercises that are more difficult than average.

5. Unless otherwise instructed, name your files Proof n.m, World n.m, Sentences n.m, Table n.m, or Solution n.m, where $n.m$ is the number of the exercise.

6. When using Fitch to construct Proof n.m, start with the exercise file Exercise n.m, which contains the problem setup.

7. If you use Submit to submit a text file, the Grade Grinder will not assess the file, but simply acknowledge that it has been received.

You try it sections

Throughout the book, you will find a special kind of exercise that we call **You try it** exercises. These appear as part of the text rather than in the exercise sections because they are particularly important. They either illustrate important points about logic that you will need to understand later or teach you some basic operations involving one of the computer programs that came with your book. Because of this, you shouldn't skip any of the **You try it** sections. Do these exercises as soon as you come to them, if you are in the vicinity of a computer. If you aren't in the vicinity of a computer, come back and do them as soon as you are.

Here's your first **You try it** exercise. Make sure you actually do it, right now if possible. It will teach you how to use Submit to send files to the Grade Grinder, a skill you definitely want to learn. You will need to know your email address, your instructor's name and email address, and your Book ID number before you can do the exercise. If you don't know any of these, talk to your instructor first. Your computer must be connected to the internet to submit files. If it's not, use a public computer at your school or at a public library.

You try it
. . . .

1. We're going to step you through the process of submitting a file to the ◀ Grade Grinder. The file is called World Submit Me 1. It is a Tarski's World file, but you won't have to open it using Tarski's World in order to submit it. We'll pretend that it is an exercise file that you've created while doing your homework, and now you're ready to submit it. More complete instructions on running Submit are contained in the instruction manual that came with the software.

   Find the program Submit that came with your book. Submit has an icon that looks like a cog with a capital G on it, and appears inside a folder called Submit Folder. Once you've found it, double-click on the icon to launch the program.

2. After a moment, you will see the main Submit window, which has a rotat- ◀ ing cog in the upper-left corner. The first thing you should do is fill in the requested information in the five fields. Enter your Book ID first, then your name and email address. You have to use your complete email address— for example, *claire@cs.nevada-state.edu*, not just *claire* or *claire@cs*—since the Grade Grinder will need the full address to send its response back to you. Also, if you have more than one email address, you have to use the same one every time you submit files, since your email address and Book ID together are how Grade Grinder will know that it is really you submitting files. Finally, fill in your instructor's name and complete email address. Be very careful to enter the correct and complete email addresses!

3. We're now ready to specify the file to submit. Click on the button **Choose** ◀ **Files To Submit** in the lower-left corner. This opens a window showing two file lists. The list on the left shows files on your computer, while the one on the right (which is currently empty) will list files you want to submit. We need to locate the file World Submit Me 1 on the left and copy it over to the right. By the way, our software uses files with the extensions.sen, .wld, .prf, .tt, but we don't mention these in this book when referring to files. our informal name for the file we are looking for is World Submit Me 1.

   The file World Submit Me 1 is located in the Tarski's World exercise files folder. To find this folder you will have to navigate among folders until it appears in the file list on the left. Start by clicking once on the **Submit Folder** button above the left-hand list. A menu will appear and you can then move up to higher folders by choosing their names (the higher folders appear lower on this menu). Move to the next folder up from the **Submit Folder**, which should be called **LPL Software**. When you choose this folder, the list of files will change. On the new list, find the folder **Tarski's World Folder** and double-click on its name to see the contents of the folder. The list will again change and you should now be able to see the folder **TW Exercise Files**. Double-click on this folder and the file list will show the contents of this folder. Toward the bottom of the list (you will have to scroll down the list by clicking on the scroll buttons), you will find **World Submit Me 1**. Double-click on this file and its name will move to the list on the right.

▶ 4. When you have successfully gotten the file **World Submit Me 1** on the righthand list, click the **Done** button underneath the list. This should bring you back to the original Submit window, only now the file you want to submit appears in the list of files. (Macintosh users can get to this point quickly by dragging the files they want to submit onto the Submit icon in the Finder. This will launch Submit and put those files in the submission list. If you drag a folder of files, it will put all the files in the folder onto the list.)

▶ 5. When you have the correct file on the submission list, click on the **Submit Files** button under this list. As this is your very first submission, Submit requires you to confirm your email address, in addition to asking you to confirm that you want to submit **World Submit Me 1**. The confirmation of your email address will only happen this time. You also have an opportunity to saywhether you want to send the results just to you or also to your instructor. In this case, select **Just Me**. When you are submitting finished homework exercises, you should select **Instructor Too**. Once you've chosen who the results should go to, click the **Proceed** button and your submission will be sent. (With real homework, you can always do a trial submission to see if you got the answers right, asking that the results be sent just to you. When you are satisfied with your solutions, submit the files again, asking that the results be sent to the instructor too. But don't forget the second submission!)

▶ 6. In a moment, you will get a dialog box that will tell you if your submission has been successful. If so, it will give you a "receipt" message that you can save, if you like. If you do not get this receipt, then your submission has not gone through and you will have to try again.

7. A few minutes after the Grade Grinder receives your file, you should get ◄ an email message saying that it has been received. If this were a real homework exercise, it would also tell you if the Grade Grinder found any errors in your homework solutions. You won't get an email report if you put in the wrong, or a misspelled, email address. If you don't get a report, try submitting again with the right address.

8. Quit from Submit when you are done. Congratulations on submitting your ◄ first file.

. . . . *Congratulations*

Here's an important thing for you to know: when you submit files to the Grade Grinder, Submit sends a copy of the files. The original files are still *what gets sent* on the disk where you originally saved them. If you saved them on a public computer, it is best not to leave them lying around. Put them on a thumb drive that you can take with you, and delete any copies from the public computer's hard disk.

You should carefully read the email that you receive from the Grade Grinder since it contains information concerning the errors that the Grade Grinder found in your work. Even if there are no errors, you should keep the email that you receive as a reminder that you have submitted the work. In addition, if you log in at our web site, `https://www.gradegrinder.net/`, you will be able to see the complete history of your submissions to the Grade Grinder.

To the instructor

Students, you may skip this section. It is a personal note from us, the authors, to instructors planning to use this package in their logic courses.

Practical matters

We use the *Language, Proof and Logic* package (LPL) in two very different sorts of courses. One is a first course in logic for undergraduates with no previous background in logic, philosophy, mathematics, or computer science. This important course, sometimes disparagingly referred to as "baby logic," is often an undergraduate's first and only exposure to the rigorous study of reasoning. When we teach this course, we cover much of the first two parts of the book, leaving out many of the sections indicated as optional in the table of contents. Although some of the material in these two parts may seem more advanced than is usually covered in a traditional introductory course, we find that the software makes it completely accessible to even the relatively unprepared student.

At the other end of the spectrum, we use LPL in an introductory graduate-level course in metatheory, designed for students who have already had some exposure to logic. In this course, we quickly move through the first two parts, thereby giving the students both a review and a common framework for use in the discussions of soundness and completeness. Using the Grade Grinder, students can progress through much of the early material at their own pace, doing only as many exercises as is needed to demonstrate competence.

There are no doubt many other courses for which the package would be suitable. Though we have not had the opportunity to use it this way, it would be ideally suited for a two-term course in logic and its metatheory.

Our courses are typically listed as philosophy courses, though many of the students come from other majors. Since LPL is designed to satisfy the logical needs of students from a wide variety of disciplines, it fits naturally into logic courses taught in other departments, most typically mathematics and computer science. Instructors in different departments may select different parts of the optional material. For example, computer science instructors may want to cover the sections on resolution in Part III, though philosophy instructors generally do not cover this material.

If you have not used software in your teaching before, you may be concerned about how to incorporate it into your class. Again, there is a spectrum of possibilities. At one end is to conduct your class exactly the way you always do, letting the students use the software on their own to complete homework assignments. This is a perfectly fine way to use the package, and the students will still benefit significantly from the suite of software tools.

At the other end are courses given in computer labs or classrooms, where the instructor is more a mentor offering help to students as they proceed at their own pace, a pace you can keep in step with periodic quizzes and exams. Here the student becomes a more active participant in the learning. For a class of 30 or fewer students, this can be a very effective way to teach a beginning logic course.

In between, and the style we typically use, is to give reasonably traditional presentations, but to bring a laptop to class from time to time to illustrate important material using the programs. This requires some sort of projection system, but also allows you to ask the students to do some of the computer problems in class. We encourage you to get students to operate the computer themselves in front of the class, since they thereby learn from one another, both about strategies for solving problems and constructing proofs, and about different ways to use the software. A variant of this is to schedule a weekly lab session as part of the course.

The book contains an extremely wide variety of exercises. There are far more of them than you can expect your students to do in a single quarter or semester. Beware that many exercises, especially those using Tarski's World, should be thought of as exercise sets. They may, for example, involve translating ten or twenty sentences, or transforming several sentences into conjunctive normal form. Students can find hints and solutions to selected exercises on our web site. You can download a list of these exercises from the same site.

Although there are more exercises than you can reasonably assign in a semester, and so you will have to select those that best suit your course, we do urge you to assign all of the **You try it** exercises. These are not difficult and do not test students' knowledge. Instead, they are designed to illustrate important logical concepts, to introduce students to important features of the programs, or both. The Grade Grinder will check any files that the students create in these sections.

We should say a few words about the Grade Grinder, since it is a truly innovative feature of this package. Most important, the Grade Grinder will free you from the most tedious aspect of teaching logic, namely, grading those kinds of problems whose assessment can be mechanized. These include formal proofs, translation into FOL, truth tables, and various other kinds of exercises. This will allow you to spend more time on the more rewarding parts of teaching the material.

That said, it is important to emphasize two points. The first is that the Grade Grinder is not limited in the way that most computerized grading programs are. It uses sophisticated techniques, including a powerful first-order theorem prover, in assessing student answers and providing intelligent reports on those answers. Second, in designing this package, we have not fallen into the trap of tailoring the material to what can be mechanically assessed. We firmly believe that computer-assisted learning has an important but limited role to play in logic instruction. Much of what we teach goes beyond what can be assessed automatically. This is why about half of the exercises in the book still require human attention.

It is a bit misleading to say that the Grade Grinder "grades" the homework. The Grade Grinder simply reports to you any errors in the students' solutions, leaving the decision to you what weight to give to individual problems and whether partial credit is appropriate for certain mistakes. A more detailed explanation of what the Grade Grinder does and what grade reports look like can be found at the web address given on page 16.

registering with
the Grade Grinder

Before your students can request that their Grade Grinder results be sent to you, you will have to register with the Grade Grinder as an instructor. This can be done by going to the LPL web site and following the Instructor links.

Philosophical remarks

This book, and the supporting software that comes with it, grew out of our own dissatisfaction with beginning logic courses. It seems to us that students all too often come away from these courses with neither of the things we want them to have. They do not understand the first-order language or the rationale for it, and they are unable to explain why or even whether one claim follows logically from another. Worse, they often come away with a complete misconception about logic. They leave their first (and only) course in logic having learned what seem like a bunch of useless formal rules. They gain little if any understanding about why those rules, rather than some others, were chosen, and they are unable to take any of what they have learned and apply it in other fields of rational inquiry or in their daily lives. Indeed, many come away convinced that logic is both arbitrary and irrelevant. Nothing could be further from the truth.

The real problem, as we see it, is a failure on the part of logicians to find a simple way to explain the relationship between meaning and the laws of logic. In particular, we do not succeed in conveying to students what sentences in FOL mean, or in conveying how the meanings of sentences govern which methods of inference are valid and which are not. It is this problem we set out to solve with LPL.

There are two ways to learn a second language. One is to learn how to translate sentences of the language to and from sentences of your native language. The other is to learn by using the language directly. In teaching FOL, the first way has always been the prevailing method of instruction. There are serious problems with this approach. Some of the problems, oddly enough, stem from the simplicity, precision, and elegance of FOL. This results in a distracting mismatch between the student's native language and FOL. It forces students trying to learn FOL to be sensitive to subtleties of their native language that normally go unnoticed. While this is useful, it often interferes with the learning of FOL. Students mistake complexities of their native tongue for complexities of the new language they are learning.

In LPL, we adopt the second method for learning FOL. Students are given many tasks involving the language, tasks that help them understand the meanings of sentences in FOL. Only then, after learning the basics of the symbolic language, are they asked to translate between English and FOL. Correct translation involves finding a sentence in the target language whose meaning approximates, as closely as possible, the meaning of the sentence being translated. To do this well, a translator must already be fluent in both languages.

We have been using this approach for several years. What allows it to work is Tarski's World, one of the computer programs in this package. Tarski's World provides a simple environment in which FOL can be used in many of the ways that we use our native language. We provide a large number of problems and exercises that walk students through the use of the language in this setting. We build on this in other problems where they learn how to put the language to more sophisticated uses.

As we said earlier, besides teaching the language FOL, we also discuss basic methods of proof and how to use them. In this regard, too, our approach is somewhat unusual. We emphasize both informal and formal methods of proof. We first discuss and analyze informal reasoning methods, the kind used in everyday life, and then formalize these using a Fitch-style natural deduction system. The second piece of software that comes with the book, which we call Fitch, makes it easy for students to learn this formal system and to understand its relation to the crucial informal methods that will assist them in other disciplines and in any walk of life.

A word is in order about why we chose a Fitch-style system of deduction, rather than a more semantically based method like truth trees or semantic tableau. In our experience, these semantic methods are easy to teach, but are only really applicable to arguments in formal languages. In contrast, the important rules in the Fitch system, those involving subproofs, correspond closely to essential methods of reasoning and proof, methods that can be used in virtually any context: formal or informal, deductive or inductive, practical or theoretical. The point of teaching a formal system of deduction is not so students will use the specific system later in life, but rather to foster an understanding of the most basic methods of reasoning—methods that they *will* use—and to provide a precise model of reasoning for use in discussions of soundness and completeness.

Tarski's World also plays a significant role in our discussion of proof, along with Fitch, by providing an environment for showing that one claim does not follow from another. With LPL, students learn not just how to prove consequences of premises, but also the equally important technique of showing that a given claim does *not* follow logically from its premises. To do this, they learn how to give counterexamples, which are really proofs of *nonconsequence*. These will often be given using Tarski's World.

The approach we take in LPL is also unusual in two other respects. One is our emphasis on languages in which all the basic symbols are assumed to be meaningful. This is in contrast to the so-called "uninterpreted languages" (surely an oxymoron) so often found in logic textbooks. Another is the inclusion of various topics not usually covered in introductory logic books. These include the theory of conversational implicature, material on generalized quantifiers, and most of the material in Part III. We believe that even if these topics are not covered, their presence in the book illustrates to the student the richness and open-endedness of the discipline of logic.

Web address

In addition to the book, software, and grading service, additional material can be found on the Web at the following address:

```
https://www.gradegrinder.net
```

At the web site you will find hints and solutions to selected exercises, an online version of the software manuals, support pages where you can browse our list of frequently asked questions, and directly request technical support and submit bug reports. In addition registered users may log in at the site. Students can view their history of submissions to the Grade Grinder, and download the latest versions of the software, while instructors can view the history of submissions by all of their students.

PART I

Propositional Logic

CHAPTER 1

Atomic Sentences

In the Introduction, we talked about FOL as though it were a single language. Actually, it is more like a family of languages, all having a similar grammar and sharing certain important vocabulary items, known as the connectives and quantifiers. Languages in this family can differ, however, in the specific vocabulary used to form their most basic sentences, the so-called atomic sentences.

Atomic sentences correspond to the most simple sentences of English, sentences consisting of some names connected by a predicate. Examples are *Max ran, Max saw Claire,* and *Claire gave Scruffy to Max.* Similarly, in FOL atomic sentences are formed by combining names (or individual constants, as they are often called) and predicates, though the way they are combined is a bit different from English, as you will see.

atomic sentences

Different versions of FOL have available different names and predicates. We will frequently use a first-order language designed to describe blocks arranged on a chessboard, arrangements that you will be able to create in the program Tarski's World. This language has names like b, e, and n_2, and predicates like Cube, Larger, and Between. Some examples of atomic sentences in this language are Cube(b), Larger(c, f), and Between(b, c, d). These sentences say, respectively, that b is a cube, that c is larger than f, and that b is between c and d.

names and predicates

Later in this chapter, we will look at the atomic sentences used in two other versions of FOL, the first-order languages of set theory and arithmetic. In the next chapter, we begin our discussion of the connectives and quantifiers common to all first-order languages.

Individual constants

Individual constants are simply symbols that are used to refer to some fixed individual object. They are the FOL analogue of names, though in FOL we generally don't capitalize them. For example, we might use max as an individual constant to denote a particular person, named Max, or 1 as an individual constant to denote a particular number, the number one. In either case, they would basically work exactly the way names work in English. Our blocks

names in FOL

language takes the letters a through f plus n_1, n_2, ... as its names.

The main difference between names in English and the individual constants of FOL is that we require the latter to refer to exactly one object. Obviously, the name *Max* in English can be used to refer to many different people, and might even be used twice in a single sentence to refer to two different people. Such wayward behavior is frowned upon in FOL.

There are also names in English that do not refer to any actually existing object. For example *Pegasus*, *Zeus*, and *Santa Claus* are perfectly fine names in English; they just fail to refer to anything or anybody. We don't allow such names in FOL.[1] What we do allow, though, is for one object to have more than one name; thus the individual constants matthew and max might both refer to the same individual. We also allow for nameless objects, objects that have no name at all.

> **Remember**
>
> In FOL,
>
> o Every individual constant must name an (actually existing) object.
>
> o No individual constant can name more than one object.
>
> o An object can have more than one name, or no name at all.

SECTION 1.2
Predicate symbols

predicate or relation symbols

Predicate symbols are symbols used to express some property of objects or some relation between objects. Because of this, they are also sometimes called relation symbols. As in English, predicates are expressions that, when combined with names, form atomic sentences. But they don't correspond exactly to the predicates of English grammar.

Consider the English sentence *Max likes Claire*. In English grammar, this is analyzed as a subject-predicate sentence. It consists of the subject *Max* followed by the predicate *likes Claire*. In FOL, by contrast, we usually view

logical subjects

this as a claim involving two "logical subjects," the names *Max* and *Claire*, and

---

[1] There is, however, a variant of first-order logic called *free logic* in which this assumption is relaxed. In free logic, there can be individual constants without referents. This yields a language more appropriate for mythology and fiction.

a predicate, *likes,* that expresses a relation between the referents of the names. Thus, atomic sentences of FOL often have two or more logical subjects, and the predicate is, so to speak, whatever is left. The logical subjects are called the "arguments" of the predicate. In this case, the predicate is said to be binary, since it takes two arguments.

arguments of a predicate

In English, some predicates have optional arguments. Thus you can say *Claire gave, Claire gave Scruffy,* or *Claire gave Scruffy to Max.* Here the predicate *gave* is taking one, two, and three arguments, respectively. But in FOL, each predicate has a fixed number of arguments, a fixed *arity* as it is called. This is a number that tells you how many individual constants the predicate symbol needs in order to form a sentence. The term "arity" comes from the fact that predicates taking one argument are called *unary,* those taking two are *binary,* those taking three are *ternary,* and so forth.

arity of a predicate

If the arity of a predicate symbol Pred is 1, then Pred will be used to express some property of objects, and so will require exactly one argument (a name) to make a claim. For example, we might use the unary predicate symbol Home to express the property of being at home. We could then combine this with the name max to get the expression Home(max), which expresses the claim that Max is at home.

If the arity of Pred is 2, then Pred will be used to represent a relation between two objects. Thus, we might use the expression Taller(claire, max) to express a claim about Max and Claire, the claim that Claire is taller than Max. In FOL, we can have predicate symbols of any arity. However, in the blocks language used in Tarski's World we restrict ourselves to predicates with arities 1, 2, and 3. Here we list the predicates of that language, this time with their arity.

Arity 1: Cube, Tet, Dodec, Small, Medium, Large

Arity 2: Smaller, Larger, LeftOf, RightOf, BackOf, FrontOf, SameSize, Same-Shape, SameRow, SameCol, Adjoins, =

Arity 3: Between

Tarski's World assigns each of these predicates a fixed interpretation, one reasonably consistent with the corresponding English verb phrase. For example, Cube corresponds to *is a cube,* BackOf corresponds to *is in back of,* and so forth. You can get the hang of them by working through the first set of exercises given below. To help you learn exactly what the predicates mean, Table 1.1 lists atomic sentences that use these predicates, together with their interpretations.

In English, predicates are sometimes vague. It is often unclear whether

vagueness

Table 1.1: Blocks language predicates.

Atomic Sentence	Interpretation
Tet(a)	*a* is a tetrahedron
Cube(a)	*a* is a cube
Dodec(a)	*a* is a dodecahedron
Small(a)	*a* is small
Medium(a)	*a* is medium
Large(a)	*a* is large
SameSize(a, b)	*a* is the same size as *b*
SameShape(a, b)	*a* is the same shape as *b*
Larger(a, b)	*a* is larger than *b*
Smaller(a, b)	*a* is smaller than *b*
SameCol(a, b)	*a* is in the same column as *b*
SameRow(a, b)	*a* is in the same row as *b*
Adjoins(a, b)	*a* and *b* are located on adjacent (but not diagonally) squares
LeftOf(a, b)	*a* is located nearer to the left edge of the grid than *b*
RightOf(a, b)	*a* is located nearer to the right edge of the grid than *b*
FrontOf(a, b)	*a* is located nearer to the front of the grid than *b*
BackOf(a, b)	*a* is located nearer to the back of the grid than *b*
Between(a, b, c)	*a*, *b* and *c* are in the same row, column, or diagonal, and *a* is between *b* and *c*

an individual has the property in question or not. For example, Claire, who is sixteen, is young. She will not be young when she is 96. But there is no determinate age at which a person stops being young: it is a gradual sort of thing. FOL, however, assumes that every predicate is interpreted by a determinate property or relation. By a *determinate* property, we mean a property for which, given any object, there is a definite fact of the matter whether or not the object has the property.

determinate property

This is one of the reasons we say that the blocks language predicates are *somewhat* consistent with the corresponding English predicates. Unlike the English predicates, they are given very precise interpretations, interpretations that are suggested by, but not necessarily identical with, the meanings of the corresponding English phrases. The case where the discrepancy is probably the greatest is between Between and *is between.*

> ### Remember
>
> In FOL,
>
> o Every predicate symbol comes with a single, fixed "arity," a number that tells you how many names it needs to form an atomic sentence.
>
> o Every predicate is interpreted by a determinate property or relation of the same arity as the predicate.

Atomic sentences

In FOL, the simplest kinds of claims are those made with a single predicate and the appropriate number of individual constants. A sentence formed by a predicate followed by the right number of names is called an *atomic* sentence. For example Taller(claire, max) and Cube(a) are atomic sentences, provided the names and predicate symbols in question are part of the vocabulary of our language. In the case of the identity symbol, we put the two required names on either side of the predicate, as in a = b. This is called "infix" notation, since the predicate symbol = appears in between its two arguments. With the other predicates we use "prefix" notation: the predicate precedes the arguments.

atomic sentence

infix vs. prefix notation

The order of the names in an atomic sentence is quite important. Just as *Claire is taller than Max* means something different from *Max is taller than Claire,* so too Taller(claire, max) means something completely different than Taller(max, claire). We have set things up in our blocks language so that the order of the arguments of the predicates is like that in English. Thus LeftOf(b, c) means more or less the same thing as the English sentence *b is left of c,* and Between(b, c, d) means roughly the same as the English *b is between c and d.*

Predicates and names designate properties and objects, respectively. What

claims

truth value

makes sentences special is that they make claims (or express propositions). A claim is something that is either true or false; which of these it is we call its *truth value*. Thus Taller(claire, max) expresses a claim whose truth value is TRUE, while Taller(max, claire) expresses a claim whose truth value is FALSE. (You probably didn't know that, but now you do.) Given our assumption that predicates express determinate properties and that names denote definite individuals, it follows that each atomic sentence of FOL must express a claim that is either true or false.

You try it
. . . .

▶   1. It is time to try your hand using Tarski's World. In this exercise, you will use Tarski's World to become familiar with the interpretations of the atomic sentences of the blocks language. Before starting, though, you need to learn how to launch Tarski's World and perform some basic operations. Read the appropriate sections of the user's manual describing Tarski's World before going on.

▶   2. Launch Tarski's World and open the files called Wittgenstein's World and Wittgenstein's Sentences. You will find these in the folder TW Exercises. In these files, you will see a blocks world and a list of atomic sentences. (We have added comments to some of the sentences. Comments are prefaced by a semicolon (";"), which tells Tarski's World to ignore the rest of the line.)

▶   3. Move through the sentences using the arrow keys on your keyboard, mentally assessing the truth value of each sentence in the given world. Use the **Verify Sentence** button to check your assessments. This button is on the left of the group of three colored buttons on the toolbar (the one which has **T/F** written on it). (Since the sentences are all atomic sentences the **Game** button, on the right of the same group, will not be helpful.) If you are surprised by any of the evaluations, try to figure out how your interpretation of the predicate differs from the correct interpretation.

▶   4. Next change Wittgenstein's World in many different ways, seeing what happens to the truth of the various sentences. The main point of this is to help you figure out how Tarski's World interprets the various predicates. For example, what does BackOf(d, c) mean? Do two things have to be in the same column for one to be in back of the other?

▶   5. Play around as much as you need until you are sure you understand the meanings of the atomic sentences in this file. For example, in the original world none of the sentences using Adjoins comes out true. You should try to modify the world to make some of them true. As you do this, you will notice that large blocks cannot adjoin other blocks.

6. In doing this exercise, you will no doubt notice that Between does not mean ◄ exactly what the English *between* means. This is due to the necessity of interpreting Between as a determinate predicate. For simplicity, we insist that in order for b to be between c and d, all three must be in the same row, column, or diagonal.

7. When you are finished, close the files, but do not save the changes you ◄ have made to them.
. . . . *Congratulations*

Remember

In FOL,

- Atomic sentences are formed by putting a predicate of arity n in front of n names (enclosed in parentheses and separated by commas).

- Atomic sentences are built from the identity predicate, =, using infix notation: the arguments are placed on either side of the predicate.

- The order of the names is crucial in forming atomic sentences.

Exercises

You will eventually want to read the entire chapter of the user's manual on how to use Tarski's World. To do the following problems, you will need to read at least the first four sections. Also, if you don't remember how to name and submit your solution files, you should review the section on essential instructions in the Introduction, starting on page 6.

1.1 If you skipped the **You try it** section, go back and do it now. This is an easy but crucial exercise that will familiarize you with the atomic sentences of the blocks language. There is nothing you need to turn in or submit, but don't skip the exercise!

1.2 (Copying some atomic sentences) This exercise will give you some practice with the Tarski's
✒ World keyboard window, as well as with the syntax of atomic sentences. The following are all atomic sentences of our language. Start a new sentence file and copy them into it. Have Tarski's World check each formula after you write it to see that it is a sentence. If you make a mistake, edit it before going on. Make sure you use the **Add Sentence** command between sentences,

not the return key. If you've done this correctly, the sentences in your list will be numbered and separated by horizontal lines.

1. Tet(a)
2. Medium(a)
3. Dodec(b)
4. Cube(c)
5. FrontOf(a, b)
6. Between(a, b, c)
7. a = d
8. Larger(a, b)
9. Smaller(a, c)
10. LeftOf(b, c)

Remember, you should save these sentences in a file named Sentences 1.2. When you've finished your first assignment, submit all of your solution files using the Submit program.

1.3   (Building a world) Build a world in which all the sentences in Exercise 1.2 are simultaneously true. Remember to name and submit your world file as World 1.3.

1.4   (Translating atomic sentences) Here are some simple sentences of English. Start a new sentence file and translate them into FOL.

1. *a is a cube.*
2. *b is smaller than a.*
3. *c is between a and d.*
4. *d is large.*
5. *e is larger than a.*
6. *b is a tetrahedron.*
7. *e is a dodecahedron.*
8. *e is right of b.*
9. *a is smaller than e.*
10. *d is in back of a.*
11. *b is in the same row as d.*
12. *b is the same size as c.*

After you've translated the sentences, build a world in which all of your translations are true. Submit your sentence and world files as Sentences 1.4 and World 1.4.

1.5   (Naming objects) Open Lestrade's Sentences and Lestrade's World. You will notice that none of the objects in this world has a name. Your task is to assign the objects names in such a way that all the sentences in the list come out true. Remember to save your solution in a file named World 1.5. Be sure to use **Save World As...**, not **Save World**.

1.6
✦★   (Naming objects, continued) Not all of the choices in Exercise 1.5 were forced on you. That is, you could have assigned the names differently and still had the sentences come out true. Change the assignment of as many names as possible while still making all the sentences true, and submit the changed world as World 1.6. In order for us to compare your files, you must submit both World 1.5 and World 1.6 at the same time.

1.7
✦|✎   (Context sensitivity of predicates) We have stressed the fact that FOL assumes that every predicate is interpreted by a determinate relation, whereas this is not the case in natural languages like English. Indeed, even when things seem quite determinate, there is often some form of context sensitivity. In fact, we have built some of this into Tarski's World. Consider, for example, the difference between the predicates Larger and BackOf. Whether or not cube a is larger than cube b is a determinate matter, and also one that does not vary depending on your perspective on the world. Whether or not a is back of b is also determinate, but in this case it does depend on your perspective. If you rotate the world by 90°, the answer might change.

Open Austin's Sentences and Wittgenstein's World. Evaluate the sentences in this file and tabulate the resulting truth values in a table like the one below. We've already filled in the first column, showing the values in the original world. Rotate the world 90° clockwise and evaluate the sentences again, adding the results to the table. Repeat until the world has come full circle.

	Original	Rotated 90°	Rotated 180°	Rotated 270°
1.	FALSE			
2.	FALSE			
3.	TRUE			
4.	FALSE			
5.	TRUE			
6.	FALSE			

You should be able to think of an atomic sentence in the blocks language that would produce a row across the table with the following pattern:

TRUE   FALSE   TRUE   FALSE

Add a seventh sentence to Austin's Sentences that would display the above pattern.

Are there any atomic sentences in the language that would produce a row with this pattern?

FALSE   TRUE   FALSE   FALSE

If so, add such a sentence as sentence eight in Austin's Sentences. If not, leave sentence eight blank.

Are there any atomic sentences that would produce a row in the table containing exactly three TRUE's? If so, add such a sentence as number nine. If not, leave sentence nine blank.

Submit your modified sentence file as Sentences 1.7. Turn in your completed table to your instructor.

General first-order languages

First-order languages differ in the names and predicates they contain, and so in the atomic sentences that can be formed. What they share are the connectives and quantifiers that enable us to build more complex sentences from these simpler parts. We will get to those common elements in later chapters.

translation

When you translate a sentence of English into FOL, you will sometimes have a "predefined" first-order language that you want to use, like the blocks language of Tarski's World, or the language of set theory or arithmetic described later in this chapter. If so, your goal is to come up with a translation that captures the meaning of the original English sentence as nearly as possible, given the names and predicates available in your predefined first-order language.

designing languages

Other times, though, you will not have a predefined language to use for your translation. If not, the first thing you have to do is decide what names and predicates you need for your translation. In effect, you are designing, on the fly, a new first-order language capable of expressing the English sentence you want to translate. We've been doing this all along, for example when we introduced Home(max) as the translation of *Max is at home* and Taller(claire, max) as the translation of *Claire is taller than Max.*

When you make these decisions, there are often alternative ways to go. For example, suppose you were asked to translate the sentence *Claire gave Scruffy to Max.* You might introduce a binary predicate GaveScruffy(x, y), meaning *x gave Scruffy to y*, and then translate the original sentence as GaveScruffy(claire, max). Alternatively, you might introduce a three-place predicate Gave(x, y, z), meaning *x gave y to z*, and then translate the sentence as Gave(claire, scruffy, max).

choosing predicates

There is nothing wrong with either of these predicates, or their resulting translations, so long as you have clearly specified what the predicates mean. Of course, they may not be equally useful when you go on to translate other sentences. The first predicate will allow you to translate sentences like *Max gave Scruffy to Evan* and *Evan gave Scruffy to Miles.* But if you then run into the sentence *Max gave Carl to Claire,* you would be stuck, and would have to introduce an entirely new predicate, say, GaveCarl(x, y). The three-place predicate is thus more flexible. A first-order language that contained it (plus the relevant names) would be able to translate any of these sentences.

In general, when designing a first-order language we try to economize on the predicates by introducing more flexible ones, like Gave(x, y, z), rather than less flexible ones, like GaveScruffy(x, y) and GaveCarl(x, y). This produces a more expressive language, and one that makes the logical relations between various claims more perspicuous.

Names can be introduced into a first-order language to refer to anything *objects* that can be considered an object. But we construe the notion of an "object" pretty flexibly—to cover anything that we can make claims about. We've already seen languages with names for people and the blocks of Tarski's World. Later in the chapter, we'll introduce languages with names for sets and numbers. Sometimes we will want to have names for still other kinds of "objects," like days or times. Suppose, for example, that we want to translate the sentences:

> *Claire gave Scruffy to Max on Saturday.*
> *Sunday, Max gave Scruffy to Evan.*

Here, we might introduce a four-place predicate Gave(w, x, y, z), meaning *w gave x to y on day z*, plus names for particular days, like last Saturday and last Sunday. The resulting translations would look something like this:

> Gave(claire, scruffy, max, saturday)
> Gave(max, scruffy, evan, sunday)

Designing a first-order language with just the right names and predicates requires some skill. Usually, the overall goal is to come up with a language that can say everything you want, but that uses the smallest "vocabulary" possible. Picking the right names and predicates is the key to doing this.

Exercises

1.8 Suppose we have two first-order languages: the first contains the binary predicates
✎ GaveScruffy(x, y) and GaveCarl(x, y), and the names max and claire; the second contains the ternary predicate Gave(x, y, z) and the names max, claire, scruffy, and carl.

  1. List all of the atomic sentences that can be expressed in the first language. (Some of these may say weird things like GaveScruffy(claire, claire), but don't worry about that.)
  2. How many atomic sentences can be expressed in the second language? (Count all of them, including odd ones like Gave(scruffy, scruffy, scruffy).)
  3. How many names and binary predicates would a language like the first need in order to say everything you can say in the second?

Table 1.2: Names and predicates for a language.

English	FOL	Comment
Names:		
Max	max	
Claire	claire	
Folly	folly	The name of a certain dog.
Carl	carl	The name of another dog.
Scruffy	scruffy	The name of a certain cat.
Pris	pris	The name of another cat.
2 pm, Jan 2, 2011	2:00	The name of a time.
2:01 pm, Jan 2, 2011	2:01	One minute later.
$\vdots$	$\vdots$	Similarly for other times.
Predicates:		
x is a pet	Pet(x)	
x is a person	Person(x)	
x is a student	Student(x)	
x is at home	Home(x)	
x is happy	Happy(x)	
t is earlier than t′	$t < t'$	Earlier-than for times.
x was hungry at time t	Hungry(x, t)	
x was angry at time t	Angry(x, t)	
x owned y at time t	Owned(x, y, t)	
x gave y to z at t	Gave(x, y, z, t)	
x fed y at time t	Fed(x, y, t)	

1.9 We will be giving a number of problems that use the symbols explained in Table 1.2. Start a new sentence file in Tarski's World and translate the following into FOL, using the names and predicates listed in the table. (You can switch to the Pets language in the Sentence toolbar. When you type, make sure they appear exactly as in the table; for example, use 2:00, not 2:00 pm or 2 pm.) All references to times are assumed to be to times on January 2, 2011.

1. *Claire owned Folly at 2 pm.*
2. *Claire gave Pris to Max at 2:05 pm.*
3. *Max is a student.*
4. *Claire fed Carl at 2 pm.*
5. *Folly belonged to Max at 3:05 pm.*
6. *2:00 pm is earlier than 2:05 pm.*

Name and submit your file in the usual way.

1.10 Translate the following into natural sounding, colloquial English, consulting Table 1.2.

1. Owned(max, scruffy, 2:00)
2. Fed(max, scruffy, 2:30)
3. Gave(max, scruffy, claire, 3:00)
4. 2:00 < 2:00

1.11 For each sentence in the following list, suggest a translation into an atomic sentence of FOL. In addition to giving the translation, explain what kinds of objects your names refer to and the intended meaning of the predicate you use.

1. *Max shook hands with Claire.*
2. *Max shook hands with Claire yesterday.*
3. *AIDS is less contagious than influenza.*
4. *Spain is between France and Portugal in size.*
5. *Misery loves company.*

SECTION 1.5

Function symbols

Some first-order languages have, in addition to names and predicates, other expressions that can appear in atomic sentences. These expressions are called *function symbols*. Function symbols allow us to form name-like terms from names and other name-like terms. They allow us to express, using atomic sentences, complex claims that could not be perspicuously expressed using just names and predicates. Some English examples will help clarify this.

function symbols

English has many sorts of noun phrases, expressions that can be combined with a verb phrase to get a sentence. Besides names like *Max* and *Claire*, other noun phrases include expressions like *Max's father, Claire's mother, Every girl who knows Max, No boy who knows Claire, Someone* and so forth. Each of these combines with a singular verb phrase such as *likes unbuttered popcorn* to make a sentence. But notice that the sentences that result have very different logical properties. For example,

Claire's mother likes unbuttered popcorn

implies that someone likes unbuttered popcorn, while

No boy who knows Claire likes unbuttered popcorn

does not.

Since these noun phrases have such different logical properties, they are treated differently in FOL. Those that intuitively refer to an individual are

terms

called "terms," and behave like the individual constants we have already discussed. In fact, individual constants are the simplest terms, and more complex terms are built from them using function symbols. Noun phrases like *No boy who knows Claire* are handled with very different devices, known as quantifiers, which we will discuss later.

complex terms

The FOL analog of the noun phrase *Max's father* is the term father(max). It is formed by putting a function symbol, father, in front of the individual constant max. The result is a complex term that we use to refer to the father of the person referred to by the name max. Similarly, we can put the function symbol mother together with the name claire and get the term mother(claire), which functions pretty much like the English term *Claire's mother*.

We can repeat this construction as many times as we like, forming more and more complex terms:

$$\text{father(father(max))}$$
$$\text{mother(father(claire))}$$
$$\text{mother(mother(mother(claire)))}$$

The first of these refers to Max's paternal grandfather, the second to Claire's paternal grandmother, and so forth.

These function symbols are called unary function symbols, because, like unary predicates, they take one argument. The resulting terms function just like names, and can be used in forming atomic sentences. For instance, the FOL sentence

$$\text{Taller(father(max), max)}$$

says that Max's father is taller than Max. Thus, in a language containing function symbols, the definition of atomic sentence needs to be modified to allow complex terms to appear in the argument positions in addition to names.

function symbols vs. predicates

Students often confuse function symbols with predicates, because both take terms as arguments. But there is a big difference. When you combine a unary function symbol with a term you do not get a sentence, but another term: something that refers (or should refer) to an object of some sort. This is why function symbols can be reapplied over and over again. As we have seen, the following makes perfectly good sense:

$$\text{father(father(max))}$$

This, on the other hand, is total nonsense:

$$\text{Dodec(Dodec(a))}$$

To help prevent this confusion, we will always capitalize predicates of FOL and leave function symbols and names in lower case.

Besides unary function symbols, FOL allows function symbols of any arity. Thus, for example, we can have binary function symbols. Simple English counterparts of binary function symbols are hard to come up with, but they are quite common in mathematics. For instance, we might have a function symbol sum that combines with two terms, t_1 and t_2, to give a new term, $sum(t_1, t_2)$, which refers to the sum of the numbers referred to by t_1 and t_2. Then the complex term $sum(3, 5)$ would give us another way of referring to 8. In a later section, we will introduce a function symbol to denote addition, but we will use infix notation, rather than prefix notation. Thus $3 + 5$ will be used instead of $sum(3, 5)$.

arity of function symbols

In FOL, just as we assume that every name refers to an actual object, we also assume that every complex term refers to exactly one object. This is a somewhat artificial assumption, since many function-like expressions in English don't always work this way. Though we may assume that

$$mother(father(father(max)))$$

refers to an actual (deceased) individual—one of Max's great-grandmothers— there may be other uses of these function symbols that don't seem to give us genuinely referring expressions. For example, perhaps the complex terms mother(adam) and mother(eve) fail to refer to any individuals, if Adam and Eve were in fact the first people. And certainly the complex term mother(3) doesn't refer to anything, since the number three has no mother. When designing a first-order language with function symbols, you should try to ensure that your complex terms always refer to unique, existing individuals.

The blocks world language as it is implemented in Tarski's World does not contain function symbols, but we could easily extend the language to include some. Suppose for example we introduced the function expressions fm, bm, lm and rm, that allowed us to form complex terms like:

functions symbols for blocks language

$$fm(a)$$
$$lm(bm(c))$$
$$rm(rm(fm(d)))$$

We could interpret these function symbols so that, for example, fm(a) refers to the frontmost block in the same column as a. Thus, if there are several blocks in the column with a, then fm(a) refers to whichever one is nearest the front. (Notice that fm(a) may not itself have a name; fm(a) may be our only way to refer to it.) If a is the only block in the column, or is the frontmost in its column, then fm(a) would refer to a. Analogously, bm, lm and rm could be interpreted to mean *backmost*, *leftmost* and *rightmost*, respectively.

With this interpretation, the term lm(bm(c)) would refer to the leftmost block in the same row as the backmost block in the same column as c. The atomic sentence Larger(lm(bm(c)), c) would then be true if and only if this block is larger than c.

Notice that in this expanded language, the sentence lm(bm(c)) = bm(lm(c)) is not always true. (Can you think of an example where it is false?) On the other hand, fm(fm(a)) = fm(a) is always true. Can you think of any other atomic sentences using these function symbols that are always true? How about sentences that are always false?

Remember

In a language with function symbols,

- Complex terms are typically formed by putting a function symbol of arity n in front of n terms (simple or complex).

- Complex terms are used just like names (simple terms) in forming atomic sentences.

- In FOL, complex terms are assumed to refer to one and only one object.

Exercises

1.12 Express in English the claims made by the following sentences of FOL as clearly as you can. You should try to make your English sentences as natural as possible. All the sentences are, by the way, true.
1. Taller(father(claire), father(max))
2. john = father(max)
3. Taller(claire, mother(mother(claire)))
4. Taller(mother(mother(max)), mother(father(max)))
5. mother(melanie) = mother(claire)

1.13 Assume that we have expanded the blocks language to include the function symbols fm, bm, lm and rm described earlier. Then the following formulas would all be sentences of the language:
1. Tet(lm(e))
2. fm(c) = c
3. bm(b) = bm(e)
4. FrontOf(fm(e), e)
5. LeftOf(fm(b), b)

    6. SameRow(rm(c), c)
    7. bm(lm(c)) = lm(bm(c))
    8. SameShape(lm(b), bm(rm(e)))
    9. d = lm(fm(rm(bm(d))))
   10. Between(b, lm(b), rm(b))

Fill in the following table with TRUE's and FALSE's according to whether the indicated sentence is true or false in the indicated world. Since Tarski's World does not understand the function symbols, you will not be able to check your answers. We have filled in a few of the entries for you. Turn in the completed table to your instructor.

	Leibniz's	Bolzano's	Boole's	Wittgenstein's
1.	TRUE			
2.				
3.				
4.				
5.	FALSE			
6.		TRUE		
7.				
8.			FALSE	
9.				
10.				

1.14 As you probably noticed in doing Exercise 1.13, three of the sentences came out true in all four worlds. It turns out that one of these three cannot be falsified in any world, because of the meanings of the predicates and function symbols it contains. Your goal in this problem is to build a world in which all of the other sentences in Exercise 1.13 come out false. When you have found such a world, submit it as World 1.14.

1.15 Suppose we have two first-order languages for talking about fathers. The first, which we'll call the functional language, contains the names claire, melanie, and jon, the function symbol father, and the predicates = and Taller. The second language, which we will call the relational language, has the same names, no function symbols, and the binary predicates =, Taller, and FatherOf, where FatherOf(c, b) means that c is the father of b. Translate the following atomic sentences from the relational language into the functional language. Be careful. Some atomic sentences, such as claire = claire, are in both languages! Such a sentence counts as a translation of itself.

    1. FatherOf(jon, claire)
    2. FatherOf(jon, melanie)

3. Taller(claire, melanie)

Which of the following atomic sentences of the functional language can be translated into atomic sentences of the relational language? Translate those that can be and explain the problem with those that can't.

4. father(melanie) = jon
5. father(melanie) = father(claire)
6. Taller(father(claire), father(jon))

When we add connectives and quantifiers to the language, we will be able to translate freely back and forth between the functional and relational languages.

1.16 Let's suppose that everyone has a favorite movie star. Given this assumption, make up a first-order language for talking about people and their favorite movie stars. Use a function symbol that allows you to refer to an individual's favorite actor, plus a relation symbol that allows you to say that one person is a better actor than another. Explain the interpretation of your function and relation symbols, and then use your language to express the following claims:

1. *Harrison is Nancy's favorite actor.*
2. *Nancy's favorite actor is better than Sean.*
3. *Nancy's favorite actor is better than Max's.*
4. *Claire's favorite actor's favorite actor is Brad.*
5. *Sean is his own favorite actor.*

1.17 Make up a first-order language for talking about people and their relative heights. Instead of using relation symbols like Taller, however, use a function symbol that allows you to refer to people's heights, plus the relation symbols = and <. Explain the interpretation of your function symbol, and then use your language to express the following two claims:

1. *George is taller than Sam.*
2. *Sam and Mary are the same height.*

Do you see any problem with this function symbol? If so, explain the problem. [Hint: What happens if you apply the function symbol twice?]

1.18 For each sentence in the following list, suggest a translation into an atomic sentence of FOL. In addition to giving the translation, explain what kinds of objects your names refer to and the intended meaning of the predicates and function symbols you use.

1. *Indiana's capital is larger than California's.*
2. *Hitler's mistress died in 1945.*
3. *Max shook Claire's father's hand.*
4. *Max is his father's son.*
5. *John and Nancy's eldest child is younger than Jon and Mary Ellen's.*

The first-order language of set theory

FOL was initially developed for use in mathematics, and consequently the most familiar first-order languages are those associated with various branches of mathematics. One of the most common of these is the language of set theory. This language has only two predicates, both binary. The first is the identity symbol, $=$, which we have already encountered, and the second is the symbol $\in$, for set membership.

predicates of set theory

It is standard to use infix notation for both of these predicates. Thus, in set theory, atomic sentences are always formed by placing individual constants on either side of one of the two predicates. This allows us to make identity claims, of the form $a = b$, and membership claims, of the form $a \in b$ (where a and b are individual constants).

A sentence of the form $a \in b$ is true if and only if the thing named by b is a set, and the thing named by a is a member of that set. For example, suppose a names the number 2 and b names the set $\{2, 4, 6\}$. Then the following table tells us which membership claims made up using these names are true and which are false.[2]

membership ($\in$)

$a \in a$	FALSE
$a \in b$	TRUE
$b \in a$	FALSE
$b \in b$	FALSE

Notice that there is one striking difference between the atomic sentences of set theory and the atomic sentences of the blocks language. In the blocks language, you can have a sentence, like LeftOf(a, b), that is true in a world, but which can be made false simply by moving one of the objects. Moving an object does not change the way the name works, but it can turn a true sentence into a false one, just as the sentence *Claire is sitting down* can go from true to false in virtue of Claire's standing up.

In set theory, we won't find this sort of thing happening. Here, the analog of a world is just a domain of objects and sets. For example, our domain might consist of all natural numbers, sets of natural numbers, sets of sets of natural numbers, and so forth. The difference between these "worlds" and those of Tarski's World is that the truth or falsity of the atomic sentences is determined entirely once the reference of the names is fixed. There is nothing that corresponds to moving the blocks around. Thus if the universe contains

---

[2]For the purposes of this discussion we are assuming that numbers are not sets, and that sets can contain either numbers or other sets as members.

the objects 2 and $\{2, 4, 6\}$, and if the names a and b are assigned to them, then the atomic sentences must get the values indicated in the previous table. The only way those values can change is if the names name different things. Identity claims also work this way, both in set theory and in Tarski's World.

Exercises

1.19 Which of the following atomic sentences in the first-order language of set theory are true
✎ and which are false? We use, in addition to a and b as above, the name c for 6 and d for
$\{2, 7, \{2, 4, 6\}\}$.

    1. $a \in c$
    2. $a \in d$
    3. $b \in c$
    4. $b \in d$
    5. $c \in d$
    6. $c \in b$

To answer this exercise, submit a Tarski's World sentence file with an uppercase **T** or **F** in each sentence slot to indicate your assessment.

SECTION 1.7
The first-order language of arithmetic

---

predicates ($=, <$) and functions ($+, \times$) of arithmetic

While neither the blocks language as implemented in Tarski's World nor the language of set theory has function symbols, there are languages that use them extensively. One such first-order language is the language of arithmetic. This language allows us to express statements about the natural numbers $0, 1, 2, 3, \ldots$, and the usual operations of addition and multiplication.

There are several more or less equivalent ways of setting up this language. The one we will use has two names, 0 and 1, two binary relation symbols, $=$ and $<$, and two binary function symbols, $+$ and $\times$. The atomic sentences are those that can be built up out of these symbols. We will use infix notation both for the relation symbols and the function symbols.

Notice that there are infinitely many different terms in this language (for example, $0, 1, (1 + 1), ((1 + 1) + 1), (((1 + 1) + 1) + 1), \ldots$), and so an infinite number of atomic sentences. Our list also shows that every natural number is named by some term of the language. This raises the question of how we can specify the set of terms in a precise way. We can't list them all explicitly, since there are too many. The way we get around this is by using what is known as an *inductive* definition.

Definition The terms of first-order arithmetic are formed in the following way:

    terms of arithmetic

1. The names $0, 1$ are terms.

2. If t_1, t_2 are terms, then the expressions $(t_1 + t_2)$ and $(t_1 \times t_2)$ are also terms.

3. Nothing is a term unless it can be obtained by repeated application of (1) and (2).

We should point out that this definition does indeed allow the function symbols to be applied over and over. Thus, $(1 + 1)$ is a term by clause 2 and the fact that 1 is a term. In which case $((1 + 1) \times (1 + 1))$ is also a term, again by clause 2. And so forth.

The third clause in the above definition is not as straightforward as one might want, since the phrase "can be obtained by repeated application of" is a bit vague. In Chapter 16, we will see how to give definitions like the above in a more satisfactory way, one that avoids this vague clause.

The atomic sentences in the language of first-order arithmetic are those that can be formed from the terms and the two binary predicate symbols, $=$ and $<$. So, for example, the FOL version of *1 times 1 is less than 1 plus 1* is the following:

    atomic sentences of arithmetic

$$(1 \times 1) < (1 + 1)$$

Exercises

1.20 ✎   Show that the following expressions are terms in the first-order language of arithmetic. Do this by explaining which clauses of the definition are applied and in what order. What numbers do they refer to?

    1. $(0 + 0)$
    2. $(0 + (1 \times 0))$
    3. $((1 + 1) + ((1 + 1) \times (1 + 1)))$
    4. $(((1 \times 1) \times 1) \times 1)$

1.21 ✎   Find a way to express the fact that three is less than four using the first-order language of arithmetic.

1.22 ✎⋆   Show that there are infinitely many terms in the first-order language of arithmetic referring to the number one.

SECTION 1.8
Alternative notation

As we said before, FOL is like a family of languages. But, as if that were not enough diversity, even the very same first-order language comes in a variety of dialects. Indeed, almost no two logic books use exactly the same notational conventions in writing first-order sentences. For this reason, it is important to have some familiarity with the different dialects—the different notational conventions—and to be able to translate smoothly between them. At the end of most chapters, we discuss common notational differences that you are likely to encounter.

Some notational differences, though not many, occur even at the level of atomic sentences. For example, some authors insist on putting parentheses around atomic sentences whose binary predicates are in infix position. So $(a = b)$ is used rather than $a = b$. By contrast, some authors omit parentheses surrounding the argument positions (and the commas between them) when the predicate is in prefix position. These authors use Rab instead of $R(a, b)$. We have opted for the latter simply because we use predicates made up of several letters, and the parentheses make it clear where the predicate ends and the arguments begin: Cubed is not nearly as perspicuous as Cube(d).

What is important in these choices is that sentences should be unambiguous and easy to read. Typically, the first aim requires parentheses to be used in one way or another, while the second suggests using no more than is necessary.

CHAPTER 2

The Logic of Atomic Sentences

A major concern in logic is the concept of *logical consequence*: When does one sentence, statement, or claim follow logically from others? In fact, one of the main motivations in the design of FOL was to make logical consequence as perspicuous as possible. It was thought that by avoiding the ambiguities and complexities of ordinary language, we would be able to recognize the consequences of our claims more easily. This is, to a certain extent, true; but it is also true that we should be able to recognize the consequences of our claims whether or not they are expressed in FOL.

In this chapter, we will explain what we mean by "logical consequence," or equivalently, what we mean when we say that an argument is "logically valid." This is a fairly simple concept to understand, but it can also be devilishly difficult to apply in specific cases. Indeed, in mathematics there are many, many examples where we do not know whether a given claim is a consequence of other known truths. Mathematicians may work for years or decades trying to answer such questions. After explaining the notion of consequence, we will describe the principal techniques for showing that a claim is or is not a consequence of other claims, and begin presenting what is known as a formal system of deduction, a system that allows us to show that a sentence of FOL is a consequence of others. We will continue developing this system as we learn more about FOL in later chapters.

Valid and sound arguments

Just what do we mean by logical consequence? Or rather, since this phrase is sometimes used in quite different contexts, what does a *logician* mean by logical consequence?

A few examples will help. First, let's say that an *argument* is any series of statements in which one (called the *conclusion*) is meant to follow from, or be supported by, the others (called the *premises*). Don't think of two people arguing back and forth, but of one person trying to convince another of some conclusion on the basis of mutually accepted premises. Arguments in our sense may appear as part of the more disagreeable sort of "arguments"— the kind parents have with their children—but *our* arguments also appear

arguments, premises, and conclusions

in newspaper editorials, in scholarly books, and in all forms of scientific and rational discourse. Name calling doesn't count.

There are many devices in ordinary language for indicating premises and conclusions of arguments. Words like *hence*, *thus*, *so*, and *consequently* are used to indicate that what follows is the conclusion of an argument. The words *because*, *since*, *after all*, and the like are generally used to indicate premises. Here are a couple of examples of arguments:

All men are mortal. Socrates is a man. So, Socrates is mortal.

Lucretius is a man. After all, Lucretius is mortal and all men are mortal.

One difference between these two arguments is the placement of the conclusion. In the first argument, the conclusion comes at the end, while in the second, it comes at the start. This is indicated by the words *so* and *after all*, respectively. A more important difference is that the first argument is good, while the second is bad. We will say that the first argument is *logically valid*, or that its conclusion is a *logical consequence* of its premises. The reason we say this is that it is impossible for this conclusion to be false if the premises are true. In contrast, our second conclusion might be false (suppose Lucretius is my pet goldfish), even though the premises are true (goldfish are notoriously mortal). The second conclusion is not a logical consequence of its premises.

Roughly speaking, an argument is logically valid if and only if the conclusion must be true on the assumption that the premises are true. Notice that this does not mean that an argument's premises have to be true in order for it to be valid. When we give arguments, we naturally intend the premises to be true, but sometimes we're wrong about that. We'll say more about this possibility in a minute. In the meantime, note that our first example above would be a valid argument even if it turned out that we were mistaken about one of the premises, say if Socrates turned out to be a robot rather than a man. It would still be impossible for the premises to be true and the conclusion false. In that eventuality, we would still say that the argument was logically valid, but since it had a false premise, we would not be guaranteed that the conclusion was true. It would be a valid argument with a false premise.

Here is another example of a valid argument, this time one expressed in the blocks language. Suppose we are told that Cube(c) and that c = b. Then it certainly follows that Cube(b). Why? Because there is no possible way for the premises to be true—for c to be a cube and for c to be the very same object as b—without the conclusion being true as well. Note that we can recognize that the last statement is a consequence of the first two without knowing that the premises are actually, as a matter of fact, true. For the crucial observation is that *if* the premises are true, *then* the conclusion must also be true.

A valid argument is one that guarantees the truth of its conclusion on the assumption that the premises are true. Now, as we said before, when we actually present arguments, we want them to be more than just valid: we also want the premises to be true. If an argument is valid and the premises are also true, then the argument is said to be *sound*. Thus a sound argument insures the truth of its conclusion. The argument about Socrates given above was not only valid, it was sound, since its premises were true. (He was not, contrary to rumors, a robot.) But here is an example of a valid argument that is not sound:

sound arguments

> All rich actors are good actors. Brad Pitt is a rich actor. So he must
> be a good actor.

The reason this argument is unsound is that its first premise is false. Because of this, although the argument is indeed valid, we are not assured that the conclusion is true. It may be, but then again it may not. We in fact think that Brad Pitt is a good actor, but the present argument does not show this.

Logic focuses, for the most part, on the validity of arguments, rather than their soundness. There is a simple reason for this. The truth of an argument's premises is generally an issue that is none of the logician's business: the truth of "Socrates is a man" is something historians had to ascertain; the falsity of "All rich actors are good actors" is something a movie critic might weigh in about. What logicians can tell you is how to reason correctly, *given* what you know or believe to be true. Making sure that the premises of your arguments are true is something that, by and large, we leave up to you.

In this book, we often use a special format to display arguments, which we call "Fitch format" after the logician Frederic Fitch. The format makes clear which sentences are premises and which is the conclusion. In Fitch format, we would display the above, unsound argument like this:

Fitch format

> All rich actors are good actors.
> Brad Pitt is a rich actor.
>
> Brad Pitt is a good actor.

Here, the sentences above the short, horizontal line are the premises, and the sentence below the line is the conclusion. We call the horizontal line the Fitch bar. Notice that we have omitted the words "So ... must be ..." in the conclusion, because they were in the original only to make clear which sentence was supposed to be the conclusion of the argument. In our conventional

Fitch bar

format, the Fitch bar gives us this information, and so these words are no longer needed.

Remember

1. An *argument* is a series of statements in which one, called the *conclusion*, is meant to be a consequence of the others, called the *premises*.

2. An argument is *valid* if the conclusion must be true in any circumstance in which the premises are true. We say that the conclusion of a logically valid argument is a *logical consequence* of its premises.

3. An argument is *sound* if it is valid and the premises are all true.

Exercises

2.1 (Classifying arguments) Open the file **Socrates' Sentences**. This file contains eight arguments ✦|✎ separated by dashed lines, with the premises and conclusion of each labeled.

1. In the first column of the following table, classify each of these arguments as valid or invalid. In making these assessments, you may presuppose any general features of the worlds that can be built in Tarski's World (for example, that two blocks cannot occupy the same square on the grid).

Argument	Valid?	Sound in Socrates' World?	Sound in Wittgenstein's World?
1.			
2.			
3.			
4.			
5.			
6.			
7.			
8.			

2. Now open **Socrates' World** and evaluate each sentence. Use the results of your evaluation to enter *sound* or *unsound* in each row of the second column in the table, depending on whether the argument is sound or unsound in this world. (Remember that only valid arguments can be sound; invalid arguments are automatically unsound.)

3. Open **Wittgenstein's World** and fill in the third column of the table.

4. For each argument that you have marked invalid in the table, construct a world in which the argument's premises are all true but the conclusion is false. Submit the world as World 2.1.x, where x is the number of the argument. (If you have trouble doing this, you may want to rethink your assessment of the argument's validity.) Turn in your completed table to your instructor.

This problem makes a very important point, one that students of logic sometimes forget. The point is that the validity of an argument depends only on the argument, not on facts about the specific world the statements are about. The soundness of an argument, on the other hand, depends on both the argument and the world.

By the way, the Grade Grinder will only tell you that the files that you submit are or are not counterexamples. For obvious reasons, if there is a counterexample to an argument but you don't submit one, the Grade Grinder will not complain (to you, but it will tell the instructor).

2.2 (Classifying arguments) For each of the arguments below, identify the premises and conclusion by putting the argument into Fitch format. Then say whether the argument is valid. For the first five arguments, also give your opinion about whether they are sound. (Remember that only valid arguments can be sound.) If your assessment of an argument depends on particular interpretations of the predicates, explain these dependencies.

1. Anyone who wins an academy award is famous. Meryl Streep won an academy award. Hence, Meryl Streep is famous.
2. Harrison Ford is not famous. After all, actors who win academy awards are famous, and he has never won one.
3. The right to bear arms is the most important freedom. Charlton Heston said so, and he's never wrong.
4. Al Gore must be dishonest. After all, he's a politician and hardly any politicians are honest.
5. Mark Twain lived in Hannibal, Missouri, since Sam Clemens was born there, and Mark Twain *is* Sam Clemens.
6. No one under 21 bought beer here last night, officer. Geez, we were closed, so no one bought anything last night.
7. Claire must live on the same street as Laura, since she lives on the same street as Max and he and Laura live on the same street.

2.3 For each of the arguments below, identify the premises and conclusion by putting the argument into Fitch format, and state whether the argument is valid. If your assessment of an argument depends on particular interpretations of the predicates, explain these dependencies.

1. Many of the students in the film class attend film screenings. Consequently, there must be many students in the film class.
2. There are few students in the film class, but many of them attend the film screenings. So there are many students in the film class.

3. There are many students in the film class. After all, many students attend film screenings and only students in the film class attend screenings.

4. There are thirty students in my logic class. Some of the students turned in their homework on time. Most of the students went to the all-night party. So some student who went to the party managed to turn in the homework on time.

5. There are thirty students in my logic class. Some student who went to the all-night party must have turned in the homework on time. Some of the students turned in their homework on time, and they all went to the party.

6. There are thirty students in my logic class. Most of the students turned in their homework on time. Most of the students went to the all-night party. Thus, some student who went to the party turned in the homework on time.

2.4 (Validity and truth) Can a valid argument have false premises and a false conclusion? False premises and a true conclusion? True premises and a false conclusion? True premises and a true conclusion? If you answer *yes* to any of these, give an example of such an argument. If your answer is *no*, explain why.

SECTION 2.2
Methods of proof

Our description of the logical consequence relation is fine, as far as it goes. But it doesn't give us everything we would like. In particular, it does not tell us how to *show* that a given conclusion S follows, or does not follow, from some premises P, Q, R, ... In the examples we have looked at, this may not seem very problematic, since the answers are fairly obvious. But when we are dealing with more complicated sentences or more subtle reasoning, things are sometimes far from simple.

In this course you will learn the fundamental methods of showing when claims follow from other claims and when they do not. The main technique for doing the latter, for showing that a given conclusion does *not* follow from some premises, is to find a possible circumstance in which the premises are true but the conclusion false. In fact we have already used this method to show that the argument about Lucretius was invalid. We will use the method repeatedly, and introduce more precise versions of it as we go on.

What methods are available to us for showing that a given claim *is* a logical consequence of some premises? Here, the key notion is that of a *proof*.
proof   A proof is a step-by-step demonstration that a conclusion (say S) follows from some premises (say P, Q, R). The way a proof works is by establishing a

series of intermediate conclusions, each of which is an obvious consequence of the original premises and the intermediate conclusions previously established. The proof ends when we finally establish S as an obvious consequence of the original premises and the intermediate conclusions. For example, from P, Q, R it might be obvious that S_1 follows. And from all of these, including S_1, it might be obvious that S_2 follows. Finally, from all these together we might be able to draw our desired conclusion S. If our individual steps are correct, then the proof shows that S is indeed a consequence of P, Q, R. After all, if the premises are all true, then our intermediate conclusions must be true as well. And in that case, our final conclusion must be true, too.

Consider a simple, concrete example. Suppose we want to show that *Socrates sometimes worries about dying* is a logical consequence of the four premises *Socrates is a man*, *All men are mortal*, *No mortal lives forever*, and *Everyone who will eventually die sometimes worries about it*. A proof of this conclusion might pass through the following intermediate steps. First we note that from the first two premises it follows that Socrates is mortal. From this intermediate conclusion and the third premise (that no mortal lives forever), it follows that Socrates will eventually die. But this, along with the fourth premise, gives us the desired conclusion, that Socrates sometimes worries about dying.

By the way, when we say that S is a logical consequence of premises P, Q, ..., we do not insist that each of the premises really play an essential role. So, for example, if S is a logical consequence of P then it is also a logical consequence of P and Q. This follows immediately from the definition of logical consequence. But it has a corollary for our notion of proof: We do not insist that each of the premises in a proof actually be used in the proof.

A proof that S follows from premises $P_1, \ldots, P_n$ may be quite long and complicated. But each step in the proof is supposed to provide absolutely incontrovertible evidence that the intermediate conclusion follows from things already established. Here, the logician's standards of rigor are extreme. It is not enough to show that each step in a purported proof almost certainly follows from the ones that come before. That may be good enough for getting around in our daily life, but it is not good enough if our aim is to demonstrate that S *must* be true provided $P_1, \ldots, P_n$ are all true.

demand for rigor

There is a practical reason for this demand for rigor. In ordinary life, we frequently reason in a step-by-step fashion, without requiring absolute certainty at each step. For most purposes, this is fine, since our everyday "proofs" generally traverse only a small number of intermediate conclusions. But in many types of reasoning, this is not the case.

Think of what you did in high school geometry. First you started with a small number of axioms that stated the basic premises of Euclidean geometry.

You then began to prove conclusions, called theorems, from these axioms. As you went on to prove more interesting theorems, your proofs would cite earlier theorems. These earlier theorems were treated as intermediate conclusions in justifying the new results. What this means is that the complete proofs of the later theorems really include the proofs of the earlier theorems that they presuppose. Thus, if they were written out in full, they would contain hundreds or perhaps thousands of steps. Now suppose we only insisted that each step show with probability .99 that the conclusion follows from the premises. Then each step in such a proof would be a pretty good bet, but given a long enough proof, the proof would carry virtually no weight at all about the truth of the conclusion.

This demand for certainty becomes even more important in proofs done by computers. Nowadays, theorems are sometimes proven by computers, and the proofs can be millions of steps long. If we allowed even the slightest uncertainty in the individual steps, then this uncertainty would multiply until the alleged "proof" made the truth of the conclusion no more likely than its falsity.

methods of proof

Each time we introduce new types of expressions into our language, we will discuss new methods of proof supported by those expressions. We begin by discussing the main informal methods of proof used in mathematics, science, and everyday life, emphasizing the more important methods like indirect and conditional proof. Following this discussion we will "formalize" the methods by incorporating them into what we call a formal system of deduction. A formal system of deduction uses a fixed set of rules specifying what counts as an acceptable step in a proof.

formal systems

informal proofs

The difference between an informal proof and a formal proof is not one of rigor, but of style. An informal proof of the sort used by mathematicians is every bit as rigorous as a formal proof. But it is stated in English and is usually more free-wheeling, leaving out the more obvious steps. For example, we could present our earlier argument about Socrates in the form of the following informal proof:

> **Proof:** Since Socrates is a man and all men are mortal, it follows that Socrates is mortal. But all mortals will eventually die, since that is what it means to be mortal. So Socrates will eventually die. But we are given that everyone who will eventually die sometimes worries about it. Hence Socrates sometimes worries about dying.

formal proofs

A formal proof, by contrast, employs a fixed stock of rules and a highly stylized method of presentation. For example, the simple argument from Cube(c) and c = b to Cube(b) discussed in the last section will, in our formal system, take the following form:

> 1. Cube(c)
> 2. c = b
>
> 3. Cube(b)   = Elim: 1, 2

As you can see, we use an extension of the Fitch format as a way of presenting formal proofs. The main difference is that a formal proof will usually have more than one step following the Fitch bar (though not in this example), and each of these steps will be justified by citing a rule of the formal system. We will explain later the various conventions used in formal proofs.

In the course of this book you will learn how to give both informal and formal proofs. We do not want to give the impression that formal proofs are somehow better than informal proofs. On the contrary, for purposes of proving things for ourselves, or communicating proofs to others, informal methods are usually preferable. Formal proofs come into their own in two ways. One is that they display the logical structure of a proof in a form that can be mechanically checked. There are advantages to this, if you are a logic teacher grading lots of homework, a computer, or not inclined to think for some other reason. The other is that they allow us to prove things about provability itself, such as Gödel's Completeness Theorem and Incompleteness Theorems, discussed in the final section of this book.

formal vs. informal proofs

Remember

1. A proof of a statement S from premises $P_1, \ldots, P_n$ is a step-by-step demonstration which shows that S *must* be true in any circumstances in which the premises $P_1, \ldots, P_n$ are all true.

2. Informal and formal proofs differ in style, not in rigor.

Proofs involving the identity symbol

We have already seen one example of an important method of proof. If we can prove, from whatever our premises happen to be, that $b = c$, then we know that anything that is true of b is also true of c. After all, b *is* c. In philosophy, this simple observation sometimes goes by the fancy name *the indiscernibility of identicals* and sometimes by the less pretentious name *substitution*. Shakespeare no doubt had this principle in mind when he wrote "A rose, by any other name, would smell as sweet."

indiscernibility of identicals

We will call the formal rule corresponding to this principle *Identity Elimi-*

identity elimination

nation, abbreviated = **Elim**. The reason for this name is that an application of this rule "eliminates" a use of the identity symbol when we move from the premises of the argument to its conclusion. We will have another rule that introduces the identity symbol.

The principle of identity elimination is used repeatedly in mathematics. For example, the following derivation uses the principle in conjunction with the well-known algebraic identity $x^2 - 1 = (x - 1)(x + 1)$:

$$x^2 > x^2 - 1$$

so

$$x^2 > (x - 1)(x + 1)$$

We are all familiar with reasoning that uses such substitutions repeatedly.

reflexivity of identity or identity introduction

Another principle, so simple that one often overlooks it, is the so-called reflexivity of identity. The formal rule corresponding to it is called *Identity Introduction*, or = **Intro**, since it allows us to introduce identity statements into proofs. It tells us that any sentence of the form a = a can be validly inferred from whatever premises are at hand, or from no premises at all. This is because of the assumption made in FOL that names always refer to one and only one object. This is not true about English, as we have noted before. But it is in FOL, which means that in a proof you can always take any name a that is in use and assert a = a, if it suits your purpose for some reason. (As a matter of fact, it is rarely of much use.) Gertrude Stein was surely referring to this principle when she observed "A rose is a rose is a rose."

symmetry of identity

Another principle, a bit more useful, is that of the symmetry of identity. It allows us to conclude b = a from a = b. Actually, if we wanted, we could derive this as a consequence of our first two principles, by means of the following proof.

> **Proof:** Suppose that a = b. We know that a = a, by the reflexivity of identity. Now substitute the name b for the first use of the name a in a = a, using the indiscernibility of identicals. We come up with b = a, as desired.

The previous paragraph is another example of an informal proof. In an informal proof, we often begin by stating the premises or assumptions of the proof, and then explain in a step-by-step fashion how we can get from these assumptions to the desired conclusion. There are no strict rules about how detailed the explanation needs to be. This depends on the sophistication of the intended audience for the proof. But each step must be phrased in clear and unambiguous English, and the validity of the step must be apparent. In the next section, we will see how to formalize the above proof.

A third principle about identity that bears noting is its so-called transiti- *transitivity of identity*
vity. If $a = b$ and $b = c$ are both true, then so is $a = c$. This is so obvious
that there is no particular need to prove it, but it can be proved using the
indiscernibility of identicals. (See Exercise 2.5.)

If you are using a language that contains function symbols (introduced
in the optional Section 1.5), the identity principles we've discussed also hold
for complex terms built up using function symbols. For example, if you know
that Happy(john) and john = father(max), you can use identity elimination to
conclude Happy(father(max)), even though father(max) is a complex term, not
a name. In fact, the example where we substituted $(x - 1)(x + 1)$ for $x^2 - 1$
also applied the indiscernibility of identicals to complex terms.

Remember

There are four important principles that hold of the identity relation:

1. **= Elim:** If $b = c$, then whatever holds of b holds of c. This is also
   known as the **indiscernibility of identicals.**

2. **= Intro:** Sentences of the form $b = b$ are always true (in FOL). This
   is also known as the **reflexivity of identity.**

3. **Symmetry of Identity:** If $b = c$, then $c = b$.

4. **Transitivity of Identity:** If $a = b$ and $b = c$, then $a = c$.

The latter two principles follow from the first two.

Proofs involving other predicates and relations

Sometimes there will be logical dependencies among the predicate symbols in
a first-order language, dependencies similar to those just discussed involving
the identity symbol. This is the case, for example, in the blocks language.
When this is so, proofs may need to exploit such relationships. For example,
the sentence Larger(a, c) is a consequence of Larger(a, b) and Larger(b, c). This *other transitive*
is because the larger-than relation, like the identity relation, is transitive. It is *relations*
because of this that any world where the latter two sentences are true will also
be one in which the first is true. Similar examples are given in the problems.

Another example of this sort that is used frequently in mathematics involves the transitivity of the less-than relation. You frequently encounter

proofs written in the following form:

$$k_1 \ < \ k_2$$
$$k_2 \ < \ k_3$$
$$k_3 \ < \ k_4$$

so

$$k_1 \ < \ k_4$$

This proof contains two implicit uses of the transitivity of $<$.

There is no way to catalog all the legitimate inferences involving predicate and relation symbols in all the languages we might have occasion to deal with. But the example of identity gives us a few things to look for. Many relations besides identity are transitive: *larger than* and *less than* are just two examples. *reflexive and symmetric relations* And many are reflexive and/or symmetric: *being the same size as* and *being in the same row as* are both. But you will run across other logical dependencies that don't fall under these headings. For instance, you might be told that b is larger than c and want to infer that c is smaller than b. This holds because *inverse relations* *larger than* and *smaller than* are what is known as "inverses": they refer to the same relation but point, so to speak, in opposite directions. Usually, you will have no trouble spotting the logical dependencies among predicates, but in giving a proof, you need to make explicit the ones you are assuming.

Let's look at one final example before trying our hand at some exercises. Suppose we were asked to give an informal proof of the following argument:

> RightOf(b, c)
> LeftOf(d, e)
> b = d
> ——
> LeftOf(c, e)

Our informal proof might run like this:

> **Proof:** We are told that b is to the right of c. So c must be to the left of b, since *right of* and *left of* are inverses of one another. And since $b = d$, c is left of d, by the indiscernibility of identicals. But we are also told that d is left of e, and consequently c is to the left of e, by the transitivity of *left of*. This is our desired conclusion.

Exercises

2.5 (Transitivity of Identity) Give an informal proof of the following argument using only indiscernibility of identicals. Make sure you say which name is being substituted for which, and in what sentence.

$$
\begin{array}{|l}
b = c \\
a = b \\
\hline
a = c
\end{array}
$$

2.6 Give an informal proof that the following argument is valid. If you proved the transitivity of identity by doing Exercise 2.5, you may use this principle; otherwise, use only the indiscernibility of identicals.

$$
\begin{array}{|l}
\text{SameRow}(a, a) \\
a = b \\
b = c \\
\hline
\text{SameRow}(c, a)
\end{array}
$$

2.7 Consider the following sentences.

1. *Max and Claire are not related.*
2. *Nancy is Max's mother.*
3. *Nancy is not Claire's mother.*

Does (3) follow from (1) and (2)? Does (2) follow from (1) and (3)? Does (1) follow from (2) and (3)? In each case, if your answer is *no*, describe a possible circumstance in which the premises are true and the conclusion false.

Given the meanings of the atomic predicates in the blocks language, assess the following arguments for validity. (You may again assume any general facts about the worlds that can be built in Tarski's World.) If the argument is valid, give an informal proof of its validity and turn it in on paper to your instructor. If the conclusion is not a consequence of the premises, submit a world in which the premises are true and the conclusion false.

2.8
$$
\begin{array}{|l}
\text{Large}(a) \\
\text{Larger}(a, c) \\
\hline
\text{Small}(c)
\end{array}
$$

2.9
$$
\begin{array}{|l}
\text{LeftOf}(a, b) \\
b = c \\
\hline
\text{RightOf}(c, a)
\end{array}
$$

2.10
$$
\begin{array}{|l}
\text{SameSize}(b, c) \\
\text{SameShape}(b, c) \\
\hline
b = c
\end{array}
$$

2.11
$$
\begin{array}{|l}
\text{LeftOf}(a, b) \\
\text{RightOf}(c, a) \\
\hline
\text{LeftOf}(b, c)
\end{array}
$$

2.12
$$
\begin{array}{|l}
\text{BackOf}(a, b) \\
\text{FrontOf}(a, c) \\
\hline
\text{FrontOf}(b, c)
\end{array}
$$

2.13
$$
\begin{array}{|l}
\text{SameSize}(a, b) \\
\text{Larger}(a, c) \\
\text{Smaller}(d, c) \\
\hline
\text{Smaller}(d, b)
\end{array}
$$

2.14

Between(b, a, c)
LeftOf(a, c)

LeftOf(a, b)

SECTION 2.3
Formal proofs

deductive systems

the system $\mathcal{F}$

In this section we will begin introducing our system for presenting formal proofs, what is known as a "deductive system." There are many different styles of deductive systems. The system we present in the first two parts of the book, which we will call $\mathcal{F}$, is a "Fitch-style" system, so called because Frederic Fitch first introduced this format for giving proofs. We will look at a very different deductive system in Part III, one known as the resolution method, which is of considerable importance in computer science.

In the system $\mathcal{F}$, a proof of a conclusion S from premises P, Q, and R, looks very much like an argument presented in Fitch format. The main difference is that the proof displays, in addition to the conclusion S, all of the intermediate conclusions $S_1, \ldots, S_n$ that we derive in getting from the premises to the conclusion S:

P
Q
R

S_1 **Justification 1**
⋮   ⋮
S_n **Justification n**
S   **Justification n+1**

There are two graphical devices to notice here, the vertical and horizontal lines. The vertical line that runs on the left of the steps draws our attention to the fact that we have a single purported proof consisting of a sequence of several steps. The horizontal Fitch bar indicates the division between the claims that are assumed and those that allegedly follow from them. Thus the fact that P, Q, and R are above the bar shows that these are the premises of our proof, while the fact that $S_1, \ldots, S_n$, and S are below the bar shows that these sentences are supposed to follow logically from the premises.

Notice that on the right of every step below the Fitch bar, we give a *justification* of the step. In our deductive system, a justification indicates which rule allows us to make the step, and which earlier steps (if any) the rule is applied to. In giving an actual formal proof, we will number the steps, so we can refer to them in justifying later steps.

We already gave one example of a formal proof in the system $\mathcal{F}$, back on page 48. For another example, here is a formalization of our informal proof of the symmetry of identity.

$$
\begin{array}{ll}
\quad 1.\ a = b & \\
\quad 2.\ a = a & =\textbf{ Intro} \\
\quad 3.\ b = a & =\textbf{ Elim: } 2,\ 1
\end{array}
$$

In the right hand margin of this proof you find a justification for each step below the Fitch bar. These are applications of rules we are about to introduce. The numbers at the right of step 3 show that this step follows from steps 2 and 1 by means of the rule cited.

The first rule we use in the above proof is **Identity Introduction**. This rule allows you to introduce, for any name (or complex term) n in use in the proof, the assertion n = n. You are allowed to do this at any step in the proof, and need not cite any earlier step as justification. We will abbreviate our statement of this rule in the following way:

Identity Introduction (= Intro):

$$
\triangleright \quad \big|\ n = n
$$

We have used an additional graphical device in stating this rule. This is the symbol $\triangleright$. We will use it in stating rules to indicate which step is being licensed by the rule. In this example there is only one step mentioned in the rule, but in other examples there will be several steps.

The second rule of $\mathcal{F}$ is **Identity Elimination**. It tells us that if we have proven a sentence containing n (which we indicate by writing P(n)) and a sentence of the form n = m, then we are justified in asserting any sentence which results from P(n) by replacing some or all of the occurrences of n by m.

Identity Elimination (= Elim):

$$
\begin{array}{l}
\quad P(n) \\
\quad \vdots \\
\quad n = m \\
\quad \vdots \\
\rhd \quad P(m)
\end{array}
$$

When we apply this rule, it does not matter which of $P(n)$ and $n = m$ occurs first in the proof, as long as they both appear before $P(m)$, the inferred step. In justifying the step, we cite the name of the rule, followed by the steps in which $P(n)$ and $n = m$ occur, in that order.

We could also introduce rules justified by the meanings of other predicates besides = into the system $\mathcal{F}$. For example, we could introduce a formal rule of the following sort:

Bidirectionality of Between:

$$
\begin{array}{l}
\quad \text{Between}(a, b, c) \\
\quad \vdots \\
\rhd \quad \text{Between}(a, c, b)
\end{array}
$$

We don't do this because there are just too many such rules. We could state them for a few predicates, but certainly not all of the predicates you will encounter in first-order languages.

Reiteration

There is one rule that is not technically necessary, but which will make some proofs look more natural. This rule is called **Reiteration**, and simply allows you to repeat an earlier step, if you so desire.

Reiteration (Reit):

$$
\begin{array}{l}
\quad P \\
\quad \vdots \\
\rhd \quad P
\end{array}
$$

To use the Reiteration rule, just repeat the sentence in question and, on the right, write "**Reit:** x," where x is the number of the earlier occurrence of the sentence.

Reiteration is obviously a valid rule of inference, since any sentence is a logical consequence of itself. The reason for having the rule will become clear as proofs in the system $\mathcal{F}$ become more complicated. For now, let's just say that it is like remarking, in the course of giving an informal proof, "we have already shown that P." This is often a helpful reminder to the person reading the proof.

Now that we have the first three rules of $\mathcal{F}$, let's try our hand constructing a formal proof. Suppose we were asked to prove SameRow(b, a) from the premises SameRow(a, a) and b = a. We might begin by writing down the premises and the conclusion, leaving space in between to fill in the intermediate steps in our proof.

> 1. SameRow(a, a)
> 2. b = a
> ⋮
> ?. SameRow(b, a)

It might at first seem that this proof should be a one step application of = **Elim**. But notice that the way we have stated this rule requires that we replace the first name in the identity sentence, b, for the second, a, but we want to substitute the other way around. So we need to derive a = b as an intermediate conclusion before we can apply = **Elim**.

> 1. SameRow(a, a)
> 2. b = a
> ⋮
> ?. a = b
> ?. SameRow(b, a)   = **Elim:** 1, ?

Since we have already seen how to prove the symmetry of identity, we can now fill in all the steps of the proof. The finished proof looks like this. Make sure you understand why all the steps are there and how we arrived at them.

> 1. SameRow(a, a)
> 2. b = a
>
> 3. b = b   = **Intro**
> 4. a = b   = **Elim:** 3, 2
> 5. SameRow(b, a)   = **Elim:** 1, 4

Constructing proofs in Fitch

the program Fitch

Writing out a long formal proof in complete detail, let alone reading or check-ing it, can be a pretty tedious business. The system $\mathcal{F}$ makes this less painful than many formal systems, but it's still not easy. This book comes with a sec-ond program, Fitch, that makes constructing formal proofs much less painful. Fitch can also check your proof, telling you whether it is correct, and if it isn't, which step or steps are mistaken. This means you will never be in any doubt about whether your formal proofs meet the standard of rigor demanded of them. And, as a practical matter, you can make sure they are correct before submitting them.

Fitch vs. $\mathcal{F}$

There are other ways in which Fitch makes life simpler, as well. One is that Fitch is more flexible than the system $\mathcal{F}$. It lets you take certain shortcuts that are logically correct but do not, strictly speaking, fall under the rules of $\mathcal{F}$. You can always go back and expand a proof in Fitch to a formally correct $\mathcal{F}$ proof, but we won't often insist on this.

Let us now use Fitch to construct a simple formal proof. Before going on, you will want to read the first few sections of the chapter on how to use Fitch in the manual.

You try it
. . . .

▶   1. We are going to use Fitch to construct the formal proof of SameRow(b, a) from premises SameRow(a, a) and b = a. Launch Fitch and open the file Identity 1. Here we have the beginnings of the formal proof. The premises appear above the Fitch bar. It may look slightly different from the proofs we have in the book, since in Fitch the steps don't have to be numbered, for reasons we'll soon find out. (If you would like to have numbered steps, you can choose **Show Step Numbers** from the **Proof** menu. But don't try this yet.)

▶   2. Before we start to construct the proof, notice that at the bottom of the proof window there is a separate panel called the "goal strip," contain-ing the goal of the proof. In this case the goal is to prove the sentence SameRow(b, a). If we successfully satisfy this goal, we will be able to get Fitch to put a checkmark to the right of the goal.

▶   3. Let's construct the proof. What we need to do is fill in the steps needed to complete the proof, just as we did at the end of the last section. Add a new step to the proof by choosing **Add Step After** from the **Proof** menu. In the new step, enter the sentence a = b, either by typing it in or by using the toolbar at the top of the proof window. We will first use this step to get our conclusion and then go back and prove this step.

4. Once you have entered a = b, add another step below this and enter the goal sentence SameRow(b, a). Use the mouse to click on the word **Rule?** that appears to the right of SameRow(b, a). In the menu that pops up, go to the Elimination Rules and select =. If you did this right, the rule name should now say = **Elim**. If not, try again.

5. Next cite the first premise and the intermediate sentence you first entered. You do this in Fitch by clicking on the two sentences, in either order. If you click on the wrong one, just click again and it will be un-cited. Once you have the right sentences cited, choose **Verify Proof** from the **Proof** menu. The last step should now check out, as it is a valid instance of = **Elim**. The step containing a = b will not check out, since we haven't yet indicated what it follows from. Nor will the goal check out, since we don't yet have a complete proof of SameRow(b, a). All in good time.

6. Now add a step before the first introduced step (the one containing a = b), and enter the sentence b = b. Do this by moving the focus slider (the triangle in the left margin) to the step containing a = b and choosing **Add Step Before** from the **Proof** menu. (If the new step appears in the wrong place, choose **Delete Step** from the **Proof** menu.) Enter the sentence b = b and justify it by using the rule = **Intro**. Check the step.

7. Finally, justify the step containing a = b by using the = **Elim** rule. You will need to move the focus slider to this step, and then cite the second premise and the sentence b = b. Now the whole proof, including the goal, should check out. To find out if it does, choose **Verify Proof** from the **Proof** menu. The proof should look like the completed proof on page 57, except for the absence of numbers on the steps. (Try out **Show Step Numbers** from the **Proof** menu now. The highlighting on support steps will go away and numbers will appear, just like in the book.)

8. We mentioned earlier that Fitch lets you take some shortcuts, allowing you to do things in one step that would take several if we adhered strictly to $\mathcal{F}$. This proof is a case in point. We have constructed a proof that falls under $\mathcal{F}$ but Fitch actually has symmetry of identity built into = **Elim**. So we could prove the conclusion directly from the two premises, using a single application of the rule = **Elim**. We'll do this next.

▶ 9. Add another step at the very end of your proof. Here's a trick you will find handy: Click on the goal sentence at the very bottom of the window. This puts the focus on the goal sentence. Choose **Copy** from the **Edit** menu, and then click back on the empty step at the end of your proof. Choose **Paste** from the **Edit** menu and the goal sentence will be entered into this step. This time, justify the new step using = **Elim** and citing just the two premises. You will see that the step checks out.

▶ 10. Save your proof as Proof Identity 1.

. . . . *Congratulations*

Since the proof system $\mathcal{F}$ does not have any rules for atomic predicates other than identity, neither does Fitch. However, Fitch does have a mechanism that, among other things, lets you check for consequences among atomic sentences that involve many of the predicates in the blocks world language.[1]

Analytic Consequence   This is a rule we call **Analytic Consequence** or **Ana Con** for short. **Ana Con** is not restricted to atomic sentences, but that is the only application of the rule we will discuss at the moment. This rule allows you to cite some sentences in support of a claim if any world that makes the cited sentences true also makes the conclusion true, given the meaning of the predicates as used in Tarski's World. Let's get a feeling for **Ana Con** with some examples.

You try it
. . . .

▶ 1. Use Fitch to open the file Ana Con 1. In this file you will find nine premises followed by six conclusions that are consequences of these premises. Indeed, each of the conclusions follows from three or fewer of the premises.

▶ 2. Position the focus slider (the little triangle) at the first conclusion following the Fitch bar, SameShape(c, b). We have invoked the rule **Ana Con** but we have not cited any sentences. This conclusion follows from Cube(b) and Cube(c). Cite these sentences and check the step.

▶ 3. Now move the focus slider to the step containing SameRow(b, a). Since the relation of being in the same row is symmetric and transitive, this follows from SameRow(b, c) and SameRow(a, c). Cite these two sentences and check the step.

---

[1]This mechanism does not handle the predicates Adjoins and Between, due to the complexity of the ways the meanings of these predicates interact with the others.

4. The third conclusion, BackOf(e, c), follows from three of the premises. See ◀
if you can find them. Cite them. If you get it wrong, Fitch will give you
an X when you try to check the step.

5. Now fill in the citations needed to make the fourth and fifth conclusions ◀
check out. For these, you will have to invoke the **Ana Con** rule yourself. (You will find the rule on the **Con** submenu of the **Rule?** popup.)
Remember, you may only cite the premises, not previous conclusions.

6. The final conclusion, SameCol(b, b), does not require that any premises be ◀
cited in support. It is simply an analytic truth, that is, true in virtue of
its meaning. Specify the rule and check this step.

7. When you are done, choose **Verify Proof** to see that all the goals check ◀
out. Save your work as Proof Ana Con 1.

. . . . *Congratulations*

The **Ana Con** mechanism is not really a rule, technically speaking, though
we will continue to call it that since it appears on the **Rule?** menu in Fitch.
This mechanism, along with the two others appearing on the **Con** submenu,
apply complicated procedures to see whether the sentence in question follows
from the cited sentences. As we will explain later, these three items try to find
proofs of the sentence in question "behind the scenes," and then give you a
checkmark if they succeed. The proof they find may in fact apply many, many
different rules in getting from the cited steps to the target sentence.

*rules vs. **Con**
mechanisms*

The main difference you will run into between the genuine rules in Fitch
and the mechanisms appearing on the **Con** menu is that the latter "rules"
will sometimes fail even though your step is actually correct. With the genuine
rules, Fitch will always give your step either a checkmark or an X, depending
on whether the rule is applied correctly. But with the **Con** mechanisms, Fitch
will sometimes try to find a proof of the target sentence but fail. In these
cases, Fitch will give the step a question mark rather than a check or an X,
since there might be a complicated proof that it just couldn't find.

To mark the difference between the genuine rules of $\mathcal{F}$ and the three consequence mechanisms, Fitch displays the rule names in green and the consequence mechanisms in blue. Because the **Con** mechanisms look for a proof
behind the scenes, we will often ask you not to use them in giving solutions to
homework problems. After all, the point is not to have Fitch do your homework for you! In the following problems, you should only use the **Ana Con**
rule if we explicitly say you can. To see whether a problem allows you to use

any of the **Con** mechanisms, double click on the goal or choose **View Goal Constraints** from the **Goal** menu.

> ### Remember
>
> The deductive system you will be learning is a Fitch-style deductive system, named $\mathcal{F}$. The computer application that assists you in constructing proofs in $\mathcal{F}$ is therefore called Fitch. If you write out your proofs on paper, you are using the system $\mathcal{F}$, but not the program Fitch.

Exercises

2.15 If you skipped the **You try it** sections, go back and do them now. Submit the files Proof Identity 1 and Proof Ana Con 1.

2.16 Use Fitch to give a formal version of the informal proof you gave in Exercise 2.5. Remember, you will find the problem setup in the file Exercise 2.16. You should begin your proof from this saved file. Save your completed proof as Proof 2.16.

In the following exercises, use Fitch to construct a formal proof that the conclusion is a consequence of the premises. Remember, begin your proof by opening the corresponding file, Exercise 2.x, *and save your solution as* Proof 2.x. *We're going to stop reminding you.*

2.17
| SameCol(a, b)
| b = c
| c = d
|——
| SameCol(a, d)

2.18
| Between(a, d, b)
| a = c
| e = b
|——
| Between(c, d, e)

2.19
| Smaller(a, b)
| Smaller(b, c)
|——
| Smaller(a, c)

You will need to use **Ana Con** in this proof. This proof shows that the predicate Smaller in the blocks language is transitive.

2.20
| RightOf(b, c)
| LeftOf(d, e)
| b = d
|——
| LeftOf(c, e)

Make your proof parallel the informal proof we gave on page 52, using both an identity rule and **Ana Con** (where necessary).

Demonstrating nonconsequence

Proofs come in a variety of different forms. When a mathematician proves a theorem, or when a prosecutor proves a defendant's guilt, they are showing that a particular claim follows from certain accepted information, the information they take as given. This kind of proof is what we call a proof of *consequence*, a proof that a particular piece of information must be true if the given information, the premises of the argument, are correct.

proofs of consequence

A very different, but equally important kind of proof is a proof of *nonconsequence*. When a defense attorney shows that the crime might have been committed by someone other than the client, say by the butler, the attorney is trying to prove that the client's guilt *does not* follow from the evidence in the case. When mathematicians show that the parallel postulate is not a consequence of the other axioms of Euclidean geometry, they are doing the same thing: they are showing that it would be possible for the claim in question (the parallel postulate) to be false, even if the other information (the remaining axioms) is true.

proofs of nonconsequence

We have introduced a few methods for demonstrating the validity of an argument, for showing that its conclusion is a consequence of its premises. We will be returning to this topic repeatedly in the chapters that follow, adding new tools for demonstrating consequence as we add new expressions to our language. In this section, we discuss the most important method for demonstrating nonconsequence, that is, for showing that some purported conclusion is not a consequence of the premises provided in the argument.

Recall that logical consequence was defined in terms of the validity of arguments. An argument is valid if every possible circumstance that makes the premises of the argument true also makes the conclusion true. Put the other way around, the argument is *invalid* if there is some circumstance that makes the premises true but the conclusion false. Finding such a circumstance is the key to demonstrating nonconsequence.

To show that a sentence Q is not a consequence of premises $P_1, \ldots, P_n$, we must show that the argument with premises $P_1, \ldots, P_n$ and conclusion Q is invalid. This requires us to demonstrate that it is possible for $P_1, \ldots, P_n$ to be true while Q is simultaneously false. That is, we must show that there is a possible situation or circumstance in which the premises are all true while the conclusion is false. Such a circumstance is said to be a *counterexample* to the argument.

counterexamples

Informal proofs of nonconsequence can resort to many ingenious ways for showing the existence of a counterexample. We might simply describe what is clearly a possible situation, one that makes the premises true and the conclusion false. This is the technique used by defense attorneys, who hope to create a reasonable doubt that their client is guilty (the prosecutor's conclusion) in spite of the evidence in the case (the prosecution's premises). We might draw a picture of such a situation or build a model out of Lego blocks or clay. We might act out a situation. Anything that clearly shows the existence of a counterexample is fair game.

Recall the following argument from an earlier exercise.

> Al Gore is a politician.
> Hardly any politicians are honest.
>
> Al Gore is dishonest.

If the premises of this argument are true, then the conclusion is likely. But still the argument is not valid: the conclusion is not a logical consequence of the premises. How can we see this? Well, imagine a situation where there are 10,000 politicians, and that Al Gore is the only honest one of the lot. In such circumstances both premises would be true but the conclusion would be false. Such a situation is a counterexample to the argument; it demonstrates that the argument is invalid.

What we have just given is an informal proof of nonconsequence. Are there such things as formal proofs of nonconsequence, similar to the formal proofs of validity constructed in $\mathcal{F}$? In general, no. But we will define the notion of a formal proof of nonconsequence for the blocks language used in Tarski's World. These formal proofs of nonconsequence are simply stylized counterparts of informal counterexamples.

For the blocks language, we will say that a formal proof that Q is not a consequence of $P_1, \ldots, P_n$ consists of a sentence file with $P_1, \ldots, P_n$ labeled as premises, Q labeled as conclusion, and a world file that makes each of $P_1, \ldots, P_n$ true and Q false. The world depicted in the world file will be called the counterexample to the argument in the sentence file.

You try it

. . . .

▶ 1. Launch Tarski's World and open the sentence file Bill's Argument. This argument claims that Between(b, a, d) follows from these three premises: Between(b, c, d), Between(a, b, d), and LeftOf(a, c). Do you think it does?

▶ 2. Start a new world and put four blocks, labeled a, b, c, and d on one row of the grid.

3. Arrange the blocks so that the conclusion is false. Check the premises. If ◀
   any of them are false, rearrange the blocks until they are all true. Is the
   conclusion still false? If not, keep trying.

4. If you have trouble, try putting them in the order *d*, *a*, *b*, *c*. Now you will ◀
   find that all the premises are true but the conclusion is false. This world is
   a counterexample to the argument. Thus we have demonstrated that the
   conclusion does not follow from the premises.

5. Save your counterexample as World Counterexample 1. ◀
   . . . . *Congratulations*

Remember

To demonstrate the invalidity of an argument with premises $P_1, \ldots, P_n$
and conclusion Q, find a counterexample: a possible circumstance that
makes $P_1, \ldots, P_n$ all true but Q false. Such a counterexample shows that
Q is not a consequence of $P_1, \ldots, P_n$.

Exercises

2.21 If you have skipped the **You try it** section, go back and do it now. Submit the world file World
✎ Counterexample 1.

2.22 Is the following argument valid? Sound? If it is valid, give an informal proof of it. If it is not
✎ valid, give an informal counterexample to it.

> All computer scientists are rich. Anyone who knows how to program a computer is a
> computer scientist. Bill Gates is rich. Therefore, Bill Gates knows how to program a
> computer.

2.23 Is the following argument valid? Sound? If it is valid, give an informal proof of it. If it is not
✎ valid, give an informal counterexample to it.

> Philosophers have the intelligence needed to be computer scientists. Anyone who be-
> comes a computer scientist will eventually become wealthy. Anyone with the intelli-
> gence needed to be a computer scientist will become one. Therefore, every philosopher
> will become wealthy.

Each of the following problems presents a formal argument in the blocks language. If the argument is valid, submit a proof of it using Fitch. (You will find Exercise files for each of these in the usual place.) Important: if you use **Ana Con** *in your proof, cite at most two sentences in each application. If the argument is not valid, submit a counterexample world using Tarski's World.*

2.24

> Larger(b, c)
> Smaller(b, d)
> SameSize(d, e)
>
> Larger(e, c)

2.25

> FrontOf(a, b)
> LeftOf(a, c)
> SameCol(a, b)
>
> FrontOf(c, b)

2.26

> SameRow(b, c)
> SameRow(a, d)
> SameRow(d, f)
> LeftOf(a, b)
>
> LeftOf(f, c)

2.27

> SameRow(b, c)
> SameRow(a, d)
> SameRow(d, f)
> FrontOf(a, b)
>
> FrontOf(f, c)

SECTION 2.6

Alternative notation

You will often see arguments presented in the following way, rather than in Fitch format. The symbol ∴ (read "therefore") is used to indicate the conclusion:

> All men are mortal.
> Socrates is a man.
> ∴ Socrates is mortal.

There is a huge variety of formal deductive systems, each with its own notation. We can't possibly cover all of these alternatives, though we describe one, the resolution method, in Chapter 17.

CHAPTER 3

The Boolean Connectives

So far, we have discussed only atomic claims. To form complex claims, FOL provides us with connectives and quantifiers. In this chapter we take up the three simplest connectives: conjunction, disjunction, and negation, corresponding to simple uses of the English *and*, *or*, and *it is not the case that*. Because they were first studied systematically by the English logician George Boole, they are called the Boolean operators or Boolean connectives.

Boolean connectives

The Boolean connectives are also known as *truth-functional* connectives. There are additional truth-functional connectives which we will talk about later. These connectives are called "truth functional" because the truth value of a complex sentence built up using these connectives depends on nothing more than the truth values of the simpler sentences from which it is built. Because of this, we can explain the meaning of a truth-functional connective in a couple of ways. Perhaps the easiest is by constructing a *truth table*, a table that shows how the truth value of a sentence formed with the connective depends on the truth values of the sentence's immediate parts. We will give such tables for each of the connectives we introduce. A more interesting way, and one that can be particularly illuminating, is by means of a game, sometimes called the Henkin-Hintikka game, after the logicians Leon Henkin and Jaakko Hintikka.

truth-functional connectives

truth table

Henkin-Hintikka game

Imagine that two people, say Max and Claire, disagree about the truth value of a complex sentence. Max claims it is true, Claire claims it is false. The two repeatedly challenge one another to justify their claims in terms of simpler claims, until finally their disagreement is reduced to a simple atomic claim, one involving an atomic sentence. At that point they can simply examine the world to see whether the atomic claim is true—at least in the case of claims about the sorts of worlds we find in Tarski's World. These successive challenges can be thought of as a game where one player will win, the other will lose. The legal moves at any stage depend on the form of the sentence. We will explain them below. The one who can ultimately justify his or her claims is the winner.

When you play this game in Tarski's World, the computer takes the side opposite you, even if it knows you are right. If you are mistaken in your initial assessment, the computer will be sure to win the game. If you are right, though, the computer plugs away, hoping you will blunder. If you slip up, the computer will win the game. We will use the game rules as a second way of explaining the meanings of the truth-functional connectives.

67

Section 3.1

Negation symbol: ¬

The symbol ¬ is used to express negation in our language, the notion we commonly express in English using terms like *not*, *it is not the case that*, *non-* and *un-*. In first-order logic, we always apply this symbol to the front of a sentence to be negated, while in English there is a much more subtle system for expressing negative claims. For example, the English sentences *John isn't home* and *It is not the case that John is home* have the same first-order translation:

$$\neg Home(john)$$

This sentence is true if and only if Home(john) isn't true, that is, just in case John isn't home.

In English, we generally avoid double negatives—negatives inside other negatives. For example, the sentence *It doesn't make no difference* is problematic. If someone says it, they usually mean that it doesn't make any difference. In other words, the second negative just functions as an intensifier of some sort. On the other hand, this sentence could be used to mean just what it says, that it does not make *no* difference, it makes *some* difference.

FOL is much more systematic. You can put a negation symbol in front of any sentence whatsoever, and it always negates it, no matter how many other negation symbols the sentence already contains. For example, the sentence

$$\neg\neg Home(john)$$

negates the sentence

$$\neg Home(john)$$

and so is true if and only if John is home.

literals

The negation symbol, then, can apply to complex sentences as well as to atomic sentences. We will say that a sentence is a *literal* if it is either atomic or the negation of an atomic sentence. This notion of a literal will be useful later on.

nonidentity symbol (≠)

We will abbreviate negated identity claims, such as ¬(b = c), using ≠, as in b ≠ c. The symbol ≠ is available on the keyboard palettes in both Tarski's World and Fitch.

Semantics and the game rule for negation

Given any sentence P of FOL (atomic or complex), there is another sentence ¬P. This sentence is true if and only if P is false. This can be expressed in terms of the following truth table.

P	¬P
TRUE	FALSE
FALSE	TRUE

The game rule for negation is very simple, since you never have to *do* anything. Once you commit yourself to the truth of ¬P this is the same as committing yourself to the falsity of P. Similarly, if you commit yourself to the falsity of ¬P, this is tantamount to committing yourself to the truth of P. So in either case Tarski's World simply replaces your commitment about the more complex sentence by the opposite commitment about the simpler sentence.

You try it
. . . .

1. Open Wittgenstein's World. Start a new sentence file and write the following sentence. ◄

$$\neg\neg\neg\neg\neg\text{Between}(e, d, f)$$

2. Use the **Verify Sentence** button to check the truth value of the sentence. ◄

3. Now play the game, choosing whichever commitment you please. What happens to the number of negation symbols as the game proceeds? What happens to your commitment? ◄

4. Now play the game again with the opposite commitment. If you won the first time, you should lose this time, and vice versa. Don't feel bad about losing. ◄

5. There is no need to save the sentence file when you are done. ◄
. . . . *Congratulations*

Remember

1. If P is a sentence of FOL, then so is ¬P.

2. The sentence ¬P is true if and only if P is not true.

3. A sentence that is either atomic or the negation of an atomic sentence is called a *literal*.

Exercises

3.1   If you skipped the **You try it** section, go back and do it now. There are no files to submit, but you wouldn't want to miss it.

3.2   (Assessing negated sentences) Open Boole's World and Brouwer's Sentences. In the sentence file you will find a list of sentences built up from atomic sentences using only the negation symbol. Read each sentence and decide whether you think it is true or false. Check your assessment. If the sentence is false, make it true by adding or deleting a negation sign. When you have made all the sentences in the file true, submit the modified file as Sentences 3.2

3.3   (Building a world) Start a new sentence file. Write the following sentences in your file and save the file as Sentences 3.3.

1. $\neg$Tet(f)
2. $\neg$SameCol(c, a)
3. $\neg\neg$SameCol(c, b)
4. $\neg$Dodec(f)
5. c $\neq$ b
6. $\neg$(d $\neq$ e)
7. $\neg$SameShape(f, c)
8. $\neg\neg$SameShape(d, c)
9. $\neg$Cube(e)
10. $\neg$Tet(c)

Now start a new world file and build a world where all these sentences are true. As you modify the world to make the later sentences true, make sure that you have not accidentally falsified any of the earlier sentences. When you are done, submit both your sentences and your world.

3.4   Let P be a true sentence, and let Q be formed by putting some number of negation symbols in front of P. Show that if you put an even number of negation symbols, then Q is true, but that if you put an odd number, then Q is false. [Hint: A complete proof of this simple fact would require what is known as "mathematical induction." If you are familiar with proof by induction, then go ahead and give a proof. If you are not, just explain as clearly as you can why this is true.]

Now assume that P is atomic but of unknown truth value, and that Q is formed as before. No matter how many negation symbols Q has, it will always have the same truth value as a literal, namely either the literal P or the literal $\neg$P. Describe a simple procedure for determining which.

Section 3.2
Conjunction symbol: ∧

The symbol ∧ is used to express conjunction in our language, the notion we normally express in English using terms like *and, moreover*, and *but*. In first-order logic, this connective is always placed between two sentences, whereas in English we can also conjoin other parts of speech, such as nouns. For example, the English sentences *John and Mary are home* and *John is home and Mary is home* have the same first-order translation:

$$\text{Home(john)} \wedge \text{Home(mary)}$$

This sentence is read aloud as "Home John and home Mary." It is true if and only if John is home and Mary is home.

In English, we can also conjoin verb phrases, as in the sentence *John slipped and fell*. But in FOL we must translate this the same way we would translate *John slipped and John fell*:

$$\text{Slipped(john)} \wedge \text{Fell(john)}$$

This sentence is true if and only if the atomic sentences Slipped(john) and Fell(john) are both true.

A lot of times, a sentence of FOL will contain ∧ when there is no visible sign of conjunction in the English sentence at all. How, for example, do you think we might express the English sentence *d is a large cube* in FOL? If you guessed

$$\text{Large(d)} \wedge \text{Cube(d)}$$

you were right. This sentence is true if and only if *d* is large and *d* is a cube—that is, if *d* is a large cube.

Some uses of the English *and* are not accurately mirrored by the FOL conjunction symbol. For example, suppose we are talking about an evening when Max and Claire were together. If we were to say *Max went home and Claire went to sleep*, our assertion would carry with it a temporal implication, namely that Max went home *before* Claire went to sleep. Similarly, if we were to reverse the order and assert *Claire went to sleep and Max went home* it would suggest a very different sort of situation. By contrast, no such implication, implicit or explicit, is intended when we use the symbol ∧. The sentence

$$\text{WentHome(max)} \wedge \text{FellAsleep(claire)}$$

is true in exactly the same circumstances as

$$\text{FellAsleep(claire)} \wedge \text{WentHome(max)}$$

Semantics and the game rule for ∧

Just as with negation, we can put complex sentences as well as simple ones together with ∧. A sentence P ∧ Q is true if and only if both P and Q are true. Thus P ∧ Q is false if either or both of P or Q is false. This can be summarized by the following truth table.

truth table for ∧

P	Q	P ∧ Q
TRUE	TRUE	TRUE
TRUE	FALSE	FALSE
FALSE	TRUE	FALSE
FALSE	FALSE	FALSE

game rule for ∧

The Tarski's World game is more interesting for conjunctions than negations. The way the game proceeds depends on whether you have committed to TRUE or to FALSE. If you commit to the truth of P ∧ Q then you have implicitly committed yourself to the truth of each of P and Q. Thus, Tarski's World gets to choose either one of these simpler sentences and hold you to the truth of it. (Which one will Tarski's World choose? If one or both of them are false, it will choose a false one so that it can win the game. If both are true, it will choose at random, hoping that you will make a mistake later on.)

If you commit to the falsity of P ∧ Q, then you are claiming that at least one of P or Q is false. In this case, Tarski's World will ask *you* to choose one of the two and thereby explicitly commit to its being false. The one you choose had better be false, or you will eventually lose the game.

You try it
. . . .

▶   1. Open Claire's World. Start a new sentence file and enter the sentence

$$\neg\text{Cube}(a) \wedge \neg\text{Cube}(b) \wedge \neg\text{Cube}(c)$$

▶   2. Notice that this sentence is false in this world, since c is a cube. Play the game committed (mistakenly) to the truth of the sentence. You will see that Tarski's World immediately zeros in on the false conjunct. Your commitment to the truth of the sentence guarantees that you will lose the game, but along the way, the reason the sentence is false becomes apparent.

▶   3. Now begin playing the game committed to the falsity of the sentence. When Tarski's World asks you to choose a conjunct you think is false, pick the first sentence. This is not the false conjunct, but select it anyway and see what happens after you choose **OK**.

4. Play until Tarski's World says that you have lost. Then click on **Back** a couple of times, until you are back to where you are asked to choose a false conjunct. This time pick the false conjunct and resume the play of the game from that point. This time you will win. ◄

5. Notice that you can lose the game even when your original assessment is correct, if you make a bad choice along the way. But Tarski's World always allows you to back up and make different choices. If your original assessment is correct, there will always be a way to win the game. If it is impossible for you to win the game, then your original assessment was wrong. ◄

6. Save your sentence file as **Sentences Game 1** when you are done. ◄
. . . . *Congratulations*

> **Remember**
>
> 1. If P and Q are sentences of FOL, then so is P ∧ Q.
>
> 2. The sentence P ∧ Q is true if and only if both P and Q are true.

Exercises

3.5 If you skipped the **You try it** section, go back and do it now. Make sure you follow all the instructions. Submit the file **Sentences Game 1**.

3.6 Start a new sentence file and open **Wittgenstein's World**. Write the following sentences in the sentence file.

       1. Tet(f) ∧ Small(f)
       2. Tet(f) ∧ Large(f)
       3. Tet(f) ∧ ¬Small(f)
       4. Tet(f) ∧ ¬Large(f)
       5. ¬Tet(f) ∧ ¬Small(f)
       6. ¬Tet(f) ∧ ¬Large(f)
       7. ¬(Tet(f) ∧ Small(f))
       8. ¬(Tet(f) ∧ Large(f))

9. ¬(¬Tet(f) ∧ ¬Small(f))
10. ¬(¬Tet(f) ∧ ¬Large(f))

Once you have written these sentences, decide which you think are true. Record your evaluations, to help you remember. Then go through and use Tarski's World to evaluate your assessments. Whenever you are wrong, play the game to see where you went wrong.

If you are never wrong, playing the game will not be very instructive. Play the game a couple times anyway, just for fun. In particular, try playing the game committed to the falsity of sentence 9. Since this sentence is true in Wittgenstein's World, Tarski's World should be able to beat you. Make sure you understand everything that happens as the game proceeds.

Next, change the size or shape of block f, predict how this will affect the truth values of your ten sentences, and see if your prediction is right. What is the maximum number of these sentences that you can get to be true in a single world? Build a world in which the maximum number of sentences are true. Submit both your sentence file and your world file, naming them as usual.

3.7 (Building a world) Open Max's Sentences. Build a world where all these sentences are true. You should start with a world with six blocks and make changes to it, trying to make all the sentences true. Be sure that as you make a later sentence true you do not inadvertently falsify an earlier sentence.

Section 3.3
Disjunction symbol: ∨

The symbol ∨ is used to express disjunction in our language, the notion we express in English using *or*. In first-order logic, this connective, like the conjunction sign, is always placed between two sentences, whereas in English we can also disjoin nouns, verbs, and other parts of speech. For example, the English sentences *John or Mary is home* and *John is home or Mary is home* both have the same first-order translation:

$$\text{Home(john)} \lor \text{Home(mary)}$$

This FOL sentence is read "Home John or home Mary."

exclusive vs. inclusive disjunction

Although the English *or* is sometimes used in an "exclusive" sense, to say that *exactly* one (i.e., one but no more than one) of the two disjoined sentences is true, the first-order logic ∨ is always given an "inclusive" interpretation: it means that at least one and possibly both of the two disjoined sentences is true. Thus, our sample sentence is true if John is home but Mary is not, if Mary is home but John is not, or if both John and Mary are home.

If we wanted to express the exclusive sense of *or* in the above example, we could do it as follows:

[Home(john) ∨ Home(mary)] ∧ ¬[Home(john) ∧ Home(mary)]

As you can see, this sentence says that John or Mary is home, but it is not the case that they are both home.

Many students are tempted to say that the English expression *either ... or* expresses exclusive disjunction. While this is sometimes the case (and indeed the simple *or* is often used exclusively), it isn't always. For example, suppose Pris and Scruffy are in the next room and the sound of a cat fight suddenly breaks out. If we say *Either Pris bit Scruffy or Scruffy bit Pris*, we would not be wrong if each had bit the other. So this would be translated as

Bit(pris, scruffy) ∨ Bit(scruffy, pris)

We will see later that the expression *either* sometimes plays a different logical function.

Another important English expression that we can capture without introducing additional symbols is *neither ... nor*. Thus *Neither John nor Mary is at home* would be expressed as:

¬(Home(john) ∨ Home(mary))

This says that it's not the case that at least one of them is at home, i.e., that neither of them is home.

Semantics and the game rule for ∨

Given any two sentences P and Q of FOL, atomic or not, we can combine them using ∨ to form a new sentence P ∨ Q. The sentence P ∨ Q is true if at least one of P or Q is true. Otherwise it is false. Here is the truth table.

P	Q	P ∨ Q
TRUE	TRUE	TRUE
TRUE	FALSE	TRUE
FALSE	TRUE	TRUE
FALSE	FALSE	FALSE

truth table for ∨

The game rules for ∨ are the "duals" of those for ∧. If you commit yourself to the truth of P ∨ Q, then Tarski's World will make you live up to this by committing yourself to the truth of one or the other. If you commit yourself to the falsity of P ∨ Q, then you are implicitly committing yourself to the falsity

game rule for ∨

of each, so Tarski's World will choose one and hold you to the commitment that it is false. (Tarski's World will, of course, try to win by picking a true one, if it can.)

You try it
. . . .

▶ 1. Open the file Ackermann's World. Start a new sentence file and enter the sentence

$$\text{Cube(c)} \lor \neg(\text{Cube(a)} \lor \text{Cube(b)})$$

Make sure you get the parentheses right!

▶ 2. Play the game committed (mistakenly) to this sentence being true. Since the sentence is a disjunction, and you are committed to TRUE, you will be asked to pick a disjunct that you think is true. Since the first one is obviously false, pick the second.

▶ 3. You now find yourself committed to the falsity of a (true) disjunction. Hence you are committed to the falsity of each disjunct. Tarski's World will then point out that you are committed to the falsity of Cube(b). But this is clearly wrong, since b is a cube. Continue until Tarski's World says you have lost.

▶ 4. Play the game again, this time committed to the falsity of the sentence. You should be able to win the game this time. If you don't, back up and try again.

▶ 5. Save your sentence file as Sentences Game 2
. . . . *Congratulations*

Remember

1. If P and Q are sentences of FOL, then so is P ∨ Q.

2. The sentence P ∨ Q is true if and only if P is true or Q is true (or both are true).

Exercises

3.8 If you skipped the **You try it** section, go back and do it now. You'll be glad you did. Well, maybe. Submit the file Sentences Game 2.

3.9
✎
Open Wittgenstein's World and the sentence file Sentences 3.6 that you created for Exercise 3.6. Edit the sentences by replacing ∧ by ∨ throughout, saving the edited list as Sentences 3.9. Once you have changed these sentences, decide which you think are true. Again, record your evaluations to help you remember them. Then go through and use Tarski's World to evaluate your assessment. Whenever you are wrong, play the game to see where you went wrong. If you are never wrong, then play the game anyway a couple times, knowing that you should win. As in Exercise 3.6, find the maximum number of sentences you can make true by changing the size or shape (or both) of block f. Submit both your sentences and world.

3.10
✎
Open Ramsey's World and start a new sentence file. Type the following four sentences into the file:

1. Between(a, b, c) ∨ Between(b, a, c)
2. FrontOf(a, b) ∨ FrontOf(c, b)
3. ¬SameRow(b, c) ∨ LeftOf(b, a)
4. RightOf(b, a) ∨ Tet(a)

Assess each of these sentences in Ramsey's World and check your assessment. Then make a single change to the world that makes all four of the sentences come out false. Save the modified world as World 3.10. Submit both files.

SECTION 3.4

Remarks about the game

We summarize the game rules for the three connectives, ¬, ∧, and ∨, in Table 3.1. The first column indicates the form of the sentence in question, and the second indicates your current commitment, TRUE or FALSE. Which player moves depends on this commitment, as shown in the third column. The goal of that player's move is indicated in the final column. Notice that although the player to move depends on the commitment, the goal of that move does not depend on the commitment. You can see why this is so by thinking about the first row of the table, the one for P ∨ Q. When you are committed to TRUE, it is clear that your goal should be to choose a true disjunct. But when you are committed to FALSE, Tarski's World is committed to TRUE, and so also has the same goal of choosing a true disjunct.

commitment and rules

There is one somewhat subtle point that should be made about our way of describing the game. We have said, for example, that when you are committed to the truth of the disjunction P ∨ Q, you are committed to the truth of one of the disjuncts. This of course is true, but does not mean you necessarily know which of P or Q is true. For example, if you have a sentence of the form

Table 3.1: Game rules for ∧, ∨, and ¬

FORM	YOUR COMMITMENT	PLAYER TO MOVE	GOAL
P ∨ Q	TRUE	you	Choose one of P, Q that
	FALSE	Tarski's World	is true.
P ∧ Q	TRUE	Tarski's World	Choose one of P, Q that
	FALSE	you	is false.
¬P	either	—	Replace ¬P by P and switch commitment.

P ∨ ¬P, then you know that it is true, no matter how the world is. After all, if P is not true, then ¬P will be true, and vice versa; in either event P ∨ ¬P will be true. But if P is quite complex, or if you have imperfect information about the world, you may not know which of P or ¬P is true. Suppose P is a sentence like *There is a whale swimming below the Golden Gate Bridge right now.* In such a case you would be willing to commit to the truth of the disjunction (since either there *is* or there *isn't*) without knowing just how to play the game and win. You know that there is a winning strategy for the game, but just don't know what it is.

Since there is a moral imperative to live up to one's commitments, the use of the term "commitment" in describing the game is a bit misleading. You are perfectly justified in asserting the truth of P ∨ ¬P, even if you do not happen to know your winning strategy for playing the game. Indeed, it would be foolish to claim that the sentence is *not* true. But if you do claim that P ∨ ¬P is true, and then play the game, you will be asked to say which of P or ¬P you think is true. With Tarski's World, unlike in real life, you can always get complete information about the world by going to the 2D view, and so always live up to such commitments.

Exercises

Here is a problem that illustrates the remarks we made about sometimes being able to tell that a sentence is true, without knowing how to win the game.

3.11 Make sure Tarski's World is set to display the world in 3D. Then open Kleene's World and Kleene's Sentences. Some objects are hidden behind other objects, thus making it impossible to assess the truth of some of the sentences. Each of the six names a, b, c, d, e, and f are in use, naming some object. Now even though you cannot see all the objects, some of the sentences in the list can be evaluated with just the information at hand. Assess the truth of each claim, if you can, without recourse to the 2-D view. Then play the game. If your initial commitment is right, but you lose the game, back up and play over again. Then go through and add comments to each sentence explaining whether you can assess its truth in the world as shown, and why. Finally, display the 2-D view and check your work. We have annotated the first sentence for you to give you the idea. (The semicolon ";" tells Tarski's World that what follows is a comment.) When you are done, print out your annotated sentences to turn in to your instructor.

SECTION 3.5

Ambiguity and parentheses

When we first described FOL, we stressed the lack of ambiguity of this language as opposed to ordinary languages. For example, English allows us to say things like *Max is home or Claire is home and Carl is happy*. This sentence can be understood in two quite different ways. One reading claims that either Claire is home and Carl is happy, or Max is home. On this reading, the sentence would be true if Max was home, even if Carl was unhappy. The other reading claims both that Max or Claire is home and that Carl is happy.

FOL avoids this sort of ambiguity by requiring the use of parentheses, much the way they are used in algebra. So, for example, FOL would not have one sentence corresponding to the ambiguous English sentence, but two:

$$\text{Home(max)} \lor (\text{Home(claire)} \land \text{Happy(carl)})$$
$$(\text{Home(max)} \lor \text{Home(claire)}) \land \text{Happy(carl)}$$

The parentheses in the first indicate that it is a disjunction, whose second disjunct is itself a conjunction. In the second, they indicate that the sentence is a conjunction whose first conjunct is a disjunction. As a result, the truth conditions for the two are quite different. This is analogous to the difference in algebra between the expressions $2 + (x \times 3)$ and $(2 + x) \times 3$. This analogy between logic and algebra is one we will come back to later.

scope of negation   Parentheses are also used to indicate the "scope" of a negation symbol when it appears in a complex sentence. So, for example, the two sentences

$$\neg\mathsf{Home(claire)} \land \mathsf{Home(max)}$$
$$\neg(\mathsf{Home(claire)} \land \mathsf{Home(max)})$$

mean quite different things. The first is a conjunction of literals, the first of which says Claire is not home, the second of which says that Max is home. By contrast, the second sentence is a negation of a sentence which itself is a conjunction: it says that they are not both home. You have already encountered this use of parentheses in earlier exercises.

Many logic books require that you always put parentheses around any pair of sentences joined by a binary connective (such as $\land$ or $\lor$). These books do not allow sentences of the form:

$$\mathsf{P} \land \mathsf{Q} \land \mathsf{R}$$

but instead require one of the following:

$$((\mathsf{P} \land \mathsf{Q}) \land \mathsf{R})$$
$$(\mathsf{P} \land (\mathsf{Q} \land \mathsf{R}))$$

leaving out parentheses   The version of FOL that we use in this book is not so fussy, in a couple of ways. First of all, it allows you to conjoin any number of sentences without using parentheses, since the result is not ambiguous, and similarly for disjunctions. Second, it allows you to leave off the outermost parentheses, since they serve no useful purpose. You can also add extra parentheses (or brackets or braces) if you want to for the sake of readability. For the most part, all we will require is that your expression be unambiguous.

> **Remember**
>
> Parentheses must be used whenever ambiguity would result from their omission. In practice, this means that conjunctions and disjunctions must be "wrapped" in parentheses whenever combined by means of some other connective.

You try it
. . . .

▶ 1. Let's try our hand at evaluating some sentences built up from atomic sentences using all three connectives $\land, \lor, \neg$. Open Boole's Sentences and Wittgenstein's World. If you changed the size or shape of f while doing Exercises 3.6 and 3.9, make sure that you change it back to a large tetrahedron.

2. Evaluate each sentence in the file and check your assessment. If your assessment is wrong, play the game to see why. Don't go from one sentence to the next until you understand why it has the truth value it does. ◀

3. Do you see the importance of parentheses? After you understand all the sentences, go back and see which of the false sentences you can make true just by adding, deleting, or moving parentheses, but without making any other changes. Save your file as **Sentences Ambiguity 1**. ◀

. . . .*Congratulations*

Exercises

To really master a new language, you have to use it, not just read about it. The exercises and problems that follow are intended to let you do just that.

3.12 If you skipped the **You try it** section, go back and do it now. Submit the file **Sentences Ambiguity 1**.

3.13 (Building a world) Open **Schröder's Sentences**. Build a single world where all the sentences in this file are true. As you work through the sentences, you will find yourself successively modifying the world. Whenever you make a change in the world, be careful that you don't make one of your earlier sentences false. When you are finished, verify that all the sentences are really true. Submit your world as **World 3.13**.

3.14 (Parentheses) Show that the sentence

¬(Small(a) ∨ Small(b))

is not a consequence of the sentence

¬Small(a) ∨ Small(b)

You will do this by submitting a counterexample world in which the second sentence is true but the first sentence is false.

3.15 (More parentheses) Show that

Cube(a) ∧ (Cube(b) ∨ Cube(c))

is not a consequence of the sentence

(Cube(a) ∧ Cube(b)) ∨ Cube(c)

You will do this by submitting a counterexample world in which the second sentence is true but the first sentence is false.

3.16 (DeMorgan Equivalences) Open the file **DeMorgan's Sentences**. Construct a world where all the *odd* numbered sentences are true. Notice that no matter how you do this, the even numbered sentences also come out true. Submit this as **World 3.16.1**. Next build a world where all the odd numbered sentences are *false*. Notice that no matter how you do it, the even numbered sentences also come out false. Submit this as **World 3.16.2**.

3.17 In Exercise 3.16, you noticed an important fact about the relation between the even and odd numbered sentences in DeMorgan's Sentences. Try to explain why each even numbered sentence always has the same truth value as the odd numbered sentence that precedes it.

Section 3.6

Equivalent ways of saying things

Every language has many ways of saying the same thing. This is particularly true of English, which has absorbed a remarkable number of words from other languages in the course of its history. But in any language, speakers always have a choice of many synonymous ways of getting across their point. The world would be a boring place if there were just one way to make a given claim.

FOL is no exception, even though it is far less rich in its expressive capacities than English. In the blocks language, for example, none of our predicates is synonymous with another predicate, though it is obvious that we could do without many of them without cutting down on the claims expressible in the language. For instance, we could get by without the predicate RightOf by expressing everything we need to say in terms of the predicate LeftOf, systematically reversing the order of the names to get equivalent claims. This is not to say that RightOf means the same thing as LeftOf—it obviously does not—but just that the blocks language offers us a simple way to construct equivalent claims using these predicates. In the exercises at the end of this section, we explore a number of equivalences made possible by the predicates of the blocks language.

Some versions of FOL are more parsimonious with their basic predicates than the blocks language, and so may not provide equivalent ways of expressing atomic claims. But even these languages cannot avoid multiple ways of expressing more complex claims. For example, $P \wedge Q$ and $Q \wedge P$ express the same claim in any first-order language. More interesting, because of the superficial differences in form, are the equivalences illustrated in Exercise 3.16, known as *DeMorgan's laws*. The first of DeMorgan's laws tells us that the negation of a conjunction, $\neg(P \wedge Q)$, is logically equivalent to the disjunction of the negations of the original conjuncts: $\neg P \vee \neg Q$. The other tells us that the negation of a disjunction, $\neg(P \vee Q)$, is equivalent to the conjunction of the negations of the original disjuncts: $\neg P \wedge \neg Q$. These laws are simple consequences of the meanings of the Boolean connectives. Writing $S_1 \Leftrightarrow S_2$ to indicate that S_1 and S_2 are logically equivalent, we can express DeMorgan's

DeMorgan's laws

laws in the following way:

$$\neg(P \wedge Q) \Leftrightarrow (\neg P \vee \neg Q)$$
$$\neg(P \vee Q) \Leftrightarrow (\neg P \wedge \neg Q)$$

There are many other equivalences that arise from the meanings of the Boolean connectives. Perhaps the simplest is known as the principle of *double negation*. Double negation says that a sentence of the form $\neg\neg P$ is equivalent to the sentence P. We will systematically discuss these and other equivalences in the next chapter. In the meantime, we simply note these important equivalences before going on. Recognizing that there is more than one way of expressing a claim is essential before we tackle complicated claims involving the Boolean connectives.

double negation

Remember

(Double negation and DeMorgan's Laws) For any sentences P and Q:

1. Double negation: $\neg\neg P \Leftrightarrow P$

2. DeMorgan: $\neg(P \wedge Q) \Leftrightarrow (\neg P \vee \neg Q)$

3. DeMorgan: $\neg(P \vee Q) \Leftrightarrow (\neg P \wedge \neg Q)$

Exercises

3.18 (Equivalences in the blocks language) In the blocks language used in Tarski's World there are a number of equivalent ways of expressing some of the predicates. Open Bernays' Sentences. You will find a list of atomic sentences, where every other sentence is left blank. In each blank, write a sentence that is equivalent to the sentence above it, but does not use the predicate used in that sentence. (In doing this, you may presuppose any general facts about Tarski's World, for example that blocks come in only three shapes.) If your answers are correct, the odd numbered sentences will have the same truth values as the even numbered sentences in every world. Check that they do in Ackermann's World, Bolzano's World, Boole's World, and Leibniz's World. Submit the modified sentence file as Sentences 3.18.

3.19 (Equivalences in English) There are also equivalent ways of expressing predicates in English. For each of the following sentences of FOL, find an atomic sentence in English that expresses the same thing. For example, the sentence Man(max) $\wedge$ ¬Married(max) could be expressed in

English by means of the atomic sentence *Max is a bachelor*.

1. FatherOf(chris, alex) ∨ MotherOf(chris, alex)
2. BrotherOf(chris, alex) ∨ SisterOf(chris, alex)
3. Human(chris) ∧ Adult(chris) ∧ ¬Woman(chris)
4. Number(4) ∧ ¬Odd(4)
5. Person(chris) ∧ ¬Odd(chris)
6. mother(mother(alex)) = mary ∨ mother(father(alex)) = mary [Notice that mother and father are function symbols. If you did not cover Section 1.5, you may skip this sentence.]

SECTION 3.7
Translation

An important skill that you will want to master is that of translating from English to FOL, and vice versa. But before you can do that, you need to know how to express yourself in both languages. The problems below are designed to help you learn these related skills.

correct translation

How do we know if a translation is correct? Intuitively, a correct translation is a sentence with the same meaning as the one being translated. But what is the meaning? FOL finesses this question, settling for "truth conditions." What we require of a correct translation in FOL is that it be true in the same circumstances as the original sentence. If two sentences are true in exactly the same circumstances, we say that they have the same *truth conditions*. For sentences of Tarski's World, this boils down to being true in the very same worlds.

truth conditions

Note that it is not sufficient that the two sentences have the same truth value in some *particular* world. If that were so, then any true sentence of English could be translated by any true sentence of FOL. So, for example, if Claire and Max are both at home, we could translate *Max is at home* by means of Home(claire). No, having the same actual truth value is not enough. They have to have the same truth values in all circumstances.

Remember

In order for an FOL sentence to be a good translation of an English sentence, it is sufficient that the two sentences have the same truth values in all possible circumstances, that is, that they have the same *truth conditions*.

In general, this is all we require of translations into and out of FOL. Thus, given an English sentence S and a good FOL translation of it, say S, any other sentence S' that is equivalent to S will also count as an acceptable translation of it, since S and S' have the same truth conditions. But there is a matter of style. Some good translations are better than others. You want sentences that are easy to understand. But you also want to keep the FOL connectives close to the English, if possible.

For example, a good translation of *It is not true that Claire and Max are both at home* would be given by

$$\neg(\mathsf{Home(claire)} \wedge \mathsf{Home(max)})$$

This is equivalent to the following sentence (by the first DeMorgan law), so we count it too as an acceptable translation:

$$\neg\mathsf{Home(claire)} \vee \neg\mathsf{Home(max)}$$

But there is a clear stylistic sense in which the first is a better translation, since it conforms more closely to the form of the original. There are no hard and fast rules for determining which among several logically equivalent sentences is the best translation of a given sentence.

Many stylistic features of English have nothing to do with the truth conditions of a sentence, and simply can't be captured in an FOL translation. For example, consider the English sentence *Pris is hungry but Carl is not*. This sentence tells us two things, that Pris is hungry and that Carl is not hungry. So it would be translated into FOL as

$$\mathsf{Hungry(pris)} \wedge \neg\mathsf{Hungry(carl)}$$

When it comes to truth conditions, *but* expresses the same truth function as *and*. Yet it is clear that *but* carries an additional suggestion that *and* does not, namely, that the listener may find the sentence following the *but* a bit surprising, given the expectations raised by the sentence preceding it. The words *but, however, yet, nonetheless*, and so forth, all express ordinary conjunction, and so are translated into FOL using $\wedge$. The fact that they also communicate a sense of unexpectedness is just lost in the translation. FOL, as much as we love it, sometimes sacrifices style for clarity.

but, however, yet, nonetheless

In Exercise 3.21, sentences 1, 8, and 10, you will discover an important function that the English phrases *either... or* and *both... and* sometimes play. *Either* helps disambiguate the following *or* by indicating how far to the left its scope extends; similarly *both* indicates how far to the left the following *and* extends. For example, *Either Max is home and Claire is home or Carl*

either... or, both... and

is happy is unambiguous, whereas it would be ambiguous without the *either*. What it means is that

$$[\text{Home(max)} \wedge \text{Home(claire)}] \vee \text{Happy(carl)}$$

In other words, *either* and *both* can sometimes act as left parentheses act in FOL. The same list of sentences demonstrates many other uses of *either* and *both*.

Remember

1. The English expression *and* sometimes suggests a temporal ordering; the FOL expression $\wedge$ never does.

2. The English expressions *but, however, yet, nonetheless,* and *moreover* are all stylistic variants of *and*.

3. The English expressions *either* and *both* are often used like parentheses to clarify an otherwise ambiguous sentence.

Exercises

3.20 (Describing a simple world) Open Boole's World. Start a new sentence file, named Sentences 3.20, where you will describe some features of this world. Check each of your sentences to see that it is indeed a sentence and that it is true in this world.

1. Notice that f (the large dodecahedron in the back) is not in front of a. Use your first sentence to say this.
2. Notice that f is to the right of a and to the left of b. Use your second sentence to say this.
3. Use your third sentence to say that f is either in back of or smaller than a.
4. Express the fact that both e and d are between c and a.
5. Note that neither e nor d is larger than c. Use your fifth sentence to say this.
6. Notice that e is neither larger than nor smaller than d. Use your sixth sentence to say this.
7. Notice that c is smaller than a but larger than e. State this fact.
8. Note that c is in front of f; moreover, it is smaller than f. Use your eighth sentence to state these things.

9. Notice that b is in the same row as a but is not in the same column as f. Use your ninth sentence to express this fact.

10. Notice that e is not in the same column as either c or d. Use your tenth sentence to state this.

Now let's change the world so that none of the above mentioned facts hold. We can do this as follows. First move f to the front right corner of the grid. (Be careful not to drop it off the edge. You might find it easier to make the move from the 2-D view. If you accidentally drop it, just open Boole's World again.) Then move e to the back left corner of the grid and make it large. Now none of the facts hold; if your answers to 1–10 are correct, all of the sentences should now be false. Verify that they are. If any are still true, can you figure out where you went wrong? Submit your sentences when you think they are correct. There is no need to submit the modified world file.

3.21 (Some translations) Tarski's World provides you with a very useful way to check whether your translation of a given English sentence is correct. If it is correct, then it will always have the same truth value as the English sentence, no matter what world the two are evaluated in. So when you are in doubt about one of your translations, simply build some worlds where the English sentence is true, others where it is false, and check to see that your translation has the right truth values in these worlds. You should use this technique frequently in all of the translation exercises.

Start a new sentence file, and use it to enter translations of the following English sentences into first-order logic. You will only need to use the connectives $\land$, $\lor$, and $\neg$.

1. *Either a is small or both c and d are large.*
2. *d and e are both in back of b.*
3. *d and e are both in back of b and larger than it.*
4. *Both d and c are cubes, however neither of them is small.*
5. *Neither e nor a is to the right of c and to the left of b.*
6. *Either e is not large or it is in back of a.*
7. *c is neither between a and b, nor in front of either of them.*
8. *Either both a and e are tetrahedra or both a and f are.*
9. *Neither d nor c is in front of either c or b.*
10. *c is either between d and f or smaller than both of them.*
11. *It is not the case that b is in the same row as c.*
12. *b is in the same column as e, which is in the same row as d, which in turn is in the same column as a.*

Before you submit your sentence file, do the next exercise.

3.22 (Checking your translations) Open Wittgenstein's World. Notice that all of the English sentences from Exercise 3.21 are true in this world. Thus, if your translations are accurate, they will also be true in this world. Check to see that they are. If you made any mistakes, go back and fix them. But as we have stressed, even if one of your sentences comes out true in Wittgenstein's World, it does not mean that it is a proper translation of the corresponding English sentence. All you know for sure is that your translation and the original sentence have the same truth value in this particular world. If the translation is correct, it will have the same truth value as the English sentence in *every* world. Thus, to have a better test of your translations, we will examine them in a number of worlds, to see if they have the same truth values as their English counterparts in all of these worlds.

Let's start by making modifications to Wittgenstein's World. Make all the large or medium objects small, and the small objects large. With these changes in the world, the English sentences 1, 3, 4, and 10 become false, while the rest remain true. Verify that the same holds for your translations. If not, correct your translations. Next, rotate your modified Wittgenstein's World 90° clockwise. Now sentences 5, 6, 8, 9, and 11 should be the only true ones that remain.

Let's check your translations in another world. Open Boole's World. The only English sentences that are true in this world are sentences 6 and 11. Verify that all of your translations except 6 and 11 are false. If not, correct your translations.

Now modify Boole's World by exchanging the positions of b and c. With this change, the English sentences 2, 5, 6, 7, and 11 come out true, while the rest are false. Check that the same is true of your translations.

There is nothing to submit except Sentences 3.21.

3.23 Start a new sentence file and translate the following into FOL. Use the names and predicates presented in Table 1.2 on page 30.
1. *Max is a student, not a pet.*
2. *Claire fed Folly at 2 pm and then ten minutes later gave her to Max.*
3. *Folly belonged to either Max or Claire at 2:05 pm.*
4. *Neither Max nor Claire fed Folly at 2 pm or at 2:05 pm.*
5. *2:00 pm is between 1:55 pm and 2:05 pm.*
6. *When Max gave Folly to Claire at 2 pm, Folly wasn't hungry, but she was an hour later.*

3.24 Referring again to Table 1.2, page 30, translate the following into natural, colloquial English. Turn in your translations to your instructor.
1. Student(claire) ∧ ¬Student(max)
2. Pet(pris) ∧ ¬Owned(max, pris, 2:00)
3. Owned(claire, pris, 2:00) ∨ Owned(claire, folly, 2:00)
4. ¬(Fed(max, pris, 2:00) ∧ Fed(max, folly, 2:00))

5. $((\text{Gave}(\text{max}, \text{pris}, \text{claire}, 2\!:\!00) \wedge \text{Hungry}(\text{pris}, 2\!:\!00)) \vee$
   $(\text{Gave}(\text{max}, \text{folly}, \text{claire}, 2\!:\!00) \wedge \text{Hungry}(\text{folly}, 2\!:\!00))) \wedge$
   $\text{Angry}(\text{claire}, 2\!:\!05)$

3.25
✎⋆ Translate the following into FOL, introducing names, predicates, and function symbols as needed. Explain the meaning of each predicate and function symbol, unless it is completely obvious.

1. *AIDS is less contagious than influenza, but more deadly.*
2. *Abe fooled Stephen on Sunday, but not on Monday.*
3. *Sean or Brad admires Meryl and Harrison.*
4. *Daisy is a jolly miller, and lives on the River Dee.*
5. *Polonius's eldest child was neither a borrower nor a lender.*

3.26
✎ (Boolean solids) Many of you know how to do a "Boolean search" on the Web or on your computer. When we do a Boolean search, we are really using a generalization of the Boolean truth functions. We specify a Boolean combination of words as a criterion for finding documents that contain (or do not contain) those words. Another generalization of the Boolean operations is to spatial objects. In Figure 3.1 we show four ways to combine a vertical cylinder (A) with a horizontal cylinder (B) to yield a new solid. Give an intuitive explanation of how the Boolean connectives are being applied in this example. Then describe what the object $\neg(A \wedge B)$ would be like and explain why we didn't give you a picture of this solid.

Figure 3.1: Boolean combinations of solids: $A \vee B$, $A \wedge \neg B$, $\neg A \wedge B$, and $A \wedge B$.

SECTION 3.8

Alternative notation

As we mentioned in Chapter 2, there are various dialect differences among users of FOL. It is important to be aware of these so that you will not be stymied by superficial differences. In fact, you will run into alternate symbols being used for each of the three connectives studied in this chapter.

The most common variant of the negation sign, $\neg$, is the symbol known as the tilde, $\sim$. Thus you will frequently encounter $\sim P$ where we would write $\neg P$. A more old-fashioned alternative is to draw a bar completely across the negated sentence, as in $\overline{P}$. This has one advantage over $\neg$, in that it allows you to avoid certain uses of parentheses, since the bar indicates its own scope by what lies under it. For example, where we have to write $\neg(P \wedge Q)$, the bar equivalent would simply be $\overline{P \wedge Q}$. None of these symbols are available on all keyboards, a serious problem in some contexts, such as programming languages. Because of this, many programming languages use an exclamation point to indicate negation. In the Java programming language, for example, $\neg P$ would be written !P.

There are only two common variants of $\wedge$. By far the most common is &, or sometimes (as in Java), &&. An older notation uses a centered dot, as in multiplication. To make things more confusing still, the dot is sometimes omitted, again as in multiplication. Thus, for $P \wedge Q$ you might see any of the following: P&Q, P&&Q, P $\cdot$ Q, or just PQ.

Happily, the symbol $\vee$ is pretty standard. The only exception you may encounter is a single or double vertical line, used in programming languages. So if you see P | Q or P || Q, what is meant is probably $P \vee Q$. Unfortunately, though, some old textbooks use P | Q to express *not both P and Q*.

Alternatives to parentheses

dot notation

There are ways to get around the use of parentheses in FOL. At one time, a common alternative to parentheses was a system known as dot notation. This system involved placing little dots next to connectives indicating their relative "power" or scope. In this system, the two sentences we write as $P \vee (Q \wedge R)$ and $(P \vee Q) \wedge R$ would have been written $P \vee . Q \wedge R$ and $P \vee Q . \wedge R$, respectively. With more complex sentences, multiple dots were used. Fortunately, this notation has just about died out, and the present authors never speak to anyone who uses it.

Polish notation

Another approach to parentheses is known as Polish notation. In Polish notation, the usual infix notation is replaced by prefix notation, and this makes parentheses unnecessary. Thus the distinction between our $\neg(P \lor Q)$ and $(\neg P \lor Q)$ would, in prefix form, come out as $\neg \lor PQ$ and $\lor \neg PQ$, the order of the connectives indicating which includes the other in its scope.

Besides prefix notation, Polish notation uses certain capital letters for connectives (N for $\neg$, K for $\land$, and A for $\lor$), and lower case letters for its atomic sentences (to distinguish them from connectives). So an actual sentence of the Polish dialect would look like this:

$$\text{ApKNqr}$$

Since this expression starts with A, we know right away that it is a disjunction. What follows must be its two disjuncts, in sequence. So the first disjunct is p and the second is KNqr, that is, the conjunction of the negation of q and of r. So this is the Polish version of

$$P \lor (\neg Q \land R)$$

Though Polish notation may look hard to read, many of you have already mastered a version of it. Calculators use two styles for entering formulas. One is known as algebraic style, the other as RPN style. The RPN stands for "reverse Polish notation." If you have a calculator that uses RPN, then to calculate the value of, say, $(7 \times 8) + 3$ you enter things in this order: 7, 8, $\times$, 3, $+$. This is just the reverse of the Polish, or prefix, ordering.

reverse Polish notation

In order for Polish notation to work without parentheses, the connectives must all have a fixed arity. If we allowed conjunction to take an arbitrary number of sentences as arguments, rather than requiring exactly two, a sentence like KpNKqrs would be ambiguous. It could either mean $P \land \neg(Q \land R) \land S$ or $P \land \neg(Q \land R \land S)$, and these aren't equivalent.

Remember

The following table summarizes the alternative notations discussed so far.

Our notation	Common equivalents
$\neg P$	$\sim P$, $\overline{P}$, !P, Np
$P \land Q$	P&Q, P&&Q, P$\cdot$Q, PQ, Kpq
$P \lor Q$	P $\mid$ Q, P $\parallel$ Q, Apq

Exercises

3.27 (Overcoming dialect differences) The following are all sentences of FOL. But they're in different dialects. Submit a sentence file in which you've translated them into our dialect.
1. $\overline{P\&\overline{Q}}$
2. !(P || (Q&&P))
3. $(\sim P \lor Q) \cdot P$
4. $P(\sim Q \lor RS)$

3.28 (Translating from Polish) Try your hand at translating the following sentences from Polish notation into our dialect. Submit the resulting sentence file.
1. NKpq
2. KNpq
3. NAKpqArs
4. NAKpAqrs
5. NAKApqrs

CHAPTER 4

The Logic of Boolean Connectives

The connectives ∧, ∨, and ¬ are truth-functional connectives. Recall what this means: the truth value of a complex sentence built by means of one of these symbols can be determined simply by looking at the truth values of the sentence's immediate constituents. So to know whether P ∨ Q is true, we need only know the truth values of P and Q. This particularly simple behavior is what allows us to capture the meanings of truth-functional connectives using truth tables.

Other connectives we could study are not this simple. Consider, the sentence *it is necessarily the case that S*. Since some true claims are necessarily true, that is, could not have been false (for instance, a = a), while other true claims are *not* necessarily true (for instance, Cube(a)), we can't figure out the truth value of the original sentence if we are only told the truth value of its constituent sentence *S*. *It is necessarily the case*, unlike *it is not the case*, is not truth-functional.

truth-functional vs. non-truth-functional operators

The fact that the Boolean connectives are truth functional makes it very easy to explain their meanings. It also provides us with a simple but powerful technique to study their logic. The technique is an extension of the truth tables used to present the meanings of the connectives. It turns out that we can often calculate the logical properties of complex sentences by constructing truth tables that display all possible assignments of truth values to the atomic constituents from which the sentences are built. The technique can, for example, tell us that a particular sentence S is a logical consequence of some premises $P_1, \ldots, P_n$. And since logical consequence is one of our main concerns, the technique is an important one to learn.

In this chapter we will discuss what truth tables can tell us about three related logical notions: the notions of *logical consequence*, *logical equivalence*, and *logical truth*. Although we've already discussed logical consequence at some length, we'll tackle these in reverse order, since the related truth table techniques are easier to understand in that order.

Section 4.1
Tautologies and logical truth

We said that a sentence S is a logical consequence of a set of premises $P_1, \ldots, P_n$ if it is impossible for the premises all to be true while the conclusion S is false. That is, the conclusion *must* be true if the premises are true.

Notice that according to this definition there are some sentences that are logical consequences of *any* set of premises, even the empty set. This will be true of any sentence whose truth is itself a logical necessity. For example, given our assumptions about FOL, the sentence $a = a$ is necessarily true. So of course, no matter what your initial premises may be, it will be impossible for those premises to be true and for $a = a$ to be false—simply because it is impossible for $a = a$ to be false! We will call such logically necessary sentences *logical truths.*

logical truth

The intuitive notions of logical possibility and logical necessity have already come up several times in this book in characterizing valid arguments and the consequence relation. But this is the first time we have applied them to individual sentences. Intuitively, a sentence is logically possible if it could be (or could have been) true, at least on logical grounds. There might be some other reasons, say physical, why the statement could not be true, but there are no logical reasons preventing it. For example, it is not physically possible to go faster than the speed of light, though it is logically possible: they do it on Star Trek all the time. On the other hand, it is not even logically possible for an object not to be identical to itself. That would simply violate the meaning of identity. The way it is usually put is that a claim is logically possible if there is some logically possible circumstance (or situation or world) in which the claim is true. Similarly, a sentence is logically necessary if it is true in every logically possible circumstance.

logical possibility and necessity

These notions are quite important, but they are also annoyingly vague. As we proceed through this book, we will introduce several precise concepts that help us clarify these notions. The first of these precise concepts, which we introduce in this section, is the notion of a *tautology.*

tautology

How can a precise concept help clarify an imprecise, intuitive notion? Let's think for a moment about the blocks language and the intuitive notion of logical possibility. Presumably, a sentence of the blocks language is logically possible if there could be a blocks world in which it is true. Clearly, if we can construct a world in Tarski's World that makes it true, then this demonstrates that the sentence is indeed logically possible. On the other hand, there are logically possible sentences that can't be made true in the worlds you can build with Tarski's World. For example, the sentence

$$\neg(\mathsf{Tet(b)} \lor \mathsf{Cube(b)} \lor \mathsf{Dodec(b)})$$

is surely *logically* possible, say if b were a sphere or an icosahedron. You can't build such a world with Tarski's World, but that is not logic's fault, just as it's not logic's fault that you can't travel faster than the speed of light. Tarski's World has its non-logical laws and constraints just like the physical world.

The Tarski's World program gives rise to a precise notion of possibility for sentences in the blocks language. We could say that a sentence is TW-possible if it is true in some world that can be built using the program. Our observations in the preceding paragraph could then be rephrased by saying that every TW-possible sentence is logically possible, but that the reverse is not in general true. Some logically possible sentences are not TW-possible.

TW-possible

It may seem surprising that we can make such definitive claims involving a vague notion like logical possibility. But really, it's no more surprising than the fact that we can say with certainty that a particular apple is red, even though the boundaries of the color red are vague. There may be cases where it is hard to decide whether something is red, but this doesn't mean there aren't many perfectly clear-cut cases.

Tarski's World gives us a precise method for showing that a sentence of the blocks language is logically possible, since whatever is possible in Tarski's World is logically possible. In this section, we will introduce another precise method, one that can be used to show that a sentence built up using truth-functional connectives is logically necessary. The method uses truth tables to show that certain sentences cannot possibly be false, due simply to the meanings of the truth-functional connectives they contain. Like the method given to us by Tarski's World, the truth table method works only in one direction: when it says that a sentence is logically necessary, then it definitely is. On the other hand, some sentences are logically necessary for reasons that the truth table method cannot detect.

truth table method

Suppose we have a complex sentence S with n atomic sentences, $\mathsf{A}_1, \ldots, \mathsf{A}_n$. To build a truth table for S, one writes the atomic sentences $\mathsf{A}_1, \ldots, \mathsf{A}_n$ across the top of the page, with the sentence S to their right. It is customary to draw a double line separating the atomic sentences from S. Your truth table will have one row for every possible way of assigning TRUE and FALSE to the atomic sentences. Since there are two possible assignments to each atomic sentence, there will be 2^n rows. Thus if $n = 1$ there will be two rows, if $n = 2$ there will be four rows, if $n = 3$ there will be eight rows, if $n = 4$ there will be sixteen rows, and so forth. It is customary to make the leftmost column have the top half of the rows marked TRUE, the second half FALSE. The next

number of rows in a truth table

column splits each of these, marking the first and third quarters of the rows with TRUE, the second and fourth quarters with FALSE, and so on. This will result in the last column having TRUE and FALSE alternating down the column.

Let's start by looking at a very simple example of a truth table, one for the sentence Cube(a) ∨ ¬Cube(a). Since this sentence is built up from one atomic sentence, our truth table will contain two rows, one for the case where Cube(a) is true and one for when it is false.

Cube(a)	Cube(a) ∨ ¬Cube(a)
T	
F	

reference columns

In a truth table, the column or columns under the atomic sentences are called *reference columns*. Once the reference columns have been filled in, we are ready to fill in the remainder of the table. To do this, we construct columns of **T**'s and **F**'s beneath each connective of the target sentence S. These columns are filled in one by one, using the truth tables for the various connectives. We start by working on connectives that apply only to atomic sentences. Once this is done, we work on connectives that apply to sentences whose main connective has already had its column filled in. We continue this process until the main connective of S has had its column filled in. This is the column that shows how the truth of S depends on the truth of its atomic parts.

Our first step in filling in this truth table, then, is to calculate the truth values that should go in the column under the innermost connective, which in this case is the ¬. We do this by referring to the truth values in the reference column under Cube(a), switching values in accord with the meaning of ¬.

Cube(a)	Cube(a) ∨ ¬Cube(a)
T	**F**
F	**T**

Once this column is filled in, we can determine the truth values that should go under the ∨ by looking at the values under Cube(a) and those under the negation sign, since these correspond to the values of the two disjuncts to which ∨ is applied. (Do you understand this?) Since there is at least one **T** in each row, the final column of the truth table looks like this.

Cube(a)	Cube(a) ∨ ¬Cube(a)
T	**T** F
F	**T** T

Not surprisingly, our table tells us that the sentence Cube(a) ∨ ¬Cube(a) cannot be false. It is what we will call a *tautology*, an especially simple kind of logical truth. We will give a precise definition of tautologies later. Our sentence is in fact an instance of a principle, P ∨ ¬P, that is known as the law of the excluded middle. Every instance of this principle is a tautology.

law of excluded middle

Let's next look at a more complex truth table, one for a sentence built up from three atomic sentences.

$$(\text{Cube}(a) \wedge \text{Cube}(b)) \vee \neg \text{Cube}(c)$$

In order to make our table easier to read, we will abbreviate the atomic sentences by A, B, and C. Since there are three atomic sentences, our table will have eight (2^3) rows. Look carefully at how we've arranged the T's and F's and convince yourself that every possible assignment is represented by one of the rows.

A	B	C	(A ∧ B) ∨ ¬C
T	T	T	
T	T	F	
T	F	T	
T	F	F	
F	T	T	
F	T	F	
F	F	T	
F	F	F	

Since two of the connectives in the target sentence apply to atomic sentences whose values are specified in the reference column, we can fill in these columns using the truth tables for ∧ and ¬ given earlier.

A	B	C	(A ∧ B)	∨	¬C
T	T	T	T		F
T	T	F	T		T
T	F	T	F		F
T	F	F	F		T
F	T	T	F		F
F	T	F	F		T
F	F	T	F		F
F	F	F	F		T

This leaves only one connective, the main connective of the sentence. We fill in the column under it by referring to the two columns just completed, using the truth table for ∨.

A	B	C	(A ∧ B)	∨	¬C
T	T	T	T	**T**	F
T	T	F	T	**T**	T
T	F	T	F	**F**	F
T	F	F	F	**T**	T
F	T	T	F	**F**	F
F	T	F	F	**T**	T
F	F	T	F	**F**	F
F	F	F	F	**T**	T

When we inspect the final column of this table, the one beneath the connective ∨, we see that the sentence will be false in any circumstance where Cube(c) is true and one of Cube(a) or Cube(b) is false. This table shows that our sentence is not a tautology. Furthermore, since there clearly are blocks worlds in which c is a cube and either a or b is not, the claim made by our original sentence is not logically necessary.

Let's look at one more example, this time for a sentence of the form

$$¬(A ∧ (¬A ∨ (B ∧ C))) ∨ B$$

This sentence, though it has the same number of atomic constituents, is considerably more complex than our previous example. We begin the truth table by filling in the columns under the two connectives that apply directly to atomic sentences.

A	B	C	¬(A ∧ (¬A ∨ (B ∧ C)))	∨ B
T	T	T	**F**	**T**
T	T	F	**F**	**F**
T	F	T	**F**	**F**
T	F	F	**F**	**F**
F	T	T	**T**	**T**
F	T	F	**T**	**F**
F	F	T	**T**	**F**
F	F	F	**T**	**F**

We can now fill in the column under the ∨ that connects ¬A and B ∧ C by referring to the columns just filled in. This column will have an **F** in it if and only if both of the constituents are false.

A	B	C	¬(A ∧ (¬A ∨ (B ∧ C))) ∨ B
T	T	T	F **T** T
T	T	F	F **F** F
T	F	T	F **F** F
T	F	F	F **F** F
F	T	T	T **T** T
F	T	F	T **T** F
F	F	T	T **T** F
F	F	F	T **T** F

We now fill in the column under the remaining ∧. To do this, we need to refer to the reference column under A, and to the just completed column. The best way to do this is to run two fingers down the relevant columns and enter a **T** in only those rows where both your fingers are pointing to **T**'s.

A	B	C	¬(A ∧ (¬A ∨ (B ∧ C))) ∨ B
T	T	T	**T** F T T
T	T	F	**F** F F F
T	F	T	**F** F F F
T	F	F	**F** F F F
F	T	T	**F** T T T
F	T	F	**F** T T F
F	F	T	**F** T T F
F	F	F	**F** T T F

We can now fill in the column for the remaining ¬ by referring to the previously completed column. The ¬ simply reverses **T**'s and **F**'s.

A	B	C	¬(A ∧ (¬A ∨ (B ∧ C))) ∨ B
T	T	T	**F** T F T T
T	T	F	**T** F F F F
T	F	T	**T** F F F F
T	F	F	**T** F F F F
F	T	T	**T** F T T T
F	T	F	**T** F T T F
F	F	T	**T** F T T F
F	F	F	**T** F T T F

Finally, we can fill in the column under the main connective of our sentence. We do this with the two-finger method: running our fingers down the reference column for B and the just completed column, entering **T** whenever at least one finger points to a **T**.

A	B	C	¬(A ∧ (¬A ∨ (B ∧ C))) ∨ B
T	T	T	F T F T T **T**
T	T	F	T F F F F **T**
T	F	T	T F F F F **T**
T	F	F	T F F F F **T**
F	T	T	T F T T T **T**
F	T	F	T F T T F **T**
F	F	T	T F T T F **T**
F	F	F	T F T T F **T**

tautology

We will say that a *tautology* is any sentence whose truth table has only **T**'s in the column under its main connective. Thus, we see from the final column of the above table that any sentence of the form

$$\neg(A \wedge (\neg A \vee (B \wedge C))) \vee B$$

is a tautology.

You try it
. . . .

▶ 1. Open the program Boole from the software that came with the book. We will use Boole to reconstruct the truth table just discussed. The first thing to do is enter the sentence ¬(A ∧ (¬A ∨ (B ∧ C))) ∨ B at the top, right of the table. To do this, use the toolbar to enter the logical symbols and the keyboard to type the letters A, B, and C. (You can also enter the logical symbols from the keyboard by typing &, |, and ∼ for ∧, ∨, and ¬, respectively. If you enter the logical symbols from the keyboard, make sure you add spaces before and after the binary connectives so that the columns under them will be reasonably spaced out.) If your sentence is ill-formed, part of the sentence will be displayed in red. The point at which the color changes is Boole's best guess about where the error is.

▶ 2. To build the reference columns, click in the top left portion of the table to move your insertion point to the top of the first reference column. Enter C in this column. Then choose **Add Column Before** from the **Table** menu and enter B. Repeat this procedure and add a column headed by A. To fill in the reference columns, click under each of them in turn, and type the desired pattern of **T**'s and **F**'s.

▶ 3. Click under the various connectives in the target sentence, and notice that green squares appear in the columns whose values the connective depends upon. Select a column so that the highlighted columns are already filled in, and fill in that column with the appropriate truth values. Continue this process until your table is complete. When you are done use the **Verify** item from the **Table** menu to see if all the values are correct and your table complete. You can also verify your table using the colored button on the toolbar (just to the left of the print button). If you have filled the table correctly, green check marks should appear to the left of each row, and next to the target sentence. Red crosses indicate that you have made a mistake, and you should fix these now.

4. Once you have a correct and complete truth table, click on the **Assessment** button in the pink area under the toolbar. This will allow you to say whether you think the sentence is a tautology. Say that it is (since it is), and check your assessment by again selecting **Verify** from the **Table** menu (or by using the toolbar button). You should now see a green check mark next to the word "Tautology" on the assessment panel. Save your table as Table Tautology 1.

◀

. . . . *Congratulations*

There is a slight problem with our definition of a tautology, in that it assumes that every sentence has a main connective. This is almost always the case, but not in sentences like:

$$P \wedge Q \wedge R$$

main connectives

For purposes of constructing truth tables, we will assume that the main connective in conjunctions with more than two conjuncts is always the rightmost $\wedge$. That is to say, we will construct a truth table for $P \wedge Q \wedge R$ the same way we would construct a truth table for:

$$(P \wedge Q) \wedge R$$

More generally, we construct the truth table for:

$$P_1 \wedge P_2 \wedge P_3 \wedge \ldots \wedge P_n$$

as if it were "punctuated" like this:

$$(((P_1 \wedge P_2) \wedge P_3) \wedge \ldots) \wedge P_n$$

We treat long disjunctions similarly.

Any tautology is logically necessary. After all, its truth is guaranteed simply by its structure and the meanings of the truth-functional connectives. Tautologies are logical necessities in a very strong sense. Their truth is independent of both the way the world happens to be and even the meanings of the atomic sentences out of which they are composed.

tautologies and logical necessity

It should be clear, however, that not all logically necessary claims are tautologies. The simplest example of a logically necessary claim that is not a tautology is the FOL sentence a = a. Since this is an atomic sentence, its truth table would contain one **T** and one **F**. The truth table method is too coarse to recognize that the row containing the **F** does not represent a genuine possibility.

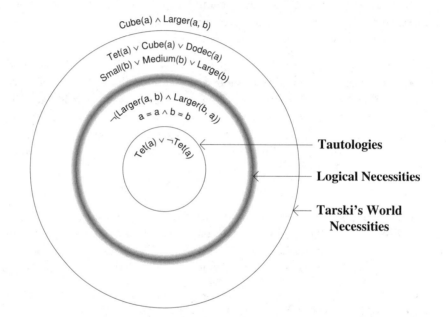

Figure 4.1: The relation between tautologies, logical truths, and TW-necessities.

You should be able to think of any number of sentences that are not tautological, but which nonetheless seem logically necessary. For example, the sentence

$$\neg(\mathsf{Larger}(a, b) \wedge \mathsf{Larger}(b, a))$$

cannot possibly be false, yet a truth table for the sentence will not show this. The sentence will be false in the row of the truth table that assigns **T** to both Larger(a, b) and Larger(b, a).

We now have two methods for exploring the notions of logical possibility and necessity, at least for the blocks language. First, there are the blocks worlds that can be constructed using Tarski's World. If a sentence is true in some such world, we have called it TW-possible. Similarly, if a sentence is true in every world in which it has a truth value (that is, in which its names all have referents), we can call it TW-necessary. The second method is that of truth tables. If a sentence comes out true in every row of its truth table, we could call it TT-necessary or, more traditionally, tautological. If a sentence is true in at least one row of its truth table, we will call it TT-possible.

TT-possible

None of these concepts correspond exactly to the vague notions of logical possibility and necessity. But there are clear and important relationships between the notions. On the necessity side, we know that all tautologies are logically necessary, and that all logical necessities are TW-necessary. These relationships are depicted in the "Euler circle" diagram in Figure 4.1, where we have represented the set of logical necessities as the interior of a circle with a fuzzy boundary. The set of tautologies is represented by a precise circle contained inside the fuzzy circle, and the set of Tarski's World necessities is represented by a precise circle containing both these circles.

There is, in fact, another method for showing that a sentence is a logical truth, one that uses the technique of proofs. If you can prove a sentence using no premises whatsoever, then the sentence is logically necessary. In the following chapters, we will give you some more methods for giving proofs. Using these, you will be able to prove that sentences are logically necessary without constructing their truth tables. When we add quantifiers to our language, the gap between tautologies and logical truths will become very apparent, making the truth table method less useful. By contrast, the methods of proof that we discuss later will extend naturally to sentences containing quantifiers.

proof and logical truth

Remember

Let S be a sentence of FOL built up from atomic sentences by means of truth-functional connectives alone. A truth table for S shows how the truth of S depends on the truth of its atomic parts.

1. S is a tautology if and only if every row of the truth table assigns TRUE to S.

2. If S is a tautology, then S is a logical truth (that is, is logically necessary).

3. Some logical truths are not tautologies.

4. S is TT-possible if and only if at least one row of the truth table assigns TRUE to S.

Exercises

In this chapter, you will often be using Boole to construct truth tables. Although Boole has the capability of building and filling in reference columns for you, do not use this feature. To understand truth tables, you need to be able to do this yourself. In later chapters, we will let you use the feature, once you've learned how to do it yourself. The Grade Grinder will, by the way, be able to tell if Boole constructed the reference columns.

4.1 If you skipped the **You try it** section, go back and do it now. Submit the file **Table Tautology 1**.

4.2 Assume that A, B, and C are atomic sentences. Use Boole to construct truth tables for each of the following sentences and, based on your truth tables, say which are tautologies. Name your tables **Table 4.2.x**, where x is the number of the sentence.
1. $(A \land B) \lor (\neg A \lor \neg B)$
2. $(A \land B) \lor (A \land \neg B)$
3. $\neg(A \land B) \lor C$
4. $(A \lor B) \lor \neg(A \lor (B \land C))$

4.3 In Exercise 4.2 you should have discovered that two of the four sentences are tautologies, and hence logical truths.
1. Suppose you are told that the atomic sentence A is in fact a logical truth (for example, $a = a$). Can you determine whether any additional sentences in the list (1)-(4) are logically necessary based on this information?
2. Suppose you are told that A is in fact a logically false sentence (for example, $a \neq a$). Can you determine whether any additional sentences in the list (1)-(4) are logical truths based on this information?

In the following four exercises, use Boole to construct truth tables and indicate whether the sentence is TT-*possible and whether it is a tautology. Remember how you should treat long conjunctions and disjunctions.*

4.4 $\neg(B \land \neg C \land \neg B)$

4.5 $A \lor \neg(B \lor \neg(C \land A))$

4.6 $\neg[\neg A \lor \neg(B \land C) \lor (A \land B)]$

4.7 $\neg[(\neg A \lor B) \land \neg(C \land D)]$

4.8 Make a copy of the Euler circle diagram on page 102 and place the numbers of the following sentences in the appropriate region.
1. $a = b$
2. $a = b \lor b = b$

3. $a = b \wedge b = b$
4. $\neg(\text{Large}(a) \wedge \text{Large}(b) \wedge \text{Adjoins}(a, b))$
5. $\text{Larger}(a, b) \vee \neg\text{Larger}(a, b)$
6. $\text{Larger}(a, b) \vee \text{Smaller}(a, b)$
7. $\neg\text{Tet}(a) \vee \neg\text{Cube}(b) \vee a \neq b$
8. $\neg(\text{Small}(a) \wedge \text{Small}(b)) \vee \text{Small}(a)$
9. $\text{SameSize}(a, b) \vee \neg(\text{Small}(a) \wedge \text{Small}(b))$
10. $\neg(\text{SameCol}(a, b) \wedge \text{SameRow}(a, b))$

4.9 (Logical dependencies) Use Tarski's World to open Weiner's Sentences.

1. For each of the ten sentences in this file, construct a truth table in Boole and assess whether the sentence is TT-possible. Name your tables Table 4.9.x, where x is the number of the sentence in question. Use the results to fill in the first column of the following table:

Sentence	TT-possible	TW-possible
1		
2		
3		
⋮		
10		

2. In the second column of the table, put *yes* if you think the sentence is TW-possible, that is, if it is possible to make the sentence true by building a world in Tarski's World, and *no* otherwise. For each sentence that you mark TW-possible, actually build a world in which it is true and name it World 4.9.x, where x is the number of the sentence in question. The truth tables you constructed before may help you build these worlds.

3. Are any of the sentences TT-possible but not TW-possible? Explain why this can happen. Are any of the sentences TW-possible but not TT-possible? Explain why not. Submit the files you created and turn in the table and explanations to your instructor.

4.10 Draw an Euler circle diagram similar to the diagram on page 102, but this time showing the relationship between the notions of logical possibility, TW-possibility, and TT-possibility. For each region in the diagram, indicate an example sentence that would fall in that region. Don't forget the region that falls outside all the circles.

All necessary truths are obviously possible: since they are true in *all* possible circumstances, they are surely true in *some* possible circumstances. Given this reflection, where would the sentences from our previous diagram on page 102 fit into the new diagram?

4.11 Suppose that S is a tautology, with atomic sentences A, B, and C. Suppose that we replace all occurrences of A by another sentence P, possibly complex. Explain why the resulting sentence is still a tautology. This is expressed by saying that substitution preserves tautologicality. Explain why substitution of atomic sentences does not always preserve logical truth, even though it preserves tautologies. Give an example.

Logical and tautological equivalence

In the last chapter, we introduced the notion of logically equivalent sentences, sentences that have the same truth values in every possible circumstance. When two sentences are logically equivalent, we also say they have the same truth conditions, since the conditions under which they come out true or false are identical.

logical equivalence

The notion of logical equivalence, like logical necessity, is somewhat vague, but not in a way that prevents us from studying it with precision. For here too we can introduce precise concepts that bear a clear relationship to the intuitive notion we aim to understand better. The key concept we will introduce in this section is that of *tautological equivalence*. Two sentences are tautologically equivalent if they can be seen to be equivalent simply in virtue of the meanings of the truth-functional connectives. As you might expect, we can check for tautological equivalence using truth tables.

tautological equivalence

Suppose we have two sentences, S and S′, that we want to check for tautological equivalence. What we do is construct a truth table with a reference column for each of the atomic sentences that appear in either of the two sentences. To the right, we write both S and S′, with a vertical line separating them, and fill in the truth values under the connectives as usual. We call this a *joint* truth table for the sentences S and S′. When the joint truth table is completed, we compare the column under the main connective of S with the column under the main connective of S′. If these columns are identical, then we know that the truth conditions of the two sentences are the same.

joint truth tables

Let's look at an example. Using A and B to stand for arbitrary atomic sentences, let us test the first DeMorgan law for tautological equivalence. We would do this by means of the following joint truth table.

A	B	¬(A ∧ B)		¬A ∨ ¬B		
T	T	**F**	T	F	**F**	F
T	F	**T**	F	F	**T**	T
F	T	**T**	F	T	**T**	F
F	F	**T**	F	T	**T**	T

In this table, the columns in bold correspond to the main connectives of the two sentences. Since these columns are identical, we know that the sentences must have the same truth values, no matter what the truth values of their atomic constituents may be. This holds simply in virtue of the structure of the two sentences and the meanings of the Boolean connectives. So, the two sentences are indeed tautologically equivalent.

Let's look at a second example, this time to see whether the sentence $\neg((A \vee B) \wedge \neg C)$ is tautologically equivalent to $(\neg A \wedge \neg B) \vee C$. To construct a truth table for this pair of sentences, we will need eight rows, since there are three atomic sentences. The completed table looks like this.

A	B	C	$\neg((A \vee B) \wedge \neg C)$	$(\neg A \wedge \neg B) \vee C$
T	T	T	**T** T F F	F F F **T**
T	T	F	**F** T T T	F F F **F**
T	F	T	**T** T F F	F F T **T**
T	F	F	**F** T T T	F F T **F**
F	T	T	**T** T F F	T F F **T**
F	T	F	**F** T T T	T F F **F**
F	F	T	**T** F F F	T T T **T**
F	F	F	**T** F F T	T T T **T**

Once again, scanning the final columns under the two main connectives reveals that the sentences are tautologically equivalent, and hence logically equivalent.

All tautologically equivalent sentences are logically equivalent, but the reverse does not in general hold. Indeed, the relationship between these notions is the same as that between tautologies and logical truths. Tautological equivalence is a strict form of logical equivalence, one that won't apply to some logically equivalent pairs of sentences. Consider the pair of sentences:

tautological vs. logical equivalence

$$a = b \wedge \text{Cube}(a)$$
$$a = b \wedge \text{Cube}(b)$$

These sentences are logically equivalent, as is demonstrated in the following informal proof.

Proof: Suppose that the sentence $a = b \wedge \text{Cube}(a)$ is true. Then $a = b$ and $\text{Cube}(a)$ are both true. Using the indiscernibility of identicals (Identity Elimination), we know that $\text{Cube}(b)$ is true, and hence that $a = b \wedge \text{Cube}(b)$ is true. So the truth of $a = b \wedge \text{Cube}(a)$ logically implies the truth of $a = b \wedge \text{Cube}(b)$.

The reverse holds as well. For suppose that $a = b \wedge \text{Cube}(b)$ is true. Then by symmetry of identity, we also know $b = a$. From this and $\text{Cube}(b)$ we can conclude $\text{Cube}(a)$, and hence that $a = b \wedge \text{Cube}(a)$

is true. So the truth of $a = b \wedge Cube(b)$ implies the truth of $a = b \wedge Cube(a)$.

Thus $a = b \wedge Cube(a)$ is true if and only if $a = b \wedge Cube(b)$ is true.

This proof shows that these two sentences have the same truth values in any possible circumstance. For if one were true and the other false, this would contradict the conclusion of one of the two parts of the proof. But consider what happens when we construct a joint truth table for these sentences. Three atomic sentences appear in the pair of sentences, so the joint table will look like this. (Notice that the ordinary truth table for either of the sentences alone would have only four rows, but that the joint table must have eight. Do you understand why?)

number of rows in joint table

$a = b$	$Cube(a)$	$Cube(b)$	$a = b \wedge Cube(a)$	$a = b \wedge Cube(b)$
T	T	T	**T**	**T**
T	T	F	**T**	**F**
T	F	T	**F**	**T**
T	F	F	**F**	**F**
F	T	T	**F**	**F**
F	T	F	**F**	**F**
F	F	T	**F**	**F**
F	F	F	**F**	**F**

This table shows that the two sentences are not tautologically equivalent, since it assigns the sentences different values in the second and third rows. Look closely at those two rows to see what's going on. Notice that in both of these rows, $a = b$ is assigned **T** while $Cube(a)$ and $Cube(b)$ are assigned different truth values. Of course, *we* know that neither of these rows corresponds to a logically possible circumstance, since if a and b are identical, the truth values of $Cube(a)$ and $Cube(b)$ must be the same. But the truth table method doesn't detect this, since it is sensitive only to the meanings of the truth-functional connectives.

As we expand our language to include quantifiers, we will find many logical equivalences that are not tautological equivalences. But this is not to say there aren't a lot of important and interesting tautological equivalences. We've already highlighted three in the last chapter: double negation and the two DeMorgan equivalences. We leave it to you to check that these principles are, in fact, tautological equivalences. In the next section, we will introduce other principles and see how they can be used to simplify sentences of FOL.

> **Remember**
>
> Let S and S′ be a sentences of FOL built up from atomic sentences by means of truth-functional connectives alone. To test for tautological equivalence, we construct a *joint* truth table for the two sentences.
>
> 1. S and S′ are *tautologically equivalent* if and only if every row of the joint truth table assigns the same values to S and S′.
>
> 2. If S and S′ are tautologically equivalent, then they are logically equivalent.
>
> 3. Some logically equivalent sentences are not tautologically equivalent.

Exercises

In Exercises 4.12-4.18, use Boole to construct joint truth tables showing that the pairs of sentences are logically (indeed, tautologically) equivalent. To add a second sentence to your joint truth table, choose **Add Column After** *from the* **Table** *menu. Don't forget to specify your assessments, and remember, you should build and fill in your own reference columns.*

4.12 (DeMorgan)
¬(A ∨ B) and ¬A ∧ ¬B

4.13 (Associativity)
(A ∧ B) ∧ C and A ∧ (B ∧ C)

4.14 (Associativity)
(A ∨ B) ∨ C and A ∨ (B ∨ C)

4.15 (Idempotence)
A ∧ B ∧ A and A ∧ B

4.16 (Idempotence)
A ∨ B ∨ A and A ∨ B

4.17 (Distribution)
A ∧ (B ∨ C) and (A ∧ B) ∨ (A ∧ C)

4.18 (Distribution)
A ∨ (B ∧ C) and (A ∨ B) ∧ (A ∨ C)

4.19 (TW-equivalence) Suppose we introduced the notion of TW-equivalence, saying that two sentences of the blocks language are TW-equivalent if and only if they have the same truth value in every world that can be constructed in Tarski's World.
  1. What is the relationship between TW-equivalence, tautological equivalence and logical equivalence?
  2. Give an example of a pair of sentences that are TW-equivalent but not logically equivalent.

SECTION 4.3

Logical and tautological consequence

Our main concern in this book is with the logical consequence relation, of which logical truth and logical equivalence can be thought of as very special cases: A logical truth is a sentence that is a logical consequence of any set of premises, and logically equivalent sentences are sentences that are logical consequences of one another.

As you've probably guessed, truth tables allow us to define a precise notion of *tautological* consequence, a strict form of logical consequence, just as they allowed us to define tautologies and tautological equivalence, strict forms of logical truth and logical equivalence.

Let's look at the simple case of two sentences, P and Q, both built from atomic sentences by means of truth-functional connectives. Suppose you want to know whether Q is a consequence of P. Create a joint truth table for P and Q, just like you would if you were testing for tautological equivalence. After you fill in the columns for P and Q, scan the columns under the main connectives for these sentences. In particular, look at every row of the table in which P is true. If each such row is also one in which Q is true, then Q is said to be a *tautological consequence* of P. The truth table shows that if P is true, then Q must be true as well, and that this holds simply due to the meanings of the truth-functional connectives.

tautological consequence

Just as tautologies are logically necessary, so too any tautological consequence Q of a sentence P must also be a logical consequence of P. We can see this by proving that if Q is *not* a logical consequence of P, then it can't possibly pass our truth table test for tautological consequence.

> **Proof:** Suppose Q is not a logical consequence of P. Then by our definition of logical consequence, there must be a possible circumstance in which P is true but Q is false. This circumstance will determine truth values for the atomic sentences in P and Q, and these values will correspond to a row in the joint truth table for P and Q, since all possible assignments of truth values to the atomic sentences are represented in the truth table. Further, since P and Q are built up from the atomic sentences by truth-functional connectives, and since the former is true in the original circumstance and the latter false, P will be assigned **T** in this row and Q will be assigned **F**. Hence, Q is not a tautological consequence of P.

Let's look at a very simple example. Suppose we wanted to check to see whether A ∨ B is a consequence of A ∧ B. The joint truth table for these sentences looks like this.

A	B	A ∧ B	A ∨ B
T	T	**T**	**T**
T	F	**F**	**T**
F	T	**F**	**T**
F	F	**F**	**F**

When you compare the columns under these two sentences, you see that the sentences are most definitely *not* tautologically equivalent. No surprise. But we are interested in whether A ∧ B logically implies A ∨ B, and so the only rows we care about are those in which the former sentence is true. A ∧ B is only true in the first row, and A ∨ B is also true in that row. So this table shows that A ∨ B is a tautological consequence (and hence a logical consequence) of A ∧ B.

Notice that our table also shows that A ∧ B is not a tautological consequence of A ∨ B, since there are rows in which the latter is true and the former false. Does this show that A ∧ B is not a *logical* consequence of A ∨ B? Well, we have to be careful. A ∧ B is not in general a logical consequence of A ∨ B, but it might be in certain cases, depending on the sentences A and B. We'll ask you to come up with an example in the exercises.

Not every logical consequence of a sentence is a tautological consequence of that sentence. For example, the sentence a = c is a logical consequence of the sentence (a = b ∧ b = c), but it is not a tautological consequence of it. Think about the row that assigns **T** to the atomic sentences a = b and b = c, but **F** to the sentence a = c. Clearly this row, which prevents a = c from being a tautological consequence of (a = b ∧ b = c), does not respect the meanings of the atomic sentences out of which the sentences are built. It does not correspond to a genuinely possible circumstance, but the truth table method does not detect this.

logical vs. tautological consequence

The truth table method of checking tautological consequence is not restricted to just one premise. You can apply it to arguments with any number of premises $P_1, \ldots, P_n$ and conclusion Q. To do so, you have to construct a joint truth table for all of the sentences $P_1, \ldots, P_n$ and Q. Once you've done this, you need to check every row in which the premises all come out true to see whether the conclusion comes out true as well. If so, the conclusion is a tautological consequence of the premises.

Let's try this out on a couple of simple examples. First, suppose we want to check to see whether B is a consequence of the two premises A ∨ B and ¬A. The joint truth table for these three sentences comes out like this. (Notice that since one of our target sentences, the conclusion B, is atomic, we have simply repeated the reference column when this sentence appears again on the right.)

A	B	A ∨ B	¬A	B
T	T	**T**	**F**	**T**
T	F	**T**	**F**	**F**
F	T	**T**	**T**	**T**
F	F	**F**	**T**	**F**

Scanning the columns under our two premises, A ∨ B and ¬A, we see that there is only one row where both premises come out true, namely the third. And in the third row, the conclusion B also comes out true. So B is indeed a tautological (and hence logical) consequence of these premises.

In both of the examples we've looked at so far, there has been only one row in which the premises all came out true. This makes the arguments easy to check for validity, but it's not at all something you can count on. For example, suppose we used the truth table method to check whether A ∨ C is a consequence of A ∨ ¬B and B ∨ C. The joint truth table for these three sentences looks like this.

A	B	C	A ∨ ¬B	B ∨ C	A ∨ C
T	T	T	**T** F	**T**	**T**
T	T	F	**T** F	**T**	**T**
T	F	T	**T** T	**T**	**T**
T	F	F	**T** T	**F**	**T**
F	T	T	**F** F	**T**	**T**
F	T	F	**F** F	**T**	**F**
F	F	T	**T** T	**T**	**T**
F	F	F	**T** T	**F**	**F**

Here, there are four rows in which the premises, A ∨ ¬B and B ∨ C, are both true: the first, second, third, and seventh. But in each of these rows the conclusion, A ∨ C, is also true. The conclusion is true in other rows as well, but we don't care about that. This inference, from A ∨ ¬B and B ∨ C to A ∨ C, is logically valid, and is an instance of an important pattern known in computer science as *resolution*.

We should look at an example where the truth table method reveals that the conclusion is not a tautological consequence of the premises. Actually, the last truth table will serve this purpose. For this table also shows that the sentence A ∨ ¬B is *not* a tautological consequence of the two premises B ∨ C and A ∨ C. Can you find the row that shows this? (Hint: It's got to be the first, second, third, fifth, or seventh, since these are the rows in which B ∨ C and A ∨ C are both true.)

> **Remember**
>
> Let $P_1, \ldots, P_n$ and Q be sentences of FOL built up from atomic sentences by means of truth functional connectives alone. Construct a joint truth table for all of these sentences.
>
> 1. Q is a tautological consequence of $P_1, \ldots, P_n$ if and only if every row that assigns **T** to each of $P_1, \ldots, P_n$ also assigns **T** to Q.
>
> 2. If Q is a tautological consequence of $P_1, \ldots, P_n$, then Q is also a logical consequence of $P_1, \ldots, P_n$.
>
> 3. Some logical consequences are not tautological consequences.

Exercises

For each of the arguments below, use the truth table method to determine whether the conclusion is a tautological consequence of the premises. Your truth table for Exercise 4.24 will be fairly large. It's good for the soul to build a large truth table every once in a while. Be thankful you have Boole to help you. (But make sure you build your own reference columns!)

4.20

$(\text{Tet}(a) \wedge \text{Small}(a)) \vee \text{Small}(b)$

$\text{Small}(a) \vee \text{Small}(b)$

4.21

$\text{Taller}(\text{claire}, \text{max}) \vee \text{Taller}(\text{max}, \text{claire})$

$\text{Taller}(\text{claire}, \text{max})$

$\neg\text{Taller}(\text{max}, \text{claire})$

4.22

$\text{Large}(a)$

$\text{Cube}(a) \vee \text{Dodec}(a)$

$(\text{Cube}(a) \wedge \text{Large}(a)) \vee (\text{Dodec}(a) \wedge \text{Large}(a))$

4.23

$A \vee \neg B$

$B \vee C$

$C \vee D$

$A \vee \neg D$

4.24

$\neg A \vee B \vee C$

$\neg C \vee D$

$\neg(B \wedge \neg E)$

$D \vee \neg A \vee E$

4.25 Give an example of two different sentences A and B in the blocks language such that $A \wedge B$ is a logical consequence of $A \vee B$. [Hint: Note that $A \wedge A$ is a logical consequence of $A \vee A$, but here we insist that A and B be distinct sentences.]

SECTION 4.4

Tautological consequence in Fitch

We hope you solved Exercise 4.24, because the solution gives you a sense of both the power and the drawbacks of the truth table method. We were tempted to ask you to construct a table requiring 64 rows, but thought better of it. Constructing large truth tables may build character, but like most things that build character, it's a drag.

Checking to see if Q is a tautological consequence of $P_1, \ldots, P_n$ is a mechanical procedure. If the sentences are long it may require a lot of tedious work, but it doesn't take any originality. This is just the sort of thing that computers are good at. Because of this, we have built a mechanism into Fitch,

Taut Con mechanism

called **Taut Con**, that is similar to **Ana Con** but checks to see whether a sentence is a tautological consequence of the sentences cited in support. Like **Ana Con**, **Taut Con** is not really an inference rule (we will introduce inference rules for the Boolean connectives in Chapter 6), but is useful for quickly testing whether one sentence follows tautologically from others.

You try it
. . . .

▶ 1. Launch Fitch and open the file Taut Con 1. In this file you will find an argument that has the same form as the argument in Exercise 4.23. (Ignore the two goal sentences. We'll get to them later.) Move the focus slider to the last step of the proof. From the **Rule?** menu, go down to the **Con** submenu and choose **Taut Con**.

▶ 2. Now cite the three premises as support for this sentence and check the step. The step will not check out since this sentence is not a tautological consequence of the premises, as you discovered if you did Exercise 4.23, which has the same form as this inference.

▶ 3. Edit the step that did not check out to read:

$$\text{Home(max)} \lor \text{Home(carl)}$$

This sentence is a tautological consequence of two of the premises. Figure out which two and cite just them. If you cited the right two, the step should check out. Try it.

▶ 4. Add another step to the proof and enter the sentence:

$$\text{Home(carl)} \lor (\text{Home(max)} \land \text{Home(pris)})$$

Use **Taut Con** to see if this sentence follows tautologically from the three premises. Choose **Verify Proof** from the **Proof** menu. You will find that although the step checks out, the goal does not. This is because we have put a special constraint on your use of **Taut Con** in this exercise.

5. Choose **View Goal Constraints** from the **Goal** menu. You will find that ◄ in this proof, you are allowed to use **Taut Con**, but can only cite two or fewer support sentences when you use it. Close the goal window to get back to the proof.

6. The sentence you entered also follows from the sentence immediately above ◄ it plus just one of the three premises. Uncite the three premises and see if you can get the step to check out citing just *two* sentences in support. Once you succeed, verify the proof and save it as Proof Taut Con 1. Do not close the proof, since it will be needed in the next "You Try It".
. . . . *Congratulations*

You are probably curious about the relationship between **Taut Con** and **Ana Con**—and for that matter, what the other mysterious item on the **Con** menu, **FO Con**, might do. These are in fact three increasingly strong methods that Fitch uses to test for logical consequence. **Taut Con** is the weakest. It checks to see whether the current step follows from the cited sentences in virtue of the meanings of the truth-functional connectives. It ignores the meanings of any predicates that appear in the sentence and, when we introduce quantifiers into the language, it will ignore those as well.

Taut Con, FO Con, and Ana Con

To help you keep track of the information that **Taut Con** considers when checking a step, Fitch has special "goggles" which obscure the information that will not be considered when checking the step. As we said above, the only things that matter when checking a step justified by **Taut Con** step are the propositional connectives in the formulae, and the pattern of occurrence of the atomic formulae.

What this means is that the meanings of the predicate symbols and names in the formulae do not matter, and Fitch's goggles obscure this information. When you put the goggles on, every individual atomic formula involved in the step appears as a block of color, hiding the particular atomic formula that is present. Every occurrence of the same atomic formula will be represented by the same color, and different formulae will have different colors.

You try it
. . . .

1. Return to the file Taut Con 1 again that you made in the previous "You ◄ Try It" section, and focus on the last step of the proof, which contains an

application of the **Taut Con** rule.

▶ 2. Click on the picture of a pair of goggles that appears to the right of the rule name, and notice how the conclusion and the cited sentences change into blocks of color.

▶ 3. You should see a more colorful version of this

                                                                 Taut Con

Be sure to understand how the atomic formulas relate to their corresponding colors.

▶ 4. There is nothing to save.

. . . . *Congratulations*

FO Con, which stands for "first-order consequence," pays attention to the truth-functional connectives, the quantifiers, and the identity predicate when it checks for consequence. **FO Con** would, for example, identify a = c as a consequence of a = b ∧ b = c. It is stronger than **Taut Con** in the sense that any consequence that **Taut Con** recognizes as valid will also be recognized by **FO Con**. But it may take longer since it has to apply a more complex procedure, thanks to identity and the quantifiers. After we get to quantifiers, we'll talk more about the procedure it is applying.

The strongest rule of the three is **Ana Con**, which tries to recognize consequences due to truth-functional connectives, quantifiers, identity, *and* most of the blocks language predicates. (**Ana Con** ignores Between and Adjoins, simply for practical reasons.) Any inference that checks out using either **Taut Con** or **FO Con** should, in principle, check out using **Ana Con** as well. In practice, though, the procedure that **Ana Con** uses may bog down or run out of memory in cases where the first two have no trouble.

As we said before, you should only use a procedure from the **Con** menu when the exercise makes clear that the procedure is allowed in the solution. Moreover if an exercise asks you to use **Taut Con**, don't use **FO Con** or **Ana Con** instead, even if these more powerful rules seem to work just as well. If you are in doubt about which rules you are allowed to use, choose **View Goal Constraints** from the **Goal** menu.

You try it

. . . .

1. Open the file Taut Con 2. You will find a proof containing ten steps whose rules have not been specified. ◄

2. Focus on each step in turn. You will find that the supporting steps have already been cited. Convince yourself that the step follows from the cited sentences. Is it a tautological consequence of the sentences cited? If so, change the rule to **Taut Con** and see if you were right. If not, change it to **Ana Con** and see if it checks out. (If **Taut Con** will work, make sure you use it rather than the stronger **Ana Con**.) ◄

3. When all of your steps check out using **Taut Con** or **Ana Con**, go back and find the one step whose rule can be changed from **Ana Con** to the weaker **FO Con**. ◄

4. When each step checks out using the weakest **Con** rule possible, save your proof as Proof Taut Con 2. ◄

. . . . *Congratulations*

Exercises

4.26 If you skipped the **You try it** sections, go back and do them now. Submit the files Proof Taut
✎ Con 1 and Proof Taut Con 2.

For each of the following arguments, decide whether the conclusion is a tautological consequence of the premises. If it is, submit a proof that establishes the conclusion using one or more applications of **Taut Con**. *Do not cite more than two sentences at a time for any of your applications of* **Taut Con**. *If the conclusion is not a consequence of the premises, submit a counterexample world showing that the argument is not valid.*

4.27
✎

| Cube(a) ∨ Cube(b) |
| Dodec(c) ∨ Dodec(d) |
¬Cube(a) ∨ ¬Dodec(c)
Cube(b) ∨ Dodec(d)

4.28
✎

| Large(a) ∨ Large(b) |
Large(a) ∨ Large(c)
Large(a) ∧ (Large(b) ∨ Large(c))

4.29

Small(a) ∨ Small(b)
Small(b) ∨ Small(c)
Small(c) ∨ Small(d)
Small(d) ∨ Small(e)
¬Small(c)

Small(a) ∨ Small(e)

4.30

Tet(a) ∨ ¬(Tet(b) ∧ Tet(c))
¬(¬Tet(b) ∨ ¬Tet(d))
(Tet(e) ∧ Tet(c)) ∨ (Tet(c) ∧ Tet(d))

Tet(a)

Section 4.5

Pushing negation around

When two sentences are logically equivalent, each is a logical consequence of the other. As a result, in giving an informal proof, you can always go from an established sentence to one that is logically equivalent to it. This fact makes observations like the DeMorgan laws and double negation quite useful in giving informal proofs.

substitution of logical equivalents

What makes these equivalences even more useful is the fact that logically equivalent sentences can be substituted for one another in the context of a larger sentence and the resulting sentences will also be logically equivalent. An example will help illustrate what we mean. Suppose we start with the sentence:

$$\neg(\text{Cube}(a) \land \neg\neg\text{Small}(a))$$

By the principle of double negation, we know that Small(a) is logically equivalent to ¬¬Small(a). Since these have exactly the same truth conditions, we can substitute Small(a) for ¬¬Small(a) in the context of the above sentence, and the result,

$$\neg(\text{Cube}(a) \land \text{Small}(a))$$

will be logically equivalent to the original, a fact that you can check by constructing a joint truth table for the two sentences.

We can state this important fact in the following way. Let's write S(P) for an FOL sentence that contains the (possibly complex) sentence P as a component part, and S(Q) for the result of substituting Q for P in S(P). Then if P and Q are logically equivalent:

$$P \Leftrightarrow Q$$

it follows that S(P) and S(Q) are also logically equivalent:

$$S(P) \Leftrightarrow S(Q)$$

This is known as the principle of *substitution of logical equivalents*.

We won't prove this principle at the moment, because it requires a proof by induction, a style of proof we get to in a later chapter. But the observation allows us to use a few simple equivalences to do some pretty amazing things. For example, using only the two DeMorgan laws and double negation, we can take any sentence built up with $\wedge$, $\vee$, and $\neg$, and transform it into one where $\neg$ applies only to atomic sentences. Another way of expressing this is that any sentence built out of atomic sentences using the three connectives $\wedge$, $\vee$, and $\neg$ is logically equivalent to one built from literals using just $\wedge$ and $\vee$.

To obtain such a sentence, you simply drive the $\neg$ in, switching $\wedge$ to $\vee$, $\vee$ to $\wedge$, and canceling any pair of $\neg$'s that are right next to each other, not separated by any parentheses. Such a sentence is said to be in *negation normal form* or NNF. Here is an example of a derivation of the negation normal form of a sentence. We use A, B, and C to stand for any *atomic* sentences of the language.

negation normal form (NNF)

$$
\begin{aligned}
\neg((A \vee B) \wedge \neg C) &\Leftrightarrow \neg(A \vee B) \vee \neg\neg C \\
&\Leftrightarrow \neg(A \vee B) \vee C \\
&\Leftrightarrow (\neg A \wedge \neg B) \vee C
\end{aligned}
$$

In reading and giving derivations of this sort, remember that the symbol $\Leftrightarrow$ is not itself a symbol of the first-order language, but a shorthand way of saying that two sentences are logically equivalent. In this derivation, the first step is an application of the first DeMorgan law to the whole sentence. The second step applies double negation to the component $\neg\neg C$. The final step is an application of the second DeMorgan law to the component $\neg(A \vee B)$. The sentence we end up with is in negation normal form, since the negation signs apply only to atomic sentences.

We end this section with a list of some additional logical equivalences that allow us to simplify sentences in useful ways. You already constructed truth tables for most of these equivalences in Exercises 4.13-4.16 at the end of Section 4.2.

1. (Associativity of $\wedge$) An FOL sentence $P \wedge (Q \wedge R)$ is logically equivalent to $(P \wedge Q) \wedge R$, which is in turn equivalent to $P \wedge Q \wedge R$. That is,

associativity

$$P \wedge (Q \wedge R) \Leftrightarrow (P \wedge Q) \wedge R \Leftrightarrow P \wedge Q \wedge R$$

2. (Associativity of $\vee$) An FOL sentence $P \vee (Q \vee R)$ is logically equivalent to $(P \vee Q) \vee R$, which is in turn equivalent to $P \vee Q \vee R$. That is,

$$P \vee (Q \vee R) \Leftrightarrow (P \vee Q) \vee R \Leftrightarrow P \vee Q \vee R$$

commutativity

3. (Commutativity of $\wedge$) A conjunction $P \wedge Q$ is logically equivalent to $Q \wedge P$. That is,

$$P \wedge Q \Leftrightarrow Q \wedge P$$

As a result, any rearrangement of the conjuncts of an FOL sentence is logically equivalent to the original. For example, $P \wedge Q \wedge R$ is equivalent to $R \wedge Q \wedge P$.

4. (Commutativity of $\vee$) A disjunction $P \vee Q$ is logically equivalent to $Q \vee P$. That is,

$$P \vee Q \Leftrightarrow Q \vee P$$

As a result, any rearrangement of the disjuncts of an FOL sentence is logically equivalent to the original. For example, $P \vee Q \vee R$ is equivalent to $R \vee Q \vee P$.

idempotence

5. (Idempotence of $\wedge$) A conjunction $P \wedge P$ is equivalent to P. That is,

$$P \wedge P \Leftrightarrow P$$

More generally (given Commutativity), any conjunction with a repeated conjunct is equivalent to the result of removing all but one occurrence of that conjunct. For example, $P \wedge Q \wedge P$ is equivalent to $P \wedge Q$.

6. (Idempotence of $\vee$) A disjunction $P \vee P$ is equivalent to P. That is,

$$P \vee P \Leftrightarrow P$$

More generally (given Commutativity), any disjunction with a repeated disjunct is equivalent to the result of removing all but one occurrence of that disjunct. For example, $P \vee Q \vee P$ is equivalent to $P \vee Q$.

Here is an example where we use some of these laws to show that the first sentence in the following list is logically equivalent to the last. Once again (as in what follows), we use A, B, and C to stand for arbitrary atomic sentences of FOL. Thus the result is in negation normal form.

$$
\begin{aligned}
(A \vee B) \wedge C \wedge (\neg(\neg B \wedge \neg A) \vee B) \ &\Leftrightarrow\ (A \vee B) \wedge C \wedge ((\neg\neg B \vee \neg\neg A) \vee B) \\
&\Leftrightarrow\ (A \vee B) \wedge C \wedge ((B \vee A) \vee B) \\
&\Leftrightarrow\ (A \vee B) \wedge C \wedge (B \vee A \vee B) \\
&\Leftrightarrow\ (A \vee B) \wedge C \wedge (B \vee A) \\
&\Leftrightarrow\ (A \vee B) \wedge C \wedge (A \vee B) \\
&\Leftrightarrow\ (A \vee B) \wedge C
\end{aligned}
$$

We call a demonstration of this sort a *chain of equivalences*. The first step in this chain is justified by one of the DeMorgan laws. The second step involves two applications of double negation. In the next step we use associativity to remove the unnecessary parentheses. In the fourth step, we use idempotence of ∨. The next to the last step uses commutativity of ∨, while the final step uses idempotence of ∧.

chain of equivalences

Remember

1. *Substitution of equivalents:* If P and Q are logically equivalent:

$$P \Leftrightarrow Q$$

   then the results of substituting one for the other in the context of a larger sentence are also logically equivalent:

$$S(P) \Leftrightarrow S(Q)$$

2. A sentence is in *negation normal form* (NNF) if all occurrences of ¬ apply directly to atomic sentences.

3. Any sentence built from atomic sentences using just ∧, ∨, and ¬ can be put into negation normal form by repeated application of the DeMorgan laws and double negation.

4. Sentences can often be further simplified using the principles of associativity, commutativity, and idempotence.

Exercises

4.31 (Negation normal form) Use Tarski's World to open **Turing's Sentences**. You will find the following five sentences, each followed by an empty sentence position.

   1. ¬(Cube(a) ∧ Larger(a, b))
   3. ¬(Cube(a) ∨ ¬Larger(b, a))
   5. ¬(¬Cube(a) ∨ ¬Larger(a, b) ∨ a ≠ b)
   7. ¬(Tet(b) ∨ (Large(c) ∧ ¬Smaller(d, e)))
   9. Dodec(f) ∨ ¬(Tet(b) ∨ ¬Tet(f) ∨ ¬Dodec(f))

In the empty positions, write the negation normal form of the sentence above it. Then build any world where all of the names are in use. If you have gotten the negation normal forms correct, each even numbered sentence will have the same truth value in your world as the odd numbered sentence above it. Verify that this is so in your world. Submit the modified sentence file as Sentences 4.31.

4.32 (Negation normal form) Use Tarski's World to open the file Sextus' Sentences. In the odd
✎ numbered slots, you will find the following sentences.

1. ¬(Home(carl) ∧ ¬Home(claire))
3. ¬[Happy(max) ∧ (¬Likes(carl, claire) ∨ ¬Likes(claire, carl))]
5. ¬¬¬[(Home(max) ∨ Home(carl)) ∧ (Happy(max) ∨ Happy(carl))]

Use Double Negation and DeMorgan's laws to put each sentence into negation normal form in the slot below it. Submit the modified file as Sentences 4.32.

In each of the following exercises, use associativity, commutativity, and idempotence to simplify the sentence as much as you can using just these rules. Your answer should consist of a chain of logical equivalences like the chain given on page 120. At each step of the chain, indicate which principle you are using.

4.33 (A ∧ B) ∧ A
✎

4.34 (B ∧ (A ∧ B ∧ C))
✎

4.35 (A ∨ B) ∨ (C ∧ D) ∨ A
✎

4.36 (¬A ∨ B) ∨ (B ∨ C)
✎

4.37 (A ∧ B) ∨ C ∨ (B ∧ A) ∨ A
✎

SECTION 4.6

Conjunctive and disjunctive normal forms

We have seen that with a few simple principles of Boolean logic, we can start with a sentence and transform it into a logically equivalent sentence in negation normal form, one where all negations occur in front of atomic sentences. We can improve on this by introducing the so-called distributive laws. These additional equivalences will allow us to transform sentences into what are known as *conjunctive normal form* (CNF) and *disjunctive normal form* (DNF). These normal forms are quite important in certain applications of logic in computer science, as we discuss in Chapter 17. We will also use disjunctive normal form to demonstrate an important fact about the Boolean connectives in Chapter 7.

distribution   Recall that in algebra you learned that multiplication distributes over addition: $a \times (b+c) = (a \times b) + (a \times c)$. The distributive laws of logic look formally much the same. One version tells us that $P \wedge (Q \vee R)$ is logically equivalent to $(P \wedge Q) \vee (P \wedge R)$. That is, $\wedge$ distributes over $\vee$. To see that this is so, notice that the first sentence is true if and only if P plus at least one of Q or R is true. But a moment's thought shows that the second sentence is true in exactly the same circumstances. This can also be confirmed by constructing a joint truth table for the two sentences, which you've already done if you did Exercise 4.17.

In arithmetic, $+$ does not distribute over $\times$. However, $\vee$ does distribute over $\wedge$. That is to say, $P \vee (Q \wedge R)$ is logically equivalent to $(P \vee Q) \wedge (P \vee R)$, as you also discovered in Exercise 4.18.

Remember

(The distributive laws) For any sentences P, Q, and R:

1. Distribution of $\wedge$ over $\vee$: $P \wedge (Q \vee R) \Leftrightarrow (P \wedge Q) \vee (P \wedge R)$

2. Distribution of $\vee$ over $\wedge$: $P \vee (Q \wedge R) \Leftrightarrow (P \vee Q) \wedge (P \vee R)$

As you may recall from algebra, the distributive law for $\times$ over $+$ is incredibly useful. It allows us to transform any algebraic expression involving $+$ and $\times$, no matter how complex, into one that is just a sum of products. For example, the following transformation uses distribution three times.

$$
\begin{aligned}
(a + b)(c + d) &= (a + b)c + (a + b)d \\
&= ac + bc + (a + b)d \\
&= ac + bc + ad + bd
\end{aligned}
$$

In exactly the same way, the distribution of $\wedge$ over $\vee$ allows us to transform any sentence built up from literals by means of $\wedge$ and $\vee$ into a logically equivalent sentence that is a disjunction of (one or more) conjunctions of (one or more) literals. That is, using this first distributive law, we can turn any sentence in negation normal form into a sentence that is a disjunction of conjunctions of literals. A sentence in this form is said to be in *disjunctive normal form*.

disjunctive normal form (DNF)

Here is an example that parallels our algebraic example. Notice that, as in the algebraic example, we are distributing in from the right as well as the left, even though our statement of the rule only illustrates distribution from the left.

$$
\begin{aligned}
(A \vee B) \wedge (C \vee D) &\Leftrightarrow [(A \vee B) \wedge C] \vee [(A \vee B) \wedge D] \\
&\Leftrightarrow (A \wedge C) \vee (B \wedge C) \vee [(A \vee B) \wedge D] \\
&\Leftrightarrow (A \wedge C) \vee (B \wedge C) \vee (A \wedge D) \vee (B \wedge D)
\end{aligned}
$$

As you can see, distribution of ∧ over ∨ lets us drive conjunction signs deeper and deeper, just as the DeMorgan laws allow us to move negations deeper. Thus, if we take any sentence and first use DeMorgan (and double negation) to get a sentence in negation normal form, we can then use this first distribution law to get a sentence in disjunctive normal form, one in which all the conjunction signs apply to literals.

Likewise, using distribution of ∨ over ∧, we can turn any negation normal form sentence into one that is a *conjunction* of one or more sentences, each of which is a *disjunction* of one or more literals. A sentence in this form is said to be in *conjunctive normal form (CNF)*. Here's an example, parallel to the one given above but with ∧ and ∨ interchanged:

conjunctive normal form (CNF)

$$(A \land B) \lor (C \land D) \quad \Leftrightarrow \quad [(A \land B) \lor C] \land [(A \land B) \lor D]$$
$$\Leftrightarrow \quad (A \lor C) \land (B \lor C) \land [(A \land B) \lor D]$$
$$\Leftrightarrow \quad (A \lor C) \land (B \lor C) \land (A \lor D) \land (B \lor D)$$

On page 119, we showed how to transform the sentence $\neg((A \lor B) \land \neg C)$ into one in negation normal form. The result was $(\neg A \land \neg B) \lor C$. This sentence just happens to be in disjunctive normal form. Let us repeat our earlier transformation, but continue until we get a sentence in conjunctive normal form.

$$\neg((A \lor B) \land \neg C) \quad \Leftrightarrow \quad \neg(A \lor B) \lor \neg\neg C$$
$$\Leftrightarrow \quad \neg(A \lor B) \lor C$$
$$\Leftrightarrow \quad (\neg A \land \neg B) \lor C$$
$$\Leftrightarrow \quad (\neg A \lor C) \land (\neg B \lor C)$$

It is important to remember that a sentence can count as being in both conjunctive and disjunctive normal forms at the same time. For example, the sentence

$$\text{Home(claire)} \land \neg\text{Home(max)}$$

is in both DNF and CNF. On the one hand, it is in disjunctive normal form since it is a disjunction of one sentence (itself) which is a conjunction of two literals. On the other hand, it is in conjunctive normal form since it is a conjunction of two sentences, each of which is a disjunction of one literal.

In case you find this last remark confusing, here are simple tests for whether sentences are in disjunctive normal form and conjunctive normal form. The tests assume that the sentence has no unnecessary parentheses and contains only the connectives ∧, ∨, and ¬.

test for DNF

To check whether a sentence is in DNF, ask yourself whether all the negation signs apply directly to atomic sentences and whether all the conjunction signs apply directly to literals. If both answers are *yes*, then the sentence is in disjunctive normal form.

To check whether a sentence is in CNF, ask yourself whether all the negation signs apply directly to atomic sentences and all the disjunction signs apply directly to literals. If both answers are *yes*, then the sentence is in conjunctive normal form.

test for CNF

Now look at the above sentence again and notice that it passes both of these tests (in the CNF case because it has no disjunction signs).

> **Remember**
>
> 1. A sentence is in *disjunctive normal form* (DNF) if it is a disjunction of one or more conjunctions of one or more literals.
>
> 2. A sentence is in *conjunctive normal form* (CNF) if it is a conjunction of one or more disjunctions of one or more literals.
>
> 3. Distribution of ∧ over ∨ allows you to transform any sentence in negation normal form into disjunctive normal form.
>
> 4. Distribution of ∨ over ∧ allows you to transform any sentence in negation normal form into conjunctive normal form.
>
> 5. Some sentences are in both CNF and DNF.

You try it
. . . .

1. Use Tarski's World to open the file **DNF Example**. In this file you will find two sentences. The second sentence is the result of putting the first into disjunctive normal form, so the two sentences are logically equivalent. ◀

2. Build a world in which the sentences are true. Since they are equivalent, you could try to make either one true, but you will find the second one easier to work on. ◀

3. Play the game for each sentence, committed correctly to the truth of the sentence. You should be able to win both times. Count the number of steps it takes you to win. ◀

▶ 4. In general it is easier to evaluate the truth value of a sentence in disjunctive normal form. This comes out in the game, which takes at most three steps for a sentence in DNF, one each for ∨, ∧, and ¬, in that order. There is no limit to the number of steps a sentence in other forms may take.

▶ 5. Save the world you have created as World DNF 1.

. . . . *Congratulations*

Exercises

4.38 If you skipped the **You try it** section, go back and do it now. Submit the file World DNF 1.

4.39 Open CNF Sentences. In this file you will find the following conjunctive normal form sentences in the odd numbered positions, but you will see that the even numbered positions are blank.
1. (LeftOf(a, b) ∨ BackOf(a, b)) ∧ Cube(a)
3. Larger(a, b) ∧ (Cube(a) ∨ Tet(a) ∨ a = b)
5. (Between(a, b, c) ∨ Tet(a) ∨ ¬Tet(b)) ∧ Dodec(c)
7. Cube(a) ∧ Cube(b) ∧ (¬Small(a) ∨ ¬Small(b))
9. (Small(a) ∨ Medium(a)) ∧ (Cube(a) ∨ ¬Dodec(a))

In the even numbered positions you should fill in a DNF sentence logically equivalent to the sentence above it. Check your work by opening several worlds and checking to see that each of your sentences has the same truth value as the one above it. Submit the modified file as Sentences 4.39.

4.40 Open More CNF Sentences. In this file you will find the following sentences in every third position.
1. ¬[(Cube(a) ∧ ¬Small(a)) ∨ (¬Cube(a) ∧ Small(a))]
4. ¬[(Cube(a) ∨ ¬Small(a)) ∧ (¬Cube(a) ∨ Small(a))]
7. ¬(Cube(a) ∧ Larger(a, b)) ∧ Dodec(b)
10. ¬(¬Cube(a) ∧ Tet(b))
13. ¬¬Cube(a) ∨ Tet(b)

The two blanks that follow each sentence are for you to first transform the sentence into negation normal form, and then put that sentence into CNF. Again, check your work by opening several worlds to see that each of your sentences has the same truth value as the original. When you are finished, submit the modified file as Sentences 4.40.

In Exercises 4.41-4.43, use a chain of equivalences to convert each sentence into an equivalent sentence in disjunctive normal form. Simplify your answer as much as possible using the laws of associativity, commutativity, and idempotence. At each step in your chain, indicate which principle you are applying. Assume that A, B, C, *and* D *are literals.*

4.41 $C \land (A \lor (B \land C))$

4.42 $B \land (A \land B \land (A \lor B \lor (B \land C)))$

4.43 $A \land (A \land (B \lor (A \land C)))$

CHAPTER 5

Methods of Proof for Boolean Logic

limitations of truth table methods

Truth tables give us powerful techniques for investigating the logic of the Boolean operators. But they are by no means the end of the story. Truth tables are fine for showing the validity of simple arguments that depend only on truth-functional connectives, but the method has two very significant limitations.

First, truth tables get extremely large as the number of atomic sentences goes up. An argument involving seven atomic sentences is hardly unusual, but testing it for validity would call for a truth table with $2^7 = 128$ rows. Testing an argument with 14 atomic sentences, just twice as many, would take a table containing over 16 thousand rows. You could probably get a Ph.D. in logic for building a truth table that size. This exponential growth severely limits the practical value of the truth table method.

The second limitation is, surprisingly enough, even more significant. Truth table methods can't be easily extended to reasoning whose validity depends on more than just truth-functional connectives. As you might guess from the artificiality of the arguments looked at in the previous chapter, this rules out most kinds of reasoning you'll encounter in everyday life. Ordinary reasoning relies heavily on the logic of the Boolean connectives, make no mistake about that. But it also relies on the logic of other kinds of expressions. Since the truth table method detects only tautological consequence, we need a method of applying Boolean logic that can work along with other valid principles of reasoning.

Methods of proof, both formal and informal, give us the required extensibility. In this chapter we will discuss legitimate patterns of inference that arise when we introduce the Boolean connectives into a language, and show how to apply the patterns in informal proofs. In Chapter 6, we'll extend our formal system with corresponding rules. The key advantage of proof methods over truth tables is that we'll be able to use them even when the validity of our proof depends on more than just the Boolean operators.

The Boolean connectives give rise to many valid patterns of inference. Some of these are extremely simple, like the entailment from the sentence $P \wedge Q$ to P. These we will refer to as *valid inference steps*, and will discuss them briefly in the first section. Much more interesting are two new *methods of proof* that are allowed by the new expressions: proof by cases and proof by contradiction. We will discuss these later, one at a time.

Valid inference steps

Here's an important rule of thumb: In an informal proof, it is always legitimate to move from a sentence P to another sentence Q if both you and your "audience" (the person or people you're trying to convince) already know that Q is a logical consequence of P. The main exception to this rule is when you give informal proofs to your logic instructor: presumably, your instructor knows the assigned argument is valid, so in these circumstances, you have to pretend you're addressing the proof to someone who doesn't already know that. What you're really doing is convincing your instructor that *you* see that the argument is valid and that you could prove it to someone who did not.

important rule of thumb

The reason we start with this rule of thumb is that you've already learned several well-known logical equivalences that you should feel free to use when giving informal proofs. For example, you can freely use double negation or idempotence if the need arises in a proof. Thus a chain of equivalences of the sort we gave on page 120 is a legitimate component of an informal proof. Of course, if you are asked to prove one of the named equivalences, say one of the distribution or DeMorgan laws, then you shouldn't presuppose it in your proof. You'll have to figure out a way to prove it to someone who doesn't already know that it is valid.

A special case of this rule of thumb is the following: If you already know that a sentence Q is a logical truth, then you may assert Q at any point in your proof. We already saw this principle at work in Chapter 2, when we discussed the reflexivity of identity, the principle that allowed us to assert a sentence of the form $a = a$ at any point in a proof. It also allows us to assert other simple logical truths, like excluded middle ($P \vee \neg P$), at any point in a proof. Of course, the logical truths have to be simple enough that you can be sure your audience will recognize them.

There are three simple inference steps that we will mention here that don't involve logical equivalences or logical truths, but that are clearly supported by the meanings of $\wedge$ and $\vee$. First, suppose we have managed to prove a conjunction, say $P \wedge Q$, in the course of our proof. The individual conjuncts P and Q are clearly consequences of this conjunction, because there is no way for the conjunction to be true without each conjunct being true. Thus, we are justified in asserting either. More generally, we are justified in inferring, from a conjunction of any number of sentences, any one of its conjuncts. This inference pattern is sometimes called *conjunction elimination* or *simplification*, when it is presented in the context of a formal system of deduction. When it is used in informal proofs, however, it usually goes by without comment, since it is so obvious.

Only slightly more interesting is the converse. Given the meaning of ∧, it is clear that P ∧ Q is a logical consequence of the pair of sentences P and Q: there is no way the latter could be true without former also being true. Thus if we have managed to prove P and to prove Q from the same premises, then

we are entitled to infer the conjunction P ∧ Q. More generally, if we want to prove a conjunction of a bunch of sentences, we may do so by proving each conjunct separately. In a formal system of deduction, steps of this sort are sometimes called *conjunction introduction* or just *conjunction*. Once again, in real life reasoning, these steps are too simple to warrant mention. In our informal proofs, we will seldom point them out explicitly.

Finally, let us look at one valid inference pattern involving ∨. It is a simple step, but one that strikes students as peculiar. Suppose that you have proven

Cube(b). Then you can conclude Cube(a) ∨ Cube(b) ∨ Cube(c), if you should want to for some reason, since the latter is a consequence of the former. More generally, if you have proven some sentence P then you can infer any disjunction that has P as one of its disjuncts. After all, if P is true, so is any such disjunction.

What strikes newcomers to logic as peculiar about such a step is that using it amounts to throwing away information. Why in the world would you want to conclude P ∨ Q when you already know the more informative claim P? But as we will see, this step is actually quite useful when combined with some of the methods of proof to be discussed later. Still, in mathematical proofs, it generally goes by unnoticed. In formal systems, it is dubbed *disjunction introduction*, or (rather unfortunately) *addition*.

Matters of style

Informal proofs serve two purposes. On the one hand, they are a method of discovery; they allow us to extract new information from information already obtained. On the other hand, they are a method of communication; they allow us to convey our discoveries to others. As with all forms of communication, this can be done well or done poorly.

When we learn to write, we learn certain basic rules of punctuation, capitalization, paragraph structure and so forth. But beyond the basic rules, there are also matters of style. Different writers have different styles. And it is a good thing, since we would get pretty tired of reading if everyone wrote with the very same style. So too in giving proofs. If you go on to study mathematics, you will read lots of proofs, and you will find that every writer has his or her own style. You will even develop a style of your own.

Every step in a "good" proof, besides being correct, should have two properties. It should be *easily understood* and *significant*. By "easily understood" we mean that other people should be able to follow the step without undue difficulty: they should be able to see that the step is valid without having to engage in a piece of complex reasoning of their own. By "significant" we mean that the step should be informative, not a waste of the reader's time.

These two criteria pull in opposite directions. Typically, the more significant the step, the harder it is to follow. Good style requires a reasonable balance between the two. And that in turn requires some sense of who your audience is. For example, if you and your audience have been working with logic for a while, you will recognize a number of equivalences that you will want to use without further proof. But if you or your audience are beginners, the same inference may require several steps.

knowing your audience

Remember

1. In giving an informal proof from some premises, if Q is already known to be a logical consequence of sentences $P_1, \ldots, P_n$ and each of $P_1, \ldots, P_n$ has been proven from the premises, then you may assert Q in your proof.

2. Each step in an informal proof should be *significant* but *easily understood*.

3. Whether a step is significant or easily understood depends on the audience to whom it is addressed.

4. The following are valid patterns of inference that generally go unmentioned in informal proofs:

   ○ From $P \wedge Q$, infer P.

   ○ From P and Q, infer $P \wedge Q$.

   ○ From P, infer $P \vee Q$.

Exercises

In the following exercises we list a number of patterns of inference, only some of which are valid. For each pattern, determine whether it is valid. If it is, explain why it is valid, appealing to the truth tables for the connectives involved. If it is not, give a specific example of how the step could be used to get from true premises to a false conclusion.

5.1 From P ∨ Q and ¬P, infer Q.

5.2 From P ∨ Q and Q, infer ¬P.

5.3 From ¬(P ∨ Q), infer ¬P.

5.4 From ¬(P ∧ Q) and P, infer ¬Q.

5.5 From ¬(P ∧ Q), infer ¬P.

5.6 From P ∧ Q and ¬P, infer R.

SECTION 5.2
Proof by cases

The simple forms of inference discussed in the last section are all instances of the principle that you can use already established cases of logical consequence in informal proofs. But the Boolean connectives also give rise to two entirely new methods of proof, methods that are explicitly applied in all types of rigorous reasoning. The first of these is the method of *proof by cases*. In our formal system $\mathcal{F}$, this method will be called *disjunction elimination*, but don't be misled by the ordinary sounding name: it is far more significant than, say, disjunction introduction or conjunction elimination.

We begin by illustrating proof by cases with a well-known piece of mathematical reasoning. The reasoning proves that there are irrational numbers b and c such that b^c is rational. First, let's review what this means. A number is said to be *rational* if it can be expressed as a fraction n/m, for integers n and m. If it can't be so expressed, then it is irrational. Thus 2 is rational $(2 = 2/1)$, but $\sqrt{2}$ is irrational. (We will prove this latter fact in the next section, to illustrate proof by contradiction; for now, just take it as a well-known truth.) Here now is our proof:

Proof: To show that there are irrational numbers b and c such that b^c is rational, we will consider the number $\sqrt{2}^{\sqrt{2}}$. We note that this number is either rational or irrational.

If $\sqrt{2}^{\sqrt{2}}$ is rational, then we have found our b and c; namely, we take $b = c = \sqrt{2}$.

Suppose, on the other hand, that $\sqrt{2}^{\sqrt{2}}$ is irrational. Then we take $b = \sqrt{2}^{\sqrt{2}}$ and $c = \sqrt{2}$ and compute b^c:

$$
\begin{aligned}
b^c &= (\sqrt{2}^{\sqrt{2}})^{\sqrt{2}} \\
&= \sqrt{2}^{(\sqrt{2} \cdot \sqrt{2})} \\
&= \sqrt{2}^2 \\
&= 2
\end{aligned}
$$

Thus, we see that in this case, too, b^c is rational.

Consequently, whether $\sqrt{2}^{\sqrt{2}}$ is rational or irrational, we know that there are irrational numbers b and c such that b^c is rational.

What interests us here is not the result itself but the general structure of the argument. We begin with a desired goal that we want to prove, say S, and a disjunction we already know, say P $\lor$ Q. We then show two things: that S follows if we assume that P is the case, and that S follows if we assume that Q is the case. Since we know that one of these must hold, we then conclude that S must be the case. This is the pattern of reasoning known as proof by cases.

proof by cases

In proof by cases, we aren't limited to breaking into just two cases, as we did in the example. If at any stage in a proof we have a disjunction containing n disjuncts, say $P_1 \lor \ldots \lor P_n$, then we can break into n cases. In the first we assume P_1, in the second P_2, and so forth for each disjunct. If we are able to prove our desired result S in each of these cases, we are justified in concluding that S holds.

Let's look at an even simpler example of proof by cases. Suppose we want to prove that Small(c) is a logical consequence of

$$(\text{Cube}(c) \land \text{Small}(c)) \lor (\text{Tet}(c) \land \text{Small}(c))$$

This is pretty obvious, but the proof involves breaking into cases, as you will notice if you think carefully about how you recognize this. For the record, here is how we would write out the proof.

Proof: We are given

$$(\text{Cube}(c) \land \text{Small}(c)) \lor (\text{Tet}(c) \land \text{Small}(c))$$

as a premise. We will break into two cases, corresponding to the two disjuncts. First, assume that Cube(c) $\land$ Small(c) holds. But then (by conjunction elimination, which we really shouldn't even mention) we have Small(c). But likewise, if we assume Tet(c) ∧ Small(c), then it follows that Small(c). So, in either case, we have Small(c), as desired.

Our next example shows how the odd step of disjunction introduction (from P infer P ∨ Q) can be used fruitfully with proof by cases. Suppose we know that either Max is home and Carl is happy, or Claire is home and Scruffy is happy, i.e.,

$$(\text{Home}(\text{max}) \wedge \text{Happy}(\text{carl})) \vee (\text{Home}(\text{claire}) \wedge \text{Happy}(\text{scruffy}))$$

We want to prove that either Carl or Scruffy is happy, that is,

$$\text{Happy}(\text{carl}) \vee \text{Happy}(\text{scruffy})$$

A rather pedantic, step-by-step proof would look like this:

Proof: Assume the disjunction:

$$(\text{Home}(\text{max}) \wedge \text{Happy}(\text{carl})) \vee (\text{Home}(\text{claire}) \wedge \text{Happy}(\text{scruffy}))$$

Then either:
$$\text{Home}(\text{max}) \wedge \text{Happy}(\text{carl})$$

or:
$$\text{Home}(\text{claire}) \wedge \text{Happy}(\text{scruffy}).$$

If the first alternative holds, then Happy(carl), and so we have

$$\text{Happy}(\text{carl}) \vee \text{Happy}(\text{scruffy})$$

by disjunction introduction. Similarly, if the second alternative holds, we have Happy(scruffy), and so

$$\text{Happy}(\text{carl}) \vee \text{Happy}(\text{scruffy})$$

So, in either case, we have our desired conclusion. Thus our conclusion follows by proof by cases.

Arguing by cases is extremely useful in everyday reasoning. For example, one of the authors (call him J) and his wife recently realized that their parking meter had expired several hours earlier. J argued in the following way that there was no point in rushing back to the car (logicians argue this way; don't marry one):

Proof: At this point, either we've already gotten a ticket or we haven't. If we've gotten a ticket, we won't get another one in the time it takes us to get to the car, so rushing would serve no purpose. If we haven't gotten a ticket in the past several hours, it is extremely unlikely that we will get one in the next few minutes, so again, rushing would be pointless. In either event, there's no need to rush.

J's wife responded with the following counterargument (showing that many years of marriage to a logician has an impact):

Proof: Either we are going to get a ticket in the next few minutes or we aren't. If we are, then rushing might prevent it, which would be a good thing. If we aren't, then it will still be good exercise and will also show our respect for the law, both of which are good things. So in either event, rushing back to the car is a good thing to do.

J's wife won the argument.

The validity of proof by cases cannot be demonstrated by the simple truth table method introduced in Chapter 4. The reason is that we infer the conclusion S from the fact that S is *provable* from each of the disjuncts P and Q. It relies on the principle that if S is a logical consequence of P, and also a logical consequence of Q, then it is a logical consequence of P $\vee$ Q. This holds because any circumstance that makes P $\vee$ Q true must make at least one of P or Q true, and hence S as well, by the fact that S is a consequence of both.

Remember

Proof by cases: To prove S from $P_1 \vee \ldots \vee P_n$ using this method, prove S from each of $P_1, \ldots, P_n$.

Exercises

The next two exercises present valid arguments. Turn in informal proofs of the arguments' validity. Your proofs should be phrased in complete, well-formed English sentences, making use of first-order sentences as convenient, much in the style we have used above. Whenever you use proof by cases, say so. You don't have to be explicit about the use of simple proof steps like conjunction elimination. By the way, there is typically more than one way to prove a given result.

5.7

> Home(max) ∨ Home(claire)
> ¬Home(max) ∨ Happy(carl)
> ¬Home(claire) ∨ Happy(carl)
>
> Happy(carl)

5.8

> LeftOf(a, b) ∨ RightOf(a, b)
> BackOf(a, b) ∨ ¬LeftOf(a, b)
> FrontOf(b, a) ∨ ¬RightOf(a, b)
> SameCol(c, a) ∧ SameRow(c, b)
>
> BackOf(a, b)

5.9 Assume the same four premises as in Exercise 5.8. Is LeftOf(b, c) a logical consequence of these premises? If so, turn in an informal proof of the argument's validity. If not, submit a counterexample world.

5.10 Suppose Max's favorite basketball team is the Chicago Bulls and favorite football team is the Denver Broncos. Max's father John is returning from Indianapolis to San Francisco on United Airlines, and promises that he will buy Max a souvenir from one of his favorite teams on the way. Explain John's reasoning, appealing to the annoying fact that all United flights between Indianapolis and San Francisco stop in either Denver or Chicago. Make explicit the role proof by cases plays in this reasoning.

5.11 Suppose the police are investigating a burglary and discover the following facts. All the doors to the house were bolted from the inside and show no sign of forced entry. In fact, the only possible ways in and out of the house were a small bathroom window on the first floor that was left open and an unlocked bedroom window on the second floor. On the basis of this, the detectives rule out a well-known burglar, Julius, who weighs two hundred and fifty pounds and is arthritic. Explain their reasoning.

5.12 In our proof that there are irrational numbers b and c where b^c is rational, one of our steps was to assert that $\sqrt{2}^{\sqrt{2}}$ is either rational or irrational. What justifies the introduction of this claim into our proof?

5.13 Describe an everyday example of reasoning by cases that you have performed in the last few days.

5.14 Give an informal proof that if S is a tautological consequence of P and a tautological consequence of Q, then S is a tautological consequence of P ∨ Q. Remember that the joint truth table for P ∨ Q and S may have more rows than either the joint truth table for P and S, or the joint truth table for Q and S. [Hint: Assume you are looking at a single row of the joint truth table for P ∨ Q and S in which P ∨ Q is true. Break into cases based on whether P is true or Q is true and prove that S must be true in either case.]

SECTION 5.3

Indirect proof: proof by contradiction

One of the most important methods of proof is known as *proof by contradiction*. It is also called *indirect proof* or *reductio ad absurdum*. Its counterpart in $\mathcal{F}$ is called *negation introduction*.

indirect proof or proof by contradiction

The basic idea is this. Suppose you want to prove a negative sentence, say ¬S, from some premises, say $P_1, \ldots, P_n$. One way to do this is by temporarily assuming S and showing that a contradiction follows from this assumption. If you can show this, then you are entitled to conclude that ¬S is a logical consequence of the original premises. Why? Because your proof of the contradiction shows that S, $P_1, \ldots, P_n$ cannot all be true simultaneously. (If they were, the contradiction would have to be true, and it can't be.) Hence if $P_1, \ldots, P_n$ are true in any set of circumstances, then S must be false in those circumstances. Which is to say, if $P_1, \ldots, P_n$ are all true, then ¬S must be true as well.

Let's look at a simple indirect proof. Assume Cube(c) ∨ Dodec(c) and Tet(b). Let us prove ¬(b = c).

> **Proof:** In order to prove ¬(b = c), we assume b = c and attempt to get a contradiction. From our first premise we know that either Cube(c) or Dodec(c). If the first is the case, then we conclude Cube(b) by the indiscernibility of identicals, which contradicts Tet(b). But similarly, if the second is the case, we get Dodec(b) which contradicts Tet(b). So neither case is possible, and we have a contradiction. Thus our initial assumption that b = c must be wrong. So proof by contradiction gives us our desired conclusion, ¬(b = c). (Notice that this argument also uses the method of proof by cases.)

Let us now give a more interesting and famous example of this method of proof. The Greeks were shocked to discover that the square root of 2 could not be expressed as a fraction, or, as we would put it, is irrational. The proof of this fact proceeds via contradiction. Before we go through the proof, let's review some simple numerical facts that were well known to the Greeks. The first is that any rational number can be expressed as a fraction p/q where at least one of p and q is odd. (If not, keep dividing both the numerator and denominator by 2 until one of them is odd.) The other fact follows from the observation that when you square an odd number, you always get an odd number. So if n^2 is an even number, then so is n. And from this, we see that if n^2 is even, it must be divisible by 4.

Now we're ready for the proof that $\sqrt{2}$ is irrational.

Proof: With an eye toward getting a contradiction, we will assume that $\sqrt{2}$ is rational. Thus, on this assumption, $\sqrt{2}$ can be expressed in the form p/q, where at least one of p and q is odd. Since $p/q = \sqrt{2}$ we can square both sides to get:

$$\frac{p^2}{q^2} = 2$$

Multiplying both sides by q^2, we get $p^2 = 2q^2$. But this shows that p^2 is an even number. As we noted before, this allows us to conclude that p is even and that p^2 is divisible by 4. Looking again at the equation $p^2 = 2q^2$, we see that if p^2 is divisible by 4, then $2q^2$ is divisible by 4 and hence q^2 must be divisible by 2. In which case, q is even as well. So both p and q are even, contradicting the fact that at least one of them is odd. Thus, our assumption that $\sqrt{2}$ is rational led us to a contradiction, and so we conclude that it is irrational.

In both of these examples, we used the method of indirect proof to prove a sentence that begins with a negation. (Remember, "irrational" simply means *not rational*.) You can also use this method to prove a sentence S that does not begin with a negation. In this case, you would begin by assuming ¬S, obtain a contradiction, and then conclude that ¬¬S is the case, which of course is equivalent to S.

contradiction

In order to apply the method of proof by contradiction, it is important that you understand what a contradiction is, since that is what you need to prove from your temporary assumption. Intuitively, a contradiction is any claim that cannot possibly be true, or any set of claims which cannot all be true simultaneously. Examples are a sentence Q and its negation ¬Q, a pair of inconsistent claims like Cube(c) and Tet(c) or $x < y$ and $y < x$, or a single sentence of the form $a \neq a$. We can take the notion of a contradictory or inconsistent set of sentences to be any set of sentences that could not all be true in any single situation.

contradiction symbol (⊥)

The symbol ⊥ is often used as a short-hand way of saying that a contradiction has been obtained. Different people read ⊥ as "contradiction," "the absurd," and "the false," but what it means is that a conclusion has been reached which is logically impossible, or that several conclusions have been derived which, taken together, are impossible.

Notice that a sentence S is a logical impossibility if and only if its negation ¬S is logically necessary. This means that any method we have of demonstrating that a sentence is logically necessary also demonstrates that its negation is logically impossible, that is, a contradiction. For example, if a truth table shows that ¬S is a tautology, then we know that S is a contradiction.

Similarly, the truth table method gives us a way of showing that a collection of sentences are mutually contradictory. Construct a joint truth table for $P_1, \ldots, P_n$. These sentences are TT-contradictory if every row has an **F** assigned to at least one of the sentences. If the sentences are TT-contradictory, we know they cannot all be true at once, simply in virtue of the meanings of the truth functional connectives out of which they are built. We have already mentioned one such example: any pair of sentences, one of which is the negation of the other.

TT-*contradictory*

The method of proof by contradiction, like proof by cases, is often encountered in everyday reasoning, though the derived contradiction is sometimes left implicit. People will often assume a claim for the sake of argument and then show that the assumption leads to something else that is known to be false. They then conclude the negation of the original claim. This sort of reasoning is in fact an indirect proof: the inconsistency becomes explicit if we add the known fact to our set of premises.

Let's look at an example of this kind of reasoning. Imagine a defense attorney presenting the following summary to the jury:

> The prosecution claims that my client killed the owner of the KitKat Club. Assume that they are correct. You've heard their own experts testify that the murder took place at 5:15 in the afternoon. We also know the defendant was still at work at City Hall at 4:45, according to the testimony of five co-workers. It follows that my client had to get from City Hall to the KitKat Club in 30 minutes or less. But to make that trip takes 35 minutes under the best of circumstances, and police records show that there was a massive traffic jam the day of the murder. I submit that my client is innocent.

Clearly, reasoning like this is used all the time: whenever we assume something and then rule out the assumption on the basis of its consequences. Sometimes these consequences are not contradictions, or even things that we know to be false, but rather future consequences that we consider unacceptable. You might for example assume that you will go to Hawaii for spring break, calculate the impact on your finances and ability to finish the term papers coming due, and reluctantly conclude that you can't make the trip. When you reason like this, you are using the method of indirect proof.

Remember

Proof by contradiction: To prove $\neg S$ using this method, assume S and prove a contradiction $\bot$.

Exercises

In the following exercises, decide whether the displayed argument is valid. If it is, turn in an informal proof, phrased in complete, well-formed English sentences, making use of first-order sentences as convenient. Whenever you use proof by cases or proof by contradiction, say so. You don't have to be explicit about the use of simple proof steps like conjunction elimination. If the argument is invalid, construct a counterexample world in Tarski's World. (Argument 5.16 is valid, and so will not require a counterexample.)

5.15

> b is a tetrahedron.
> c is a cube.
> Either c is larger than b or else they are identical.
>
> b is smaller than c.

5.16

> Max or Claire is at home but either Scruffy or Carl is unhappy.
> Either Max is not home or Carl is happy.
> Either Claire is not home or Scruffy is unhappy.
>
> Scruffy is unhappy.

5.17

> Cube(a) ∨ Tet(a) ∨ Large(a)
> ¬Cube(a) ∨ a = b ∨ Large(a)
> ¬Large(a) ∨ a = c
> ¬(c = c ∧ Tet(a))
>
> a = b ∨ a = c

5.18

> Cube(a) ∨ Tet(a) ∨ Large(a)
> ¬Cube(a) ∨ a = b ∨ Large(a)
> ¬Large(a) ∨ a = c
> ¬(c = c ∧ Tet(a))
>
> ¬(Large(a) ∨ Tet(a))

5.19 Consider the following sentences.

1. *Folly was Claire's pet at 2 pm or at 2:05 pm.*
2. *Folly was Max's pet at 2 pm.*
3. *Folly was Claire's pet at 2:05 pm.*

Does (3) follow from (1) and (2)? Does (2) follow from (1) and (3)? Does (1) follow from (2) and (3)? In each case, give either a proof of consequence, or describe a situation that makes the premises true and the conclusion false. You may assume that Folly can only be one person's pet at any given time.

5.20 Suppose it is Friday night and you are going out with your boyfriend. He wants to see a romantic comedy, while you want to see the latest Wes Craven slasher movie. He points out that if he watches the Wes Craven movie, he will not be able to sleep because he can't stand the sight of blood, and he has to take the MCAT test tomorrow. If he does not do well on the MCAT, he won't get into medical school. Analyze your boyfriend's argument, pointing out where indirect proof is being used. How would you rebut his argument?

5.21 Describe an everyday example of an indirect proof that you have used in the last few days.

5.22 Prove that indirect proof is a tautologically valid method of proof. That is, show that if $P_1, \ldots, P_n, S$ is TT-contradictory, then $\neg S$ is a tautological consequence of $P_1, \ldots, P_n$.

In the next three exercises we ask you to prove simple facts about the natural numbers. We do not expect you to phrase the proofs in FOL. You will have to appeal to basic facts of arithmetic plus the definitions of even and odd number. This is OK, but make these appeals explicit. Also make explicit any use of proof by contradiction.

5.23 Assume that n^2 is odd. Prove that n is odd.

5.24 Assume that $n + m$ is odd. Prove that $n \times m$ is even.

5.25 Assume that n^2 is divisible by 3. Prove that n^2 is divisible by 9.

5.26 A good way to make sure you understand a proof is to try to generalize it. Prove that $\sqrt{3}$ is irrational. [Hint: You will need to figure out some facts about divisibility by 3 that parallel the facts we used about even and odd, for example, the fact expressed in Exercise 5.25.] Can you generalize these two results?

SECTION 5.4

Arguments with inconsistent premises

What follows from an inconsistent set of premises? If you look back at our definition of logical consequence, you will see that every sentence is a consequence of such a set. After all, if the premises are contradictory, then there are no circumstances in which they are all true. Thus, there are no circumstances in which the premises are true and the conclusion is false. Which is to say, in any situation in which the premises are all true (there aren't any of these!), the conclusion will be true as well. Hence any argument with an inconsistent set of premises is trivially valid. In particular, if one can establish a contradiction ⊥ on the basis of the premises, then one is entitled to assert any sentence at all.

always valid

This often strikes students as a very odd method of reasoning, and for very good reason. For recall the distinction between a valid argument and a sound one. A *sound* argument is a valid argument with true premises. Even though any argument with an inconsistent set of premises is valid, no such argument is sound, since there is no way the premises of the argument can all be true. For this reason, an argument with an inconsistent set of premises is not worth

never sound

much on its own. After all, the reason we are interested in logical consequence is because of its relation to truth. If the premises can't possibly be true, then even knowing that the argument is valid gives us no clue as to the truth or falsity of the conclusion. An unsound argument gives no more support for its conclusion than an invalid one.

In general, methods of proof don't allow us to show that an argument is unsound. After all, the truth or falsity of the premises is not a matter of logic, but of how the world happens to be. But in the case of arguments with inconsistent premises, our methods of proof do give us a way to show that at least one of the premises is false (though we might not know which one), and hence that the argument is unsound. To do this, we prove that the premises are inconsistent by deriving a contradiction.

Suppose, for example, you are given a proof that the following argument is valid:

$$\begin{array}{|l} \text{Home(max)} \lor \text{Home(claire)} \\ \neg\text{Home(max)} \\ \neg\text{Home(claire)} \\ \hline \text{Home(max)} \land \text{Happy(carl)} \end{array}$$

While it is true that this conclusion is a consequence of the premises, your reaction should not be to believe the conclusion. Indeed, using proof by cases we can show that the premises are inconsistent, and hence that the argument is unsound. There is no reason to be convinced of the conclusion of an unsound argument.

Remember

A proof of a contradiction $\bot$ from premises $P_1, \ldots, P_n$ (without additional assumptions) shows that the premises are inconsistent. An argument with inconsistent premises is always valid, but more importantly, always unsound.

Exercises

5.27 Give two different proofs that the premises of the above argument are inconsistent. Your first should use proof by cases but not DeMorgan's law, while your second can use DeMorgan but not proof by cases.

CHAPTER 6

Formal Proofs and Boolean Logic

The deductive system $\mathcal{F}$ is what is known as a system of *natural deduction.* Such systems are intended to be models of the valid principles of reasoning used in informal proofs. In this chapter, we will present the inference rules of $\mathcal{F}$ that correspond to the informal principles of Boolean reasoning discussed in the previous chapter. You will easily recognize the rules as formal counterparts of some of the principles we've already discussed.

natural deduction

Although natural deduction systems like $\mathcal{F}$ are meant to model informal reasoning, they are also designed to be relatively spare or "stripped down" versions of such reasoning. For example, we told you that in giving an informal proof, you can always presuppose steps that you and your audience already know to be logically valid. So if one of the equivalence laws is not at issue in a proof, you can simply apply it in a single step of your informal proof. However, in $\mathcal{F}$ we will give you a very elegant but restricted collection of inference rules that you must apply in constructing a formal proof. Many of the valid inference steps that we have seen (like the DeMorgan Laws) are not allowed as single steps; they must be justified in terms of more basic steps. The advantage to this "lean and mean" approach is that it makes it easier to prove results *about* the deductive system, since the fewer the rules, the simpler the system. For example, one of the things we can prove is that anything you could demonstrate with a system that contained rules for all of the named logical equivalences of Chapter 4 can be proved in the leaner system $\mathcal{F}$.

Systems of natural deduction like $\mathcal{F}$ use two rules for each connective, one that allows us to prove statements containing the symbol, and one that allows us to prove things *from* statements containing the symbol. The former are called *introduction* rules since they let us introduce these symbols into proofs. By contrast, the latter are called *elimination* rules. This is similar to our treatment of the identity predicate in Chapter 2. If you go on to study proof theory in more advanced logic courses, you will see that that this elegant pairing of rules has many advantages over systems that include more inference steps as basic.

introduction and elimination rules

The formal rules of $\mathcal{F}$ are all implemented in the program Fitch, allowing you to construct formal proofs much more easily than if you had to write them out by hand. Actually, Fitch's interpretation of the introduction and

rule defaults

elimination rules is a bit more generous in spirit than $\mathcal{F}$. It doesn't allow you to do anything that $\mathcal{F}$ wouldn't permit, but there are cases where Fitch will let you do in one step what might take several in $\mathcal{F}$. Also, many of Fitch's rules have "default applications" that can save you a lot of time. If you want the default use of some rule, all you have to do is specify the rule and cite the step or steps you are applying it to; Fitch will then fill in the appropriate conclusion for you. Similarly, if you have filled in the formula and rule, Fitch can sometimes add appropriate support steps for you via the **Add Support Steps** command. At the end of each section below we'll explain the default uses of the rules introduced in that section.

Section 6.1

Conjunction rules

The simplest principles to formalize are those that involve the conjunction symbol $\wedge$. These are the rules of conjunction elimination and conjunction introduction.

Conjunction elimination

The rule of conjunction elimination allows you to assert any conjunct P_i of a conjunctive sentence $P_1 \wedge \ldots \wedge P_i \wedge \ldots \wedge P_n$ that you have already derived in the proof. (P_i can, by the way, be any conjunct, including the first or the last.) You justify the new step by citing the step containing the conjunction. We abbreviate this rule with the following schema:

Conjunction Elimination ($\wedge$ Elim):

$$
\begin{array}{l|l}
 & P_1 \wedge \ldots \wedge P_i \wedge \ldots \wedge P_n \\
 & \vdots \\
\triangleright & P_i
\end{array}
$$

You try it
. . . .

▶ 1. Open the file Conjunction 1. There are three sentences that you are asked to prove. They are shown in the goal strip at the bottom of the proof window as usual.

▶ 2. The first sentence you are to prove is Tet(a). To do this, first add a new step to the proof and write the sentence Tet(a).

3. Next, go to the popup **Rule?** menu and under the Elimination Rules,   ◀
   choose ∧.

4. If you try to check this step, you will see that it fails, because you have   ◀
   have not yet cited any sentences in support of the step. In this example,
   you need to cite the single premise in support. Do this and then check the
   step.

5. You should be able to prove each of the other sentences similarly, by means   ◀
   of a single application of ∧ **Elim**. When you have proven these sentences,
   check your goals and save the proof as Proof Conjunction 1.

. . . . *Congratulations*

Conjunction introduction

The corresponding introduction rule, conjunction introduction, allows you to
assert a conjunction $P_1 \wedge \ldots \wedge P_n$ provided you have already established each
of its constituent conjuncts P_1 through P_n. We will symbolize this rule in the
following way:

Conjunction Introduction (∧ Intro):

$$
\begin{array}{|l}
P_1 \\
\Downarrow \\
P_n \\
 \vdots \\
\triangleright \quad P_1 \wedge \ldots \wedge P_n
\end{array}
$$

In this rule, we have used the notation:

$$
\begin{array}{c}
P_1 \\
\Downarrow \\
P_n
\end{array}
$$

to indicate that each of P_1 through P_n must appear in the proof before you
can assert their conjunction. The order in which they appear does not matter,
and they do not have to appear one right after another. They just need to
appear somewhere earlier in the proof.

Here is a simple example of our two conjunction rules at work together. It
is a proof of $C \wedge B$ from $A \wedge B \wedge C$.

$$
\begin{array}{lll}
& 1.\ A \wedge B \wedge C & \\
\hline
& 2.\ B & \wedge \ \textbf{Elim: } 1 \\
& 3.\ C & \wedge \ \textbf{Elim: } 1 \\
& 4.\ C \wedge B & \wedge \ \textbf{Intro: } 3,\ 2
\end{array}
$$

Let's try our hand using both conjunction rules in Fitch.

You try it

▶ 1. Open the file Conjunction 2. We will help you prove the two sentences requested in the goals. You will need to use both of the conjunction rules in each case.

▶ 2. The first goal is Medium(d) ∧ ¬Large(c). Add a new step and enter this sentence. (Remember that you can copy the sentence from the goal strip and paste it into the new step. It's faster than typing it in.)

▶ 3. Above the step you just created, add two more steps, typing one of the conjuncts in each. If you can prove these, then the conclusion will follow by ∧ **Intro**. Show this by choosing this rule at the conjunction step and citing the two conjuncts in support.

▶ 4. Now all you need to do is prove each of the conjuncts. This is easily done using the rule ∧ **Elim** at each of these steps. Do this, cite the appropriate support sentences, and check the proof. The first goal should check out.

▶ 5. Prove the second goal sentence similarly. Once both goals check out, save your proof as Proof Conjunction 2.

. . . . *Congratulations*

Default and generous uses of the ∧ rules

As we said, Fitch is generous in its interpretation of the inference rules of $\mathcal{F}$. For example, Fitch considers the following to be an acceptable use of ∧ **Elim**:

$$
\begin{array}{lll}
& 17.\ \text{Tet}(a) \wedge \text{Tet}(b) \wedge \text{Tet}(c) \wedge \text{Tet}(d) & \\
& \vdots & \\
& 26.\ \text{Tet}(d) \wedge \text{Tet}(b) & \wedge \ \textbf{Elim: } 17
\end{array}
$$

What we have done here is pick two of the conjuncts from step 17 and assert the conjunction of these in step 26. Technically, $\mathcal{F}$ would require us to derive the two conjuncts separately and, like Humpty Dumpty, put them back together again. Fitch does this for us.

Since Fitch lets you take any collection of conjuncts in the cited sentence and assert their conjunction in any order, Fitch's interpretation of ∧ **Elim** allows you to prove that conjunction is "commutative." In other words, you can use it to take a conjunction and reorder its conjuncts however you please:

> 13. Tet(a) ∧ Tet(b)
> ⋮
> 21. Tet(b) ∧ Tet(a)   ∧ **Elim**: 13

You try it
. . . .

1. Open the file Conjunction 3. Notice that there are two goals. The first goal   ◀
   asks you to prove Tet(c) ∧ Tet(a) from the premise. Strictly speaking, this
   would take two uses of ∧ **Elim** followed by one use of ∧ **Intro**. However,
   Fitch lets you do this with a single use of ∧ **Elim**. Try this and then check
   the step.

2. Verify that the second goal sentence also follows by a single application of   ◀
   Fitch's rule of ∧ **Elim**. When you have proven these sentences, check your
   goals and save the proof as Proof Conjunction 3.

3. Next try out other sentences to see whether they follow from the given   ◀
   sentence by ∧ **Elim**. For example, does Tet(c) ∧ Small(a) follow? Should
   it?

4. When you are satisfied you understand conjunction elimination, close the   ◀
   file, but don't save the changes you made in step 3.
. . . . *Congratulations*

The ∧ **Intro** rule implemented in Fitch is also less restrictive than our discussion of the formal rule might suggest. First of all, Fitch does not care about the order in which you cite the supporting sentences. Second, if you cite a sentence, that sentence can appear more than once as a conjunct in the concluding sentence. For example, you can use this rule to conclude Cube(a) ∧ Cube(a) from the sentence Cube(a), if you want to for some reason.

default uses of conjunction rules

Both of the conjunction rules have default uses. If at a new step you cite a conjunction and specify the rule as ∧ **Elim**, then when you check the step (or choose **Check Proof**), Fitch will fill in the blank step with the leftmost conjunct in the cited sentence. If you cite several sentences and apply ∧ **Intro**, Fitch will fill in the conjunction of those steps, ordering conjuncts in the same order they were cited.

You try it

▶   1. Open the file Conjunction 4.

▶   2. Move the focus to the first blank step, the one immediately following the premises. Notice that this step has a rule specified, as well as a support sentence cited. Check the step to see what default Fitch generates.

▶   3. Then, focus on each successive step, try to predict what the default will be, and check the step. (The last two steps give different results because we entered the support steps in different orders.)

▶   4. When you have checked all the steps, save your proof as Proof Conjunction 4.

▶   5. Feel free to experiment with the rule defaults some more, to see when they are useful.

. . . . *Congratulations*

You can use the **Add Support Steps** command (found on the **Proof** menu) with either of the conjunction rules. In either case you must have chosen a rule, and have entered a formula in the focus step. In the case of ∧ **Elim**, a single support step will be created, and this step will contain the formula at the focus step, followed by a conjunction symbol to indicate that you must enter more conjuncts to complete the support formula. If the **Add Support Steps** is used with the ∧ **Intro rule**, and the focus formula is a conjunction, then one support step is introduced for each conjunct of the focus formula.

parentheses and conjunction rules

One final point: In applying conjunction introduction, you will sometimes have to be careful about parentheses, due to our conventions about dropping outermost parentheses. If one of the conjuncts is itself a conjunction, then of course there is no need to add any parentheses before forming the larger conjunction, unless you want to. For example, the following are both correct applications of the rule. (The first is what Fitch's default mechanism would give you.)

Correct:
> 1. A ∧ B
> 2. C
>
> 3. (A ∧ B) ∧ C   ∧ **Intro**: 1, 2

Correct:
> 1. A ∧ B
> 2. C
>
> 3. A ∧ B ∧ C   ∧ **Intro**: 1, 2

However, if one of the conjuncts is a disjunction (or some other complex sentence), to prevent ambiguity you may need to reintroduce the parentheses that you omitted before. Thus the first of the following is a correct proof, but the second contains a faulty application of conjunction introduction, since it concludes with an ambiguous sentence.

Correct:
> 1. A ∨ B
> 2. C
>
> 3. (A ∨ B) ∧ C   ∧ **Intro**: 1, 2

Wrong:
> 1. A ∨ B
> 2. C
>
> 3. A ∨ B ∧ C   ∧ **Intro**: 1, 2

Disjunction rules

We know: the conjunction rules were boring. Not so the disjunction rules, particularly disjunction elimination.

Disjunction introduction

The rule of disjunction introduction allows you to go from a sentence P_i to any disjunction that has P_i among its disjuncts, say $P_1 \lor \ldots \lor P_i \lor \ldots \lor P_n$. In schematic form:

Disjunction Introduction (∨ Intro):

> P_i
> $\vdots$
> ▷ $P_1 \lor \ldots \lor P_i \lor \ldots \lor P_n$

Once again, we stress that P_i may be the first or last disjunct of the conclusion. Further, as with conjunction introduction, some thought ought to be given to whether parentheses must be added to P_i to prevent ambiguity.

As we explained in Chapter 5, disjunction introduction is a less peculiar rule than it may at first appear. But before we look at a sensible example of how it is used, we need to have at our disposal the second disjunction rule.

Disjunction elimination

We now come to the first rule that corresponds to what we called a method of proof in the last chapter. This is the rule of disjunction elimination, the formal counterpart of proof by cases. Recall that proof by cases allows you to conclude a sentence S from a disjunction $P_1 \vee \ldots \vee P_n$ if you can prove S from each of P_1 through P_n individually. The form of this rule requires us to discuss an important new structural feature of the Fitch-style system of deduction. This is the notion of a *subproof*.

subproofs

A subproof, as the name suggests, is a proof that occurs within the context of a larger proof. As with any proof, a subproof generally begins with an assumption, separated from the rest of the subproof by the Fitch bar. But

temporary assumptions

the assumption of a subproof, unlike a premise of the main proof, is only temporarily assumed. Throughout the course of the subproof itself, the assumption acts just like an additional premise. But after the subproof, the assumption is no longer in force.

Before we give the schematic form of disjunction elimination, let's look at a particular proof that uses the rule. This will serve as a concrete illustration of how subproofs appear in $\mathcal{F}$.

1. $(A \wedge B) \vee (C \wedge D)$		
2. $A \wedge B$		
3. B	$\wedge$ **Elim**: 2	
4. $B \vee D$	$\vee$ **Intro**: 3	
5. $C \wedge D$		
6. D	$\wedge$ **Elim**: 5	
7. $B \vee D$	$\vee$ **Intro**: 6	
8. $B \vee D$	$\vee$ **Elim**: 1, 2–4, 5–7	

With appropriate replacements for A, B, C, and D, this is a formalization of the proof given on page 134. It contains two subproofs. One of these runs from line 2 to 4, and shows that $B \vee D$ follows if we (temporarily) assume $A \wedge B$. The other runs from line 5 to 7, and shows that the same conclusion follows from the assumption $C \wedge D$. These two proofs, together with the premise $(A \wedge B) \vee (C \wedge D)$, are just what we need to apply the method of proof by cases—or as we will now call it, the rule of disjunction elimination.

Look closely at this proof and compare it to the informal proof given on page 134 to see if you can understand what is going on. Notice that the assumption steps of our two subproofs do not have to be justified by a rule any more than the premise of the larger "parent" proof requires a justification. This is because we are not claiming that these assumptions follow from what comes before, but are simply assuming them to show what follows from their supposition. Notice also that we have used the rule $\vee$ **Intro** twice in this proof, since that is the only way we can derive the desired sentence in each subproof. Although it seems like we are throwing away information when we infer $B \vee D$ from the stronger claim B, when you consider the overall proof, it is clear that $B \vee D$ is the strongest claim that follows from the original premise.

We can now state the schematic version of disjunction elimination.

Disjunction Elimination ($\vee$ Elim):

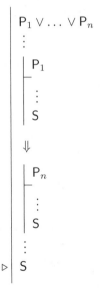

What this says is that if you have established a disjunction $P_1 \vee \ldots \vee P_n$, and you have also shown that S follows from each of the disjuncts P_1 through P_n, then you can conclude S. Again, it does not matter what order the subproofs appear in, or even that they come after the disjunction. When applying the rule, you will cite the step containing the disjunction, plus each of the required subproofs.

Let's look at another example of this rule, to emphasize how justifications involving subproofs are given. Here is a proof showing that A follows from the sentence (B ∧ A) ∨ (A ∧ C).

1. (B ∧ A) ∨ (A ∧ C)		
2. B ∧ A		
3. A	∧ **Elim**: 2
4. A ∧ C		
5. A	∧ **Elim**: 4
6. A	∨ **Elim**: 1, 2–3, 4–5

The citation for step 6 shows the form we use when citing subproofs. The citation "*n–m*" is our way of referring to the subproof that begins on line *n* and ends on line *m*.

Sometimes, in using disjunction elimination, you will find it natural to use the reiteration rule introduced in Chapter 3. For example, suppose we modify the above proof to show that A follows from (B ∧ A) ∨ A.

1. (B ∧ A) ∨ A		
2. B ∧ A		
3. A	∧ **Elim**: 2
4. A		
5. A	**Reit**: 4
6. A	∨ **Elim**: 1, 2–3, 4–5

Here, the assumption of the second subproof is A, exactly the sentence we want to prove. So all we need to do is repeat that sentence to get the subproof into the desired form. (We could also just give a subproof with one step, but it is more natural to use reiteration in such cases.)

You try it
. . . .

▶ 1. Open the file **Disjunction 1**. In this file, you are asked to prove

$$\mathsf{Medium}(c) \lor \mathsf{Large}(c)$$

from the sentence

$$(\mathsf{Cube}(c) \land \mathsf{Large}(c)) \lor \mathsf{Medium}(c)$$

We are going to step you through the construction of the following proof:

1. $(\mathsf{Cube}(c) \land \mathsf{Large}(c)) \lor \mathsf{Medium}(c)$	
2. $\mathsf{Cube}(c) \land \mathsf{Large}(c)$	
3. $\mathsf{Large}(c)$	$\land$ **Elim**: 2
4. $\mathsf{Medium}(c) \lor \mathsf{Large}(c)$	$\lor$ **Intro**: 3
5. $\mathsf{Medium}(c)$	
6. $\mathsf{Medium}(c) \lor \mathsf{Large}(c)$	$\lor$ **Intro**: 5
7. $\mathsf{Medium}(c) \lor \mathsf{Large}(c)$	$\lor$ **Elim**: 1, 2–4, 5–6

2. To use $\lor$ **Elim** in this case, we need to get two subproofs, one for each ◄ of the disjuncts in the premise. It is a good policy to begin by specifying both of the necessary subproofs before doing anything else. To start a subproof, add a new step and choose **New Subproof** from the **Proof** menu. Fitch will indent the step and allow you to enter the sentence you want to assume. Enter the first disjunct of the premise, $\mathsf{Cube}(c) \land \mathsf{Large}(c)$, as the assumption of this subproof.

3. Rather than work on this subproof now, let's specify the second case before ◄ we forget what we're trying to do. To do this, we need to end the first subproof and start a second subproof after it. You end the current subproof by choosing **End Subproof** from the **Proof** menu. This will give you a new step outside of, but immediately following the subproof.

4. Start your second subproof at this new step by choosing **New Subproof** ◄ from the **Proof** menu. This time type the other disjunct of the premise, $\mathsf{Medium}(c)$. We have now specified the assumptions of the two cases we need to consider. Our goal is to prove that the conclusion follows in both of these cases.

5. Go back to the first subproof and add a step following the assumption. (Fo- ◄ cus on the assumption step of the subproof and choose **Add Step After** from the **Proof** menu.) In this step use $\land$ **Elim** to prove $\mathsf{Large}(c)$. Then add another step to that subproof and prove the goal sentence, using $\lor$ **Intro**. In both steps, you will have to cite the necessary support sentences.

▶   6. After you've finished the first subproof and all the steps check out, move the focus slider to the assumption step of the second subproof and add a new step. Use ∨ **Intro** to prove the goal sentence from your assumption.

▶   7. We've now derived the goal sentence in both of the subproofs, and so are ready to add the final step of our proof. While focussed on the last step of the second subproof, choose **End Subproof** from the **Proof** menu. Enter the goal sentence into this new step.

▶   8. Specify the rule in the final step as ∨ **Elim**. For support, cite the two subproofs and the premise. Check your completed proof. If it does not check out, compare your proof carefully with the proof displayed above. Have you accidentally gotten one of your subproofs inside the other one? If so, delete the misplaced subproof by focusing on the assumption and choosing **Delete Step** from the **Proof** menu. Then try again.

▶   9. When the entire proof checks out, save it as Proof Disjunction 1.

. . . . *Congratulations*

Default and generous uses of the ∨ rules

There are a couple of ways in which Fitch is more lenient in checking ∨ **Elim** than the strict form of the rule suggests. First, the sentence S does not have to be the last sentence in the subproof, though usually it will be. S simply has to appear on the "main level" of each subproof, not necessarily as the very last step. Second, if you start with a disjunction containing more than two disjuncts, say P ∨ Q ∨ R, Fitch doesn't require three subproofs. If you have one subproof starting with P and one starting with Q ∨ R, or one starting with Q and one starting with P ∨ R, then Fitch will still be happy, as long as you've proven S in each of these cases.

default uses of
disjunction rules

Both disjunction rules have default applications, though they work rather differently. If you cite appropriate support for ∨ **Elim** (i.e., a disjunction and subproofs for each disjunct) and then check the step without typing a sentence, Fitch will look at the subproofs cited and, if they all end with the same sentence, insert that sentence into the step. If you cite a sentence and apply ∨ **Intro** without typing a sentence, Fitch will insert the cited sentence followed by ∨, leaving the insertion point after the ∨ so you can type in the rest of the disjunction you had in mind.

You try it

. . . .

1. Open the file Disjunction 2. The goal is to prove the sentence ◀

$$(\mathsf{Cube(b)} \land \mathsf{Small(b)}) \lor (\mathsf{Cube(b)} \land \mathsf{Large(b)})$$

The required proof is almost complete, though it may not look like it.

2. Focus on each empty step in succession, checking the step so that Fitch ◀
will fill in the default sentence. On the second empty step you will have to
finish the sentence by typing in the second disjunct, $(\mathsf{Cube(b)} \land \mathsf{Large(b)})$,
of the goal sentence. (If the last step does not generate a default, it is
because you have not typed the right thing in the ∨ **Intro** step.)

3. When you are finished, see if the proof checks out. Do you understand the ◀
proof? Could you have come up with it on your own?

4. Save your completed proof as Proof Disjunction 2. ◀

. . . . *Congratulations*

When you choose the ∨ **Intro** rule, and enter a disjunction at the focus
step, you can use the **Add Support Steps** command to insert an appropriate
support step. Fitch has to guess at the formula that you might want to cite as
support. Fitch chooses the first disjunct, although any disjunct of the focus
formula would be appropriate. **Add Support Steps** cannot be used with the
∨ **Elim** rule. When you use this rule, Fitch does not have enough information
to fill in the support steps, even when you have given a formula at the focus
step. You are on your own for this rule!

Exercises

6.1 If you skipped any of the **You try it** sections, go back and do them now. Submit the files Proof
Conjunction 1, Proof Conjunction 2, Proof Conjunction 3, Proof Conjunction 4, Proof Disjunction
1, and Proof Disjunction 2.

6.2 Open the file Exercise 6.2, which contains an incomplete formal proof. As it stands, none of
the steps check out, either because no rule has been specified, no support steps cited, or no
sentence typed in. Provide the missing pieces and submit the completed proof.

Use Fitch to construct formal proofs for the following arguments. You will find Exercise *files for each
argument in the usual place. As usual, name your solutions* Proof 6.x.

6.3
↗
$$a = b \land b = c \land c = d$$
$$a = c \land b = d$$

6.4
↗
$$(A \land B) \lor C$$
$$C \lor B$$

6.5
↗
$$A \land (B \lor C)$$
$$(A \land B) \lor (A \land C)$$

6.6
↗
$$(A \land B) \lor (A \land C)$$
$$A \land (B \lor C)$$

Section 6.3

Negation rules

Last but not least are the negation rules. It turns out that negation introduction is our most interesting and complex rule.

Negation elimination

The rule of negation elimination corresponds to a very trivial valid step, from $\neg\neg P$ to P. Schematically:

Negation Elimination (¬ Elim):

$$\neg\neg P$$
$$\vdots$$
$$\triangleright \quad P$$

Negation elimination gives us one direction of the principle of double negation. You might reasonably expect that our second negation rule, negation introduction, would simply give us the other direction. But if that's what you guessed, you guessed wrong.

Negation introduction

The rule of negation introduction corresponds to the method of indirect proof or proof by contradiction. Like ∨ **Elim**, it involves the use of a subproof, as will the formal analogs of all nontrivial methods of proof. The rule says that if you can prove a contradiction ⊥ on the basis of an additional assumption P, then you are entitled to infer $\neg P$ from the original premises. Schematically:

Negation Introduction (¬ Intro):

There are different ways of understanding this rule, depending on how we interpret the contradiction symbol ⊥. Some authors interpret it simply as shorthand for any contradiction of the form Q ∧ ¬Q. If we construed the schema that way, we wouldn't have to say anything more about it. But we will treat ⊥ as a symbol in its own right, to be read "contradiction."

When ⊥ appears in a truth table, we need to ensure that the column underneath it contains only the truth value FALSE. The Boole program takes this into account and will mark any row with an occurrence of TRUE in this column as being incorrect.

This treatment of ⊥ as a symbol that represent an always-false atomic sentence has several advantages that will become apparent when you use the rule. The one disadvantage is that we need to have rules about this special symbol. We introduce these rules next.

⊥ Introduction

The rule of ⊥ Introduction (⊥ **Intro**) allows us to obtain the contradiction symbol if we have established an explicit contradiction in the form of some sentence P and its negation ¬P.

⊥ Introduction (⊥ Intro):

```
 |  P
 |  ⋮
 |  ¬P
 |  ⋮
▷|  ⊥
```

Ordinarily, you will only apply ⊥ **Intro** in the context of a subproof, to show that the subproof's assumption leads to a contradiction. The only time you will be able to derive ⊥ in your main proof (as opposed to a subproof) is when the premises of your argument are themselves inconsistent. In fact, this is how we give a formal proof that a set of premises is inconsistent. A formal proof of inconsistency is a proof that derives ⊥ at the main level of the proof. Let's try out the rules of ⊥ **Intro** and ¬ **Intro** to see how they work.

You try it
. . . .

▶ 1. To illustrate these rules, we will show you how to prove ¬¬A from A. This is the other direction of double negation. Use Fitch to open the file Negation 1.

▶ 2. We will step you through the construction of the following simple proof.

| 1. A |
| 2. ¬A |
| 3. ⊥   ⊥ **Intro**: 1, 2 |
| 4. ¬¬A   ¬ **Intro**: 2–3 |

▶ 3. To construct this proof, add a step immediately after the premise. Turn it into a subproof by choosing **New Subproof** from the **Proof** menu. Enter the assumption ¬A.

▶ 4. Add a new step to the subproof and enter ⊥, changing the rule to ⊥ **Intro**. Cite the appropriate steps and check the step.

▶ 5. Now end the subproof and enter the final sentence, ¬¬A, after the subproof. Specify the rule as ¬ **Intro**, cite the preceding subproof and check the step. Your whole proof should now check out.

▶ 6. Notice that in the third line of your proof you cited a step outside the subproof, namely the premise. This is legitimate, but raises an important issue. Just what steps *can* be cited at a given point in a proof? As a first guess, you might think that you can cite any earlier step. But this turns out to be wrong. We will explain why, and what the correct answer is, in the next section.

▶ 7. Save your proof as Proof Negation 1.
. . . . *Congratulations*

The contradiction symbol ⊥ acts just like any other sentence in a proof. In particular, if you are reasoning by cases and derive ⊥ in each of your subproofs, then you can use ∨ **Elim** to derive ⊥ in your main proof. For example, here is a proof that the premises A ∨ B, ¬A, and ¬B are inconsistent.

1. A ∨ B		
2. ¬A		
3. ¬B		
4. A		
5. ⊥		⊥ **Intro**: 4, 2
6. B		
7. ⊥		⊥ **Intro**: 6, 3
8. ⊥		∨ **Elim**: 1, 4–5, 6–7

The important thing to notice here is step 8, where we have applied ∨ **Elim** to extract the contradiction symbol from our two subproofs. This is clearly justified, since we have shown that whichever of A or B holds, we immediately arrive at a contradiction. Since the premises tell us that one or the other holds, the premises are inconsistent.

Other ways of introducing ⊥

The rule of ⊥ **Intro** recognizes only the most blatant contradictions, those where you have established a sentence P and its negation ¬P. What if in the course of a proof you come across an inconsistency of some other form? For example, suppose you manage to derive a single TT-contradictory sentence like ¬(A ∨ ¬A), or the two sentences ¬A ∨ ¬B and A ∧ B, which together form a TT-contradictory set?

It turns out that if you can prove any TT-contradictory sentence or sentences, the rules we've already given you will allow you to prove ⊥. It may take a fair amount of effort and ingenuity, but it is possible. We'll eventually prove this, but for now you'll have to take our word for it.

One way to check whether some sentences are TT-contradictory is to try to derive ⊥ from them using a single application of **Taut Con**. In other words, enter ⊥, cite the sentences, and choose **Taut Con** from the **Rule?** menu. If **Taut Con** tells you that ⊥ follows from the cited sentences, then you can be sure that it is possible to prove this using just the introduction and elimination rules for ∧, ∨, ¬, and ⊥.

*introducing ⊥
with Taut Con*

Of course, there are other forms of contradiction besides TT-contradictions. For example, suppose you manage to prove the three sentences Cube(b), b = c, and ¬Cube(c). These sentences are not TT-contradictory, but you can see that a single application of = **Elim** will give you the TT-contradictory pair

Cube(c) and ¬Cube(c). If you suspect that you have derived some sentences whose inconsistency results from the Boolean connectives *plus* the identity predicate, you can check this using the **FO Con** mechanism, since **FO Con** understands the meaning of =. If **FO Con** says that ⊥ follows from the cited sentences (and if those sentences do not contain quantifiers), then you should be able to prove ⊥ using just the introduction and elimination rules for =, ∧, ∨, ¬, and ⊥.

The only time you may arrive at a contradiction but not be able to prove ⊥ using the rules of $\mathcal{F}$ is if the inconsistency depends on the meanings of predicates other than identity. For example, suppose you derived the contradiction n < n, or the contradictory pair of sentences Cube(b) and Tet(b). The rules of $\mathcal{F}$ give you no way to get from these sentences to a contradiction of the form P and ¬P, at least without some further premises.

What this means is that in Fitch, the **Ana Con** mechanism will let you establish contradictions that can't be derived in $\mathcal{F}$. Of course, the **Ana Con** mechanism only understands predicates in the blocks language (and even there, it excludes Adjoins and Between). But it will allow you to derive ⊥ from, for example, the two sentences Cube(b) and Tet(b). You can either do this directly, by entering ⊥ and citing the two sentences, or indirectly, by using **Ana Con** to prove, say, ¬Cube(b) from Tet(b).

You try it
. . . .

▶ 1. Open Negation 2 using Fitch. In this file you will find an incomplete proof. As premises, we have listed a number of sentences, several groups of which are contradictory.

▶ 2. Focus on each step that contains the ⊥ symbol. You will see that various sentences are cited in support of the step. Only one of these steps is an application of the ⊥ **Intro** rule. Which one? Specify the rule for that step as ⊥ **Intro** and check it.

▶ 3. Among the remaining steps, you will find one where the cited sentences form a TT-contradictory set of sentences. Which one? Change the justification at that step to **Taut Con** and check the step. Since it checks out, we assure you that you can derive ⊥ from these same premises using just the Boolean rules.

▶ 4. Of the remaining steps, the supports of two are contradictory in view of the meaning of the identity symbol =. Which steps? Change the justification at those step to **FO Con** and check the steps. To derive ⊥ from these premises, you would need the identity rules (in one case = **Elim**, in the other = **Intro**).

5. Verify that the remaining steps cannot be justified by any of the rules ⊥ **Intro**, **Taut Con** or **FO Con**. Change the justification at those steps to **Ana Con** and check the steps.   ◄

6. Save your proof as **Proof Negation 2**. (Needless to say, this is a formal proof of inconsistency with a vengeance!)   ◄
....*Congratulations*

⊥ Elimination

As we remarked earlier, if in a proof, or more importantly in some subproof, you are able to establish a contradiction, then you are entitled to assert any FOL sentence P whatsoever. In our formal system, this is modeled by the rule of ⊥ Elimination (⊥ **Elim**).

⊥ Elimination (⊥ Elim):

$$
\begin{array}{c|c}
 & \bot \\
 & \vdots \\
\triangleright & P
\end{array}
$$

The following **You try it** section illustrates both of the ⊥ rules. Be sure to go through it, as it presents a proof tactic you will have several occasions to use.

You try it
....

1. It often happens in giving proofs using ∨ **Elim** that one really wants to eliminate one or more of the disjuncts, because they contradict other assumptions. The form of the ∨ **Elim** rule does not permit this, though. The proof we will construct here shows how to get around this difficulty.   ◄

2. Using Fitch, open the file **Negation 3**. We will use ∨ **Elim** and the two ⊥ rules to prove P from the premises P ∨ Q and ¬Q.   ◄

3. Start two subproofs, the first with assumption P, the second with assumption Q. Our goal is to establish P in both subproofs.   ◄

4. In the first subproof, we can simply use reiteration to repeat the assumption P.   ◄

▶ 5. In the second subproof, how will we establish P? In an informal proof, we would simply eliminate this case, because the assumption contradicts one of the premises. In a formal proof, though, we must establish our goal sentence P in both subproofs, and this is where ⊥ **Elim** is useful. First use ⊥ **Intro** to show that this case is contradictory. You will cite the assumed sentence Q and the second premise ¬Q. Once you have ⊥ as the second step of this subproof, use ⊥ **Elim** to establish P in this subproof.

▶ 6. Since you now have P in both subproofs, you can finish the proof using ∨ **Elim**. Complete the proof.

▶ 7. Save your proof as Proof Negation 3.

· *Congratulations*

It turns out that we do not really need the rule of ⊥ **Elim**. You can prove any sentence from a contradiction without it; it just takes longer. Suppose, for example, that you have established a contradiction at step 17 of some proof. Here is how you can introduce P at step 21 without using ⊥ **Elim**.

17. ⊥	
18. ¬P	
19. ⊥	**Reit**: 17
20. ¬¬P	¬ **Intro**: 18–19
21. P	¬ **Elim**: 20

Still, we include ⊥ **Elim** to make our proofs shorter and more natural.

Default and generous uses of the ¬ rules

The rule of ¬ **Elim** allows you to take off two negation signs from the front of a sentence. Repeated uses of this rule would allow you to remove four, six, or indeed any even number of negation signs. For this reason, the implementation of ¬ **Elim** in Fitch allows you to remove any even number of negation signs in one step. Similarly for ¬ **Intro**, if the sentence in the assumption step of the cited subproof is a negation, ¬A, say, we allow you to deduce the unnegated sentence A, instead of ¬¬A.

default uses of negation rules

Both of the negation rules have default applications. In a default application of ¬ **Elim**, Fitch will remove as many negation signs as possible from the front of the cited sentences (the number must be even, of course) and insert the resulting sentence at the ¬ **Elim** step. In a default application of ¬ **Intro**, the inserted sentence will be the negation of the assumption step of the cited subproof.

You try it

. . . .

1. Open the file **Negation 4**. First look at the goal to see what sentence we are trying to prove. Then focus on each step in succession and check the step. Before moving to the next step, make sure you understand why the step checks out and, more important, why we are doing what we are doing at that step. At the empty steps, try to predict which sentence Fitch will provide as a default before you check the step. ◀

2. When you are done, make sure you understand the completed proof. Save your file as **Proof Negation 4**. ◀

. . . . *Congratulations*

Fitch will add a single support step if you use the **Add Support Steps** command when you have entered a formula and chosen the ¬ **Elim** rule. The support formula will be the formula from the focus step with two negation symbols preceding it. If you choose the ¬ **Intro** rule and use **Add Support Steps** then Fitch will insert a subproof as support, with the negation of the focus formula as the assumption of the subproof and ⊥ as the only other step in the subproof. You can also use **Add Support Steps** with ⊥ **Elim**. Whatever formula is present, Fitch inserts a single support step containing the support formula ⊥.

Exercises

6.7 If you skipped any of the **You try it** sections, go back and do them now. Submit the files
↗ Proof Negation 1, Proof Negation 2, Proof Negation 3, and Proof Negation 4.

6.8 (Substitution) In informal proofs, we allow you to substitute logically equivalent sentences
↗ for one another, even when they occur in the context of a larger sentence. For example, the following inference results from two uses of double negation, each applied to a part of the whole sentence:

$$\begin{array}{|l} P \wedge (Q \vee \neg\neg R) \\ \hline \neg\neg P \wedge (Q \vee R) \end{array}$$

How would we prove this using $\mathcal{F}$, which has no substitution rule? Open the file Exercise 6.8, which contains an incomplete formal proof of this argument. As it stands, none of the proof's steps check out, because no rules or support steps have been cited. Provide the missing justifications and submit the completed proof.

*Evaluate each of the following arguments. If the argument is valid, use Fitch to give a formal proof using the rules you have learned. If it not valid, use Tarski's World to construct a counterexample world. In the last two proofs you will need to use **Ana Con** to show that certain atomic sentences contradict one another to introduce ⊥. Use **Ana Con** only in this way. That is, your use of **Ana Con** should cite exactly two atomic sentences in support of an introduction of ⊥. If you have difficulty with any of these exercises, you may want to skip ahead and read Section 6.5.*

6.9

> Cube(b)
> ¬(Cube(c) ∧ Cube(b))
>> ¬Cube(c)

6.10

> Cube(a) ∨ Cube(b)
> ¬(Cube(c) ∧ Cube(b))
>> ¬Cube(c)

6.11

> Dodec(e)
> Small(e)
> ¬Dodec(e) ∨ Dodec(f) ∨ Small(e)
>> Dodec(f)

6.12

> Dodec(e)
> ¬Small(e)
> ¬Dodec(e) ∨ Dodec(f) ∨ Small(e)
>> Dodec(f)

6.13

> Dodec(e)
> Large(e)
> ¬Dodec(e) ∨ Dodec(f) ∨ Small(e)
>> Dodec(f)

6.14

> SameRow(b, f) ∨ SameRow(c, f)
>   ∨ SameRow(d, f)
> ¬SameRow(c, f)
> FrontOf(b, f)
> ¬(SameRow(d, f) ∧ Cube(f))
>> ¬Cube(f)

*In the following two exercises, determine whether the sentences are consistent. If they are, use Tarski's World to build a world where the sentences are both true. If they are inconsistent, use Fitch to give a proof that they are inconsistent (that is, derive ⊥ from them). You may use **Ana Con** in your proof, but only applied to literals (that is, atomic sentences or negations of atomic sentences).*

6.15

¬(Larger(a, b) ∧ Larger(b, a))
¬SameSize(a, b)

6.16

Smaller(a, b) ∨ Smaller(b, a)
SameSize(a, b)

The proper use of subproofs

Subproofs are the characteristic feature of Fitch-style deductive systems. It is important that you understand how to use them properly, since if you are not careful, you may "prove" things that don't follow from your premises. For example, the following formal proof looks like it is constructed according to our rules, but it purports to prove that A ∧ B follows from (B ∧ A) ∨ (A ∧ C), which is clearly not right.

1. (B ∧ A) ∨ (A ∧ C)		
2. B ∧ A		
3. B	∧ **Elim**: 2	
4. A	∧ **Elim**: 2	
5. A ∧ C		
6. A	∧ **Elim**: 5	
7. A	∨ **Elim**: 1, 2–4, 5–6	
8. A ∧ B	∧ **Intro**: 7, 3	

The problem with this proof is step 8. In this step we have used step 3, a step that occurs within an earlier subproof. But it turns out that this sort of justification—one that reaches back inside a subproof that has already ended—is not legitimate. To understand why it's not legitimate, we need to think about what function subproofs play in a piece of reasoning.

A subproof typically looks something like this:

Subproofs begin with the introduction of a new assumption, in this example R. The reasoning within the subproof depends on this new assumption, together with any other premises or assumptions of the parent proof. So in our example, the derivation of S may depend on both P and R. When the subproof ends, indicated by the end of the vertical line that ties the subproof together, the subsequent reasoning can no longer use the subproof's assumption, or anything that depends on it. We say that the assumption has been *discharged* or that the subproof has been *ended*.

When an assumption has been discharged, the individual steps of its subproof are no longer accessible. It is only the subproof as a whole that can be cited as justification for some later step. What this means is that in justifying the assertion of T in our example, we could cite P, Q, and the subproof as a whole, but we could *not* cite individual items in the subproof like R or S. For these steps rely on assumptions we no longer have at our disposal. Once the subproof has been ended, they are no longer accessible.

This, of course, is where we went wrong in step 8 of the fallacious proof given earlier. We cited a step in a subproof that had been ended, namely, step 3. But the sentence at that step, B, had been proven on the basis of the assumption B ∧ A, an assumption we only made temporarily. The assumption is no longer in force at step 8, and so cannot be used at that point.

This injunction does not prevent you from citing, from within a subproof, items that occur earlier outside the subproof, as long as they do not occur in subproofs that ended before that step. For example, in the schematic proof given above, the justification for S could well include the step that contains Q.

This observation becomes more pointed when you are working in a subproof of a subproof. We have not yet seen any examples where we needed to have subproofs within subproofs, but the following proof, of one direction of the first DeMorgan law, is one.

Notice that the subproof 2–15 contains two subproofs, 3–5 and 8–10. In step 5 of subproof 3–5, we cite step 2 from the parent subproof 2–15. Similarly, in step 10 of the subproof 8–10, we cite step 2. This is legitimate since the subproof 2–15 has not been ended by step 10. While we did not need to in this proof, we could in fact have cited step 1 in either of the sub-subproofs.

Another thing to note about this proof is the use of the Reiteration rule at step 14. We did not need to use Reiteration here, but did so just to illustrate a point. When it comes to subproofs, Reiteration is like any other rule: when you use it, you can cite steps outside of the immediate subproof, if the proofs that contain the cited steps have not yet ended. But you cannot cite a step inside a subproof that has already ended. For example, if we replaced the justification for step 15 with "**Reit**: 10," then our proof would no longer be correct.

1. ¬(P ∧ Q)		
2. ¬(¬P ∨ ¬Q)		
3. ¬P		
4. ¬P ∨ ¬Q	∨ **Intro**: 3	
5. ⊥	⊥ **Intro**: 4, 2	
6. ¬¬P	¬ **Intro**: 3–5	
7. P	¬ **Elim**: 6	
8. ¬Q		
9. ¬P ∨ ¬Q	∨ **Intro**: 8	
10. ⊥	⊥ **Intro**: 9, 2	
11. ¬¬Q	¬ **Intro**: 8–10	
12. Q	¬ **Elim**: 11	
13. P ∧ Q	∧ **Intro**: 7, 12	
14. ¬(P ∧ Q)	**Reit**: 1	
15. ⊥	⊥ **Intro**: 13, 14	
16. ¬¬(¬P ∨ ¬Q)	¬ **Intro**: 2–15	
17. ¬P ∨ ¬Q	¬ **Elim**: 16	

As you'll see, most proofs in $\mathcal{F}$ require subproofs inside subproofs—what we call *nested* subproofs. To create such a subproof in Fitch, you just choose **New Subproof** from the **Proof** menu while you're inside the first subproof. You may already have done this by accident in constructing earlier proofs. In the exercises that follow, you'll have to do it on purpose.

nested subproofs

Remember

o In justifying a step of a subproof, you may cite any earlier step contained in the main proof, or in any subproof whose assumption is still in force. You may never cite individual steps inside a subproof that has already ended.

o Fitch enforces this automatically by not permitting the citation of individual steps inside subproofs that have ended.

Exercises

6.17 Try to recreate the following "proof" using Fitch.

> 1. $(\text{Tet}(a) \wedge \text{Large}(c)) \vee (\text{Tet}(a) \wedge \text{Dodec}(b))$
>> 2. $\text{Tet}(a) \wedge \text{Large}(c)$
>> 3. $\text{Tet}(a)$ $\wedge$ **Elim**: 2
>> 4. $\text{Tet}(a) \wedge \text{Dodec}(b)$
>>> 5. $\text{Dodec}(b)$ $\wedge$ **Elim**: 4
>>> 6. $\text{Tet}(a)$ $\wedge$ **Elim**: 4
>> 7. $\text{Tet}(a)$ $\vee$ **Elim**: 1, 2–3, 4–6
>> 8. $\text{Tet}(a) \wedge \text{Dodec}(b)$ $\wedge$ **Intro**: 7, 5

What step won't Fitch let you perform? Why? Is the conclusion a consequence of the premise? Discuss this example in the form of a clear English paragraph, and turn your paragraph in to your instructor.

Use Fitch to give formal proofs for the following arguments. You will need to use subproofs within subproofs to prove these.

6.18

> $A \vee B$
>
> $A \vee \neg\neg B$

6.19

> $A \vee B$
> $\neg B \vee C$
>
> $A \vee C$

6.20

> $A \vee B$
> $A \vee C$
>
> $A \vee (B \wedge C)$

Section 6.5

Strategy and tactics

Many students try constructing formal proofs by blindly piecing together a sequence of steps permitted by the introduction and elimination rules, a process no more related to reasoning than playing solitaire. This approach occasionally works, but more often than not it will fail—or at any rate, make it harder to find a proof. In this section, we will give you some advice about how to go about finding proofs when they don't jump right out at you. The advice consists of two important strategies and an essential maxim.

an important maxim   Here is the maxim: Always keep firmly in mind what the sentences in your proof mean! Students who pay attention to the meanings of the sentences avoid innumerable pitfalls, among them the pitfall of trying to prove a sentence that doesn't really follow from the information given. Your first step in trying to construct a proof should *always* be to convince yourself that the claim made by the conclusion is a consequence of the premises. You should do this even if the exercise tells you that the argument is valid and simply asks you to find a proof. For in the process of understanding the sentences and recognizing the argument's validity, you will often get some idea how to prove it.

After you're convinced that the argument is indeed valid, the first strategy for finding a formal proof is to try giving an informal proof, the kind you might use to convince a fellow classmate. Often the basic structure of your informal reasoning can be directly formalized using the rules of $\mathcal{F}$. For example, if you find yourself using an indirect proof, then that part of the reasoning will probably require negation introduction in $\mathcal{F}$. If you use proof by cases, then you'll almost surely formalize the proof using disjunction elimination.

try informal proof

Suppose you have decided that the argument is valid, but are having trouble finding an informal proof. Or suppose you can't see how your informal proof can be converted into a proof that uses just the rules of $\mathcal{F}$. The second strategy is helpful in either of these cases. It is known as "working backwards." What you do is look at the conclusion and see what additional sentence or sentences would allow you to infer that conclusion. Then you simply insert these steps into your proof, not worrying about exactly how they will be justified, and cite them in support of your goal sentence. You then take these intermediate steps as new goals and see if you can prove them. Once you do, your proof will be complete.

working backwards

Let's work through an example that applies both of these strategies. Suppose you are asked to give a formal proof of the argument:

$$\begin{array}{|l}
\neg P \vee \neg Q \\
\hline
\neg (P \wedge Q)
\end{array}$$

You'll recognize this as an application of one of the DeMorgan laws, so you know it's valid. But when you think about it (applying our maxim) you may find that what convinces you of its validity is the following observation, which is hard to formalize: if the premise is true, then either P or Q is false, and that will make $P \wedge Q$ false, and hence the conclusion true. Though this is a completely convincing argument, it is not immediately clear how it would translate into the introduction and elimination rules of $\mathcal{F}$.

Let's try working backwards to see if we can come up with an informal proof that is easier to formalize. Since the conclusion is a negation, we could prove it by assuming $P \wedge Q$ and deriving a contradiction. So let's suppose $P \wedge Q$ and take $\bot$ as our new goal. Now things look a little clearer. For the premise tells us that either $\neg P$ or $\neg Q$ is true, but either of these cases directly contradicts one of the conjuncts of our assumption. So proof by cases will allow us to derive a contradiction. For the record, here is how we would state this as an informal proof:

> **Proof:** We are given ¬P ∨ ¬Q and want to prove ¬(P ∧ Q). For purposes of reductio, we will assume P ∧ Q and attempt to derive a contradiction. There are two cases to consider, since we are given that either ¬P or ¬Q is true. But each of these contradicts the assumption P ∧ Q: ¬P contradicts the first conjunct and ¬Q contradicts the second. Consequently, our assumption leads to a contradiction, and so our proof is complete.

In the following, we lead you through the construction of a formal proof that models this informal reasoning.

You try it
. . . .

▶ 1. Open the file **Strategy 1**. Begin by entering the desired conclusion in a new step of the proof. We will construct the proof working backwards, just like we found our informal proof. Add a step before the conclusion you've entered so that your proof looks something like this:

1. ¬P ∨ ¬Q	
2. ...	**Rule?**
3. ¬(P ∧ Q)	**Rule?**

▶ 2. The main method used in our informal proof was reductio, which corresponds to negation introduction. So change the blank step into a subproof with the assumption P ∧ Q and the contradiction symbol at the bottom. (You can also use **Add Support Steps** op achieve this.) Also add a step in between these to remind you that that's where you still need to fill things in, and enter your justification for the final step, so you remember why you added the subproof. At this point your proof should look roughly like this:

1. ¬P ∨ ¬Q	
2. P ∧ Q	
3. ...	**Rule?**
4. ⊥	**Rule?**
5. ¬(P ∧ Q)	¬ **Intro**: 2–4

3. Our informal proof showed that there was a contradiction whichever of ¬P or ¬Q was the case. The formal counterpart of proof by cases is disjunction elimination, so the next step is to fill in two subproofs, one assuming ¬P, the other assuming ¬Q, and both concluding with ⊥. Make sure you put in the justification for the step where you apply ∨ **Elim**, and it's a good idea to add empty steps to remind yourself where you need to continue working. Here's what your proof should look like now:

1. ¬P ∨ ¬Q	
2. P ∧ Q	
3. ¬P	
4. ...	**Rule?**
5. ⊥	**Rule?**
6. ¬Q	
7. ...	**Rule?**
8. ⊥	**Rule?**
9. ⊥	∨ **Elim**: 1, 3–5, 6–8
10. ¬(P ∧ Q)	¬ **Intro**: 2–9

4. Filling in the remaining steps is easy. Finish your proof as follows:

1. ¬P ∨ ¬Q	
2. P ∧ Q	
3. ¬P	
4. P	∧ **Elim**: 2
5. ⊥	⊥ **Intro**: 4, 3
6. ¬Q	
7. Q	∧ **Elim**: 2
8. ⊥	⊥ **Intro**: 7, 6
9. ⊥	∨ **Elim**: 1, 3–5, 6–8
10. ¬(P ∧ Q)	¬ **Intro**: 2–9

▶   5. Save your proof as **Proof Strategy 1**.

. . . . *Congratulations*

pitfalls of working backwards

Working backwards can be a very useful technique, since it often allows you to replace a complex goal with simpler ones or to add new assumptions from which to reason. But you should not think that the technique can be applied mechanically, without giving it any thought. Each time you add new intermediate goals, whether they are sentences or subproofs, it is essential that you stop and check whether the new goals are actually reasonable. If they don't seem plausible, you should try something else.

Here's a simple example of why this constant checking is so important. Suppose you were asked to prove the sentence $A \lor C$ from the given sentence $(A \land B) \lor (C \land D)$. Working backwards you might notice that if you could prove A, from this you could infer the desired conclusion by the rule $\lor$ **Intro**. Sketched in, your partial proof would look like this:

1. $(A \land B) \lor (C \land D)$	
2. A	**Rule?**
3. $A \lor C$	$\lor$ **Intro**

The problem with this is that A does not follow from the given sentence, and no amount of work will allow you to prove that it does. If you didn't notice this from the outset, you could spend a lot of time trying to construct an impossible proof! But if you notice it, you can try a more promising approach. (In this case, disjunction elimination is clearly the right way to go.) Working backwards, though a valuable tactic, is no replacement for good honest thinking.

checking with Con mechanisms

When you're constructing a formal proof in Fitch, you can avoid trying to prove an incorrect intermediate conclusion by checking the step with **Taut Con**. In the above example, for instance, if you use **Taut Con** at the second step, citing the premise as support, you would immediately find that it is hopeless to try to prove A from the given premise.

Many of the problems in this book ask you to determine whether an argument is valid and to back up your answer with either a proof of consequence or a counterexample, a proof of non-consequence. You will approach these problems in much the same way we've described, first trying to understand the claims involved and deciding whether the conclusion follows from the premises. If you think the conclusion does not follow, or really don't have a good hunch one way or the other, try to find a counterexample. You may succeed, in which case you will have shown the argument to be invalid. If you cannot find a counterexample, trying to find one often gives rise to insights about why the argument is valid, insights that can help you find the required proof.

We can summarize our strategy advice with a seven step procedure for approaching problems of this sort.

Remember

In assessing the validity of an argument, use the following method:

1. Understand what the sentences are saying.

2. Decide whether you think the conclusion follows from the premises.

3. If you think it does not follow, or are not sure, try to find a counterexample.

4. If you think it does follow, try to give an informal proof.

5. If a formal proof is called for, use the informal proof to guide you in finding one.

6. In giving consequence proofs, both formal and informal, don't forget the tactic of working backwards.

7. In working backwards, though, always check that your intermediate goals are consequences of the available information.

One final warning: One of the nice things about Fitch is that it will give you instant feedback about whether your proof is correct. This is a valuable learning tool, but it can be misused. You should not use Fitch as a crutch, trying out rule applications and letting Fitch tell you if they are correct. If you do this, then you are not really learning the system $\mathcal{F}$. One way to check up on yourself is to write a formal proof out on paper every now and then. If you try this and find you can't do it without Fitch's help, then you are using Fitch as a crutch, not a learning tool.

using Fitch as a crutch

Exercises

6.21 If you skipped the **You try it** section, go back and do it now. Submit the file Proof Strategy 1.
↗

6.22 Give a formal proof mirroring the informal proof on page 137 of ¬(b = c) from the premises Cube(c) ∨ Dodec(c) and Tet(b). You may apply **Ana Con** to literals in establishing ⊥.

6.23 Give an informal proof that might have been used by the authors in constructing the formal proof shown on page 167.

In each of the following exercises, give an informal proof of the validity of the indicated argument. (You should never use the principle you are proving in your informal proof, for example in Exercise 6.24, you should not use DeMorgan in your informal proof.) Then use Fitch to construct a formal proof that mirrors your informal proof as much as possible. Turn in your informal proofs to your instructor and submit the formal proof in the usual way.

6.24

¬(A ∨ B)
¬A ∧ ¬B

6.25

¬A ∧ ¬B
¬(A ∨ B)

6.26

A ∨ (B ∧ C)
¬B ∨ ¬C ∨ D
A ∨ D

6.27

(A ∧ B) ∨ (C ∧ D)
(B ∧ C) ∨ (D ∧ E)
C ∨ (A ∧ E)

*In each of the following exercises, you should assess whether the argument is valid. If it is, use Fitch to construct a formal proof. You may use **Ana Con** but only involving literals and ⊥. If it is not valid, use Tarski's World to construct a counterexample.*

6.28

Cube(c) ∨ Small(c)
Dodec(c)
Small(c)

6.29

Larger(a, b) ∨ Larger(a, c)
Smaller(b, a) ∨ ¬Larger(a, c)
Larger(a, b)

6.30

¬(¬Cube(a) ∧ Cube(b))
¬(¬Cube(b) ∨ Cube(c))
Cube(a)

6.31

Dodec(b) ∨ Cube(b)
Small(b) ∨ Medium(b)
¬(Small(b) ∧ Cube(b))
Medium(b) ∧ Dodec(b)

6.32

Dodec(b) ∨ Cube(b)
Small(b) ∨ Medium(b)
¬Small(b) ∧ ¬Cube(b)
Medium(b) ∧ Dodec(b)

Proofs without premises

Not all proofs begin with the assumption of premises. This may seem odd, but in fact it is how we use our deductive system to show that a sentence is a logical truth. A sentence that can be proven without any premises at all is necessarily true. Here's a trivial example of such a proof, one that shows that $a = a \land b = b$ is a logical truth.

demonstrating logical truth

1. $a = a$		= **Intro**
2. $b = b$		= **Intro**
3. $a = a \land b = b$		$\land$ **Intro**: 1, 2

The first step of this proof is not a premise, but an application of = **Intro**. You might think that any proof without premises would have to start with this rule, since it is the only one that doesn't have to cite any supporting steps earlier in the proof. But in fact, this is not a very representative example of such proofs. A more typical and interesting proof without premises is the following, which shows that $\neg(P \land \neg P)$ is a logical truth.

1. $P \land \neg P$	
2. P	$\land$ **Elim**: 1
3. $\neg P$	$\land$ **Elim**: 1
4. $\bot$	$\bot$ **Intro**: 2, 3
5. $\neg(P \land \neg P)$	$\neg$ **Intro**: 1–4

Notice that there are no assumptions above the first horizontal Fitch bar, indicating that the main proof has no premises. The first step of the proof is the *subproof's* assumption. The subproof proceeds to derive a contradiction, based on this assumption, thus allowing us to conclude that the negation of the subproof's assumption follows without the need of premises. In other words, it is a logical truth.

When we want you to prove that a sentence is a logical truth, we will use Fitch notation to indicate that you must prove this without assuming any premises. For example the above proof shows that the following "argument" is valid:

$$\vdash \neg(P \land \neg P)$$

We close this section with the following reminder:

> **Remember**
>
> A proof without any premises shows that its conclusion is a logical truth.

Exercises

6.33 (Excluded Middle) Open the file Exercise 6.33. This contains an incomplete proof of the law of excluded middle, P ∨ ¬P. As it stands, the proof does not check out because it's missing some sentences, some support citations, and some rules. Fill in the missing pieces and submit the completed proof as **Proof 6.33**. The proof shows that we can derive excluded middle in $\mathcal{F}$ without any premises.

*In the following exercises, assess whether the indicated sentence is a logical truth in the blocks language. If so, use Fitch to construct a formal proof of the sentence from no premises (using **Ana Con** if necessary, but only applied to literals). If not, use Tarski's World to construct a counterexample. (A counterexample here will simply be a world that makes the purported conclusion false.)*

6.34

$$\vdash \neg(a = b \land \text{Dodec}(a) \land \neg\text{Dodec}(b))$$

6.35

$$\vdash \neg(a = b \land \text{Dodec}(a) \land \text{Cube}(b))$$

6.36

$$\vdash \neg(a = b \land b = c \land a \neq c)$$

6.37

$$\vdash \neg(a \neq b \land b \neq c \land a = c)$$

6.38

$$\vdash \neg(\text{SameRow}(a, b) \land \text{SameRow}(b, c) \land \text{FrontOf}(c, a))$$

6.39

$$\vdash \neg(\text{SameCol}(a, b) \land \text{SameCol}(b, c) \land \text{FrontOf}(a, c))$$

The following sentences are all tautologies, and so should be provable in $\mathcal{F}$. Although the informal proofs are relatively simple, $\mathcal{F}$ makes fairly heavy going of them, since it forces us to prove even very obvious steps. Use Fitch to construct formal proofs. You may want to build on the proof of Excluded Middle given in Exercise 6.33. Alternatively, with the permission of your instructor, you may use **Taut Con***, but only to justify an instance of Excluded Middle. The Grade Grinder will indicate whether you used* **Taut Con** *or not.*

6.40
✎★

> A ∨ ¬(A ∧ B)

6.41
✎★

> (A ∧ B) ∨ ¬A ∨ ¬B

6.42
✎★

> ¬A ∨ ¬(¬B ∧ (¬A ∨ B))

CHAPTER 7
Conditionals

There are many logically important constructions in English besides the Boolean connectives. Even if we restrict ourselves to words and phrases that connect two simple indicative sentences, we still find many that go beyond the Boolean operators. For example, besides saying:

> *Max is home* **and** *Claire is at the library,*

and

> *Max is home* **or** *Claire is at the library,*

we can combine these same atomic sentences in the following ways, among others:

> *Max is home* **if** *Claire is at the library,*
> *Max is home* **only if** *Claire is at the library,*
> *Max is home* **if and only if** *Claire is at the library,*
> *Max is* **not** *home* **nor** *is Claire at the library,*
> *Max is home* **unless** *Claire is at the library,*
> *Max is home* **even though** *Claire is at the library,*
> *Max is home* **in spite of the fact that** *Claire is at the library,*
> *Max is home* **just in case** *Claire is at the library,*
> *Max is home* **whenever** *Claire is at the library,*
> *Max is home* **because** *Claire is at the library.*

And these are just the tip of the iceberg. There are also constructions that combine three atomic sentences to form new sentences:

> **If** *Max is home* **then** *Claire is at the library,* **otherwise** *Claire is concerned,*

and constructions that combine four:

> **If** *Max is home* **then** *Claire is at the library,* **otherwise** *Claire is concerned* **unless** *Carl is with him,*

and so forth.

Some of these constructions are truth functional, or have important truth-functional uses, while others do not. Recall that a connective is truth functional if the truth or falsity of compound statements made with it is completely

178

determined by the truth values of its constituents. Its meaning, in other words, can be captured by a truth table.

FOL does not include connectives that are not truth functional. This is not to say that such connectives aren't important, but their meanings tend to be vague and subject to conflicting interpretations. The decision to exclude them is analogous to our assumption that all the predicates of FOL have precise interpretations.

non-truth-functional connectives

Whether or not a connective in English can be, or always is, used truth functionally is a tricky matter, about which we'll have more to say later in the chapter. Of the connectives listed above, though, there is one that is very clearly not truth functional: the connective *because*. This is not hard to prove.

> **Proof:** To show that the English connective *because* is not truth functional, it suffices to find two possible circumstances in which the sentence *Max is home because Claire is at the library* would have different truth values, but in which its constituents, *Max is home* and *Claire is at the library*, have the same truth values.
>
> Why? Well, suppose that the meaning of *because* were captured by a truth table. These two circumstances would correspond to the same row of the truth table, since the atomic sentences have the same values, but in one circumstance the sentence is true and in the other it is false. So the purported truth table must be wrong, contrary to our assumption.
>
> For the first circumstance, imagine that Max learned that Claire would be at the library, hence unable to feed Carl, and so rushed home to feed him. For the second circumstance, imagine that Max is at home, expecting Claire to be there too, but she unexpectedly had to go the library to get a reference book for a report. In both circumstances the sentences *Max is home* and *Claire is at the library* are true. But the compound sentence *Max is home **because** Claire is at the library* is true in the first, false in the second.

The reason *because* is not truth functional is that it typically asserts some sort of causal connection between the facts described by the constituent sentences. This is why our compound sentence was false in the second situation: the causal connection was missing.

In this chapter, we will introduce two new truth-functional connectives, known as the material conditional and the material biconditional, both standard features of FOL. It turns out that, as we'll show at the end of the chapter, these new symbols do not actually increase the expressive power of FOL. They do, however, make it much easier to say and prove certain things, and so are valuable additions to the language.

Material conditional symbol: →

The symbol → is used to combine two sentences P and Q to form a new sentence P → Q, called a *material conditional*. The sentence P is called the *antecedent* of the conditional, and Q is called the *consequent* of the conditional. We will discuss the English counterparts of this symbol after we explain its meaning.

Semantics and the game rule for the conditional

The sentence P → Q is true if and only if either P is false or Q is true (or both). This can be summarized by the following truth table.

truth table for →

P	Q	P → Q
T	T	**T**
T	F	**F**
F	T	**T**
F	F	**T**

game rule for →

A second's thought shows that P → Q is really just another way of saying ¬P ∨ Q. Tarski's World in fact treats the former as an abbreviation of the latter. In particular, in playing the game, Tarski's World simply replaces a statement of the form P → Q by its equivalent ¬P ∨ Q.

Remember

1. If P and Q are sentences of FOL, then so is P → Q.

2. The sentence P → Q is false in only one case: if the antecedent P is true and the consequent Q is false. Otherwise, it is true.

English forms of the material conditional

if ... then

We can come fairly close to an adequate English rendering of the material conditional P → Q with the sentence *If P then Q*. At any rate, it is clear that this English conditional, like the material conditional, is false if P is true and Q is false. Thus, we will translate, for example, *If Max is home then Claire is at the library* as:

$$\text{Home(max)} \rightarrow \text{Library(claire)}$$

In this course we will always translate *if... then...* using $\rightarrow$, but there are in fact many uses of the English expression that cannot be adequately expressed with the material conditional. Consider, for example, the sentence,

> *If Max had been at home, then Carl would have been there too.*

This sentence can be false even if Max is not in fact at home. (Suppose the speaker mistakenly thought Carl was with Max, when in fact Claire had taken him to the vet.) But the first-order sentence,

$$\text{Home(max)} \rightarrow \text{Home(carl)}$$

is automatically true if Max is not at home. A material conditional with a false antecedent is always true.

We have already seen that the connective *because* is not truth functional since it expresses a causal connection between its antecedent and consequent. The English construction *if... then...* can also be used to express a sort of causal connection between antecedent and consequent. That's what seems to be going on in the above example. As a result, many uses of *if... then...* in English just aren't truth functional. The truth of the whole depends on something more than the truth values of the parts; it depends on there being some genuine connection between the subject matter of the antecedent and the consequent.

Notice that we started with the truth table for $\rightarrow$ and decided to read it as *if... then....* What if we had started the other way around, looking for a truth-functional approximation of the English conditional? Could we have found a better truth table to go with *if... then...*? The answer is clearly *no*. While the material conditional is sometimes inadequate for capturing subtleties of English conditionals, it is the best we can do with a truth-functional connective. But these are controversial matters. We will take them up further in Section 7.3.

Necessary and sufficient conditions

Other English expressions that we will translate using the material conditional $P \rightarrow Q$ include: *P only if Q, Q provided P,* and *Q if P.* Notice in particular that *P only if Q* is translated $P \rightarrow Q$, while *P if Q* is translated $Q \rightarrow P$. To

only if, provided

understand why, we need to think carefully about the difference between *only if* and *if*.

necessary condition

In English, the expression *only if* introduces what is called a *necessary condition*, a condition that must hold in order for something else to obtain. For example, suppose your instructor announces at the beginning of the course that you will pass the course only if you turn in all the homework assignments. Your instructor is telling you that turning in the homework is a necessary condition for passing: if you don't do it, you won't pass. But the instructor is not guaranteeing that you *will* pass if you *do* turn in the homework: clearly, there are other ways to fail, such as skipping the tests and getting all the homework problems wrong.

The assertion that you will pass only if you turn in all the homework really excludes just one possibility: that you pass but did not turn in all the homework. In other words, *P only if Q* is false only when P is true and Q is false, and this is just the case in which $P \rightarrow Q$ is false.

Contrast this with the assertion that you will pass the course if you turn in all the homework. Now this is a very different kettle of fish. An instructor who makes this promise is establishing a very lax grading policy: just turn in the homework and you'll get a passing grade, regardless of how well you do on the homework or whether you even bother to take the tests!

sufficient condition

In English, the expression *if* introduces what is called a *sufficient condition*, one that guarantees that something else (in this case, passing the course) will obtain. Because of this an English sentence *P if Q* must be translated as $Q \rightarrow P$. The sentence rules out Q being true (turning in the homework) and P being false (failing the course).

Other uses of $\rightarrow$

unless

In FOL we also use $\rightarrow$ in combination with $\neg$ to translate sentences of the form *Unless P, Q* or *Q unless P*. Consider, for example, the sentence *Claire is at the library unless Max is home*. Compare this with the sentence *Claire is at the library if Max is not home*. While the focus of these two sentences is slightly different, a moment's thought shows that they are false in exactly the same circumstances, namely, if Claire is not at the library, yet Max is not home (say they are both at the movies). More generally, *Unless P, Q* or *Q unless P* are true in the same circumstances as *Q if not P*, and so are translated as $\neg P \rightarrow Q$. A good way to remember this is to whisper *if not* whenever you see *unless*. If you find this translation of *unless* counterintuitive, be patient. We'll say more about it in Section 7.3.

It turns out that the most important use of $\rightarrow$ in first-order logic is not in connection with the above expressions at all, but rather with *universally quantified* sentences, sentences of the form *All A's are B's* and *Every A is a B*. The analogous first-order sentences have the form:

For every object x $(A(x) \rightarrow B(x))$

This says that any object you pick will either fail to be an A or else be a B. We'll learn about such sentences in Part II of this book.

There is one other thing we should say about the material conditional, which helps explain its importance in logic. The conditional allows us to reduce the notion of logical consequence to that of logical truth, at least in cases where we have only finitely many premises. We said that a sentence Q is a consequence of premises $P_1, \ldots, P_n$ if and only if it is impossible for all the premises to be true while the conclusion is false. Another way of saying this is that it is impossible for the single sentence $(P_1 \wedge \ldots \wedge P_n)$ to be true while Q is false.

reducing logical consequence to logical truth

Given the meaning of $\rightarrow$, we see that Q is a consequence of $P_1, \ldots, P_n$ if and only if it is impossible for the single sentence

$$(P_1 \wedge \ldots \wedge P_n) \rightarrow Q$$

to be false, that is, just in case this conditional sentence is a logical truth. Thus, one way to verify the tautological validity of an argument in propositional logic, at least in theory, is to construct a truth table for this sentence and see whether the final column contains only TRUE. This method is usually not very practical, however, since the truth tables quickly get too large to be manageable.

Remember

1. The following English constructions are all translated $P \rightarrow Q$: *If P then Q; Q if P; P only if Q*; and *Provided P, Q*.

2. *Unless P, Q* and *Q unless P* are translated $\neg P \rightarrow Q$.

3. Q is a logical consequence of $P_1, \ldots, P_n$ if and only if the sentence $(P_1 \wedge \ldots \wedge P_n) \rightarrow Q$ is a logical truth.

SECTION 7.2

Biconditional symbol: ↔

Our final connective is called the material biconditional symbol. Given any sentences P and Q there is another sentence formed by connecting these by means of the biconditional: $P \leftrightarrow Q$. A sentence of the form $P \leftrightarrow Q$ is true if

if and only if

and only if P and Q have the same truth value, that is, either they are both true or both false. In English this is commonly expressed using the expression *if and only if*. So, for example, the sentence *Max is home if and only if Claire is at the library* would be translated as:

$$\text{Home(max)} \leftrightarrow \text{Library(claire)}$$

iff

just in case

Mathematicians and logicians often write "iff" as an abbreviation for "if and only if." Upon encountering this, students and typesetters generally conclude it's a spelling mistake, to the consternation of the authors. But in fact it is shorthand for the biconditional. Mathematicians also use "just in case" as a way of expressing the biconditional. Thus the mathematical claims *n is even iff n^2 is even,* and *n is even just in case n^2 is even,* would both be translated as:

$$\text{Even(n)} \leftrightarrow \text{Even(n}^2)$$

This use of "just in case" is, we admit, one of the more bizarre quirks of mathematicians, having nothing much to do with the ordinary meaning of this phrase. In this book, we use the phrase in the mathematician's sense, just in case you were wondering.

An important fact about the biconditional symbol is that two sentences P and Q are logically equivalent if and only if the biconditional formed from them, $P \leftrightarrow Q$, is a logical truth. Another way of putting this is to say that $P \Leftrightarrow Q$ is true if and only if the FOL sentence $P \leftrightarrow Q$ is logically necessary. So, for example, we can express one of the DeMorgan laws by saying that the following sentence is a logical truth:

$$\neg(P \vee Q) \leftrightarrow (\neg P \wedge \neg Q)$$

$\leftrightarrow$ *vs.* $\Leftrightarrow$

This observation makes it tempting to confuse the symbols $\leftrightarrow$ and $\Leftrightarrow$. This temptation must be resisted. The former is a truth-functional connective of FOL, while the latter is an abbreviation of "is logically equivalent to." It is not a truth-functional connective and is not an expression of FOL.

Semantics and the game rule for $\leftrightarrow$

The semantics for the biconditional is given by the following truth table.

P	Q	$P \leftrightarrow Q$
T	T	**T**
T	F	**F**
F	T	**F**
F	F	**T**

truth table for $\leftrightarrow$

Notice that the final column of this truth table is the same as that for
(P → Q) ∧ (Q → P). (See Exercise 7.3 below.) For this reason, logicians often
treat a sentence of the form P ↔ Q as an abbreviation of (P → Q) ∧ (Q → P).
Tarski's World also uses this abbreviation in the game. Thus, the game rule   *game rule for ↔*
for P ↔ Q is simple. Whenever a sentence of this form is encountered, it is
replaced by (P → Q) ∧ (Q → P).

Remember

1. If P and Q are sentences of FOL, then so is P ↔ Q.

2. The sentence P ↔ Q is true if and only if P and Q have the same truth
   value.

Exercises

---

*For the following exercises, use Boole to determine whether the indicated pairs of sentences are tauto-
logically equivalent. Feel free to have Boole build your reference columns and fill them out for you. Don't
forget to indicate your assessment.*

7.1   A → B and ¬A ∨ B.

7.2   ¬(A → B) and A ∧ ¬B.

7.3   A ↔ B and (A → B) ∧ (B → A).

7.4   A ↔ B and (A ∧ B) ∨ (¬A ∧ ¬B).

7.5   (A ∧ B) → C and A → (B ∨ C).

7.6   (A ∧ B) → C and A → (B → C).

7.7   A → (B → (C → D)) and
         ((A → B) → C) → D.

7.8   A ↔ (B ↔ (C ↔ D)) and
         ((A ↔ B) ↔ C) ↔ D.

7.9   (Just in case) Prove that the ordinary (nonmathematical) use of *just in case* does not express
         a truth-functional connective. Use as your example the sentence *Max went home just in case
         Carl was hungry.*

7.10   (Evaluating sentences in a world) Using Tarski's World, run through Abelard's Sentences, eval-
         uating them in Wittgenstein's World. If you make a mistake, play the game to see where you
         have gone wrong. Once you have gone through all the sentences, go back and make all the false
         ones true by changing one or more names used in the sentence. Submit your edited sentences
         as Sentences 7.10.

7.11 (Describing a world) Launch Tarski's World and choose **Hide Labels** from the **World** menu. Then, with the labels hidden, open Montague's World. In this world, each object has a name, and no object has more than one name. Start a new sentence file where you will describe some features of this world. Check each of your sentences to see that it is indeed a sentence and that it is true in this world.

1. Notice that if c is a tetrahedron, then a is not a tetrahedron. (Remember, in this world each object has exactly one name.) Use your first sentence to express this fact.
2. However, note that the same is true of b and d. That is, if b is a tetrahedron, then d isn't. Use your second sentence to express this.
3. Finally, observe that if b is a tetrahedron, then c isn't. Express this.
4. Notice that if a is a cube and b is a dodecahedron, then a is to the left of b. Use your next sentence to express this fact.
5. Use your next sentence to express the fact that if b and c are both cubes, then they are in the same row but not in the same column.
6. Use your next sentence to express the fact that b is a tetrahedron only if it is small. [Check this sentence carefully. If your sentence evaluates as false, then you've got the arrow pointing in the wrong direction.]
7. Next, express the fact that if a and d are both cubes, then one is to the left of the other. [Note: You will need to use a disjunction to express the fact that one is to the left of the other.]
8. Notice that d is a cube if and only if it is either medium or large. Express this.
9. Observe that if b is neither to the right nor left of d, then one of them is a tetrahedron. Express this observation.
10. Finally, express the fact that b and c are the same size if and only if one is a tetrahedron and the other is a dodecahedron.

Save your sentences as Sentences 7.11. Now choose **Show Labels** from the **World** menu. Verify that all of your sentences are indeed true. When verifying the first three, pay particular attention to the truth values of the various constituents. Notice that sometimes the conditional has a false antecedent and sometimes a true consequent. What it never has is a true antecedent and a false consequent. In each of these three cases, play the game committed to true. Make sure you understand why the game proceeds as it does.

7.12 (Translation) Translate the following English sentences into FOL. Your translations will use all of the propositional connectives.

1. *If **a** is a tetrahedron then it is in front of **d**.*
2. ***a** is to the left of or right of **d** only if it's a cube.*
3. ***c** is between either **a** and **e** or **a** and **d**.*
4. ***c** is to the right of **a**, provided it (i.e., **c**) is small.*

5. *c is to the right of **d** only if **b** is to the right of **c** and left of **e**.*
6. *If **e** is a tetrahedron, then it's to the right of **b** if and only if it is also in front of **b**.*
7. *If **b** is a dodecahedron, then if it isn't in front of **d** then it isn't in back of **d** either.*
8. *c is in back of **a** but in front of **e**.*
9. *e is in front of **d** unless it (i.e., **e**) is a large tetrahedron.*
10. *At least one of **a**, **c**, and **e** is a cube.*
11. *a is a tetrahedron only if it is in front of **b**.*
12. *b is larger than both **a** and **e**.*
13. *a and **e** are both larger than **c**, but neither is large.*
14. *d is the same shape as **b** only if they are the same size.*
15. *a is large if and only if it's a cube.*
16. *b is a cube unless **c** is a tetrahedron.*
17. *If **e** isn't a cube, either **b** or **d** is large.*
18. *b or **d** is a cube if either **a** or **c** is a tetrahedron.*
19. *a is large just in case **d** is small.*
20. *a is large just in case **e** is.*

Save your list of sentences as **Sentences 7.12**. Before submitting the file, you should complete Exercise 7.13.

7.13 (Checking your translations) Open **Bolzano's World**. Notice that all the English sentences from Exercise 7.12 are true in this world. Thus, if your translations are accurate, they will also be true in this world. Check to see that they are. If you made any mistakes, go back and fix them.

Remember that even if one of your sentences comes out true in **Bolzano's World**, it does not mean that it is a proper translation of the corresponding English sentence. If the translation is correct, it will have the same truth value as the English sentence in *every* world. So let's check your translations in some other worlds.

Open **Wittgenstein's World**. Here we see that the English sentences 3, 5, 9, 11, 12, 13, 14, and 20 are false, while the rest are true. Check to see that the same holds of your translations. If not, correct your translations (and make sure they are still true in **Bolzano's World**).

Next open **Leibniz's World**. Here half the English sentences are true (1, 2, 4, 6, 7, 10, 11, 14, 18, and 20) and half false (3, 5, 8, 9, 12, 13, 15, 16, 17, and 19). Check to see that the same holds of your translations. If not, correct your translations.

Finally, open **Venn's World**. In this world, all of the English sentences are false. Check to see that the same holds of your translations and correct them if necessary.

There is no need to submit any files for this exercise, but don't forget to submit **Sentences 7.12**.

7.14 (Figuring out sizes and shapes) Open Euler's Sentences. The nine sentences in this file uniquely determine the shapes and sizes of blocks a, b, and c. See if you can figure out the solution just by thinking about what the sentences mean and using the informal methods of proof you've already studied. When you've figured it out, submit a world in which all of the sentences are true.

7.15 (More sizes and shapes) Start a new sentence file and use it to translate the following English sentences.

1. If *a* is a tetrahedron, then *b* is also a tetrahedron.
2. *c* is a tetrahedron if *b* is.
3. *a* and *c* are both tetrahedra only if at least one of them is large.
4. *a* is a tetrahedron but *c* isn't large.
5. If *c* is small and *d* is a dodecahedron, then *d* is neither large nor small.
6. *c* is medium only if none of *d*, *e*, and *f* are cubes.
7. *d* is a small dodecahedron unless *a* is small.
8. *e* is large just in case it is a fact that *d* is large if and only if *f* is.
9. *d* and *e* are the same size.
10. *d* and *e* are the same shape.
11. *f* is either a cube or a dodecahedron, if it is large.
12. *c* is larger than *e* only if *b* is larger than *c*.

Save these sentences as Sentences 7.15. Then see if you can figure out the sizes and shapes of a, b, c, d, e, and f. You will find it helpful to approach this problem systematically, filling in the following table as you reason about the sentences:

	a	b	c	d	e	f
Shape:						
Size:						

When you have filled in the table, use it to guide you in building a world in which the twelve English sentences are true. Verify that your translations are true in this world as well. Submit both your sentence file and your world file.

7.16 (Name that object) Open Sherlock's World and Sherlock's Sentences. You will notice that none of the objects in this world has a name. Your task is to assign the names a, b, and c in such a way that all the sentences in the list come out true. Submit the modified world as World 7.16.

7.17 (Building a world) Open Boolos' Sentences. Submit a world in which all five sentences in this file are true.

7.18 Using the symbols introduced in Table 1.2, page 30, translate the following sentences into FOL. Submit your translations as a sentence file.

1. *If Claire gave Folly to Max at 2:03 then Folly belonged to her at 2:00 and to him at 2:05.*
2. *Max fed Folly at 2:00 pm, but if he gave her to Claire then, Folly was not hungry five minutes later.*
3. *If neither Max nor Claire fed Folly at 2:00, then she was hungry.*
4. *Max was angry at 2:05 only if Claire fed either Folly or Scruffy five minutes before.*
5. *Max is a student if and only if Claire is not.*

7.19 Using Table 1.2 on page 30, translate the following into colloquial English.

1. $(\mathsf{Fed}(\mathsf{max}, \mathsf{folly}, 2{:}00) \lor \mathsf{Fed}(\mathsf{claire}, \mathsf{folly}, 2{:}00)) \rightarrow \mathsf{Pet}(\mathsf{folly})$
2. $\mathsf{Fed}(\mathsf{max}, \mathsf{folly}, 2{:}30) \leftrightarrow \mathsf{Fed}(\mathsf{claire}, \mathsf{scruffy}, 2{:}00)$
3. $\neg\mathsf{Hungry}(\mathsf{folly}, 2{:}00) \rightarrow \mathsf{Hungry}(\mathsf{scruffy}, 2{:}00)$
4. $\neg(\mathsf{Hungry}(\mathsf{folly}, 2{:}00) \rightarrow \mathsf{Hungry}(\mathsf{scruffy}, 2{:}00))$

7.20 Translate the following into FOL as best you can. Explain any predicates and function symbols you use, and any shortcomings in your first-order translations.

1. *If Abe can fool Stephen, surely he can fool Ulysses.*
2. *If you scratch my back, I'll scratch yours.*
3. *France will sign the treaty only if Germany does.*
4. *If Tweedledee gets a party, so will Tweedledum, and vice versa.*
5. *If John and Mary went to the concert together, they must like each other.*

7.21 (The monkey principle) One of the stranger uses of *if... then...* in English is as a roundabout way to express negation. Suppose a friend of yours says *If Keanu Reeves is a great actor, then I'm a monkey's uncle.* This is simply a way of denying the antecedent of the conditional, in this case that Keanu Reeves is a great actor. Explain why this works. Your explanation should appeal to the truth table for $\rightarrow$, but it will have to go beyond that. Turn in your explanation and also submit a Boole table showing that $\mathsf{A} \rightarrow \bot$ is equivalent to $\neg\mathsf{A}$.

SECTION 7.3
Conversational implicature

In translating from English to FOL, there are many problematic cases. For example, many students resist translating a sentence like *Max is home unless Claire is at the library* as:

$$\neg\mathsf{Library}(\mathsf{claire}) \rightarrow \mathsf{Home}(\mathsf{max})$$

These students usually think that the meaning of this English sentence would be more accurately captured by the biconditional claim:

$$\neg \text{Library}(\text{claire}) \leftrightarrow \text{Home}(\text{max})$$

The reason the latter seems natural is that when we assert the English sentence, there is some suggestion that if Claire *is* at at the library, then Max is *not* at home.

To resolve problematic cases like this, it is often useful to distinguish between the *truth conditions* of a sentence, and other things that in some sense follow from the assertion of the sentence. To take an obvious case, suppose someone asserts the sentence *It is a lovely day.* One thing you may conclude from this is that the speaker understands English. This is not part of what the speaker said, however, but part of what can be inferred from his saying it. The truth or falsity of the claim has nothing to do with the speaker's linguistic abilities.

The philosopher H. P. Grice developed a theory of what he called *conversational implicature* to help sort out the genuine truth conditions of a sentence from other conclusions we may draw from its assertion. These other conclusions are what Grice called *implicatures.* We won't go into this theory in detail, but knowing a little bit about it can be a great aid in translation, so we present an introduction to Grice's theory.

conversational implicatures

Suppose we have an English sentence S that someone asserts, and we are trying to decide whether a particular conclusion we draw is part of the meaning of S or, instead, one of its implicatures. Grice pointed out that if the conclusion *is* part of the meaning, then it cannot be "cancelled" by some further elaboration by the speaker. Thus, for example, the conclusion that Max is home is part of the meaning of an assertion of *Max and Claire are home*, so we can't cancel this conclusion by saying *Max and Claire are home, but Max isn't home.* We would simply be contradicting ourselves.

cancelling implicatures

Contrast this with the speaker who said *It is a lovely day.* Suppose he had gone on to say, perhaps reading haltingly from a phrase book: *Do you speak any French?* In that case, the suggestion that the speaker understands English is effectively cancelled.

A more illuminating use of Grice's cancellability test concerns the expression *either... or....* Recall that we claimed that this should be translated into FOL as an inclusive disjunction, using $\vee$. We can now see that the suggestion that this phrase expresses exclusive disjunction is generally just a conversational implicature. For example, if the waiter says *You can have either soup or salad,* there is a strong suggestion that you cannot have both. But it is clear that this is just an implicature, since the waiter could, without contradicting himself, go on to say *And you can have both, if you want.* Had the original *either...or...* expressed the exclusive disjunction, this would be like saying *You can have soup or salad but not both, and you can have both, if you want.*

Let's go back now to the sentence *Max is at home unless Claire is at the library.* Earlier we denied that the correct translation was

$$\neg\mathsf{Library}(\mathsf{claire}) \leftrightarrow \mathsf{Home}(\mathsf{max})$$

which is equivalent to the conjunction of the correct translation

$$\neg\mathsf{Library}(\mathsf{claire}) \rightarrow \mathsf{Home}(\mathsf{max})$$

with the additional claim

$$\mathsf{Library}(\mathsf{claire}) \rightarrow \neg\mathsf{Home}(\mathsf{max})$$

Is this second claim part of the meaning of the English sentence, or is it simply a conversational implicature? Grice's cancellability test shows that it is just an implicature. After all, it makes perfectly good sense for the speaker to go on to say *On the other hand, if Claire is at the library, I have no idea where Max is.* This elaboration takes away the suggestion that if Claire is at the library, then Max isn't at home.

Another common implicature arises with the phrase *only if*, which people often construe as the stronger *if and only if*. For example, suppose a father tells his son, *You can have dessert only if you eat all your lima beans.* We've seen that this is not a guarantee that if the child does eat his lima beans he will get dessert, since *only if* introduces a necessary, not sufficient, condition. Still it is clear that the father's assertion suggests that, other things equal, the child can have dessert if he eats the dreaded beans. But again, the suggestion can be cancelled. Suppose the father goes on to say: *If you eat the beans, I'll check to see if there's any ice cream left.* This cancels the suggestion that dessert is guaranteed.

Remember

If the assertion of a sentence carries with it a suggestion that could be cancelled (without contradiction) by further elaboration by the speaker, then the suggestion is a *conversational implicature*, not part of the content of the original claim.

Exercises

7.22   Suppose Claire asserts the sentence *Max managed to get Carl home.* Does this logically imply, or just conversationally implicate, that it was hard to get Carl home? Justify your answer.

7.23   Suppose Max asserts the sentence *We can walk to the movie or we can drive.* Does his assertion logically imply, or merely implicate, that we cannot both walk and drive? How does this differ from the soup or salad example?

7.24   Consider the sentence *Max is home in spite of the fact that Claire is at the library.* What would be the best translation of this sentence into FOL? Clearly, whether you would be inclined to use this sentence is not determined simply by the truth values of the atomic sentences *Max is home* and *Claire is at the library.* This may be because *in spite of the fact* is, like *because,* a non-truth-functional connective, or because it carries, like *but,* additional conversational implicatures. (See our discussion of *because* earlier in this chapter and the discussion of *but* in Chapter 3.) Which explanation do you think is right? Justify your answer.

Section 7.4

Truth-functional completeness

We now have at our disposal five truth-functional connectives, one unary ($\neg$), and four binary ($\wedge$, $\vee$, $\rightarrow$, $\leftrightarrow$). Should we introduce any more? Though we've seen a few English expressions that can't be expressed in FOL, like *because,* these have not been truth functional. We've also run into others, like *neither. . . nor. . . ,* that *are* truth functional, but which we can easily express using the existing connectives of FOL.

The question we will address in the current section is whether there are any truth-functional connectives that we need to add to our language. Is it possible that we might encounter an English construction that is truth functional but which we cannot express using the symbols we have introduced so far? If so, this would be an unfortunate limitation of our language.

How can we possibly answer this question? Well, let's begin by thinking about binary connectives, those that apply to two sentences to make a third. How many binary truth-functional connectives are possible? If we think about the possible truth tables for such connectives, we can compute the total number. First, since we are dealing with binary connectives, there are four rows in each table. Each row can be assigned either TRUE or FALSE, so there are $2^4 = 16$ ways of doing this. For example, here is the table that captures the truth function expressed by *neither. . . nor. . . .*

P	Q	Neither P nor Q
T	T	**F**
T	F	**F**
F	T	**F**
F	F	**T**

Since there are only 16 different ways of filling in the final column of such a table, there are only 16 binary truth functions, and so 16 possible binary truth-functional connectives. We could look at each of these tables in turn and show how to express the truth function with existing connectives, just as we captured *neither P nor Q* with $\neg(P \vee Q)$. But there is a more general and systematic way to show this.

Suppose we are thinking about introducing a binary truth-functional connective, say $\star$. It will have a truth table like the following, with one of the values TRUE or FALSE in each row.

P	Q	P $\star$ Q
T	T	1st *value*
T	F	2nd *value*
F	T	3rd *value*
F	F	4th *value*

If all four values are FALSE, then we can clearly express P $\star$ Q with the sentence $P \wedge \neg P \wedge Q \wedge \neg Q$. So suppose at least one of the values is TRUE. How can we express P $\star$ Q? One way would be this. Let $C_1, \ldots, C_4$ stand for the following four conjunctions:

$$
\begin{aligned}
C_1 &= (P \wedge Q) \\
C_2 &= (P \wedge \neg Q) \\
C_3 &= (\neg P \wedge Q) \\
C_4 &= (\neg P \wedge \neg Q)
\end{aligned}
$$

Notice that sentence C_1 will be TRUE if the truth values of P and Q are as specified in the first row of the truth table, and that if the values of P and Q are anything else, then C_1 will be false. Similarly with C_2 and the second row of the truth table, and so forth. To build a sentence that gets the value TRUE in exactly the same rows as P $\star$ Q, all we need do is take the disjunction of the appropriate C's. For example, if P $\star$ Q is true in rows 2 and 4, then $C_2 \vee C_4$ is equivalent to this sentence.

What this shows is that all binary truth functions are already expressible using just the connectives $\neg$, $\wedge$, and $\vee$. In fact, it shows that they can be expressed using sentences in disjunctive normal form, as described in Chapter 4.

It's easy to see that a similar procedure allows us to express all possible unary truth functions. A unary connective, say $\natural$, will have a truth table like this:

P	$\natural$P
T	*1st value*
F	*2nd value*

If both of the values under $\natural$P are FALSE, then we can express it using the sentence $P \wedge \neg P$. Otherwise, we can express $\natural$P as a disjunction of one or more of the following:

$$C_1 = P$$
$$C_2 = \neg P$$

C_1 will be included as one of the disjuncts if the first value is TRUE, and C_2 will be included if the second value is TRUE. (Of course, in only one case will there be more than one disjunct.)

Once we understand how this procedure is working, we see that it will apply equally well to truth-functional connectives of any arity. Suppose, for example, that we want to express the ternary truth-functional connective defined by the following truth table:

P	Q	R	$\clubsuit$(P, Q, R)
T	T	T	**T**
T	T	F	**T**
T	F	T	**F**
T	F	F	**F**
F	T	T	**T**
F	T	F	**F**
F	F	T	**T**
F	F	F	**F**

A fairly good English translation of $\clubsuit$(P, Q, R) is *if P then Q, else R*. When we apply the above method to express this connective, we get the following sentence:

$$(P \wedge Q \wedge R) \vee (P \wedge Q \wedge \neg R) \vee (\neg P \wedge Q \wedge R) \vee (\neg P \wedge \neg Q \wedge R)$$

More generally, if $\spadesuit$ expresses an *n*-ary connective, then we can use this procedure to get a sentence that is tautologically equivalent to $\spadesuit(P_1, \ldots, P_n)$. First, we define the conjunctions $C_1, \ldots, C_{2^n}$ that correspond to the 2^n rows of the truth table. We then form a disjunction D that contains C_k as a disjunct if and only if the kth row of the truth table has the value TRUE. (If all rows contain FALSE, then we let D be $P_1 \wedge \neg P_1$.) For reasons we've already noted, this disjunction is tautologically equivalent to $\spadesuit(P_1, \ldots, P_n)$.

We have just sketched a proof that any truth function, of any arity whatsoever, can be expressed using just the Boolean connectives $\neg$, $\wedge$, and $\vee$. This is a sufficiently interesting fact that it deserves to be highlighted. We'll say that a set of connectives is *truth-functionally complete* if the connectives in the set allow us to express any truth function. We can then highlight this important fact as a *theorem*:

truth-functional completeness

Theorem The Boolean connectives $\neg$, $\wedge$, and $\vee$ are truth-functionally complete.

A theorem is just a conclusion the author finds particularly interesting or is particularly proud of having proven.

There are other collections of operators that are truth-functionally complete. In fact, we could get rid of either $\wedge$ or $\vee$ without losing truth-functional completeness. For example, $P \vee Q$ can be expressed using just $\neg$ and $\wedge$ as follows:

$$\neg(\neg P \wedge \neg Q)$$

What this means is that we could get rid of all the occurrences of $\vee$ in our sentences in favor of $\neg$ and $\wedge$. Alternatively, we could get rid of $\wedge$ in favor of $\neg$ and $\vee$, as you'll see in Exercise 7.25. Of course either way, the resulting sentences would be much longer and harder to understand.

We could in fact be even more economical in our choice of connectives. Suppose we used $P \downarrow Q$ to express *neither P nor Q*. It turns out that the connective $\downarrow$ is, all by itself, truth-functionally complete. To see this, notice that $\neg P$ can be expressed as:

$$P \downarrow P$$

which says *neither P nor P*, and $P \wedge Q$ can be expressed as:

$$(P \downarrow P) \downarrow (Q \downarrow Q)$$

which says *neither not P nor not Q*. Thus in theory we could use just this one truth-functional connective and express anything we can now express using our current five.

There are two disadvantages to economizing on connectives. First, as we've already said, the fewer connectives we have, the harder it is to understand our sentences. But even worse, our proofs become much more complicated. For example, if we always expressed $\wedge$ in terms of $\neg$ and $\vee$, a single application of the simple rule of $\wedge$ **Intro** would have to be replaced by two uses of $\perp$ **Intro**,

disadvantages of economy

one use of ∨ **Elim**, and one use of ¬ **Intro** (see Exercise 7.26). This is why we haven't skimped on connectives.

> ### Remember
>
> 1. A set of connectives is *truth-functionally complete* if the connectives allow us to express every truth function.
>
> 2. Various sets of connectives, including the Boolean connectives, are truth-functionally complete.

Exercises

7.25 (Replacing ∧, →, and ↔) Use Tarski's World to open the file Sheffer's Sentences. In this file, you will find the following sentences in the odd-numbered positions:
  1. Tet(a) ∧ Small(a)
  3. Tet(a) → Small(a)
  5. Tet(a) ↔ Small(a)
  7. (Cube(b) ∧ Cube(c)) → (Small(b) ↔ Small(c))

In each even-numbered slot, enter a sentence that is equivalent to the one above it, but which uses only the connectives ¬ and ∨. Before submitting your solution file, you might want to try out your sentences in several worlds to make sure the new sentences have the expected truth values.

7.26 (Basic versus defined symbols in proofs) Treating a symbol as basic, with its own rules, or as a defined symbol, without its own rules, makes a big difference to the complexity of proofs. Use Fitch to open the file Exercise 7.26. In this file, you are asked to construct a proof of ¬(¬A ∨ ¬B) from the premises A and B. A proof of the equivalent sentence A ∧ B would of course take a single step.

7.27 (Simplifying *if... then... else*) Assume that P, Q, and R are atomic sentences. See if you can simplify the sentence we came up with to express ♣(P, Q, R) (*if P then Q, else R*), so that it becomes a disjunction of two sentences, each of which is a conjunction of two literals. Submit your solution as a Tarski's World sentence file.

7.28
✎⋆⋆⋆
(Expressing another ternary connective) Start a new sentence file using Tarski's World. Use the method we have developed to express the ternary connective ♡ defined in the following truth table, and enter this as the first sentence in your file. Then see if you can simplify the result as much as possible. Enter the simplified form as your second sentence. (This sentence should have no more than two occurrences each of P, Q, and R, and no more than six occurrences of the Boolean connectives, ∨, ∧ and ¬.)

P	Q	R	♡(P, Q, R)
T	T	T	**T**
T	T	F	**T**
T	F	T	**T**
T	F	F	**F**
F	T	T	**F**
F	T	F	**T**
F	F	T	**T**
F	F	F	**T**

7.29
✎⋆
(Sheffer stroke) Another binary connective that is truth-functionally complete on its own is called the Sheffer stroke, named after H. M. Sheffer, one of the logicians who discovered and studied it. It is also known as *nand* by analogy with *nor*. Here is its truth table:

P	Q	P | Q
T	T	**F**
T	F	**T**
F	T	**T**
F	F	**T**

Show how to express ¬P, P ∧ Q, and P ∨ Q using the Sheffer stroke. (We remind you that nowadays, the symbol | has been appropriated as an alternative for ∨. Don't let that confuse you.)

7.30
✎⋆
(Putting monkeys to work) Suppose we have the single binary connective →, plus the symbol for absurdity ⊥. Using just these expressions, see if you can find a way to express ¬P, P ∧ Q, and P ∨ Q. [Hint: Don't forget what you learned in Exercise 7.21.]

7.31
✎⋆
(Another non-truth-functional connective) Show that truth value at a particular time of the sentence *Max is home whenever Claire is at the library* is not determined by the truth values of the atomic sentences *Max is home* and *Claire is at the library* at that same time. That is, show that *whenever* is not truth functional.

7.32   (Exclusive disjunction) Suppose we had introduced ▽ to express exclusive disjunction. Is the
✎⋆   following a valid method of proof for this connective?

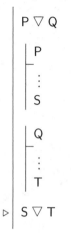

If you say *yes*, justify your answer; if *no*, give an example where the method sanctions an
invalid inference.

State valid introduction and elimination rules for ▽ using the same format we use to state
the introduction and elimination rules of $\mathcal{F}$. You may need more than one of each.

SECTION 7.5

Alternative notation

As with the other truth-functional connectives, there are alternative notations
for the material conditional and biconditional. The most common alternative
to P → Q is P ⊃ Q. Polish notation for the conditional is Cpq. The most common alternative to P ↔ Q is P ≡ Q. The Polish notation for the biconditional
is Epq.

	Remember

The following table summarizes the alternative notations discussed so far.

Our notation	Common equivalents
¬P	∼ P, $\overline{P}$, !P, Np
P ∧ Q	P&Q, P&&Q, P · Q, PQ, Kpq
P ∨ Q	P | Q, P ‖ Q, Apq
P → Q	P ⊃ Q, Cpq
P ↔ Q	P ≡ Q, Epq

CHAPTER 8

The Logic of Conditionals

One thing the theorem on page 195 tells us is that introducing the material conditional and biconditional symbols did not increase the expressive power of FOL. Since $\rightarrow$ and $\leftrightarrow$ can be defined using the Boolean connectives, we could always eliminate them from claims or proofs by means of these definitions. Thus, for example, if we wanted to prove $P \rightarrow Q$ we could just prove $\neg P \lor Q$, and then use the definition. In practice, though, this is a terrible idea. It is far more natural to use rules that involve these symbols directly, and the resulting proofs are simpler and easier to understand.[1]

The material conditional, in particular, is an extremely useful symbol to have. For example, many claims and theorems that we will run across can only be expressed naturally using the conditional. In fact, quite a few of the examples we've already used are more naturally stated as conditional claims. Thus in an earlier exercise we asked you to prove $\mathsf{Even}(n \times m)$ from the premise $\mathsf{Odd}(n + m)$. But really, the fact we were interested in was that, no matter what numbers n and m you pick, the following *conditional* claim is true:

$$\mathsf{Odd}(n + m) \rightarrow \mathsf{Even}(n \times m)$$

Given the importance of conditional claims, and the frequency you'll encounter them, we need to learn how to prove these claims.

Informal methods of proof

As before, we will first look at informal proofs involving conditionals and later incorporate the key methods into the system $\mathcal{F}$. Among the informal methods, we distinguish simple valid steps from more important methods of proof.

Valid steps

The most common valid proof step involving $\rightarrow$ goes by the Latin name *modus ponens*, or by the English *conditional elimination*. The rule says that if you

*modus ponens or
conditional elimination*

---

[1]In Exercise 8.38 we ask you to construct proofs of $(P \land Q) \rightarrow P$ and the equivalent $\neg(P \land Q) \lor P$, so that you can see for yourself how much simpler the first is than the second.

have established both P → Q and P, then you can infer Q. This rule is obviously valid, as a review of the truth table for → shows, since if P → Q and P are both true, then so must be Q.

biconditional elimination

There is a similar proof step for the biconditional, since the biconditional is logically equivalent to a conjunction of two conditionals. If you have established either P ↔ Q or Q ↔ P, then if you can establish P, you can infer Q. This is called *biconditional elimination*.

contraposition

In addition to these simple rules, there are a number of useful equivalences involving our new symbols. One of the most important is known as the Law of Contraposition. It states that P → Q is logically equivalent to ¬Q → ¬P. This latter conditional is known as the *contrapositive* of the original conditional. It is easy to see that the original conditional is equivalent to the contrapositive, since the latter is false if and only if ¬Q is true and ¬P is false, which is to say, when P is true and Q is false. Contraposition is a particularly useful equivalence since it is often easier to prove the contrapositive of a conditional than the conditional itself. We'll see an example of this in a moment.

Here are some logical equivalences to bear in mind, beginning with contraposition. Make sure you understand them all and see why they are equivalent. Use Boole to construct truth tables for any you don't immediately see.

$$P \rightarrow Q \quad \Leftrightarrow \quad \neg Q \rightarrow \neg P$$
$$P \rightarrow Q \quad \Leftrightarrow \quad \neg P \vee Q$$
$$\neg(P \rightarrow Q) \quad \Leftrightarrow \quad P \wedge \neg Q$$
$$P \leftrightarrow Q \quad \Leftrightarrow \quad (P \rightarrow Q) \wedge (Q \rightarrow P)$$
$$P \leftrightarrow Q \quad \Leftrightarrow \quad (P \wedge Q) \vee (\neg P \wedge \neg Q)$$

Remember

Let P and Q be any sentences of FOL.

1. Modus ponens: From P → Q and P, infer Q.

2. Biconditional elimination: From P and either P ↔ Q or Q ↔ P, infer Q.

3. Contraposition: P → Q ⇔ ¬Q → ¬P

The method of conditional proof

One of the most important methods of proof is that of conditional proof, a method that allows you to prove a conditional statement. Suppose you want to prove the conditional claim $P \rightarrow Q$. What you do is temporarily assume the antecedent, P, of your desired conditional. Then if, with this additional assumption, you are able to prove Q, conditional proof allows you to infer that $P \rightarrow Q$ follows from the original premises.

conditional proof

Let's look at a simple example that uses both modus ponens and conditional proof. We will show that $\text{Tet}(a) \rightarrow \text{Tet}(c)$ is a logical consequence of the two premises $\text{Tet}(a) \rightarrow \text{Tet}(b)$ and $\text{Tet}(b) \rightarrow \text{Tet}(c)$. In other words, we'll show that the operator $\rightarrow$ is transitive. This may seem obvious, but the proof is a good illustration of the method of conditional proof.

> **Proof:** We are given, as premises, $\text{Tet}(a) \rightarrow \text{Tet}(b)$ and $\text{Tet}(b) \rightarrow \text{Tet}(c)$. We want to prove $\text{Tet}(a) \rightarrow \text{Tet}(c)$. With an eye toward using conditional proof, let us assume, in addition to our premises, that $\text{Tet}(a)$ is true. Then, by applying modus ponens using our first premise, we can conclude $\text{Tet}(b)$. Using modus ponens again, this time with our second premise, we get $\text{Tet}(c)$. So we have established the consequent, $\text{Tet}(c)$, of our desired conditional on the basis of our assumption of $\text{Tet}(a)$. But then the rule of conditional proof assures us that $\text{Tet}(a) \rightarrow \text{Tet}(c)$ follows from the initial premises alone.

Conditional proof is clearly a valid form of reasoning, one that we use all the time. If Q follows logically from some premises plus the additional assumption P, then we can be sure that if those premises are true, the conditional $P \rightarrow Q$ must be true as well. After all, the conditional can only be false if P is true and Q is false, and our conditional proof shows that, given the premises, this can never happen.

Let's look at a more interesting example. This example will use both conditional proof and proof by contradiction. We will prove:

$$\text{Even}(n^2) \rightarrow \text{Even}(n)$$

> **Proof:** The method of conditional proof tells us that we can proceed by assuming $\text{Even}(n^2)$ and proving $\text{Even}(n)$. So assume that n^2 is even. To prove that n is even, we will use proof by contradiction. Thus, assume that n is not even, that is, that it is odd. Then we can express n as $2m + 1$, for some m. But then we see that:
>
> $$\begin{aligned} n^2 &= (2m+1)^2 \\ &= 4m^2 + 4m + 1 \\ &= 2(2m^2 + 2m) + 1 \end{aligned}$$
>
> But this shows that n^2 is odd, contradicting our first assumption. This contradiction shows that n is not odd, i.e., that it is even. Thus, by conditional proof, we have established $\text{Even}(n^2) \rightarrow \text{Even}(n)$.

Did you get lost? This proof has a pretty complicated structure, since we first assumed $\text{Even}(n^2)$ for the purpose of conditional proof, but then immediately assumed $\neg\text{Even}(n)$ to get an indirect proof of $\text{Even}(n)$. The contradiction that we arrived at was $\neg\text{Even}(n^2)$, which contradicted our first assumption.

Proofs of this sort are fairly common, and this is why it is often easier to prove the contrapositive of a conditional. The contrapositive of our original claim is this:

$$\neg\text{Even}(n) \to \neg\text{Even}(n^2)$$

Let's look at the proof of this contrapositive.

Proof: To prove $\neg\text{Even}(n) \to \neg\text{Even}(n^2)$, we begin by assuming $\neg\text{Even}(n)$, i.e., that n is odd. Then we can express n as $2m + 1$, for some m. But then we see that:

$$\begin{aligned} n^2 &= (2m+1)^2 \\ &= 4m^2 + 4m + 1 \\ &= 2(2m^2 + 2m) + 1 \end{aligned}$$

But this shows that n^2 is also odd, hence $\neg\text{Even}(n^2)$. Thus, by conditional proof, we have established $\neg\text{Even}(n) \to \neg\text{Even}(n^2)$.

By proving the contrapositive, we avoided the need for an indirect proof *inside* the conditional proof. This makes the proof easier to understand, and since the contrapositive is logically equivalent to our original claim, our second proof could serve as a proof of the original claim as well.

The method of conditional proof is used extensively in everyday reasoning. Some years ago Bill was trying to decide whether to take English 301, Postmodernism. His friend Sarah claimed that *if Bill takes Postmodernism, he will not get into medical school.* Sarah's argument, when challenged by Bill, took the form of a conditional proof, combined with a proof by cases.

Suppose you take Postmodernism. Then either you will adopt the postmodern disdain for rationality or you won't. If you don't, you will fail the class, which will lower your GPA so much that you will not get into medical school. But if you do adopt the postmodern contempt toward rationality, you won't be able to pass organic chemistry, and so will not get into medical school. So in either case, you will not get into medical school. Hence, if you take Postmodernism, you won't get into medical school.

Unfortunately for Bill, he had already succumbed to postmodernism, and so rejected Sarah's argument. He went ahead and took the course, failed chemistry, and did not get into medical school. He's now a wealthy lobbyist in Washington. Sarah is an executive in the computer industry in California.

Proving biconditionals

Not surprisingly, we can also use conditional proof to prove biconditionals, though we have to work twice as hard. To prove $P \leftrightarrow Q$ by conditional proof, you need to do two things: assume P and prove Q; then assume Q and prove P. This gives us both $P \rightarrow Q$ and $Q \rightarrow P$, whose conjunction is equivalent to $P \leftrightarrow Q$.

There is another form of proof involving $\leftrightarrow$ that is common in mathematics. Mathematicians are quite fond of finding results which show that several different conditions are equivalent. Thus you will find theorems that make claims like this: "The following conditions are all equivalent: Q_1, Q_2, Q_3." What they mean by this is that all of the following biconditionals hold:

$$Q_1 \leftrightarrow Q_2$$
$$Q_2 \leftrightarrow Q_3$$
$$Q_1 \leftrightarrow Q_3$$

To prove these three biconditionals in the standard way, you would have to give six conditional proofs, two for each biconditional. But we can cut our work in half by noting that it suffices to prove some cycle of results like the following:

$$Q_1 \rightarrow Q_2$$
$$Q_2 \rightarrow Q_3$$
$$Q_3 \rightarrow Q_1$$

proving a cycle of conditionals

These would be shown by three conditional proofs, rather than the six that would otherwise be required. Once we have these, there is no need to prove the reverse directions, since they follow from the transitivity of $\rightarrow$. For example, we don't need to explicitly prove $Q_2 \rightarrow Q_1$, the reverse of the first conditional, since this follows from $Q_2 \rightarrow Q_3$ and $Q_3 \rightarrow Q_1$, our other two conditionals.

When we apply this technique, we don't have to arrange the cycle in exactly the order in which the conditions are given. But we do have to make sure we have a genuine cycle, one that allows us to get from any one of our conditions to any other.

Let's give a very simple example. We will prove that the following conditions on a natural number n are all equivalent:

1. n is even
2. n^2 is even
3. n^2 is divisible by 4.

Proof: Rather than prove all six biconditionals, we prove that $(3) \rightarrow (2) \rightarrow (1) \rightarrow (3)$. Assume (3). Now clearly, if n^2 is divisible by 4, then it is divisible by 2, so we have $(3) \rightarrow (2)$. Next, we prove $(2) \rightarrow (1)$ by proving its contrapositive. Thus, assume n is odd and prove n^2 is odd. Since n is odd, we can write it in the form $2m + 1$. But then (as we've already shown) $n^2 = 2(2m^2 + 2m) + 1$ which is also odd. Finally, let us prove $(1) \rightarrow (3)$. If n is even, it can be expressed as $2m$. Thus, $n^2 = (2m)^2 = 4m^2$, which is divisible by 4. This completes the cycle, showing that the three conditions are indeed equivalent.

When you apply this method, you should look for simple or obvious implications, like $(1) \rightarrow (3)$ above, or implications that you've already established, like $(2) \rightarrow (1)$ above, and try to build them into your cycle of conditionals.

Remember

1. The method of conditional proof: To prove P → Q, assume P and prove Q.

2. To prove a number of biconditionals, try to arrange them into a cycle of conditionals.

Exercises

8.1 In the following list we give a number of inference patterns, some of which are valid, some invalid. For each pattern, decide whether you think it is valid and say so. Later, we will return to these patterns and ask you to give formal proofs for the valid ones and counterexamples for the invalid ones. But for now, just assess their validity.

1. *Affirming the Consequent:* From A → B and B, infer A.
2. *Modus Tollens:* From A → B and ¬B, infer ¬A.
3. *Strengthening the Antecedent:* From B → C, infer (A ∧ B) → C.
4. *Weakening the Antecedent:* From B → C, infer (A ∨ B) → C.
5. *Strengthening the Consequent:* From A → B, infer A → (B ∧ C).
6. *Weakening the Consequent:* From A → B, infer A → (B ∨ C).
7. *Constructive Dilemma:* From A ∨ B, A → C, and B → D, infer C ∨ D.
8. *Transitivity of the Biconditional:* From A ↔ B and B ↔ C, infer A ↔ C.

8.2 Open Conditional Sentences. Suppose that the sentences in this file are your premises. Now consider the five sentences listed below. Some of these sentences are consequences of these premises, some are not. For those that are consequences, give informal proofs and turn them in to your instructor. For those that are not consequences, submit counterexample worlds in which the premises are true but the conclusion false. Name the counterexamples World 8.2.x, where x is the number of the sentence.

1. Tet(e)
2. Tet(c) → Tet(e)
3. Tet(c) → Larger(f, e)
4. Tet(c) → LeftOf(c, f)
5. Dodec(e) → Smaller(e, f)

The following arguments are all valid. Turn in informal proofs of their validity. You may find it helpful to translate the arguments into FOL *before trying to give proofs, though that's not required. Explicitly note any inferences using modus ponens, biconditional elimination, or conditional proof.*

8.3

The unicorn, if it is not mythical, is a mammal, but if it is mythical, it is immortal.
If the unicorn is either immortal or a mammal, it is horned.
The unicorn, if horned, is magical.

The unicorn is magical.

8.4

The unicorn, if horned, is elusive and dangerous.
If elusive or mythical, the unicorn is rare.
If a mammal, the unicorn is not rare.

The unicorn, if horned, is not a mammal.

8.5

The unicorn, if horned, is elusive and magical, but if not horned, it is neither.
If the unicorn is not horned, it is not mythical.

The unicorn is horned if and only if magical or mythical.

8.6

a is a large tetrahedron or a small cube.
b is not small.
If *a* is a tetrahedron or a cube, then *b* is large or small.
a is a tetrahedron only if *b* is medium.

a is small and *b* is large.

8.7

b is small unless it's a cube.
If *c* is small, then either *d* or *e* is too.
If *d* is small, then *c* is not.
If *b* is a cube, then *e* is not small.

If *c* is small, then so is *b*.

8.8

d is in the same row as *a*, *b* or *c*.
If *d* is in the same row as *b*, then it is in the same row as *a* only if it's not in the same row as *c*.
d is in the same row as *a* if and only if it is in the same row as *c*.

d is in the same row as *a* if and only if it is not in the same row as *b*.

8.9

> *a* is either a cube, a dodecahedron, or a tetrahedron.
> *a* is small, medium, or large.
> *a* is medium if and only if it's a dodecahedron.
> *a* is a tetrahedron if and only if it is large.
>
> *a* is a cube if and only if it's small.

8.10 Open Between Sentences. Determine whether this set of sentences is satisfiable or not. If it is, submit a world in which all the sentences are true. If not, give an informal proof that the sentences are inconsistent. That is, assume all of them and derive a contradiction.

8.11 Analyze the structure of the informal proof in support of the following claim: *If the U.S. does not cut back on its use of oil soon, parts of California will be flooded within 50 years.* Are there weak points in the argument? What premises are implicitly assumed in the proof? Are they plausible?

> **Proof:** Suppose the U.S. does not cut back on its oil use soon. Then it will be unable to reduce its carbon dioxide emissions substantially in the next few years. But then the countries of China, India and Brazil will refuse to join in efforts to curb carbon dioxide emissions. As these countries develop without such efforts, the emission of carbon dioxide will get much worse, and so the greenhouse effect will accelerate. As a result the sea will get warmer, ice will melt, and the sea level will rise. In which case, low lying coastal areas in California will be subject to flooding within 50 years. So if we do not cut back on our oil use, parts of California will be flooded within 50 years.

8.12 Describe an everyday example of reasoning that uses the method of conditional proof.

8.13 Prove: $\mathsf{Odd}(n + m) \to \mathsf{Even}(n \times m)$.
[Hint: Compare this with Exercise 5.24 on page 141.]

8.14 Prove: $\mathsf{Irrational}(x) \to \mathsf{Irrational}(\sqrt{x})$.
[Hint: It is easier to prove the contrapositive.]

8.15 Prove that the following conditions on the natural number *n* are all equivalent. Use as few conditional proofs as possible.

1. *n* is divisible by 3
2. n^2 is divisible by 3
3. n^2 is divisible by 9
4. n^3 is divisible by 3
5. n^3 is divisible by 9
6. n^3 is divisible by 27

8.16 Give an informal proof that if R is a tautological consequence of $P_1, \ldots, P_n$ and Q, then $Q \to R$ is a tautological consequence of $P_1, \ldots, P_n$.

Formal rules of proof for → and ↔

We now turn to the formal analogues of the methods of proof involving the conditional and biconditional. Again, we incorporate an introduction and elimination rule for each connective into $\mathcal{F}$.

Rules for the conditional

The rule of *modus ponens* or conditional elimination is easily formalized. If you have proven both P → Q and P then you can assert Q, citing as justification these two earlier steps. Schematically:

Conditional Elimination (→ Elim):

$$
\begin{array}{c|l}
 & P \to Q \\
 & \ \ \vdots \\
 & P \\
 & \ \ \vdots \\
\rhd & Q \\
\end{array}
$$

The corresponding introduction rule is the formal counterpart of the method of conditional proof. As you would expect, it requires us to construct a subproof. To prove a statement of the form P → Q we begin our subproof with the assumption of P and try to prove Q. If we succeed, then we are allowed to discharge the assumption and conclude our desired conditional, citing the subproof as justification. Schematically:

Conditional Introduction (→ Intro):

$$
\begin{array}{c|ll}
 & & P \\
 & & \ \ \vdots \\
 & & Q \\
\rhd & P \to Q &
\end{array}
$$

Strategy and tactics

The strategy of working backwards usually works extremely well in proofs that involve conditionals, particularly when the desired conclusion is itself a

working backwards

conditional. Thus if the goal of a proof is to show that a conditional sentence P → Q is a consequence of the given information, you should sketch in a subproof with P as an assumption and Q as the final step. Use this subproof as support for an application of → **Intro**. When you check your proof, Q will become your new intermediate goal, and in proving it you can rely on the assumption P.

Let's work through a simple example that involves both of the rules for the conditional and illustrates the technique of working backwards.

You try it
. . . .

▶ 1. We will step you through a formal proof of A → C from the premise (A ∨ B) → C. Use Fitch to open the file **Conditional 1**. Notice the premise and the goal. Add a step to the proof and write in the goal sentence.

▶ 2. Start a subproof before the sentence A → C. Enter A as the assumption of the subproof.

▶ 3. Add a second step to the subproof and enter C.

▶ 4. Move the slider to the step containing the goal sentence A → C. Justify this step using the rule → **Intro**, citing the subproof for support. Check this step.

▶ 5. Now we need to go back and fill in the subproof. Add a step between the two steps of the subproof. Enter A ∨ B. Justify this step using ∨ **Intro**, citing the assumption of the subproof.

▶ 6. Now move the slider to the last step of the subproof. Justify this step using the rule → **Elim**, citing the premise and the step you just proved.

▶ 7. Verify that your proof checks out, and save it as **Proof Conditional 1**.

. . . . *Congratulations*

Once we have conditional introduction at our disposal, we can convert any proof with premises into the proof, without premises, of a corresponding conditional. For example, we showed in Chapter 6 (page 158) how to give a formal proof of ¬¬A from premise A. We can now use the earlier proof to build a proof of the logically true sentence A → ¬¬A.

1. A
2. ¬A
3. ⊥   ⊥ **Intro**: 1, 2
4. ¬¬A   ¬ **Intro**: 2–3
5. A → ¬¬A   → **Intro**: 1–4

Notice that the subproof here is identical to the original proof given on page 158. We simply embedded that proof in our new proof and applied conditional introduction to derive A → ¬¬A.

Default and generous uses of the → rules

The rule → **Elim** does not care in which order you cite the support sentences. The rule → **Intro** does not insist that the consequent be at the last step of the cited subproof, though it usually is. Also, the assumption step might be the only step in the subproof, as in a proof of a sentence of the form P → P.

The default applications of the conditional rules work exactly as you would expect. If you cite supports of the form indicated in the rule statements, Fitch will fill in the appropriate conclusion for you.

*default uses of
conditional rules*

You try it
. . . .

1. Open the file Conditional 2. Look at the goal to see what sentence we are trying to prove. Then focus on each step in succession and check the step. On the empty steps, try to predict what default Fitch will supply.

◀

2. When you are finished, make sure you understand the proof. Save the checked proof as Proof Conditional 2.

◀

. . . . *Congratulations*

You can use the **Add Support Steps** command when you are using the → **Intro** rule, and have an implication at the focus step. Fitch will insert a subproof as support. The antecedent of the implication will be the assumption of the subproof, and the consequent will appear at the last line of the subproof. You cannot use **Add Support Steps** with → **Elim**.

Rules for the biconditional

The rules for the biconditional are just what you would expect, given the rules for the conditional. The elimination rule for the biconditional can be stated schematically as follows:

Biconditional Elimination (↔ Elim):

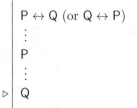

This means that you can conclude Q if you can establish P and either of the biconditionals indicated.

The introduction rule for the biconditional P ↔ Q requires that you give two subproofs, one showing that Q follows from P, and one showing that P follows from Q:

Biconditional Introduction (↔ Intro):

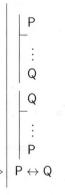

Here is a simple example of a proof using biconditional introduction. It shows how to derive the double negation law within the system $\mathcal{F}$.

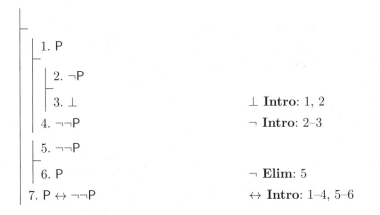

```
  | 1. P
  |   | 2. ¬P
  |   | 3. ⊥              ⊥ Intro: 1, 2
  | 4. ¬¬P                ¬ Intro: 2–3
  |   | 5. ¬¬P
  | 6. P                  ¬ Elim: 5
  | 7. P ↔ ¬¬P            ↔ Intro: 1–4, 5–6
```

Strategy and tactics

When you are constructing a proof whose conclusion is a biconditional, it is particularly important to sketch in the proof ahead of time. Add the two required subproofs and the desired conclusion, citing the subproofs in support. Then try to fill in the subproofs. This is a good idea because these proofs sometimes get quite long and involved. The sketch will help you remember what you were trying to do.

You try it

....

1. Open the file **Conditional 3**. In this file, you are asked to prove, without premises, the law of contraposition: ◀

$$(P \to Q) \leftrightarrow (\neg Q \to \neg P)$$

2. Start your proof by sketching in the two subproofs that you know you'll have to prove, plus the desired conclusion. Your partial proof will look like this: ◀

```
  |   | 1. P → Q
  |   | 2. ¬Q → ¬P          Rule?
  |   | 3. ¬Q → ¬P
  |   | 4. P → Q            Rule?
  | 5. (P → Q) ↔ (¬Q → ¬P)  ↔ Intro: 1–2, 3–4
```

▶ 3. Now that you have the overall structure, start filling in the first subproof. Since the goal of that subproof is a conditional claim, sketch in a conditional proof that would give you that claim:

1. P → Q	
2. ¬Q	
3. ¬P	**Rule?**
4. ¬Q → ¬P	→ **Intro:** 2–3
5. ¬Q → ¬P	
6. P → Q	**Rule?**
7. (P → Q) ↔ (¬Q → ¬P)	↔ **Intro:** 1–4, 5–6

▶ 4. To derive ¬P in the subsubproof, you will need to assume P and derive a contradiction. This is pretty straightforward:

1. P → Q	
2. ¬Q	
3. P	
4. Q	→ **Elim:** 1, 3
5. ⊥	⊥ **Intro:** 4, 2
6. ¬P	¬ **Intro:** 3–5
7. ¬Q → ¬P	→ **Intro:** 2–6
8. ¬Q → ¬P	
9. P → Q	**Rule?**
10. (P → Q) ↔ (¬Q → ¬P)	↔ **Intro:** 1–7, 8–9

▶ 5. This completes the first subproof. Luckily, you sketched in the second subproof so you know what you want to do next. You should be able to finish the second subproof on your own, since it is almost identical to the first.

6. When you are done, save your proof as Proof Conditional 3. ◀

. . . . *Congratulations*

The default and generous uses of the biconditional rules are exactly like those for the conditional connective, and **Add Support Steps** works exactly as you would expect.

Exercises

8.17 If you skipped any of the **You try it** sections, go back and do them now. Submit the files Proof Conditional 1, Proof Conditional 2, and Proof Conditional 3.

In the following exercises we return to the patterns of inference discussed in Exercise 8.1. Some of these are valid, some invalid. For each valid pattern, construct a formal proof in Fitch. For each invalid pattern, give a counterexample using Tarski's World. To give a counterexample in these cases, you will have to come up with sentences of the blocks language that fit the pattern, and a world that makes those specific premises true and the conclusion false. Submit both the world and the sentence file. In the sentence file, list the premises first and the conclusion last.

8.18 *Affirming the Consequent:*
From A → B and B, infer A.

8.19 *Modus Tollens:*
From A → B and ¬B, infer ¬A.

8.20 *Strengthening the Antecedent:*
From B → C, infer (A ∧ B) → C.

8.21 *Weakening the Antecedent:*
From B → C, infer (A ∨ B) → C.

8.22 *Strengthening the Consequent:*
From A → B, infer A → (B ∧ C).

8.23 *Weakening the Consequent:*
From A → B, infer A → (B ∨ C).

8.24 *Constructive Dilemma:*
From A ∨ B, A → C, and B → D, infer C ∨ D.

8.25 *Transitivity of the Biconditional:*
From A ↔ B and B ↔ C, infer A ↔ C.

*Use Fitch to construct formal proofs for the following arguments. In two cases, you may find yourself re-proving an instance of the law of Excluded Middle, P ∨ ¬P, in order to complete your proof. If you've forgotten how to do that, look back at your solution to Exercise 6.33. Alternatively, with the permission of your instructor, you may use **Taut Con** to justify an instance of Excluded Middle.*

8.26

P → (Q → P)

8.27

(P → (Q → R)) ↔ ((P ∧ Q) → R)

8.28

> P ↔ ¬P
>
> ⊥

8.29

> (P → Q) ↔ (¬P ∨ Q)

8.30

> ¬(P → Q) ↔ (P ∧ ¬Q)

*The following arguments are translations of those given in Exercises 8.3–8.9. (For simplicity we have assumed "the unicorn" refers to a specific unicorn named Charlie. This is less than ideal, but the best we can do without quantifiers.) Use Fitch to formalize the proofs you gave of their validity. You will need to use **Ana Con** to introduce ⊥ in two of your proofs.*

8.31

> (¬Mythical(c) → Mammal(c))
>   ∧ (Mythical(c) → ¬Mortal(c))
> (¬Mortal(c) ∨ Mammal(c)) → Horned(c)
> Horned(c) → Magical(c)
>
> Magical(c)

8.32

> Horned(c) → (Elusive(c)
>   ∧ Dangerous(c))
> (Elusive(c) ∨ Mythical(c)) → Rare(c)
> Mammal(c) → ¬Rare(c)
>
> Horned(c) → ¬Mammal(c)

8.33

> (Horned(c) → (Elusive(c) ∧ Magical(c)))
>   ∧ (¬Horned(c) → (¬Elusive(c)
>   ∧ ¬Magical(c)))
> ¬Horned(c) → ¬Mythical(c)
>
> Horned(c) ↔ (Magical(c) ∨ Mythical(c))

8.34

> (Tet(a) ∧ Large(a)) ∨ (Cube(a)
>   ∧ Small(a))
> ¬Small(b)
> (Tet(a) ∨ Cube(a)) → (Large(b)
>   ∨ Small(b))
> Tet(a) → Medium(b)
>
> Small(a) ∧ Large(b)

8.35

> ¬Cube(b) → Small(b)
> Small(c) → (Small(d) ∨ Small(e))
> Small(d) → ¬Small(c)
> Cube(b) → ¬Small(e)
>
> Small(c) → Small(b)

8.36

> SameRow(d, a) ∨ SameRow(d, b)
>   ∨ SameRow(d, c)
> SameRow(d, b) → (SameRow(d, a)
>   → ¬SameRow(d, c))
> SameRow(d, a) ↔ SameRow(d, c)
>
> SameRow(d, a) ↔ ¬SameRow(d, b)

8.37

Cube(a) ∨ Dodec(a) ∨ Tet(a)
Small(a) ∨ Medium(a) ∨ Large(a)
Medium(a) ↔ Dodec(a)
Tet(a) ↔ Large(a)

Cube(a) ↔ Small(a)

8.38 Use Fitch to give formal proofs of both (P ∧ Q) → P and the equivalent sentence ¬(P ∧ Q) ∨ P. (You will find the exercise files in Exercise 8.38.1 and Exercise 8.38.2.) Do you see why it is convenient to include → in FOL, rather than define it in terms of the Boolean connectives?

SECTION 8.3
Soundness and completeness

We have now introduced formal rules for all of our truth-functional connectives. Let's step back for a minute and ask two important questions about the formal system $\mathcal{F}$. The questions get at two desirable properties of a deductive system, which logicians call *soundness* and *completeness*. Don't be confused by the names, however. These uses of *sound* and *complete* are different from their use in the notions of a sound argument and a truth-functionally complete set of connectives.

Soundness

We intend our formal system $\mathcal{F}$ to be a correct system of deduction in the sense that any argument that can be proven valid in $\mathcal{F}$ should be genuinely valid. The first question that we will ask, then, is whether we have succeeded in this goal. Does the system $\mathcal{F}$ allow us to construct proofs only of genuinely valid arguments? This is known as the soundness question for the deductive system $\mathcal{F}$.

soundness of a deductive system

The answer to this question may seem obvious, but it deserves a closer look. After all, consider the rule of inference suggested in Exercise 7.32 on page 198. Probably, when you first looked at this rule, it seemed pretty reasonable, even though on closer inspection you realized it was not (or maybe you got the problem wrong). How can we be sure that something similar might not be the case for one of our official rules? Maybe there is a flaw in one of them but we just haven't thought long enough or hard enough to discover it.

Or maybe there are problems that go beyond the individual rules, something about the way the rules interact. Consider for example the following argument:

$$\begin{array}{|l} \neg(\text{Happy}(\text{carl}) \wedge \text{Happy}(\text{scruffy})) \\ \hline \neg\text{Happy}(\text{carl}) \end{array}$$

We know this argument isn't valid since it is clearly possible for the premise to be true and the conclusion false. But how do we know that the rules of proof we've introduced do not allow some very complicated and ingenious proof of the conclusion from the premise? After all, there is no way to examine all possible proofs and make sure there isn't one with this premise and conclusion: there are infinitely many proofs.

To answer our question, we need to make it more precise. We have seen that there is a certain vagueness in the notion of logical consequence. The concept of tautological consequence was introduced as a precise approximation of the informal notion. One way to make our question more precise is to ask whether the rules for the truth-functional connectives allow us to prove only arguments that are tautologically valid. This question leaves out the issue of whether the identity rules are legitimate, but we will address that question later.

$\mathcal{F}_T$

Let's introduce some new symbols to make it easier to express the claim we want to investigate. We will use $\mathcal{F}_T$ to refer to the portion of our deductive system that contains the introduction and elimination rules for $\neg, \vee, \wedge, \rightarrow, \leftrightarrow$, and $\perp$. You can think of the subscript T as standing for either "tautology" or "truth-functional." We will also write $P_1, \ldots, P_n \vdash_T S$ to indicate that there

$\vdash_T$

is a formal proof in $\mathcal{F}_T$ of S from premises $P_1, \ldots, P_n$. (The symbol $\vdash$ is commonly used in logic to indicate the provability of what's on the right from what's on the left. If you have trouble remembering what this symbol means, just think of it as a tiny Fitch bar.) We can now state our claim as follows.

Soundness of $\mathcal{F}_T$

Theorem (Soundness of $\mathcal{F}_T$) If $P_1, \ldots, P_n \vdash_T S$ then S is a tautological consequence of $P_1, \ldots, P_n$.

> **Proof:** Suppose that p is a proof constructed in the system $\mathcal{F}_T$. We will show that any sentence that occurs at any step in proof p is a tautological consequence of the assumptions in force at that step. This claim applies not just to sentences at the main level of p, but also to sentences appearing in subproofs, no matter how deeply nested. The assumptions in force at a step always include the main premises of the proof, but if we are dealing with a step inside some nested subproofs, they also include all the assumptions of these subproofs. The soundness theorem follows from our claim because if S appears

at the main level of p, then the only assumptions in force are the premises $P_1, \ldots, P_n$. So S is a tautological consequence of $P_1, \ldots, P_n$.

To prove this claim we will use proof by contradiction. Suppose that there is a step in p containing a sentence that is not a tautological consequence of the assumptions in force at that step. Call this an *invalid* step. The idea of our proof is to look at the first invalid step in p and show that none of the twelve rules of $\mathcal{F}_T$ could have justified that step. In other words, we will apply proof by cases to show that, no matter which rule of $\mathcal{F}_T$ was applied at the invalid step, we get a contradiction. (Actually, we will only look at three of the cases and leave the remaining rules as exercises.) This allows us to conclude that there can be no invalid steps in proofs in $\mathcal{F}_T$.

$\rightarrow$ **Intro**: Suppose the first invalid step derives the sentence $Q \rightarrow R$ from an application of $\rightarrow$ **Intro** to an earlier subproof with assumption Q and conclusion R.

Again let $A_1, \ldots, A_k$ be the assumptions in force at $Q \rightarrow R$. Note that the assumptions in force at R are $A_1, \ldots, A_k$ and Q. Since step R is earlier than the first invalid step, R must be a tautological consequence of $A_1, \ldots, A_k$ and Q.

Imagine constructing a joint truth table for the sentences $A_1, \ldots, A_k$, Q, $Q \rightarrow R$, and R. There must be a row h of this table in which $A_1, \ldots, A_k$ all come out true, but $Q \rightarrow R$ comes out false, by the assumption that this step is invalid. Since $Q \rightarrow R$ is false in this row, Q must be true and R must be false. But this contradicts our observation that R is a tautological consequence of $A_1, \ldots, A_k$ and Q.

→ **Elim**: Suppose the first invalid step derives the sentence R by an application of → **Elim** to sentences Q → R and Q appearing earlier in the proof. Let $A_1, \ldots, A_k$ be a list of all the assumptions in force at R. If this is an invalid step, R is not a tautological consequence of $A_1, \ldots, A_k$. But we will show that this leads us to a contradiction.

Since R is the first invalid step in p, we know that Q → R and Q are both valid steps, that is, they are tautological consequences of the assumptions in force at those steps. The crucial observation is that since $\mathcal{F}_T$ allows us to cite sentences only in the main proof or in subproofs whose assumptions are still in force, we know that the assumptions in force at steps Q → R and Q are also in force at R. Hence, the assumptions for these steps are among $A_1, \ldots, A_k$. An illustration may help. Suppose our proof takes the following form:

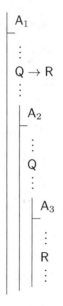

As should be clear, the restrictions on citing earlier steps guarantee that all the assumptions in force at the cited steps will still be in force at the step containing R. In the example shown, assumption A_1 is in force at the step containing Q → R, assumptions A_1 and A_2 are in force at the step containing Q, and assumptions A_1, A_2 and A_3 are in force at the step containing R.

Suppose, now, we construct a joint truth table for the sentences $A_1, \ldots, A_k$, Q, Q → R, and R. By the assumption that R is an invalid step, there must be a row h of this table in which $A_1, \ldots, A_k$ all come out true, but R comes out false. However, since Q and Q → R are tautological consequences of $A_1, \ldots, A_k$, both of these sentences are true in row h. But this contradicts the truth table for →.

⊥ **Elim**: Suppose the first invalid step derives the sentence Q from ⊥. Since this is the first invalid step, ⊥ must be a tautological consequence of the assumptions in force at ⊥. By the same considerations as in the first case, the assumptions in force at ⊥ are also in force at Q. Hence ⊥ is a tautological consequence of the assumptions $A_1, \ldots, A_k$ in force at Q. But the only way that this can be so is for $A_1, \ldots, A_k$ to be TT-contradictory. In other words, there are no rows in which all of $A_1, \ldots, A_k$ come out true. But then Q is vacuously a tautological consequence of $A_1, \ldots, A_k$.

Remaining rules: We have looked at three of the twelve rules. The remaining cases are similar to these, and so we leave them as exercises.

Once a contradiction is demonstrated in all twelve cases, we can conclude that our original assumption, that it is possible for a proof of $\mathcal{F}_T$ to contain an invalid step, must be false. This concludes the proof of soundness.

Having proven the Soundness Theorem, we can be absolutely certain that no matter how hard someone tries, no matter how ingenious they are, it will be impossible for them to produce a proof of ¬Happy(Carl) from the premise ¬(Happy(Carl) ∧ Happy(Scruffy)). Why? There is no such proof because the former is not a tautological consequence of the latter.

A corollary is a result which follows with little effort from an earlier theorem. We can state the following corollary, which simply applies the Soundness Theorem to cases in which there are no premises.

Corollary If ⊢$_T$ S, that is, if there is a proof of S in $\mathcal{F}_T$ with no premises, then S is a tautology.

The import of this corollary is illustrated in Figure 8.1. The corollary tells us that if a sentence is provable in $\mathcal{F}_T$, then it is a tautology. The Soundness Theorem assures us that this same relationship holds between provability in $\mathcal{F}_T$, with or without premises, and tautological consequence. Notice the question marks in Figure 8.1. So far we do not know whether there are any tautologies (or tautologically valid arguments) that are not provable in $\mathcal{F}_T$. This is the second question we need to address: the issue of *completeness*.

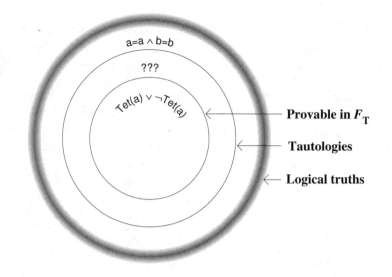

Figure 8.1: The soundness theorem for $\mathcal{F}_T$ tells us that only tautologies are provable (without premises) in $\mathcal{F}_T$.

sound deductive system vs. sound argument

Before turning to completeness, we should reiterate that our use of the term *soundness* in this section has little to do with our earlier use of the term *sound* to describe valid arguments with true premises. What *soundness* really means when applied to a deductive system is that the system only allows you to prove *valid* arguments. It would be more appropriate to call it a "validness" theorem, but that is not what it is traditionally called.

Completeness

completeness of a deductive system

Sometimes, in doing an exercise on formal proofs, you may have despaired of finding a proof, even though you could see that the conclusion followed from the premises. Our second question addresses this concern. Does our deductive system allow us to prove everything we should be able to prove?

Of course, this raises the question of what we "should" be able to prove, which again confronts us with the vagueness of the notion of logical consequence. But given the soundness theorem, we know that the most $\mathcal{F}_T$ will let us prove are tautological consequences. So we can state our question more precisely: Can we convince ourselves that given any premises $P_1, \ldots, P_n$ and any tautological consequence S of these premises, our deductive system $\mathcal{F}_T$ allows us to construct a proof of S from $P_1, \ldots, P_n$? Or could there be tautological consequences of some set of premises that are just plain out of the reach of the

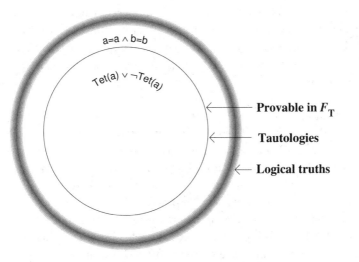

$a{=}a \wedge b{=}b$

$Tet(a) \vee \neg Tet(a)$

← **Provable in F_T**

← **Tautologies**

← **Logical truths**

Figure 8.2: Completeness and soundness of $\mathcal{F}_T$ tells us that all and only tautologies are provable (without premises) in $\mathcal{F}_T$.

deductive system $\mathcal{F}_T$? The next theorem assures us that this cannot happen.

Theorem (Completeness of $\mathcal{F}_T$) If a sentence S is a tautological consequence of $P_1, \ldots, P_n$, then $P_1, \ldots, P_n \vdash_T S$.

completeness of $\mathcal{F}_T$

The proof of this result is quite a bit more complicated than the proof of the Soundness Theorem, and requires material we have not yet introduced. Consequently, we will not be able to give the proof here, but will prove it in Chapter 17.

This result is called the *Completeness Theorem* because it tells us that the introduction and elimination rules are complete for the logic of the truth-functional connectives: anything that is a logical consequence simply in virtue of the meanings of the truth-functional connectives can be proven in $\mathcal{F}_T$. As illustrated in Figure 8.2, it assures us that *all* tautologies (and tautologically valid arguments) are provable in $\mathcal{F}_T$.

Notice, however, that the Soundness Theorem implies a kind of *incompleteness*, since it shows that the rules of $\mathcal{F}_T$ allow us to prove only *tauto*logical consequences of our premises. They do not allow us to prove any logical consequence of the premises that is not a tautological consequence of those premises. For example, it shows that there is no way to prove Dodec(c) from Dodec(b) $\wedge$ b $=$ c in $\mathcal{F}_T$, since the former is not a tautological consequence of the latter. To prove something like this, we will need the identity rules in addition to the rules for the truth-functional connectives. Similarly, to prove

soundness and incompleteness

uses of soundness and completeness

¬Larger(c, b) from Larger(b, c), we would need rules having to do with the predicate Larger. We will return to these issues in Chapter 19.

The Soundness and Completeness Theorems have practical uses that are worth keeping in mind. The Completeness Theorem gives us a method for showing that an argument *has* a proof without actually having to find such proof: just show that the conclusion is a tautological consequence of the premises. For example, it is obvious that $A \rightarrow (B \rightarrow A)$ is a tautology so by the Completeness Theorem we know it must have a proof. Similarly, the sentence $B \wedge \neg D$ is a tautological consequence of $\neg((A \wedge B) \rightarrow (C \vee D))$ so we know it must be possible to find a proof of the former from the latter.

The Soundness Theorem, on the other hand, gives us a method for telling that an argument does *not* have a proof in $\mathcal{F}_T$: show that the conclusion is not a tautological consequence of the premises. For example, $A \rightarrow (A \rightarrow B)$ is not a tautology, so it is impossible to construct a proof of it in $\mathcal{F}_T$, no matter how hard you try. Similarly, the sentence $B \wedge \neg D$ is a not tautological consequence of $\neg((A \vee B) \rightarrow (C \wedge D))$, so we know there is no proof of this in $\mathcal{F}_T$.

Recall our earlier discussion of the **Taut Con** routine in Fitch. This procedure checks to see whether a sentence is a tautological consequence of whatever sentences you cite in support. You can use the observations in the preceding paragraphs, along with **Taut Con**, to decide whether it is possible to give a proof using the rules of $\mathcal{F}_T$. If **Taut Con** says a particular sentence is a tautological consequence of the cited sentences, then you know it is possible to give a full proof of the sentence, even though you may not see exactly how the proof goes. On the other hand, if **Taut Con** says it is a not tautological consequence of the cited sentences, then there is no point in trying to find a proof in $\mathcal{F}_T$, for the simple reason that no such proof is possible.

Remember

Given an argument with premises $P_1, \ldots, P_n$ and conclusion S:

1. (Completeness of $\mathcal{F}_T$) If S is a tautological consequence of $P_1, \ldots, P_n$, then there is a proof of S from premises $P_1, \ldots, P_n$ using only the introduction and elimination rules for $\neg, \vee, \wedge, \rightarrow, \leftrightarrow,$ and $\bot$.

2. (Soundness of $\mathcal{F}_T$) If S is not a tautological consequence of $P_1, \ldots, P_n$, then there is no proof of S from premises $P_1, \ldots, P_n$ using only the rules for $\neg, \vee, \wedge, \rightarrow, \leftrightarrow,$ and $\bot$.

3. Which of these alternatives holds can be determined with the **Taut Con** procedure of Fitch.

Decide whether the following two arguments are provable in $\mathcal{F}_T$ without actually trying to find proofs. Do this by constructing a truth table in Boole to assess their tautological validity. Submit the table. Then explain clearly how you know the argument is or is not provable by applying the Soundness and Completeness results. Turn in your explanations to your instructor. (The explanations are more important than the tables, so don't forget the second part!)

8.39

A ∧ (B ∨ ¬A ∨ (C ∧ D))

E ∧ (D ∨ ¬(A ∧ (B ∨ D)))

A ∧ B

8.40

A ∧ (B ∨ ¬A ∨ (C ∧ D)) ∧ ¬(A ∧ D)

¬(E ∧ (D ∨ ¬(A ∧ (B ∨ D))))

In the proof of the Soundness Theorem, we only treated three of the twelve rules of $\mathcal{F}_T$. The next three problems ask you to treat some of the other rules.

8.41 Give the argument required for the ¬ **Elim** case of the Soundness proof. Your argument will be very similar to the one we gave for → **Elim**.

8.42 Give the argument required for the ¬ **Intro** case of the Soundness proof. Your argument will be similar to the one we gave for → **Intro**.

8.43 Give the argument required for the ∨ **Elim** case of the Soundness proof.

SECTION 8.4

Valid arguments: some review exercises

There is wisdom in the old saying "Don't lose sight of the forest for the trees." The forest in our case is an understanding of valid arguments. The trees are the various methods of proofs, formal and informal, and the notions of counterexample, tautology, and the like. The problems in this section are intended to remind you of the relationship between the forest and the trees, as well as to help you review the main ideas discussed so far.

Since you now know that our introduction and elimination rules suffice to prove any tautologically valid argument, you should feel free to use **Taut Con** in doing these exercises. In fact, you may use it in your formal proofs from now on, but with this important proviso: Make sure that you use it only in cases where the inference step is obvious and would go by without notice in an informal proof. For example, you may use it to introduce the law of excluded middle or to apply a DeMorgan equivalence. But you should still use rules like ∨ **Elim**, ¬ **Intro**, and → **Intro** when your informal proof would use proof by cases, proof by contradiction, or conditional proof. Any one-step proofs that consist of a single application of **Taut Con** will be counted as wrong!

Before doing these problems, go back and read the material in the **Remember** boxes, paying special attention to the strategy for evaluating arguments on page 173.

Remember

From this point on in the book, you may use **Taut Con** in formal proofs, but only to skip simple steps that would go unmentioned in an informal proof.

Exercises

*In the following exercises, you are given arguments in the blocks language. Evaluate each argument's validity. If it is valid, construct a formal proof to show this. If you need to use **Ana Con**, use it only to derive ⊥ from atomic sentences. If the argument is invalid, you should use Tarski's World to construct a counterexample world.*

8.44

| Adjoins(a, b) ∧ Adjoins(b, c)
| SameRow(a, c)
|
| a ≠ c

8.45

|
|
| ¬(Cube(b) ∧ b = c) ∨ Cube(c)

8.46

| Cube(a) ∨ (Cube(b) → Tet(c))
| Tet(c) → Small(c)
| (Cube(b) → Small(c)) → Small(b)
|
| ¬Cube(a) → Small(b)

8.47

| Small(a) ∧ (Medium(b) ∨ Large(c))
| Medium(b) → FrontOf(a, b)
| Large(c) → Tet(c)
|
| ¬Tet(c) → FrontOf(c, b)

8.48

| Small(a) ∧ (Medium(b) ∨ Large(c))
| Medium(b) → FrontOf(a, b)
| Large(c) → Tet(c)
|
| ¬Tet(c) → FrontOf(a, b)

8.49

| (Dodec(a) ∧ Dodec(b))
| → (SameCol(a, c) → Small(a))
| (¬SameCol(b, c) ∧ ¬Small(b))
| → (Dodec(b) ∧ ¬Small(a))
| SameCol(a, c) ∧ ¬SameCol(b, c)
|
| Dodec(a) → Small(b)

8.50
⤴

Cube(b) ↔ (Cube(a) ↔ Cube(c))

Dodec(b) → (Cube(a) ↔ ¬Cube(c))

8.51
⤴

Cube(b) ↔ (Cube(a) ↔ Cube(c))

Dodec(b) → a ≠ b

8.52
⤴

Cube(b) ↔ (Cube(a) ↔ Cube(c))

Dodec(b) → a ≠ c

8.53
⤴

Small(a) → Small(b)
Small(b) → (SameSize(b, c) → Small(c))
¬Small(a) → (Large(a) ∧ Large(c))

SameSize(b, c) → (Large(c) ∨ Small(c))

PART II

Quantifiers

CHAPTER 9

Introduction to Quantification

In English and other natural languages, basic sentences are made by combining noun phrases and verb phrases. The simplest noun phrases are names, like *Max* and *Claire*, which correspond to the constant symbols of FOL. More complex noun phrases are formed by combining common nouns with words known as *determiners*, such as *every, some, most, the, three,* and *no*, giving us noun phrases like *every cube, some man from Indiana, most children in the class, the dodecahedron in the corner, three blind mice,* and *no student of logic*.

determiners

Logicians call noun phrases of this sort *quantified expressions,* and sentences containing them *quantified sentences*. They are so called because they allow us to talk about quantities of things—every cube, most children, and so forth.

quantified sentences

The logical properties of quantified sentences are highly dependent on which determiner is used. Compare, for example, the following arguments:

> *Every rich actor is a good actor.*
> *Brad Pitt is a rich actor.*
>
> *Brad Pitt is a good actor.*

> *Many rich actors are good actors.*
> *Brad Pitt is a rich actor.*
>
> *Brad Pitt is a good actor.*

> *No rich actor is a good actor.*
> *Brad Pitt is a rich actor.*
>
> *Brad Pitt is a good actor.*

What a difference a determiner makes! The first of these arguments is obviously valid. The second is not logically valid, though the premises do make the conclusion at least plausible. The third argument is just plain dumb: in fact the premises logically imply the *negation* of the conclusion. You can hardly get a worse argument than that.

Quantification takes us out of the realm of truth-functional connectives. Notice that we can't determine the truth of quantified sentences by looking at the truth values of constituent sentences. Indeed, sentences like *Every rich actor is a good actor* and *No rich actor is a good actor* really aren't made up of simpler sentences, at least not in any obvious way. Their truth values are determined by the relationship between the collection of rich actors and the collection of good actors: by whether all of the former or none of the former are members of the latter.

hidden quantification

Various non-truth-functional constructions that we've already looked at are, in fact, hidden forms of quantification. Recall, for example, the sentence:

Max is home whenever Claire is at the library.

You saw in Exercise 7.31 that the truth of this sentence at a particular time is not a truth function of its parts at that time. The reason is that *whenever* is an implicit form of quantification, meaning *at every time that*. The sentence means something like:

Every time when Claire is at the library is a time when Max is at home.

Another example of a non-truth-functional connective that involves implicit quantification is *logically implies*. You can't tell whether P logically implies Q just by looking at the truth values of P and Q. This is because the claim means that *every logically possible circumstance that makes* P *true makes* Q *true*. The claim implicitly quantifies over possible circumstances.

quantifiers of FOL

While there are many forms of quantification in English, only two are built explicitly into FOL. This language has two quantifier symbols, ∀ and ∃, meaning *everything* and *something* respectively. This may seem like a very small number of quantifiers, but surprisingly many other forms of quantification can be defined from ∀ and ∃ using predicates and truth-functional connectives, including phrases like *every cube, three blind mice, no tall student,* and *whenever*. Some quantified expressions are outside the scope of FOL, however, including *most students, many cubes,* and *infinitely many prime numbers*. We'll discuss these issues in Chapter 14.

Section 9.1
Variables and atomic wffs

variables

Before we can show you how FOL's quantifier symbols work, we need to introduce a new type of term, called a *variable*. Variables are a kind of auxiliary symbol. In some ways they behave like individual constants, since they can appear in the list of arguments immediately following a predicate or function symbol. But in other ways they are very different from individual constants. In particular, their semantic function is not to refer to objects. Rather, they are placeholders that indicate relationships between quantifiers and the argument positions of various predicates. This will become clearer with our discussion of quantifiers.

First-order logic assumes an infinite list of variables so that we will never run out of them, no matter how complex a sentence may get. We take these variables to be any of the letters t, u, v, w, x, y, and z, with or without subscripts. So, for example, x, u_{23}, and z_6 are all variables in our dialect of FOL.

terms with variables

Adding variables expands the set of terms of the language. Up until now, individual constants (names) were the only basic terms. If the language contained function symbols, we built additional terms by repeated application of these function symbols. Now we have two types of basic terms, variables and individual constants, and can form complex terms by applying function symbols to either type of basic term. So in addition to the term father(max), we will have the term father(x), and in addition to $(0 + 1) \times 1$, we have $(y + z) \times z$.

These new terms allow us to produce expressions that look like atomic sentences, except that there are variables in place of some individual constants. For example, Home(x), Taller(max, y), and Taller(father(z), z) are such expressions. We call these expressions *atomic well-formed formulas*, or *atomic wffs*. They are not sentences, but will be used in conjunction with quantifier symbols to build sentences. The term *sentence* is reserved for well-formed formulas in which any variables that do occur are used together with quantifiers that *bind* them. We will give the definitions of sentence and bound variable in due course.

atomic wffs

Remember

1. The language FOL has an infinite number of variables, any of t, u, v, w, x, y, and z, with or without numerical subscripts.

2. Variables can occur in atomic wffs in any position normally occupied by a name.

The quantifier symbols: ∀, ∃

The quantifier symbols ∀ and ∃ let us express certain rudimentary claims about the number (or quantity) of things that satisfy some condition. Specifically, they allow us to say that *all* objects satisfy some condition, or that *at least one* object satisfies some condition. When used in conjunction with identity (=), they can also be used to express more complex numerical claims, for instance, that there are *exactly three things* that satisfy some condition.

Universal quantifier (∀)

everything, each thing, all things, anything

The symbol ∀ is used to express universal claims, those we express in English using quantified phrases like *everything, each thing, all things,* and *anything*. It is always used in connection with a variable, and so is said to be a variable binding operator. The combination ∀x is read "for every object x," or (somewhat misleadingly) "for all x."[1] If we wanted to translate the (rather unlikely) English sentence *Everything is at home* into first-order logic, we would use the FOL sentence

$$\forall x \, \text{Home}(x)$$

This says that every object x meets the following condition: x is at home. Or, to put it more naturally, it says that everything whatsoever is at home.

Of course, we rarely make such unconditional claims about absolutely everything. More common are restricted universal claims like *Every doctor is smart.* This sentence would be translated as

$$\forall x \, (\text{Doctor}(x) \rightarrow \text{Smart}(x))$$

This FOL sentence claims that given any object at all—call it x—if x is a doctor, then x is smart. To put it another way, the sentence says that if you pick anything at all, you'll find either that it is not a doctor or that it is smart (or perhaps both).

Existential quantifier (∃)

something, at least one thing, a, an

The symbol ∃ is used to express existential claims, those we express in English using such phrases as *something, at least one thing, a,* and *an*. It too is always

---

[1]We encourage students to use the first locution when reading formulas, at least for a few weeks, since we have seen many students who have misunderstood the basic function of variables as a result of reading them the second way.

used in connection with a variable, and so is a variable binding operator. The combination ∃x is read "for some object x," or (somewhat misleadingly) "for some x." If we wanted to translate the English sentence *Something is at home* into first-order logic, we would use the FOL sentence

$$\exists x \, \mathsf{Home}(x)$$

This says that some object x meets the following condition: x is at home.

While it is possible to make such claims, it is more common to assert that something of a particular kind meets some condition, say, *Some doctor is smart*. This sentence would be translated as

$$\exists x \, (\mathsf{Doctor}(x) \wedge \mathsf{Smart}(x))$$

This sentence claims that some object, call it x, meets the complex condition: x is both a doctor and smart. Or, more colloquially, it says that there is at least one smart doctor.

SECTION 9.3

Wffs and sentences

Notice that in some of the above examples, we formed *sentences* out of complex expressions that were not themselves sentences, expressions like

$$\mathsf{Doctor}(x) \wedge \mathsf{Smart}(x)$$

that contain variables with no associated quantifier. Thus, to systematically describe all the sentences of first-order logic, we first need to describe a larger class, the so-called well-formed formulas, or *wffs*.

We have already explained what an atomic wff is: any *n*-ary predicate followed by *n* terms, where terms can now contain either variables or individual constants. Using atomic wffs as our building blocks, we can construct more complicated wffs by repeatedly applying the following rules.

well-formed formula (wff)

1. If P is a wff, so is ¬P.
2. If $P_1, \ldots, P_n$ are wffs, so is $(P_1 \wedge \ldots \wedge P_n)$.
3. If $P_1, \ldots, P_n$ are wffs, so is $(P_1 \vee \ldots \vee P_n)$.
4. If P and Q are wffs, so is $(P \rightarrow Q)$.
5. If P and Q are wffs, so is $(P \leftrightarrow Q)$.
6. If P is a wff and ν is a variable (i.e., one of t, u, v, w, x, ...), then $\forall \nu \, P$ is a wff

7. If P is a wff and ν is a variable, then $\exists \nu$ P is a wff

By convention, we allow the outermost parentheses in a wff to be dropped, writing A $\wedge$ B rather than (A $\wedge$ B), but only if the parentheses enclose the whole wff.

The way these grammatical rules work is pretty straightforward. For example, starting from the atomic wffs Cube(x) and Small(x) we can apply rule 2 to get the wff:

$$(\text{Cube}(x) \wedge \text{Small}(x))$$

Similarly, starting from the atomic wff LeftOf(x, y) we can apply rule 7 to get the wff:

$$\exists y \, \text{LeftOf}(x, y)$$

The rules can also be applied to complex wffs, so from the above two wffs and rule 4 we can generate the following wff:

$$((\text{Cube}(x) \wedge \text{Small}(x)) \rightarrow \exists y \, \text{LeftOf}(x, y))$$

Some wffs have the important property that every occurrence of a variable has a quantifier associated with it, to tell us whether the variable is treated existentially or universally. When does a variable have an associated quantifier? We make this precise by defining the twin notions of *free* and *bound*

free variable
bound variable

variables.

1. Any variable in an atomic wff is *free* or *unbound*.
2. The free variables in P are also free in ¬P.
3. The free variables in $P_1, \ldots, P_n$ are all free in $(P_1 \wedge \ldots \wedge P_n)$.
4. The free variables in $P_1, \ldots, P_n$ are all free in $(P_1 \vee \ldots \vee P_n)$.
5. The free variables in P and Q are all free in $(P \rightarrow Q)$.
6. The free variables in P and Q are all free in $(P \leftrightarrow Q)$.
7. All of the free variables in P are free in $\forall \nu$ P, except for ν, and every occurrence of ν in P is said to be *bound*.
8. All of the free variables in P are free in $\exists \nu$ P, except for ν, and every occurrence of ν in P is said to be *bound*.

Look again at the wff

$$(\text{Cube}(x) \wedge \text{Small}(x)) \rightarrow \exists y \, \text{LeftOf}(x, y)$$

In this formula the variable y is bound by the quantifier $\exists y$. The variable x, on the other hand, has not been bound; it is *free*.

sentences

A *sentence* is a wff with no unbound (free) variables. None of the wffs displayed above are sentences, since they all contain free variables. To make a sentence out of the last of these, we can simply apply rule 6 to produce:

$$\forall x\,((\mathsf{Cube}(x) \land \mathsf{Small}(x)) \to \exists y\,\mathsf{LeftOf}(x, y))$$

Here all occurrences of the variable x have been bound by the quantifier ∀x. This wff is a sentence since it has no free variables. It claims that for every object x, if x is both a cube and small, then there is an object y such that x is to the left of y. Or, to put it more naturally, every small cube is to the left of something.

These rules can be applied over and over again to form more and more complex wffs. So, for example, repeated application of the first rule to the wff Home(max) will give us all of the following wffs:

$$\neg\mathsf{Home}(\mathsf{max})$$
$$\neg\neg\mathsf{Home}(\mathsf{max})$$
$$\neg\neg\neg\mathsf{Home}(\mathsf{max})$$
$$\vdots$$

Since none of these contains any variables, and so no free variables, they are all sentences. They claim, as you know, that Max is not home, that it is not the case that Max is not home, that it is not the case that it is not the case that Max is not home, and so forth.

We have said that a sentence is a wff with no free variables. However, it can sometimes be a bit tricky deciding whether a variable is free in a wff. For example, there are no free variables in the wff,

$$\exists x\,(\mathsf{Doctor}(x) \land \mathsf{Smart}(x))$$

However there *is* a free variable in the deceptively similar wff,

$$\exists x\,\mathsf{Doctor}(x) \land \mathsf{Smart}(x)$$

Here the last occurrence of the variable x is still free. We can see why this is the case by thinking about when the existential quantifier was applied in building up these two formulas. In the first one, the parentheses show that the quantifier was applied to the conjunction (Doctor(x) ∧ Smart(x)). As a consequence, all occurrences of x in the conjunction were bound by this quantifier. In contrast, the lack of parentheses show that in building up the second formula, the existential quantifier was applied to form ∃x Doctor(x), thus binding only the occurrence of x in Doctor(x). This formula was then conjoined with Smart(x), and so the latter's occurrence of x did not get bound.

scope of quantifier

Parentheses, as you can see from this example, make a big difference. They are the way you can tell what the *scope* of a quantifier is, that is, which variables fall under its influence and which don't. This example also shows that a variable can occur both free and bound in a formula. It is really an *occurrence* of a variable that is either free or bound, not the variable itself. In the formula

$$\exists x \, \text{Doctor}(x) \land \text{Smart}(x)$$

the last occurrence of x is free and the second is bound.

Remember

1. Complex wffs are built from atomic wffs by means of truth-functional connectives and quantifiers in accord with the rules on page 233.

2. When you append either quantifier $\forall x$ or $\exists x$ to a wff P, we say that the quantifier binds all the free occurrences of x in P.

3. A sentence is a wff in which no variables occur free (unbound).

Exercises

9.1 (Fixing some expressions) Open the sentence file **Bernstein's Sentences**. The expressions in this list are not quite well-formed sentences of our language, but they can all be made sentences by slight modification. Turn them into sentences *without adding or deleting any quantifier symbols*. With some of them, there is more than one way to make it a sentence. Use **Verify** to make sure your results are sentences and then submit the corrected file.

9.2 (Fixing some more expressions) Open the sentence file **Schonfinkel's Sentences**. Again, the expressions in this list are not well-formed sentences. Turn them into sentences, but this time, do it *only* by adding quantifier symbols or variables, or both. Do not add any parentheses. Use **Verify** to make sure your results are sentences and submit the corrected file.

9.3 (Making them true) Open **Bozo's Sentences** and **Leibniz's World**. Some of the expressions in this file are not wffs, some are wffs but not sentences, and one is a sentence but false. Read and assess each one. See if you can adjust each one to make it a true sentence with as little change as possible. Try to capture the intent of the original expression, if you can tell what that was (if not, don't worry). Use **Verify** to make sure your results are true sentences and then submit your file.

Semantics for the quantifiers

When we described the meanings of our various connectives, we told you how the truth value of a complex sentence, say ¬P, depends on the truth values of its constituents, in this case P. But we have not yet given you similar rules for determining the truth value of quantified sentences. The reason is simple: the expression to which we apply the quantifier in order to build a sentence is usually not itself a sentence. We could hardly tell you how the truth value of ∃x Cube(x) depends on the truth value of Cube(x), since this latter expression is not a sentence at all: it contains a free variable. Because of this, it is *neither* true *nor* false.

To describe when quantified sentences are true, we need to introduce the auxiliary notion of *satisfaction*. The basic idea is simple, and can be illustrated with a few examples. We say that an object satisfies the atomic wff Cube(x) if and only if the object is a cube. Similarly, we say an object satisfies the complex wff Cube(x) ∧ Small(x) if and only if it is both a cube and small. As a final example, an object satisfies the wff Cube(x) ∨ ¬Large(x) if and only if it is either a cube or not large (or both).

satisfaction

Different logic books treat satisfaction in somewhat different ways. We will describe the one that is built into the way that Tarski's World checks the truth of quantified sentences. Suppose S(x) is a wff containing x as its only free variable, and suppose we wanted to know whether a given object satisfies S(x). If this object has a name, say b, then form a new *sentence* S(b) by replacing all free occurrences of x by the individual constant b. If the new sentence S(b) is true, then the object satisfies the formula S(x); if the sentence is not true, then the object does not satisfy the formula.

This works fine as long as the given object has a name. But first-order logic doesn't require that every object have a name. How can we define satisfaction for objects that don't have names? It is for this reason that Tarski's World has, in addition to the individual constants a, b, c, d, e, and f, a further list $n_1, n_2, n_3, \ldots$ of individual constants. If we want to know whether a certain object without a name satisfies the formula S(x), we choose the first of these individual constants not in use, say n_6, temporarily name the given object with this symbol, and then check to see whether the sentence $S(n_6)$ is true. Thus, any small cube satisfies Cube(x) ∧ Small(x), because if we were to use n_6 as a name of such a small cube, then Cube(n_6) ∧ Small(n_6) would be a true sentence.

Once we have the notion of satisfaction, we can easily describe when a sentence of the form ∃x S(x) is true. It will be true if and only if there is at least

semantics of ∃

one object that satisfies the constituent wff $S(x)$. So $\exists x\,(\mathsf{Cube}(x) \wedge \mathsf{Small}(x))$ is true if there is at least one object that satisfies $\mathsf{Cube}(x) \wedge \mathsf{Small}(x)$, that is, if there is at least one small cube. Similarly, a sentence of the form $\forall x\,S(x)$ is true if and only if every object satisfies the constituent wff $S(x)$. Thus $\forall x\,(\mathsf{Cube}(x) \to \mathsf{Small}(x))$ is true if every object satisfies $\mathsf{Cube}(x) \to \mathsf{Small}(x)$, that is, if every object either isn't a cube or is small.

semantics of $\forall$

This approach to satisfaction is conceptually simpler than some. A more common approach is to avoid the introduction of new names by defining satisfaction for wffs with an arbitrary number of free variables. We will not need this for specifying the meaning of quantifiers, but we will need it in some of the more advanced sections. For this reason, we postpone the general discussion until later.

In giving the semantics for the quantifiers, we have implicitly assumed that there is a clear, non-empty collection of objects that we are talking about. For example, if we encounter the sentence $\forall x\,\mathsf{Cube}(x)$ in Tarski's World, we interpret this to be a claim about the objects depicted in the world window. We do not judge it to be false just because the moon is not a cube. Similarly, if we encounter the sentence $\forall x\,(\mathsf{Even}(x^2) \to \mathsf{Even}(x))$, we interpret this as a claim about the natural numbers. It is true because every object in the domain we are talking about, natural numbers, satisfies the constituent wff.

domain of discourse

In general, sentences containing quantifiers are only true or false relative to some *domain of discourse* or *domain of quantification*. Sometimes the intended domain contains all objects there are. Usually, though, the intended domain is a much more restricted collection of things, say the people in the room, or some particular set of physical objects, or some collection of numbers. In this book, we will specify the domain explicitly unless it is clear from context what domain is intended. Also, in FOL we always assume that the domain of discourse contains at least one object and that every individual constant in the language stands for an object in that domain. (We could give up these idealizations, but it would complicate things considerably without much gain in realism.)

In the above discussion, we introduced some notation that we will use a lot. Just as we often used P or Q to stand for a possibly complex sentence of propositional logic, so too we will often use $S(x)$ or $P(y)$ to stand for a possibly complex wff of first-order logic. Thus, $P(y)$ may stand for a wff like:

$$\exists x\,(\mathsf{LeftOf}(x, y) \vee \mathsf{RightOf}(x, y))$$

When we then write, say, $P(b)$, this stands for the result of replacing all the free occurrences of y by the individual constant b:

$$\exists x\,(\mathsf{LeftOf}(x, b) \vee \mathsf{RightOf}(x, b))$$

It is important to understand that the variable displayed in parentheses only stands for the free occurrences of the variable. For example, if $S(x)$ is used to refer to the wff we looked at earlier, where x appeared both free and bound:

$$\exists x\, \mathsf{Doctor}(x) \wedge \mathsf{Smart}(x)$$

then $S(c)$ would indicate the following substitution instance, where c is substituted for the *free* occurrence of x:

$$\exists x\, \mathsf{Doctor}(x) \wedge \mathsf{Smart}(c)$$

Remember

○ Quantified sentences make claims about some non-empty intended domain of discourse.

○ A sentence of the form $\forall x\, S(x)$ is true if and only if the wff $S(x)$ is satisfied by every object in the domain of discourse.

○ A sentence of the form $\exists x\, S(x)$ is true if and only if the wff $S(x)$ is satisfied by some object in the domain of discourse.

Game rules for the quantifiers

The game rules for the quantifiers are more interesting than those for the truth-functional connectives. With the connectives, moves in the game involved choosing sentences that are parts of the sentence to which you are committed. With the quantifier rules, however, moves consist in choosing objects, not sentences.

Suppose, for example, that you are committed to the truth of $\exists x\, P(x)$. This means that you are committed to there being an object that satisfies $P(x)$. Tarski's World will ask you to live up to this commitment by finding such an object. On the other hand, if you are committed to the falsity of $\exists x\, P(x)$, then you are committed to there being no object that satisfies $P(x)$. In which case, Tarski's World gets to choose: it tries to find an object that does satisfy $P(x)$, thus contradicting your commitment.

game rules for $\exists$

The rules for $\forall$ are just the opposite. If you are committed to the truth of $\forall x\, P(x)$, then you are committed to every object satisfying $P(x)$. Tarski's World will try to find an object not satisfying $P(x)$, thus contradicting your commitment. If, however, you are committed to the falsity of $\forall x\, P(x)$, then you

game rules for $\forall$

Table 9.1: Summary of the game rules

Form	Your commitment	Player to move	Goal
P ∨ Q	TRUE	you	Choose one of P, Q that is true.
	FALSE	Tarski's World	
P ∧ Q	TRUE	Tarski's World	Choose one of P, Q that is false.
	FALSE	you	
∃x P(x)	TRUE	you	Choose some **b** that satisfies the wff P(x).
	FALSE	Tarski's World	
∀x P(x)	TRUE	Tarski's World	Choose some **b** that does not satisfy P(x).
	FALSE	you	
¬P	either	—	Replace ¬P by P and switch commitment.
P → Q	either	—	Replace P → Q by ¬P ∨ Q and keep commitment.
P ↔ Q	either	—	Replace P ↔ Q by (P → Q) ∧ (Q → P) and keep commitment.

are committed to there being some object that does not satisfy P(x). Tarski's World will ask you to live up to your commitment by finding such an object.

We have now seen all the game rules. We summarize them in Table 9.1.

You try it
. . . .

▶   1. Open the files **Game World** and **Game Sentences**. Go through each sentence and see if you can tell whether it is true or false. Check your evaluation.

2. Whether you evaluated the sentence correctly or not, play the game twice ◀
   for each sentence, first committed to TRUE, then committed to FALSE.
   Make sure you understand how the game works at each step.

3. There is nothing to save except your understanding of the game. ◀
   . . . . *Congratulations*

Exercises

9.4 If you skipped the **You try it** section, go back and do it now. This is an easy but important
   exercise that will familiarize you with the game rules for the quantifiers. There is nothing you
   need to turn in or submit.

9.5 (Evaluating sentences in a world) Open Peirce's World and Peirce's Sentences. There are 30
   sentences in this file. Work through them, assessing their truth and playing the game when
   necessary. Make sure you understand why they have the truth values they do. (You may need to
   switch to the 2-D view for some of the sentences.) After you understand each of the sentences,
   go back and make the false ones true by adding or deleting a negation sign. Submit the file
   when the sentences all come out true in Peirce's World.

9.6 (Evaluating sentences in a world) Open Leibniz's World and Zorn's Sentences. The sentences
   in this file contain both quantifiers and the identity symbol. Work through them, assessing
   their truth and playing the game when necessary. After you're sure you understand why the
   sentences get the values they do, modify the false ones to make them true. But this time you
   can make any change you want *except* adding or deleting a negation sign.

9.7 In English we sometimes say things like *Every Jason is envied*, meaning that everyone named
   "Jason" is envied. For this reason, students are sometimes tempted to write expressions like
   $\forall b\, \mathsf{Cube}(b)$ to mean something like *Everything named **b** is a cube*. Explain why this is not well
   formed according to the grammatical rules on page 233.

SECTION 9.5

The four Aristotelian forms

Long before FOL was codified, Aristotle studied the kinds of reasoning associated with quantified noun phrases like *Every man*, *No man*, and *Some man*,
expressions we would translate using our quantifier symbols. The four main
sentence forms treated in Aristotle's logic were the following.

Aristotelian forms

All P's are Q's
Some P's are Q's
No P's are Q's
Some P's are not Q's

We will begin by looking at the first two of these forms, which we have already discussed to a certain extent. These forms are translated as follows. The form *All P's are Q's* is translated as:

$$\forall x\,(P(x) \rightarrow Q(x))$$

whereas the form *Some P's are Q's* is translated as:

$$\exists x\,(P(x) \wedge Q(x))$$

Beginning students are often tempted to translate the latter more like the former, namely as:

$$\exists x\,(P(x) \rightarrow Q(x))$$

This is in fact an extremely unnatural sentence of first-order logic. It is meaningful, but it doesn't mean what you might think. It is true just in case there is an object which is either not a P or else is a Q, which is something quite different than saying that some P's are Q's. We can quickly illustrate this difference with Tarski's World.

You try it
....

▶ 1. Use Tarski's World to build a world containing a single large cube and nothing else.

▶ 2. Write the sentence $\exists x\,(\mathsf{Cube}(x) \rightarrow \mathsf{Large}(x))$ in the sentence window. Check to see that the sentence is true in your world.

▶ 3. Now change the large cube into a small tetrahedron and check to see if the sentence is true or false. Do you understand why the sentence is still true? Even if you do, play the game twice, once committed to its being false, once to its being true.

▶ 4. Add a second sentence that correctly expresses the claim that there is a large cube. Make sure it is false in the current world but becomes true when you add a large cube. Save your two sentences as Sentences Quantifier 1.

....*Congratulations*

The other two Aristotelian forms are translated similarly, but using a negation. In particular *No P's are Q's* is translated

$$\forall x \, (P(x) \rightarrow \neg Q(x))$$

Many students, and one of the authors, finds it more natural to use the following, logically equivalent sentence:

$$\neg \exists x \, (P(x) \wedge Q(x))$$

Both of these assert that nothing that is a P is also a Q.

The last of the four forms, *Some P's are not Q's*, is translated by

$$\exists x \, (P(x) \wedge \neg Q(x))$$

which says there is something that is a P but not a Q.

The four Aristotelian forms are the very simplest sorts of sentences built using quantifiers. Since many of the more complicated forms we talk about later are elaborations of these, you should learn them well.

> **Remember**
>
> The four Aristotelian forms are translated as follows:
>
> | | |
> |---:|:---|
> | *All P's are Q's.* | $\forall x \, (P(x) \rightarrow Q(x))$ |
> | *Some P's are Q's.* | $\exists x \, (P(x) \wedge Q(x))$ |
> | *No P's are Q's.* | $\forall x \, (P(x) \rightarrow \neg Q(x))$ |
> | *Some P's are not Q's.* | $\exists x \, (P(x) \wedge \neg Q(x))$ |

Exercises

9.8 If you skipped the **You try it** section, go back and do it now. Submit the file Sentences Quantifier 1.

9.9 (Building a world) Open Aristotle's Sentences. Each of these sentences is of one of the four Aristotelian forms. Build a single world where all the sentences in the file are true. As you work through the sentences, you will find yourself successively modifying the world. Whenever you make a change in the world, you had better go back and check that you haven't made any of the earlier sentences false. Then, when you are finished, verify that all the sentences are really true and submit your world.

9.10   (Common translation mistakes) Open **Edgar's Sentences** and evaluate them in **Edgar's World**. Make sure you understand why each of them has the truth value it does. Play the game if any of the evaluations surprise you. Which of these sentences would be a good translation of *There is a tetrahedron that is large?* (Clearly this English sentence is false in **Edgar's World**, since there are no tetrahedra at all.) Which sentence would be a good translation of *There is a cube between **a** and **b**?* Which would be a good translation of *There is a large dodecahedron?* Express in clear English the claim made by each sentence in the file and turn in your answers to your instructor.

9.11   (Common mistakes, part 2) Open **Allan's Sentences**. In this file, sentences 1 and 4 are the correct translations of *Some dodecahedron is large* and *All tetrahedra are small*, respectively. Let's investigate the logical relations between these and sentences 2 and 3.

1. Construct a world in which sentences 2 and 4 are true, but sentences 1 and 3 are false. Save it as **World 9.11.1**. This shows that sentence 1 is not a consequence of 2, and sentence 3 is not a consequence of 4.
2. Can you construct a world in which sentence 3 is true and sentence 4 is false? If so, do so and save it as **World 9.11.2**. If not, explain why you can't and what this shows.
3. Can you construct a world in which sentence 1 is true and sentence 2 is false? If so, do so and save it as **World 9.11.3**. If not, explain why not.

Submit any world files you constructed and turn in any explanations to your instructor.

9.12   (Describing a world) Open **Reichenbach's World 1**. Start a new sentence file where you will describe some features of this world using sentences of the simple Aristotelian forms. Check each of your sentences to see that it is indeed a sentence and that it is true in this world.

1. Use your first sentence to describe the size of all the tetrahedra.
2. Use your second sentence to describe the size of all the cubes.
3. Use your third sentence to express the truism that every dodecahedron is either small, medium, or large.
4. Notice that some dodecahedron is large. Express this fact.
5. Observe that some dodecahedron is not large. Express this.
6. Notice that some dodecahedron is small. Express this fact.
7. Observe that some dodecahedron is not small. Express this.
8. Notice that some dodecahedron is neither large nor small. Express this.
9. Express the observation that no tetrahedron is large.
10. Express the fact that no cube is large.

Now change the sizes of the objects in the following way: make one of the cubes large, one of the tetrahedra medium, and all the dodecahedra small. With these changes, the following should come out false: 1, 2, 4, 7, 8, and 10. If not, then you have made an error in describing the original world. Can you figure out what it is? Try making other changes and see if your sentences have the expected truth values. Submit your sentence file.

9.13 Assume we are working in an extension of the first-order language of arithmetic with the additional predicates Even(x) and Prime(x), meaning, respectively, "x is an even number" and "x is a prime number." Create a sentence file in which you express the following claims:

1. Every even number is prime.
2. No even number is prime.
3. Some prime is even.
4. Some prime is not even.
5. Every prime is either odd or equal to 2.

[Note that you should assume your domain of discourse consists of the natural numbers, so there is no need for a predicate Number(x). Also, remember that 2 is not a constant in the language, so must be expressed using + and 1.]

9.14 (Name that object) Open Maigret's World and Maigret's Sentences. The goal is to try to figure out which objects have names, and what they are. You should be able to figure this out from the sentences, all of which are true. Once you have come to your conclusion, assign the six names to objects in the world in such a way that all the sentences do indeed evaluate as true. Submit your modified world.

Translating complex noun phrases

The first thing you have to learn in order to translate quantified English expressions is how to treat complex noun phrases, expressions like "a boy living in Omaha" or "every girl living in Duluth." In this section we will learn how to do this. We concentrate first on the former sort of noun phrase, whose most natural translation involves an existential quantifier. Typically, these will be noun phrases starting with one of the determiners *some*, *a*, and *an*, including noun phrases like *something*. These are called existential noun phrases, since they assert the existence of something or other. Of course two of our four Aristotelian forms involve existential noun phrases, so we know the general pattern: existential noun phrases are usually translated using ∃, frequently together with ∧.

*existential
noun phrases*

Let's look at a simple example. Suppose we wanted to translate the sentence *A small, happy dog is at home*. This sentence claims that there is an object which is simultaneously a small, happy dog, and at home. We would translate it as

$$\exists x\,[(\text{Small}(x) \wedge \text{Happy}(x) \wedge \text{Dog}(x)) \wedge \text{Home}(x)]$$

We have put parentheses around the first three predicates to indicate that they were all part of the translation of the subject noun phrase. But this is not really necessary.

universal
noun phrases

Universal noun phrases are those that begin with determiners like *every*, *each*, and *all*. These are usually translated with the universal quantifier. Sometimes noun phrases beginning with *no* and with *any* are also translated with the universal quantifier. Two of our four Aristotelian forms involve universal noun phrases, so we also know the general pattern here: universal noun phrases are usually translated using $\forall$, frequently together with $\rightarrow$.

Let's consider the sentence *Every small dog that is at home is happy*. This claims that everything with a complex property, that of being a small dog at home, has another property, that of being happy. This suggests that the overall sentence has the form *All A's are B's*. But in this case, to express the complex property that fills the "A" position, we will use a conjunction. Thus it would be translated as

$$\forall x\,[(\mathsf{Small}(x) \wedge \mathsf{Dog}(x) \wedge \mathsf{Home}(x)) \rightarrow \mathsf{Happy}(x)]$$

In this case, the parentheses are not optional. Without them the expression would not be well formed.

noun phrases in
non-subject positions

In both of the above examples, the complex noun phrase appeared at the beginning of the English sentence, much like the quantifier in the FOL translation. Often, however, the English noun phrase will appear somewhere else in the sentence, say as the direct object, and in these cases the FOL translation may be ordered very differently from the English sentence. For example, the sentence *Max owns a small, happy dog* might be translated:

$$\exists x\,[(\mathsf{Small}(x) \wedge \mathsf{Happy}(x) \wedge \mathsf{Dog}(x)) \wedge \mathsf{Owns}(\mathsf{max}, x)]$$

which says there is a small, happy dog that Max owns. Similarly, the English sentence *Max owns every small, happy dog* would end up turned around like this:

$$\forall x\,[(\mathsf{Small}(x) \wedge \mathsf{Happy}(x) \wedge \mathsf{Dog}(x)) \rightarrow \mathsf{Owns}(\mathsf{max}, x)]$$

You will be given lots of practice translating complex noun phrases in the exercises that follow. First, however, we discuss some troublesome cases.

> **Remember**
>
> 1. Translations of complex quantified noun phrases frequently employ conjunctions of atomic predicates.
>
> 2. The order of an English sentence may not correspond to the order of its FOL translation.

Conversational implicature and quantification

You will find that translating quantified phrases is not difficult, as long as quantifiers are not "nested" inside one another. There are, however, a couple of points that sometimes present stumbling blocks.

One thing that often puzzles students has to do with the truth value of sentences of the form

$$\forall x\,(P(x) \rightarrow Q(x))$$

in worlds where there are no objects satisfying $P(x)$. If you think about it, you will see that in such a world the sentence is true simply because there are no objects that satisfy the antecedent. This is called a *vacuously* true generalization.

vacuously true generalizations

Consider, for example, the sentence

$$\forall y(\mathsf{Tet}(y) \rightarrow \mathsf{Small}(y))$$

which asserts that every tetrahedron is small. But imagine that it has been asserted about a world in which there are no tetrahedra. In such a world the sentence is true simply because there are no tetrahedra at all, small, medium, or large. Consequently, it is impossible to find a counterexample, a tetrahedron which is not small.

What strikes students as especially odd are examples like

$$\forall y(\mathsf{Tet}(y) \rightarrow \mathsf{Cube}(y))$$

On the face of it, such a sentence looks contradictory. But we see that if it is asserted about a world in which there are no tetrahedra, then it is in fact *true*. But that is the only way it can be true: if there are no tetrahedra. In other words, the only way this sentence can be true is if it is vacuously true. Let's call generalizations with this property *inherently vacuous*. Thus, a sentence of the form $\forall x\,(P(x) \rightarrow Q(x))$ is inherently vacuous if the only worlds in which it is true are those in which $\forall x\,\neg P(x)$ is true.

inherently vacuous generalizations

You try it
. . . .

▶ 1. Open Dodgson's Sentences. Note that the first sentence says that every tetrahedron is large.

▶ 2. Open Peano's World. Sentence 1 is clearly false in this world, since the small tetrahedron is a counterexample to the universal claim. What this means is that if you play the game committed to the falsity of this claim, then when Tarski's World asks you to pick an object you will be able to pick the small tetrahedron and win the game. Try this.

▶ 3. Delete this counterexample and verify that sentence 1 is now true.

▶ 4. Now open Peirce's World. Verify that sentence 1 is again false, this time because there are three counterexamples. (Now if you play the game committed to the falsity of the sentence, you will have three different winning moves when asked to pick an object: you can pick any of the small tetrahedra and win.)

▶ 5. Delete all three counterexamples, and evaluate the claim. Is the result what you expected? The generalization is true, because there are no counterexamples to it. It is what we called a vacuously true generalization, since there are no objects that satisfy the antecedent. That is, there are no tetrahedra at all, small, medium, or large. Confirm that all of sentences 1–3 are vacuously true in the current world.

▶ 6. Two more vacuously true sentences are given in sentences 4 and 5. However, these sentences are different in another respect. Each of the first three sentences could have been non-vacuously true in a world, but these latter two can only be true in worlds containing no tetrahedra. That is, they are inherently vacuous.

▶ 7. Add a sixth generalization to the file that is vacuously true in Peirce's World but non-vacuously true in Peano's World. (In both cases, make sure you use the unmodified worlds.) Save your new sentence file as Sentences Vacuous 1.

. . . . *Congratulations*

In everyday conversation, it is rare to encounter a vacuously true generalization, let alone an inherently vacuous generalization. When we do find either of these, we feel that the speaker has misled us. For example, suppose a professor claims "Every freshman who took the class got an A," when in fact no freshman took her class. Here we wouldn't say that she lied, but we would certainly say that she misled us. Her statement typically carries the conversational implicature that there were freshmen in the class. If there were no freshmen, then that's what she would have said if she were being forthright. *Inherently* vacuous claims are true only when they are misleading, so they strike us as intuitively false.

conversational implicature

Another source of confusion concerns the relationship between the following two Aristotelian sentences:

> *Some P's are Q's*
> *All P's are Q's*

Students often have the intuition that the first should contradict the second. After all, why would you say that *some* student got an A if *every* student got an A? If this intuition were right, then the correct translation of *Some P's are Q's* would not be what we have suggested above, but rather

$$\exists x\,(P(x) \land Q(x)) \land \neg\forall x\,(P(x) \to Q(x))$$

It is easy to see, however, that the second conjunct of this sentence does not represent part of the meaning of the sentence. It is, rather, another example of a conversational implicature. It makes perfectly good sense to say "Some student got an A on the exam. In fact, every student did." If the proposed conjunction were the right form of translation, this amplification would be contradictory.

Remember

1. *All P's are Q's* does not imply, though it may conversationally suggest, that there are some P's.

2. *Some P's are Q's* does not imply, though it may conversationally suggest, that not all P's are Q's.

Exercises

9.15 If you skipped the **You try it** section, go back and do it now. Submit the file Sentences
Vacuous 1.

9.16 (Translating existential noun phrases) Start a new sentence file and enter translations of the
following English sentences. Each will use the symbol ∃ exactly once. None will use the symbol
∀. As you go, check that your entries are well-formed sentences. By the way, you will find that
many of these English sentences are translated using the same first-order sentence.

1. *Something is large.*
2. *Something is a cube.*
3. *Something is a large cube.*
4. *Some cube is large.*
5. *Some large cube is to the left of **b**.*
6. *A large cube is to the left of **b**.*
7. ***b** has a large cube to its left.*
8. ***b** is to the right of a large cube.* [Hint: This translation should be almost the same as
   the last, but it should contain the predicate symbol RightOf.]
9. *Something to the left of **b** is in back of **c**.*
10. *A large cube to the left of **b** is in back of **c**.*
11. *Some large cube is to the left of **b** and in back of **c**.*
12. *Some dodecahedron is not large.*
13. *Something is not a large dodecahedron.*
14. *It's not the case that something is a large dodecahedron.*
15. ***b** is not to the left of a cube.* [Warning: This sentence is ambiguous. Can you think of
    two importantly different translations? One starts with ∃, the other starts with ¬. Use
    the second of these for your translation, since this is the most natural reading of the
    English sentence.]

Now let's check the translations against a world. Open Montague's World.

○ Notice that all the English sentences above are true in this world. Check that all your
translations are also true. If not, you have made a mistake. Can you figure out what is
wrong with your translation?

○ Move the large cube to the back right corner of the grid. Observe that English sentences
5, 6, 7, 8, 10, 11, and 15 are now false, while the rest remain true. Check that the same
holds of your translations. If not, you have made a mistake. Figure out what is wrong
with your translation and fix it.

○ Now make the large cube small. The English sentences 1, 3, 4, 5, 6, 7, 8, 10, 11, and
15 are false in the modified world, the rest are true. Again, check that your translations
have the same truth values. If not, figure out what is wrong.

○ Finally, move *c* straight back to the back row, and make the dodecahedron large. All the
English sentences other than 1, 2, and 13 are false. Check that the same holds for your
translations. If not, figure out where you have gone wrong and fix them.

When you are satisfied that your translations are correct, submit your sentence file.

9.17 (Translating universal noun phrases) Start a new sentence file, and enter translations of the following sentences. This time each translation will contain exactly one ∀ and no ∃.

1. *All cubes are small.*
2. *Each small cube is to the right of **a**.*
3. ***a** is to the left of every dodecahedron.*
4. *Every medium tetrahedron is in front of **b**.*
5. *Each cube is either in front of **b** or in back of **a**.*
6. *Every cube is to the right of **a** and to the left of **b**.*
7. *Everything between **a** and **b** is a cube.*
8. *Everything smaller than **a** is a cube.*
9. *All dodecahedra are not small.* [Note: Most people find this sentence ambiguous. Can you find both readings? One starts with ∀, the other with ¬. Use the former, the one that means all the dodecahedra are either medium or large.]
10. *No dodecahedron is small.*
11. ***a** does not adjoin everything.* [Note: This sentence is ambiguous. We want you to interpret it as a denial of the claim that a adjoins everything.]
12. ***a** does not adjoin anything.* [Note: These last two sentences mean different things, though they can both be translated using ∀, ¬, and Adjoins.]
13. ***a** is not to the right of any cube.*
14. (⋆) *If something is a cube, then it is not in the same column as either **a** or **b**.* [Warning: While this sentence contains the noun phrase "something," it is actually making a universal claim, and so should be translated with ∀. You might first try to paraphrase it using the English phrase "every cube."]
15. (⋆) *Something is a cube if and only if it is not in the same column as either **a** or **b**.*

Now let's check the translations in some worlds.

- Open Claire's World. Check to see that all the English sentences are true in this world, then make sure the same holds of your translations. If you have made any mistakes, fix them.

- Adjust Claire's World by moving *a* directly in front of *c*. With this change, the English sentences 2, 6, and 12–15 are false, while the rest are true. Make sure that the same holds of your translations. If not, try to figure out what is wrong and fix it.

- Next, open Wittgenstein's World. Observe that the English sentences 2, 3, 7, 8, 11, 12, and 13 are true, but the rest are false. Check that the same holds for your translations. If not, try to fix them.

- Finally, open Venn's World. English sentences 2, 4, 7, and 11–14 are true; does the same hold for your translations?

When you are satisfied that your translations are correct, submit your sentence file.

9.18 (Translation) Open Leibniz's World. This time, we will translate some sentences while looking at the world they are meant to describe.

⚬ Start a new sentence file, and enter translations of the following sentences. Each of the English sentences is true in this world. As you go, check to make sure that your translation is indeed a true sentence.

1. *There are no medium-sized cubes.*
2. *Nothing is in front of* ***b***.
3. *Every cube is either in front of or in back of* ***e***.
4. *No cube is between* ***a*** *and* ***c***.
5. *Everything is in the same column as* ***a***, ***b***, *or* ***c***.

⚬ Now let's change the world so that none of the English sentences is true. We can do this as follows. First change *b* into a medium cube. Next, delete the leftmost tetrahedron and move *b* to exactly the position just vacated by the late tetrahedron. Finally, add a small cube to the world, locating it exactly where *b* used to sit. If your answers to 1–5 are correct, all of the translations should now be false. Verify that they are.

⚬ Make various changes to the world, so that some of the English sentences come out true and some come out false. Then check to see that the truth values of your translations track the truth values of the English sentences.

9.19 Start a new sentence file and translate the following into FOL using the symbols from Table 1.2, page 30. Note that all of your translations will involve quantifiers, though this may not be obvious from the English sentences. (Some of your translations will also require the identity predicate.)

1. *People are not pets.*
2. *Pets are not people.*
3. *Scruffy was not fed at either 2:00 or 2:05.* [Remember, **Fed** is a ternary predicate.]
4. *Claire fed Folly at some time between 2:00 and 3:00.*
5. *Claire gave a pet to Max at 2:00.*
6. *Claire had only hungry pets at 2:00.*
7. *Of all the students, only Claire was angry at 3:00.*
8. *No one fed Folly at 2:00.*
9. *If someone fed Pris at 2:00, they were angry.*
10. *Whoever owned Pris at 2:00 was angry five minutes later.*

9.20 Using Table 1.2, page 30, translate the following into colloquial English.
1. $\forall t \, \neg \mathsf{Gave}(\mathsf{claire}, \mathsf{folly}, \mathsf{max}, t)$
2. $\forall x \, (\mathsf{Pet}(x) \rightarrow \mathsf{Hungry}(x, 2{:}00))$

3. ∀y (Person(y) → ¬Owned(y, pris, 2:00))
4. ¬∃x (Angry(x, 2:00) ∧ Student(x) ∧ Fed(x, carl, 2:00))
5. ∀x ((Pet(x) ∧ Owned(max, x, 2:00)) → Gave(max, x, claire, 2:00))

9.21 Translate the following into FOL, introducing names, predicates, and function symbols as
✎** needed. As usual, explain your predicates and function symbols, and any shortcomings in
your translations. If you assume a particular domain of discourse, mention that as well.

1. *Only the brave know how to forgive.*
2. *No man is an island.*
3. *I care for nobody, not I,*
   If no one cares for me.
4. *Every nation has the government it deserves.*
5. *There are no certainties, save logic.*
6. *Misery (that is, a miserable person) loves company.*
7. *All that glitters is not gold.*
8. *There was a jolly miller once*
   Lived on the River Dee.
9. *If you praise everybody, you praise nobody.*
10. *Something is rotten in the state of Denmark.*

Quantifiers and function symbols

When we first introduced function symbols in Chapter 1, we presented them
as a way to form complex names from other names. Thus father(father(max))
refers to Max's father's father, and $(1 + (1 + 1))$ refers to the number 3. Now
that we have variables and quantifiers, function symbols become much more
useful than they were before. For example, they allow us to express in a very
compact way things like:

$$∀x \text{ Nicer}(\text{father}(\text{father}(x)), \text{father}(x))$$

This sentence says that everyone's paternal grandfather is nicer than their
father, a false belief held by many children.

Notice that even if our language had individual constants naming everyone's father (and their fathers' fathers and so on), we could not express the
above claim in a single sentence without using the function symbol father.
True, if we added the binary predicate FatherOf, we could get the same point
across, but the sentence would be considerably more complex. It would require

three universal quantifiers, something we haven't talked about yet:

$$\forall x \, \forall y \, \forall z \, ((\mathsf{FatherOf}(x, y) \land \mathsf{FatherOf}(y, z)) \to \mathsf{Nicer}(x, y))$$

In our informal mathematical examples, we have in fact been using function symbols along with variables throughout the book. For example in Chapter 8, we proved the conditional:

$$\mathsf{Even}(n^2) \to \mathsf{Even}(n)$$

This sentence is only partly in our official language of first-order arithmetic. Had we had quantifiers at the time, we could have expressed the intended claim using a universal quantifier and the binary function symbol $\times$:

$$\forall y \, (\mathsf{Even}(y \times y) \to \mathsf{Even}(y))$$

The blocks language does not have function symbols, though we could have introduced some. Remember the four function symbols, fm, bm, lm and rm, that we discussed in Chapter 1 (page 33). The idea was that these meant *frontmost, backmost, leftmost,* and *rightmost,* respectively, where, for instance, the complex term $\mathsf{lm}(b)$ referred to the leftmost block in the same row as b. Thus a formula like

$$\mathsf{lm}(x) = x$$

is satisfied by a block b if and only if b is the leftmost block in its row. If we append a universal quantifier to this atomic wff, we get the sentence

$$\forall x \, (\mathsf{lm}(x) = x)$$

which is true in exactly those worlds that have at most one block in each row. This claim could be expressed in the blocks language without function symbols, but again it would require a sentence with more than one quantifier.

To check if you understand these function symbols, see if you can tell which of the following two sentences is true in all worlds and which makes a substantive claim, true in some worlds and false in others:

$$\forall x \, (\mathsf{lm}(\mathsf{lm}(x)) = \mathsf{lm}(x))$$
$$\forall x \, (\mathsf{fm}(\mathsf{lm}(x)) = \mathsf{lm}(x))$$

In reading a term like $\mathsf{fm}(\mathsf{lm}(b))$, remember that you apply the *inner* function first, then the outer. That is, you first find the leftmost block in the row containing b—call it c—and then find the frontmost block in the column containing c.

Function symbols are extremely useful and important in applications of FOL. We close this chapter with some problems that use function symbols.

Exercises

9.22 Assume that we have expanded the blocks language to include the function symbols fm, bm, lm and rm described earlier. Then the following formulas would all be sentences of the language:

1. $\exists y\,(\text{fm}(y) = e)$
2. $\exists x\,(\text{lm}(x) = b \wedge x \neq b)$
3. $\forall x\,\text{Small}(\text{fm}(x))$
4. $\forall x\,(\text{Small}(x) \leftrightarrow \text{fm}(x) = x)$
5. $\forall x\,(\text{Cube}(x) \rightarrow \text{Dodec}(\text{lm}(x)))$
6. $\forall x\,(\text{rm}(\text{lm}(x)) = x)$
7. $\forall x\,(\text{fm}(\text{bm}(x)) = x)$
8. $\forall x\,(\text{fm}(x) \neq x \rightarrow \text{Tet}(\text{fm}(x)))$
9. $\forall x\,(\text{lm}(x) = b \rightarrow \text{SameRow}(x, b))$
10. $\exists y\,(\text{lm}(\text{fm}(y)) = \text{fm}(\text{lm}(y)) \wedge \neg\text{Small}(y))$

Fill in the following table with TRUE's and FALSE's according to whether the indicated sentence is true or false in the indicated world. Since Tarski's World does not understand the function symbols, you will not be able to check your answers. We have filled in a few of the entries for you. Turn in the completed table to your instructor.

	Malcev's	Bolzano's	Boole's	Wittgenstein's
1.				FALSE
2.				
3.			FALSE	
4.				
5.	TRUE			
6.				
7.				
8.		TRUE		
9.				
10.				

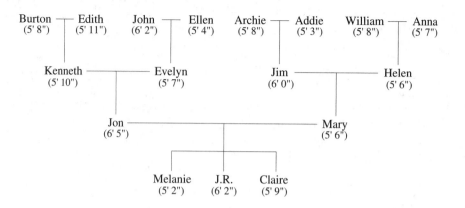

Figure 9.1: A family tree, with heights.

9.23 Consider the first-order language with function symbols mother and father, plus names for each of the people shown in the family tree in Figure 9.1. Here are some atomic wffs, each with a single free variable x. For each, pick a person for x that satisfies the wff, if you can. If there is no such person indicated in the family tree, say so.

1. mother(x) = ellen
2. father(x) = jon
3. mother(father(x)) = mary
4. father(mother(x)) = john
5. mother(father(x)) = addie
6. father(mother(father(x))) = john
7. father(father(mother(x))) = archie
8. father(father(jim)) = x
9. father(father(mother(claire))) = x
10. mother(mother(mary)) = mother(x)

9.24 Again using Figure 9.1, figure out which of the sentences listed below are true. Assume that the domain of discourse consists of the people listed in the family tree.

1. $\exists x\, \text{Taller}(x, \text{mother}(x))$
2. $\forall x\, \text{Taller}(\text{father}(x), \text{mother}(x))$
3. $\exists y\, \text{Taller}(\text{mother}(\text{mother}(y)), \text{mother}(\text{father}(y)))$
4. $\forall z\, [z \neq \text{father}(\text{claire}) \rightarrow \text{Taller}(\text{father}(\text{claire}), z)]$
5. $\forall x\, [\text{Taller}(x, \text{father}(x)) \rightarrow \text{Taller}(x, \text{claire})]$

9.25 Assume you are working in an extension of the first-order language of arithmetic with the
✎ additional predicates Even(x) and Prime(x). Express the following in this language, explicitly
using the function symbol ×, as in z × z, rather than z². Note that you do not have a predicate
Square(x).

1. No square is prime.
2. Some square is odd.
3. The square of any prime is prime.
4. The square of any prime other than 2 is odd.
5. The square of any number greater than 1 is greater than the number itself.

Submit your sentence file.

Alternative notation

The notation we have been using for the quantifiers is currently the most
popular. An older notation that is still in some use employs (x) for ∀x. Thus,
for example, in this notation our

$$\forall x\,[\mathsf{Tet}(x) \to \mathsf{Small}(x)]$$

would be written:

$$(x)\,[\mathsf{Tet}(x) \to \mathsf{Small}(x)]$$

Another notation that is occasionally used exploits the similarity between
universal quantification and conjunction by writing ⋀x instead of ∀x. In this
notation our sentence would be rendered:

$$\bigwedge x\,[\mathsf{Tet}(x) \to \mathsf{Small}(x)]$$

Finally, you will sometimes encounter the universal quantifier written Πx, as
in:

$$\Pi x\,[\mathsf{Tet}(x) \to \mathsf{Small}(x)]$$

Similar variants of ∃x are in use. One version writes (∃x) or (Ex). Other
versions write ⋁x or Σx. Thus the following are notational variants of one
another.

$$\exists x\,[\mathsf{Cube}(x) \wedge \mathsf{Large}(x)]$$
$$(Ex)\,[\mathsf{Cube}(x) \wedge \mathsf{Large}(x)]$$
$$\bigvee x\,[\mathsf{Cube}(x) \wedge \mathsf{Large}(x)]$$
$$\Sigma x\,[\mathsf{Cube}(x) \wedge \mathsf{Large}(x)]$$

Remember

The following table summarizes the alternative notations.

Our notation	Common equivalents
¬P	$\sim$ P, $\overline{P}$, !P, Np
P ∧ Q	P&Q, P&&Q, P · Q, PQ, Kpq
P ∨ Q	P | Q, P || Q, Apq
P → Q	P ⊃ Q, Cpq
P ↔ Q	P ≡ Q, Epq
∀x S(x)	(x)S(x), ⋀x S(x), Πx S(x)
∃x S(x)	(∃x)S(x), (Ex)S(x), ⋁x S(x), Σx S(x)

Exercises

9.26 ✐ (Overcoming dialect differences) The following are all sentences of FOL. But they're in different dialects. Start a new sentence file in Tarski's World and translate them into our dialect.

    1. $\sim$ (x)(P(x) ⊃ Q(x))

    2. Σy((P(y) ≡ $\overline{Q(y)}$) & R(y))

    3. $\overline{⋀x P(x)}$ ≡ ⋁x $\overline{P(x)}$

CHAPTER 10

The Logic of Quantifiers

We have now introduced all of the symbols of first-order logic, though we're nowhere near finished learning all there is to know about them. Before we go on, we should explain where the "first-order" in "first-order logic" comes from. It has to do with the kinds of things that our quantifiers quantify over. In FOL we are allowed to say things like ∃x Large(x), that is, *there is something that has the property of being large*. But we can't say things like *there is some property that Max has:* ∃P P(max).

first-order logic

First-order quantifiers allow us to make quantity claims about ordinary objects: blocks, people, numbers, sets, and so forth. (Note that we are very liberal about what an ordinary object is.) If, in addition, we want to make quantity claims about properties of the objects in our domain of discourse—say we want to claim that Max and Claire share exactly two properties—then we need what is known as *second-order* quantifiers. Since our language only has first-order quantifiers, it is known as the language of first-order logic: FOL.

second-order quantifiers

Now that we've learned the basics of how to express ourselves using first-order quantifiers, we can turn our attention to the central issues of logical consequence and logical truth:

What quantified sentences are logical truths?
What arguments involving quantification are valid?
What are the valid inference patterns involving quantifiers?
How can we formalize these valid patterns of inference?

In this chapter we take up the first two questions; the remaining two are treated in Chapters 12 and 13.

Tautologies and quantification

Introducing quantifiers required a much more radical change to the language than introducing additional truth-functional connectives. Because of the way quantifiers work, we had to introduce the notion of a well-formed formula, something very much like a sentence except that it can contain free variables. Quantifiers attach to these wffs, bind their variables, and thereby form sentences from formulas that aren't themselves sentences. This is strikingly different from the behavior of truth-functional operators.

Given how different quantified sentences are from anything we've seen before, the first thing we need to do is ask how much of the logic of truth functions applies to sentences containing quantifiers. In particular, do the notions of tautology, tautological consequence, and tautological equivalence apply to our new sentences, and if so, how?

quantified sentences and tautological consequence

The answer is that these notions do apply to quantified sentences, but they must be applied with care. Students often ignore the presence of quantifiers in sentences and try to use what they learned in propositional logic wherever it seems vaguely applicable. This can be very dangerous. For example, you might rightly notice that the following arguments are logically valid:

1. $\forall x\,(\mathsf{Cube}(x) \rightarrow \mathsf{Small}(x))$
    $\forall x\,\mathsf{Cube}(x)$
    $\overline{}$
    $\forall x\,\mathsf{Small}(x)$

2. $\forall x\,\mathsf{Cube}(x)$
    $\forall x\,\mathsf{Small}(x)$
    $\overline{}$
    $\forall x\,(\mathsf{Cube}(x) \wedge \mathsf{Small}(x))$

The first of these is valid because if every cube is small, and everything is a cube, then everything is small. The second is valid because if everything is a cube, and everything is small, then everything is a small cube. But are these arguments *tautologically* valid? Or, to put it another way, can we simply ignore the quantifiers appearing in these arguments and apply the principles of *modus ponens* and $\wedge$ **Intro**?

It doesn't take long to see that ignoring quantifiers doesn't work. For example, neither of the following arguments is valid, tautologically or otherwise:

3. $\exists x\,(\mathsf{Cube}(x) \rightarrow \mathsf{Small}(x))$
    $\exists x\,\mathsf{Cube}(x)$
    $\overline{}$
    $\exists x\,\mathsf{Small}(x)$

4. $\exists x\,\mathsf{Cube}(x)$
    $\exists x\,\mathsf{Small}(x)$
    $\overline{}$
    $\exists x\,(\mathsf{Cube}(x) \wedge \mathsf{Small}(x))$

The premises of argument 3 will be true in a world containing a large cube and a large dodecahedron, but nothing small. The premises of argument 4 will be true in a world containing a large cube and a small dodecahedron, but no small cube.

These counterexamples not only show that arguments 3 and 4 are invalid, they also show that 1 and 2 are not tautologically valid, that is, valid solely in virtue of the meanings of the truth-functional connectives. Clearly, the meaning of ∀ is an essential factor in the validity of 1 and 2, for if it were not, 3 and 4 should be valid as well. Or, to put it the other way around, if ∀ meant the same thing as ∃, then 1 and 2 would be no more valid than 3 and 4.

A similar point can be made about over-hasty applications of the notion of tautology. For example, the following sentence, which says that either there is a cube or there is something which is not a cube, is logically true:

$$\exists x\, \text{Cube}(x) \lor \exists x\, \neg\text{Cube}(x)$$

tautology and quantification

But is this sentence a tautology, true simply in virtue of the meanings of the truth-functional connectives? Again, the answer is *no*, as we can see by considering what happens when we replace the existential quantifier with a universal quantifier:

$$\forall x\, \text{Cube}(x) \lor \forall x\, \neg\text{Cube}(x)$$

This sentence says that either everything is a cube or everything is not a cube, which of course is false in any world inhabited by a mixture of cubes and non-cubes.

Are there no tautologies in a language containing quantifiers? Of course there are, but you don't find them by pretending the quantifiers simply aren't there. For example, the following sentence is a tautology:

$$\forall x\, \text{Cube}(x) \lor \neg\forall x\, \text{Cube}(x)$$

This sentence, unlike the previous one, is an instance of the law of excluded middle. It says that either everything is a cube or it's not the case that everything is a cube, and that's going to be true so long as the constituent sentence ∀x Cube(x) has a definite truth value. It would hold equally well if the constituent sentence were ∃x Cube(x), a fact you could recognize even if you didn't know exactly what this sentence meant.

Recall that if we have a tautology and replace its atomic sentences by complex sentences, the result is still a tautology, and hence also a logical truth. This holds as long as the things we are substituting are sentences that have definite truth values (whether true or not). We can use this observation to discover a large number of quantified sentences that are logical truths. Consider the following tautology:

$$(A \to B) \to (\neg B \to \neg A)$$

If we replace A and B by any quantified sentences of FOL, the result is still a tautology. For example, we might replace A by $\exists y\,(P(y) \vee R(y))$ and B by $\forall x\,(P(x) \wedge Q(x))$. The result is the following rather complex sentence, which we dub "Phred" (pronounced "Fred"):

$$(\exists y\,(P(y) \vee R(y)) \rightarrow \forall x\,(P(x) \wedge Q(x))) \rightarrow$$
$$(\neg\forall x\,(P(x) \wedge Q(x)) \rightarrow \neg\exists y\,(P(y) \vee R(y)))$$

If we just came across Phred, it would be hard to know what it even meant. But since it is a substitution instance of a tautology, we know that it is in fact a logical truth. In a similar way, we could make many other substitutions, in this and other tautologies, to discover many complex logical truths.

There is a small difficulty with this method of identifying the tautologies of FOL. A given sentence can typically be obtained by substitution from many different sentences, some tautologies, some not. Phred, for example, can be obtained from all of the following (and many others) by means of substitution:

$$A$$
$$A \rightarrow B$$
$$(A \rightarrow B) \rightarrow C$$
$$A \rightarrow (B \rightarrow C)$$
$$(A \rightarrow B) \rightarrow (C \rightarrow D)$$
$$(A \rightarrow B) \rightarrow (\neg B \rightarrow C)$$
$$(A \rightarrow B) \rightarrow (\neg B \rightarrow \neg A)$$

For instance, we can obtain Phred from the fourth of these by substituting as follows:

○ Replace A by $(\exists y\,(P(y) \vee R(y)) \rightarrow \forall x\,(P(x) \wedge Q(x)))$

○ Replace B by $\neg\forall x\,(P(x) \wedge Q(x))$

○ Replace C by $\neg\exists y\,(P(y) \vee R(y))$.

But of the seven candidates for substitution, only the last, $(A \rightarrow B) \rightarrow (\neg B \rightarrow \neg A)$, is a tautology. If a sentence can be obtained from so many different formulas by substitution, how do we know which one to look at to see if it is a tautology?

Here is a simple method for solving this problem: The basic idea is that for purposes of testing whether a sentence is a tautology, we must treat any quantified constituent of the sentence as if it is atomic. We don't look "inside" quantified parts of the sentence. But we do pay attention to all of the truth-functional connectives that are not in the scope of a quantifier. We'll describe a procedure for replacing the quantified and atomic constituents of a sentence with letters, A, B, C, ..., so that the result displays all and only the truth-functional connectives that aren't inside quantified pieces of the sentence. The result of applying this procedure is called the *truth-functional form* of the original sentence.

truth-functional form

The procedure has two main steps. The first annotates the sentence by labeling the constituents that we must treat as atomic, either because they are quantified or because they really are atomic. Applied to Phred, it yields the following result:

$$\underline{(\exists y \, (P(y) \lor R(y))}_A \to \underline{\forall x \, (P(x) \land Q(x))}_B) \to$$
$$\underline{(\neg \forall x \, (P(x) \land Q(x))}_B \to \underline{\neg \exists y \, (P(y) \lor R(y))}_A)$$

The second step replaces the underlined constituents with the sentence letters used to label them:

$$(A \to B) \to (\neg B \to \neg A)$$

We state the procedure as a little *algorithm*, a series of step-by-step instructions for finding the truth-functional form of an arbitrary sentence S of FOL.

Truth-functional form algorithm: Start at the beginning of sentence S and proceed to the right. When you come to a quantifier or an atomic sentence, begin to underline that portion of the sentence (recall that the formula a $\neq$ b is an abbreviation for ¬a = b, and is therefore not atomic). If you encountered a quantifier, underline the quantifier and the entire formula that it is applied to. (This will either be the atomic wff that immediately follows the quantifier or, if there are parentheses, the formula enclosed by the parentheses.) If you encountered an atomic sentence, just underline the atomic sentence. When you come to the end of your underline, assign the underlined constituent a sentence letter (A, B, C, ...). If an identical constituent already appears earlier in the sentence, use the same sentence letter as before; otherwise, assign the first sentence letter not yet used as a label. Once you've labeled the constituent, continue from that point. Finally, when you come to the end of the sentence, replace each underlined constituent with the sentence letter that labels it. The result is the truth-functional form of S.

truth-functional form algorithm

Let's see how we would apply this algorithm to another sentence. See if you understand each of the following steps, which begin with a quantified sentence of FOL, and end with that sentence's truth-functional form. Pay particular attention to steps 4 and 5. In step 4, do you understand why the label A is used again? In step 5, do you understand why $\forall y \, \text{Small}(y)$ gets labeled C rather than B?

1. $\neg(\text{Tet}(d) \land \forall x\, \text{Small}(x)) \to (\neg \text{Tet}(d) \lor \neg \forall y\, \text{Small}(y))$

2. $\neg(\underline{\text{Tet}(d)}_A \land \forall x\, \text{Small}(x)) \to (\neg \text{Tet}(d) \lor \neg \forall y\, \text{Small}(y))$

3. $\neg(\underline{\text{Tet}(d)}_A \land \underline{\forall x\, \text{Small}(x)}_B) \to (\neg \text{Tet}(d) \lor \neg \forall y\, \text{Small}(y))$

4. $\neg(\underline{\text{Tet}(d)}_A \land \underline{\forall x\, \text{Small}(x)}_B) \to (\neg \underline{\text{Tet}(d)}_A \lor \neg \forall y\, \text{Small}(y))$

5. $\neg(\underline{\text{Tet}(d)}_A \land \underline{\forall x\, \text{Small}(x)}_B) \to (\neg \underline{\text{Tet}(d)}_A \lor \neg \underline{\forall y\, \text{Small}(y)}_C)$

6. $\neg(A \land B) \to (\neg A \lor \neg C)$

We are now in a position to say exactly which sentences of the quantified language are tautologies.

tautologies of FOL

Definition A quantified sentence of FOL is said to be a *tautology* if and only if its truth-functional form is a tautology.

Here is a table displaying six first-order sentences and their truth-functional forms. Notice that although four of the sentences in the left column are logically true, only the first two are tautologies, as shown by their t.f. forms in the right column.

FO sentence	t.f. form
$\forall x\, \text{Cube}(x) \lor \neg \forall x\, \text{Cube}(x)$	$A \lor \neg A$
$(\exists y\, \text{Tet}(y) \land \forall z\, \text{Small}(z)) \to \forall z\, \text{Small}(z)$	$(A \land B) \to B$
$\forall x\, \text{Cube}(x) \lor \exists y\, \text{Tet}(y)$	$A \lor B$
$\forall x\, \text{Cube}(x) \to \text{Cube}(a)$	$A \to B$
$\forall x\, (\text{Cube}(x) \lor \neg \text{Cube}(x))$	A
$\forall x\, (\text{Cube}(x) \to \text{Small}(x)) \lor \exists x\, \text{Dodec}(x)$	$A \lor B$

A useful feature of the truth-functional form algorithm is that it can be applied to arguments as easily as it can be applied to sentences. All you do is continue the procedure until you come to the end of the argument, rather than stopping at the end of the first sentence. For example, applied to argument 3 on page 260, we first get the labeled argument:

$$\underline{\exists x\, (\text{Cube}(x) \to \text{Small}(x))}_A$$
$$\underline{\exists x\, \text{Cube}(x)}_B$$

$$\underline{\exists x\, \text{Small}(x)}_C$$

and then the truth-functional form:

A
B
C

This shows that argument 3 is not an instance of → **Elim**. But when we apply the algorithm to a deceptively similar argument:

$\underline{\exists x\, \mathsf{Cube}(x)}_A \to \underline{\exists x\, \mathsf{Small}(x)}_B$

$\underline{\exists x\, \mathsf{Cube}(x)}_A$

$\underline{\exists x\, \mathsf{Small}(x)}_B$

we see that this argument is indeed an instance of *modus ponens*:

A → B
A
B

The **Taut Con** procedure of Fitch uses the truth-functional form algorithm so you can use it to check whether a quantified sentence is a tautology, or whether it is a tautological consequence of other sentences.

The truth-functional form algorithm allows us to apply all of the concepts of propositional logic to sentences and arguments containing quantifiers. But we have also encountered several examples of logical truths that are not tautologies, and logically valid arguments that are not tautologically valid. In the next section we look at these.

Remember

1. Use the truth-functional form algorithm to determine the truth-functional form of a sentence or argument containing quantifiers.

2. The truth-functional form of a sentence shows how the sentence is built up from atomic and quantified sentences using truth-functional connectives.

3. A quantified sentence is a tautology if and only if its truth-functional form is a tautology.

4. Every tautology is a logical truth, but among quantified sentences there are many logical truths that are not tautologies.

5. Similarly, there are many logically valid arguments of FOL that are not tautologically valid.

Exercises

10.1 For each of the following, use the truth-functional form algorithm to annotate the sentence and determine its form. Then classify the sentence as (a) a tautology, (b) a logical truth but not a tautology, or (c) not a logical truth. (If your answer is (a), feel free to use the **Taut Con** routine in Fitch to check your answer.)

1. $\forall x\, x = x$
2. $\exists x\, \text{Cube}(x) \to \text{Cube}(a)$
3. $\text{Cube}(a) \to \exists x\, \text{Cube}(x)$
4. $\forall x\, (\text{Cube}(x) \land \text{Small}(x)) \to \forall x\, (\text{Small}(x) \land \text{Cube}(x))$
5. $\forall v\, (\text{Cube}(v) \leftrightarrow \text{Small}(v)) \leftrightarrow \neg\neg\forall v\, (\text{Cube}(v) \leftrightarrow \text{Small}(v))$
6. $\forall x\, \text{Cube}(x) \to \neg\exists x\, \neg\text{Cube}(x)$
7. $[\forall z\, (\text{Cube}(z) \to \text{Large}(z)) \land \text{Cube}(b)] \to \text{Large}(b)$
8. $\exists x\, \text{Cube}(x) \to (\exists x\, \text{Cube}(x) \lor \exists y\, \text{Dodec}(y))$
9. $(\exists x\, \text{Cube}(x) \lor \exists y\, \text{Dodec}(y)) \to \exists x\, \text{Cube}(x)$
10. $[(\forall u\, \text{Cube}(u) \to \forall u\, \text{Small}(u)) \land \neg\forall u\, \text{Small}(u)] \to \neg\forall u\, \text{Cube}(u)$

Turn in your answers by filling in a table of the following form:

	Annotated sentence	Truth-functional form	a/b/c
1.			
⋮			

In the following six exercises, use the truth-functional form algorithm to annotate the argument. Then write out its truth-functional form. Finally, assess whether the argument is (a) tautologically valid, (b) logically but not tautologically valid, or (c) invalid. Feel free to check your answers with **Taut Con***. (Exercises 10.6 and 10.7 are, by the way, particularly relevant to the proof of the Completeness Theorem for $\mathcal{F}$ given in Chapter 19.)*

10.2

$\text{Cube}(a) \land \text{Cube}(b)$
$\text{Small}(a) \land \text{Large}(b)$

$\exists x\, (\text{Cube}(x) \land \text{Small}(x)) \land \exists x\, (\text{Cube}(x) \land \text{Large}(x))$

10.3

$\forall x\, \text{Cube}(x) \to \exists y\, \text{Small}(y)$
$\neg\exists y\, \text{Small}(y)$

$\exists x\, \neg\text{Cube}(x)$

10.4

$\forall x\, \text{Cube}(x) \to \exists y\, \text{Small}(y)$
$\neg\exists y\, \text{Small}(y)$

$\neg\forall x\, \text{Cube}(x)$

10.5

$\forall x\, (\text{Tet}(x) \to \text{LeftOf}(x,b)) \lor \forall x\, (\text{Tet}(x) \to \text{RightOf}(x,b))$
$\exists x\, (\text{Tet}(x) \land \text{SameCol}(x,b)) \to \neg\forall x\, (\text{Tet}(x) \to \text{LeftOf}(x,b))$
$\forall x\, (\text{Tet}(x) \to \text{RightOf}(x,b)) \to \neg\exists x\, (\text{Tet}(x) \land \text{SameCol}(x,b))$

$\neg\exists x\, (\text{Tet}(x) \land \text{SameCol}(x,b))$

10.6

$\exists x\, (\text{Cube}(x) \land \text{Large}(x)) \to (\text{Cube}(c) \land \text{Large}(c))$
$\text{Tet}(c) \to \neg\text{Cube}(c)$
$\text{Tet}(c)$

$\forall x\, \neg(\text{Cube}(x) \land \text{Large}(x))$

10.7

$\exists x\, (\text{Cube}(x) \land \text{Large}(x)) \to (\text{Cube}(c) \land \text{Large}(c))$
$\forall x\, \neg(\text{Cube}(x) \land \text{Large}(x)) \leftrightarrow \neg\exists x\, (\text{Cube}(x) \land \text{Large}(x))$
$\text{Tet}(c) \to \neg\text{Cube}(c)$
$\text{Tet}(c)$

$\forall x\, \neg(\text{Cube}(x) \land \text{Large}(x))$

[In 10.6 and 10.7, we could think of the first premise as a way of introducing a new constant, c, by means of the assertion: *Let the constant* c *name a large cube, if there are any; otherwise, it may name any object.* Sentences of this sort are called *Henkin witnessing axioms,* and are put to important use in proving completeness for $\mathcal{F}$. The arguments show that if a constant introduced in this way ends up naming a tetrahedron, it can only be because there aren't any large cubes.]

SECTION 10.2

First-order validity and consequence

When we first discussed the intuitive notions of logical truth and logical consequence, we appealed to the idea of a logically possible circumstance. We described a logically valid argument, for example, as one whose conclusion is true in every possible circumstance in which all the premises are true. When we needed more precision than this description allowed, we introduced truth tables and the concepts of tautology and tautological consequence. These concepts add precision by modeling possible circumstances as rows of a truth table. We have seen that this move does a good job of capturing the intuitive notions of logical truth and logical consequence—provided we limit our attention to the truth-functional connectives.

Unfortunately, the concepts of tautology and tautological consequence don't get us far in first-order logic. We need a more refined method for analyzing logical truths and logically valid arguments when they depend on the quantifiers and identity. We will introduce these notions in this chapter, and develop them in greater detail in Chapter 18. The notions will give us, for first-order logic, what the concepts of tautology and tautological consequence gave us for propositional logic: precise approximations of the notions of logical truth and logical consequence.

First, a terminological point. It is a regrettable fact that there is no single term like "tautological" that logicians consistently use when applying the various logical notions to first-order sentences and arguments. That is, we don't have a uniform way of filling out the table:

Propositional logic	First-order logic	General notion
Tautology	??	*Logical truth*
Tautological consequence	??	*Logical consequence*
Tautological equivalence	??	*Logical equivalence*

One option would be to use the terms *first-order logical truth*, *first-order logical consequence*, and *first-order logical equivalence*. But these are just too much of a mouthful for repeated use, so we will abbreviate them. Instead of *first-order logical consequence*, we will use *first-order consequence* or simply *FO consequence*, and for *first-order logical equivalence*, we'll use *first-order (or FO) equivalence*. We will not, however, use *first-order truth* for *first-order logical truth*, since this might suggest that we are talking about a true (but not logically true) sentence of first-order logic.

For *first-order logical truth*, it is standard to use the term *first-order validity*. This may surprise you, since so far we've only used "valid" to apply to arguments, not sentences. This is a slight terminological inconsistency, but it shouldn't cause any problems so long as you're aware of it. In first-order logic, we use *valid* to apply to both sentences and arguments: to sentences that can't be false, and to arguments whose conclusions can't be false if their premises are true. Our completed table, then, looks like this:

first-order consequence

first-order validity

Propositional logic	First-order logic	General notion
Tautology	*FO validity*	*Logical truth*
Tautological consequence	*FO consequence*	*Logical consequence*
Tautological equivalence	*FO equivalence*	*Logical equivalence*

So what do we mean by the notions of first-order validity, first-order consequence and first-order equivalence? These concepts are meant to apply to those logical truths, consequences, and equivalences that are such solely in virtue of the truth-functional connectives, the quantifiers, and the identity symbol. Thus, for purposes of determining first-order consequence, we ignore the specific meanings of names, function symbols, and predicates other than identity.

There are two reasons for treating identity along with the quantifiers and connectives, rather than like any other predicate. The first is that almost all first-order languages use $=$. Other predicates, by contrast, vary from one first-order language to another. For example, the blocks language uses the binary predicate LeftOf, while the language of set theory uses $\in$, and the language of arithmetic uses $<$. This makes it a reasonable division of labor to try first to understand the logic implicit in the connectives, quantifiers, and identity, without regard to the meanings of the other predicates, names, and function symbols. The second reason is that the identity predicate is crucial for expressing many quantified noun phrases of English. For instance, we'll soon see how to express things like *at least three tetrahedra* and *at most four cubes*, but to express these in FOL we need identity in addition to the quantifiers $\forall$ and $\exists$. There is a sense in which identity and the quantifiers go hand in hand.

identity

If we can recognize that a sentence is logically true without knowing the meanings of the names or predicates it contains (other than identity), then we'll say the sentence is a first-order validity. Let's consider some examples from the blocks language:

$$\forall x\, \text{SameSize}(x, x)$$
$$\forall x\, \text{Cube}(x) \to \text{Cube}(b)$$
$$(\text{Cube}(b) \land b = c) \to \text{Cube}(c)$$
$$(\text{Small}(b) \land \text{SameSize}(b, c)) \to \text{Small}(c)$$

All of these are arguably logical truths of the blocks language, but only the middle two are first-order validities. One way to see this is to replace the familiar blocks language predicates with nonsensical predicates, like those used in Lewis Carroll's famous poem *Jabberwocky*.[1] The results would look

using nonsense predicates to test for FO validity

---

[1] The full text of *Jabberwocky* can be found at `http://en.wikipedia.org/wiki/Jabberwocky`. The first stanza is:

> 'Twas brillig, and the slithy toves
> Did gyre and gimble in the wabe;
> All mimsy were the borogoves,
> And the mome raths outgrabe.

"Lewis Carroll" was the pen name of the logician Charles Dodgson (after whom both Carroll's World and Dodgson's Sentences were named).

something like this:

$$\forall x \, \text{Outgrabe}(x, x)$$
$$\forall x \, \text{Tove}(x) \rightarrow \text{Tove}(b)$$
$$(\text{Tove}(b) \wedge b = c) \rightarrow \text{Tove}(c)$$
$$(\text{Slithy}(b) \wedge \text{Outgrabe}(b, c)) \rightarrow \text{Slithy}(c)$$

Notice that we can still see that the second and third sentences must be true, whatever the predicate Tove may mean. If everything is a tove, and b is an object in the domain of discourse, then b must surely be a tove. Similarly, if b is a tove, and c is the same object as b, then c is a tove as well. Contrast this with the first and fourth sentences, which no longer look logically true at all. Though we know that everything is the same size as itself, we have no idea whether everything outgrabes itself! Just so, the fact that b is slithy and outgrabes c hardly guarantees that c is slithy. Maybe it is and maybe it isn't!

The concepts of first-order consequence and first-order equivalence work similarly. For example, if you can recognize that an argument is logically valid without appealing to the meanings of the names or predicates (other than identity), then the conclusion is a first-order consequence of the premises. The following argument is an example:

> $\forall x \, (\text{Tet}(x) \rightarrow \text{Large}(x))$
> $\neg\text{Large}(b)$
>
> $\neg\text{Tet}(b)$

This argument is obviously valid. What's more, if we replace the predicates Tet and Large with nonsense predicates, say Borogove and Mimsy, the result is the following:

> $\forall x \, (\text{Borogove}(x) \rightarrow \text{Mimsy}(x))$
> $\neg\text{Mimsy}(b)$
>
> $\neg\text{Borogove}(b)$

Again, it's easy to see that if the borogoves (whatever they may be) are all mimsy (whatever that may mean), and if b is not mimsy, then it can't possibly be a borogove. So the conclusion is not just a logical consequence of the premises, it is a *first-order* consequence.

Recall that to show that a sentence was not a tautological consequence of some premises, it sufficed to find a truth-value assignment to the atomic sentences that made the premises true and the conclusion false. A similar procedure can be used to show that a conclusion is not a first-order consequence of its premises, except instead of truth-value assignments what we look for is a bit more complicated. Suppose we are given the following argument:

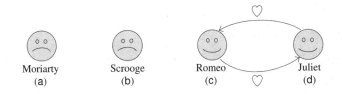

Figure 10.1: A first-order counterexample.

> ¬∃x Larger(x, a)
> ¬∃x Larger(b, x)
> Larger(c, d)
>
> Larger(a, b)

The first premise tells you that nothing is larger than *a* and the second tells you that *b* is not larger than anything. If you were trying to build a counterexample world, you might reason that *a* must be the largest object in the world (or one of them) and that *b* must be the smallest (or one of them). Since the third premise guarantees that the objects in the world aren't all the same size, the conclusion can't be falsified in a world in which the premises are true.

Is this conclusion a first-order consequence of the premises? To show that it's not, we'll do two things. First, let's replace the predicate Larger with a meaningless predicate, to help clear our minds of any constraints suggested by the predicate:

first-order counterexamples

> ¬∃x R(x, a)
> ¬∃x R(b, x)
> R(c, d)
>
> R(a, b)

Next, we'll describe a specific interpretation of R (and the names a, b, c, and d), along with a possible circumstance that would count as a counterexample to the new argument. This is easy. Suppose R means *likes*, and we are describing a situation with four individuals: Romeo and Juliet (who like each other), and Moriarty and Scrooge (who like nobody, and the feelings are mutual).

If we let a refer to Moriarty, b refer to Scrooge, c and d refer to Romeo and Juliet, then the premises of our argument are all true, though the conclusion is false. This possible circumstance, like an alternate truth assignment in propositional logic, shows that our original conclusion is not a first-order consequence of the premises. Thus we call it a *first-order counterexample*.

Let's codify this procedure.

replacement method

Replacement Method:

1. To check for first-order validity or first-order consequence, systematically replace all of the predicates, other than identity, with new, meaningless predicate symbols, making sure that if a predicate appears more than once, you replace all instances of it with the same meaningless predicate. (If there are function symbols, replace these as well.)

2. To see if S is a first-order validity, try to describe a circumstance, along with interpretations for the names, predicates, and functions in S, in which the sentence is false. If there is no such circumstance, the original sentence is a first-order validity.

3. To see if S is a first-order consequence of $P_1, \ldots, P_n$, try to find a circumstance and interpretation in which S is false while $P_1, \ldots, P_n$ are all true. If there is no such circumstance, the original inference counts as a first-order consequence.

Recognizing whether a sentence is a first-order validity, or a first-order consequence of some premises, is not as routine as with tautologies and tautological consequence. With truth tables, there may be a lot of rows to check, but at least the number is finite and known in advance. With first-order validity and consequence, the situation is much more complicated, since there are infinitely many possible circumstances that might be relevant. In fact, there is no correct and mechanical procedure, like truth tables, that always answers the question *is S a first-order validity?* But there are procedures that do a pretty good job and we have built one into Fitch; it is the procedure given as **FO Con** on the consequence menu. In checking applications of **FO Con**, Fitch will never give you the wrong answer, though sometimes it will get stuck, unable to give you any answer at all.

recognizing first-order validity

Unfortunately, judging first-order validity is no easier for you than for the computer. But unless a sentence or argument is quite complicated, the replacement method should result in a successful resolution of the issue, one that agrees with an application of **FO Con**.

As we said earlier, we will make the notions of first-order validity and first-order consequence more precise later in the book, so that we can prove theorems involving these notions. But the rough-and-ready characterizations we've given will suffice until then. Even with the current description, we can see the following:

1. If S is a tautology, then it is a first-order validity. Similarly, if S is a

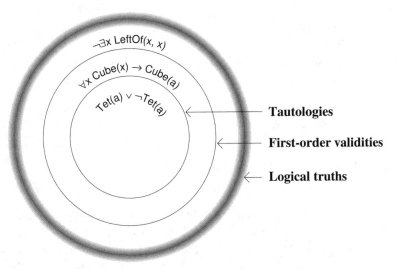

Figure 10.2: The relation between tautologies, first-order validities, and logical truths

first-order validity, it is a logical truth. The converse of neither of these statements is true, though. (See Figure 10.2.)

2. Similarly, if S is a tautological consequence of premises $P_1, \ldots, P_n$, then it is a first-order consequence of these premises. Similarly, if S is a first-order consequence of premises $P_1, \ldots, P_n$, then it is a logical consequence of these premises. Again, the converse of neither of these statements is true.

Let's try our hand at applying all of these concepts, using the various consequence mechanisms in Fitch.

You try it
. . . .

1. Open the file **FO Con 1**. Here you are given a collection of premises, plus a series of sentences that follow logically from them. Your task is to cite support sentences and specify one of the consequence rules to justify each step. But the trick is that you must use the weakest consequence mechanism possible and cite the minimal number of support sentences possible. ◀

2. Focus on the first step after the Fitch bar, $\forall x \, \text{Cube}(x) \rightarrow \text{Cube}(b)$. You will recognize this as a logical truth, which means that you should not have ◀

to cite any premises in support of this step. First, ask yourself whether the sentence is a tautology. No, it is not, so **Taut Con** will not check out. Is it a first-order validity? Yes, so change the rule to **FO Con** and see if it checks out. It would also check out using **Ana Con**, but this rule is stronger than necessary, so your answer would be counted wrong if you used this mechanism.

▶ 3. Continue through the remaining sentences, citing only necessary supporting premises and the weakest **Con** mechanism possible.

▶ 4. Save your proof as Proof FO Con 1.

. . . . *Congratulations*

Just as for **Taut Con**, **FO Con** has its own special goggles which allow you to obscure the information not considered when checking the inference rule. In this case less is obscured by the goggles, since more information is taken into account. Consider inferring ¬Tove(b) from ∀x (Tove(x) → Slithy(x)) and ¬Slithy(b).

With **FO Con**'s goggles, this inference looks like this:

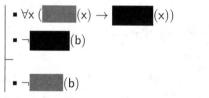

FO Con

Here the predicate symbol Slithy has been replaced by ■, and the predicate symbol Tove by ■.

FO Con goggles work exactly the same way as **Taut Con** goggles do. When focussed on a step which uses **FO Con**, the goggles are switched on by clicking on their icon, and switched off by clicking again.

You try it
. . . .

▶ 1. Open the file Goggles Example, which contains the inference described above. Use the **FO Con** goggles to verify that the effect is as we have described.

▶ 2. The particular goggles used by Fitch depend on the inference rule. To use **Taut Con** goggles on this inference, first take the **FO Con** goggles off (you can't wear both kinds at the same time), and then change the rule to **Taut Con**. Then switch on the **Taut Con** goggles.

3. Notice that you see something different now. The display should look like this:

■ ▬▬▬▬▬▬▬▬▬▬▬▬

■ ¬ ▬▬▬▬▬▬▬

■ ▬▬▬▬   **Taut Con**

Because the **Taut Con** inference rule does not take into account the meaning of quantifiers, the entire quantified formula in the initial step is treated as a single atomic formula. Since the inference contains three completely unrelated formulas, as far as **Taut Con** is concerned, this inference will not check out.

4. Take off the **Taut Con** goggles, and check the inference. As predicted, it will not check out, since the conclusion is not a tautological consequence of the premises. There is nothing to save.

. . . . *Congratulations*

Remember

1. A sentence of FOL is a first-order validity if it is a logical truth when you ignore the meanings of the names, function symbols, and predicates other than the identity symbol.

2. A sentence S is a first-order consequence of premises $P_1, \ldots, P_n$ if it a logical consequence of these premises when you ignore the meanings of the names, function symbols, and predicates other than identity.

3. The Replacement Method is useful for determining whether a sentence is a first-order validity and whether the conclusion of an argument is a first-order consequence of the premises.

4. All tautologies are first-order validities; all first-order validities are logical truths. Similarly for consequence.

Exercises

10.8 If you skipped the **You try it** section, go back and do it now. Submit the file **Proof FO Con 1**.

10.9 Open Carnap's Sentences and Bolzano's World.
1. Paraphrase each sentence in clear, colloquial English and verify that it is true in the given world.
2. For each sentence, decide whether you think it is a logical truth. If it isn't, build a world in which the sentence comes out false and save it as **World 10.9.x**, where x is the number of the sentence. [Hint: You should be able to falsify three of these sentences.]
3. Which of these sentences are first-order validities? [Hint: Three are.]
4. For the remaining four sentences (those that are logical truths but not first-order validities), apply the Replacement Method to come up with first-order counterexamples. Make sure you describe both your interpretations of the predicates and the falsifying circumstance.

Turn in your answers to parts 1, 3, and 4; submit the worlds you build in part 2.

Each of the following arguments is valid. Some of the conclusions are (a) tautological consequences of the premises, some are (b) first-order consequences that are not tautological consequences, and some are (c) logical consequences that are not first-order consequences. Use the truth-functional form algorithm and the replacement method to classify each argument. You should justify your classifications by turning in (a) the truth-functional form of the argument, (b) the truth-functional form and the argument with nonsense predicates substituted, or (c) the truth-functional form, the nonsense argument, and a first-order counterexample.

10.10
$$\text{Cube}(a) \land \text{Cube}(b)$$
$$\text{Small}(a) \land \text{Large}(b)$$

$$\exists x\,(\text{Cube}(x) \land \text{Small}(x)) \land \exists x\,(\text{Cube}(x) \land \text{Large}(x))$$

10.11
$$\text{Cube}(a) \land \text{Cube}(b)$$
$$\text{Small}(a) \land \text{Large}(b)$$

$$\exists x\,(\text{Cube}(x) \land \text{Large}(x) \land \neg\text{Smaller}(x, x))$$

10.12
$$\forall x\,\text{Cube}(x) \to \exists y\,\text{Small}(y)$$
$$\neg\exists y\,\text{Small}(y)$$

$$\exists x\,\neg\text{Cube}(x)$$

10.13
$$\forall x\,\text{Cube}(x) \to \exists y\,\text{Small}(y)$$
$$\neg\exists y\,\text{Small}(y)$$

$$\neg\forall x\,\text{Cube}(x)$$

10.14 ✎
> Cube(a)
> Dodec(b)
>
> ¬(a = b)

10.15 ✎
> Cube(a)
> ¬Cube(b)
>
> ¬(a = b)

10.16 ✎
> Cube(a)
> ¬Cube(a)
>
> ¬(a = b)

10.17 ✎
> ∀x (Dodec(x) → ¬SameCol(x, c))
>
> ¬Dodec(c)

10.18 ✎
> ∀z (Small(z) ↔ Cube(z))
> Cube(d)
>
> Small(d)

10.19 ✎
> ∀z (Small(z) → Cube(z))
> ∀w (Cube(w) → LeftOf(w, c))
>
> ¬∃y (Small(y) ∧ ¬LeftOf(y, c))

SECTION 10.3

First-order equivalence and DeMorgan's laws

There are two ways in which we can apply what we learned about tautological equivalence to first-order sentences. First of all, if you apply the truth-functional form algorithm to a pair of sentences and the resulting forms are tautologically equivalent, then of course the original sentences are first-order equivalent. For example, the sentence:

$$\neg(\exists x\, \text{Cube}(x) \land \forall y\, \text{Dodec}(y))$$

is tautologically equivalent to:

$$\neg\exists x\, \text{Cube}(x) \lor \neg\forall y\, \text{Dodec}(y)$$

When you apply the truth-functional form algorithm, you see that this is just an instance of one of DeMorgan's laws.

But it turns out that we can also apply DeMorgan, and similar principles, inside the scope of quantifiers. Let's look at an example involving the Law of Contraposition. Consider the sentences:

$$\forall x\, (\text{Cube}(x) \to \text{Small}(x))$$
$$\forall x\, (\neg\text{Small}(x) \to \neg\text{Cube}(x))$$

A moment's thought will convince you that each of these sentences is a first-order consequence of the other, and so they are first-order equivalent. But unlike the previous examples, they are not tautologically equivalent.

To see why Contraposition (and other principles of equivalence) can be applied in the scope of quantifiers, we need to consider the wffs to which the principle was applied:

$$\text{Cube}(x) \rightarrow \text{Small}(x)$$
$$\neg\text{Small}(x) \rightarrow \neg\text{Cube}(x)$$

Or, more generally, consider the wffs:

$$P(x) \rightarrow Q(x)$$
$$\neg Q(x) \rightarrow \neg P(x)$$

where $P(x)$ and $Q(x)$ may be any formulas, atomic or complex, containing the single free variable x.

Now since these formulas are not sentences, it makes no sense to say they are true in exactly the same circumstances, or that they are logical (or tautological) consequences of one another. Formulas with free variables are neither true nor false. But there is an obvious extension of the notion of logical equivalence that applies to formulas with free variables. It is easy to see that in any possible circumstance, the above two formulas will be satisfied by exactly the same objects. Here's a proof of this fact:

> **Proof:** We show this by indirect proof. Assume that in some circumstance there is an object that satisfies one but not the other of these two formulas. Let's give this object a new name, say n_1. Consider the results of replacing x by n_1 in our formulas:
>
> $$P(n_1) \rightarrow Q(n_1)$$
> $$\neg Q(n_1) \rightarrow \neg P(n_1)$$
>
> Since x was the only free variable, these are sentences. But by our assumption, one of them is true and one is false, since that is how we defined satisfaction. But this is a contradiction, since these two sentences are logically equivalent by Contraposition.

logically equivalent wffs   We will say that two wffs with free variables are logically equivalent if, in any possible circumstance, they are satisfied by the same objects.[2] Or, what

---

[2]Though we haven't discussed satisfaction for wffs with more than one free variable, a similar argument can be applied to such wffs: the only difference is that more than one name is substituted in for the free variables.

comes to the same thing, two wffs are logically equivalent if, when you replace their free variables with new names, the resulting sentences are logically equivalent.

The above proof, suitably generalized, shows that when we apply any of our principles of logical equivalence to a formula, the result is a logically equivalent formula, one that is satisfied by exactly the same objects as the original. This in turn is why the *sentence* $\forall x\,(\text{Cube}(x) \rightarrow \text{Small}(x))$ is logically equivalent to the sentence $\forall x\,(\neg\text{Small}(x) \rightarrow \neg\text{Cube}(x))$. If every object in the domain of discourse (or one object, or thirteen objects) satisfies the first formula, then every object (or one or thirteen) must satisfy the second.

Equipped with the notion of logically equivalent wffs, we can restate the principle of substitution of equivalents so that it applies to full first-order logic. Let P and Q be wffs, possibly containing free variables, and let S(P) be any sentence containing P as a component part. Then if P and Q are logically equivalent:

$$P \Leftrightarrow Q$$

substitution of equivalent wffs

then so too are S(P) and S(Q):

$$S(P) \Leftrightarrow S(Q)$$

As before, a full proof of the substitution principle requires the method of proof by induction, which we will get to in Chapter 16. But in the meantime, our current observations should make it clear why the principle still applies.

Once we see that substitution of equivalents applies to quantified sentences, the principles we have already learned yield a large number of new equivalences. For example, we can show that the sentence $\forall x\,(\text{Cube}(x) \rightarrow \text{Small}(x))$ is logically equivalent to $\forall x\,\neg(\text{Cube}(x) \wedge \neg\text{Small}(x))$ by means of the following chain of equivalences:

$$
\begin{aligned}
\forall x\,(\text{Cube}(x) \rightarrow \text{Small}(x)) \quad &\Leftrightarrow \quad \forall x\,(\neg\text{Cube}(x) \vee \text{Small}(x)) \\
&\Leftrightarrow \quad \forall x\,(\neg\text{Cube}(x) \vee \neg\neg\text{Small}(x)) \\
&\Leftrightarrow \quad \forall x\,\neg(\text{Cube}(x) \wedge \neg\text{Small}(x))
\end{aligned}
$$

Again, these sentences are not tautologically equivalent, because we are applying our principles inside a quantified sentence. But other than extending the principle of substitution, no new principles are involved. There are, however, several important logical equivalences that involve quantifiers in essential ways. Let's turn to these now.

DeMorgan laws for quantifiers

In propositional logic, the DeMorgan laws describe important relations between negation, conjunction, and disjunction. If you think about the meanings of our quantifiers, you will see that there is a strong analogy between ∀ and ∧, on the one hand, and between ∃ and ∨, on the other. For example, suppose we are talking about a world consisting of four named blocks, say a, b, c, and d. Then the sentence ∀x Cube(x) will be true if and only if the following conjunction is true:

$$\text{Cube(a)} \land \text{Cube(b)} \land \text{Cube(c)} \land \text{Cube(d)}$$

Likewise, ∃x Cube(x) will be true if and only if this disjunction is true:

$$\text{Cube(a)} \lor \text{Cube(b)} \lor \text{Cube(c)} \lor \text{Cube(d)}$$

This analogy suggests that the quantifiers may interact with negation in a way similar to conjunction and disjunction. Indeed, in our four-block world, the sentence

$$\neg\forall x\, \text{Small(x)}$$

will be true if and only if the following negation is true:

$$\neg(\text{Small(a)} \land \text{Small(b)} \land \text{Small(c)} \land \text{Small(d)})$$

which, by DeMorgan, will hold just in case the following disjunction is true:

$$\neg\text{Small(a)} \lor \neg\text{Small(b)} \lor \neg\text{Small(c)} \lor \neg\text{Small(d)}$$

which in turn will be true if and only if the following holds:

$$\exists x\, \neg\text{Small(x)}$$

DeMorgan laws for quantifiers

The DeMorgan laws for the quantifiers allow you to push a negation sign past a quantifier by switching the quantifier from ∀ to ∃ or from ∃ to ∀. So, for example, if we know that not everything has some property ($\neg\forall x\, P(x)$), then we know that something does not have the property ($\exists x\, \neg P(x)$), and vice versa. Similarly, if we know that it is not the case that something has some property ($\neg\exists x\, P(x)$), then we know that everything must fail to have it ($\forall x\, \neg P(x)$), and vice versa. We call these the DeMorgan laws for quantifiers, due to the analogy described above; they are also known as the quantifier/negation equivalences:

$$\neg\forall x\, P(x) \quad\Leftrightarrow\quad \exists x\, \neg P(x)$$
$$\neg\exists x\, P(x) \quad\Leftrightarrow\quad \forall x\, \neg P(x)$$

By applying these laws along with some earlier equivalences, we can see that there is a close relationship between certain pairs of Aristotelian sentences. In particular, the negation of *All P's are Q's* is logically equivalent to *Some P's are not Q's*. To demonstrate this equivalence, we note the following chain of equivalences. The first is the translation of *It is not true that all P's are Q's* while the last is the translation of *Some P's are not Q's*.

$$\begin{aligned}
\neg \forall x\, (P(x) \rightarrow Q(x)) \quad &\Leftrightarrow \quad \neg \forall x\, (\neg P(x) \lor Q(x)) \\
&\Leftrightarrow \quad \exists x\, \neg(\neg P(x) \lor Q(x)) \\
&\Leftrightarrow \quad \exists x\, (\neg\neg P(x) \land \neg Q(x)) \\
&\Leftrightarrow \quad \exists x\, (P(x) \land \neg Q(x))
\end{aligned}$$

The first step uses the equivalence of $P(x) \rightarrow Q(x)$ and $\neg P(x) \lor Q(x)$. The second and third steps use DeMorgan's laws, first one of the quantifier versions, and then one of the Boolean versions. The last step uses the double negation law applied to $\neg\neg P(x)$.

A similar chain of equivalences shows that the negation of *Some P's are Q's* is equivalent to *No P's are Q's*:

$$\neg \exists x\, (P(x) \land Q(x)) \quad \Leftrightarrow \quad \forall x\, (P(x) \rightarrow \neg Q(x))$$

We leave the demonstration of this as an exercise.

Remember

(DeMorgan laws for quantifiers) For any wff $P(x)$:

1. $\neg \forall x\, P(x) \Leftrightarrow \exists x\, \neg P(x)$

2. $\neg \exists x\, P(x) \Leftrightarrow \forall x\, \neg P(x)$

Exercises

10.20 Give a chain of equivalences showing that the negation of *Some P's are Q's* ($\neg \exists x\, (P(x) \land Q(x))$) is equivalent to *No P's are Q's* ($\forall x\, (P(x) \rightarrow \neg Q(x))$).

10.21 Open DeMorgan's Sentences 2. This file contains six sentences, but each of sentences 4, 5, and 6 is logically equivalent to one of the first three. Without looking at what the sentences say, see if you can figure out which is equivalent to which by opening various world files and evaluating the sentences. (You should be able to figure this out from Ackermann's, Bolzano's, and Claire's Worlds, plus what we've told you.) Once you think you've figured out which are equivalent to which, write out three equivalence chains to prove you're right. Turn these in to your instructor.

10.22 ($\forall$ versus $\wedge$) We pointed out the similarity between $\forall$ and $\wedge$, as well as that between $\exists$ and
$\mathcal{F}$ $\vee$. But we were careful not to claim that the universally quantified sentence was logically equivalent to the analogous conjunction. This problem will show you why we did not make this claim.

- Open Church's Sentences and Ramsey's World. Evaluate the sentences in this world. You will notice that the first two sentences have the same truth value, as do the second two.

- Modify Ramsey's World in any way you like, but do not add or delete objects, and do not change the names used. Verify that the first two sentences always have the same truth values, as do the last two.

- Now add one object to the world. Adjust the objects so that the first sentence is false, the second and third true, and the last false. Submit your work as World 10.22. This world shows that the first two sentences are not logically equivalent. Neither are the last two.

Section 10.4
Other quantifier equivalences

The quantifier DeMorgan laws tell us how quantifiers interact with negation. Equally important is the question of how quantifiers interact with conjunction and disjunction. The laws governing this interaction, though less interesting than DeMorgan's, are harder to remember, so you need to pay attention!

quantifiers and Boolean connectives

First of all, notice that $\forall x\,(P(x) \wedge Q(x))$, which says that everything is both P and Q, is logically equivalent to $\forall x\,P(x) \wedge \forall x\,Q(x)$, which says that everything is P and everything is Q. These are just two different ways of saying that every object in the domain of discourse has both properties P and Q. By contrast, $\forall x\,(P(x) \vee Q(x))$ is *not* logically equivalent to $\forall x\,P(x) \vee \forall x\,Q(x)$. For example, the sentence $\forall x\,(Cube(x) \vee Tet(x))$ says that everything is either a cube or a tetrahedron, but the sentence $\forall x\,Cube(x) \vee \forall x\,Tet(x)$ says that either everything is a cube or everything is a tetrahedron, clearly a very different kettle of fish. We summarize these two observations, positive and negative, as follows:

$$\forall x\,(P(x) \wedge Q(x)) \quad \Leftrightarrow \quad \forall x\,P(x) \wedge \forall x\,Q(x)$$

$$\forall x\,(P(x) \vee Q(x)) \quad \not\Leftrightarrow \quad \forall x\,P(x) \vee \forall x\,Q(x)$$

Similar observations hold with $\exists$, $\vee$, and $\wedge$, except that it works the other way around. The claim that there is some object that is either P or Q, $\exists x\,(P(x) \vee Q(x))$, is logically equivalent to the claim that something is

P or something is Q: $\exists x\, P(x) \vee \exists x\, Q(x)$. But this equivalence fails the moment we replace $\vee$ with $\wedge$. The fact that there is a cube and a tetrahedron, $\exists x\, Cube(x) \wedge \exists x\, Tet(x)$, hardly means that there is something which is both a cube and a tetrahedron: $\exists x\, (Cube(x) \wedge Tet(x))$! Again, we summarize both positive and negative observations together:

$$\exists x\, (P(x) \vee Q(x)) \quad \Leftrightarrow \quad \exists x\, P(x) \vee \exists x\, Q(x)$$

$$\exists x\, (P(x) \wedge Q(x)) \quad \not\Leftrightarrow \quad \exists x\, P(x) \wedge \exists x\, Q(x)$$

There is one circumstance when you can push a universal quantifier in past a disjunction, or move an existential quantifier out from inside a conjunction. But to explain this circumstance, we first have to talk a bit about a degenerate form of quantification. In defining the class of wffs, we did not insist that the variable being quantified actually occur free (or at all) in the wff to which the quantifier is applied. Thus, for example, the expression $\forall x\, Cube(b)$ is a wff, even though $Cube(b)$ does not contain the variable x. Similarly, the sentence $\exists y\, Small(y)$ doesn't contain any free variables. Still, we could form the sentences $\forall x\, \exists y\, Small(y)$ or even $\forall y\, \exists y\, Small(y)$. We call this sort of quantification *null quantification*. Let's think for a moment about what it might mean.

null quantification

Consider the case of the sentence $\forall x\, Cube(b)$. This sentence is true in a world if and only if every object in the domain of discourse satisfies the wff $Cube(b)$. But what does that mean, since this wff does not contain any free variables? Or, to put it another way, what does it mean to substitute a name for the (nonexistent) free variable in $Cube(b)$? Well, if you replace every occurrence of x in $Cube(b)$ with the name n_1, the result is simply $Cube(b)$. So, in a rather degenerate way, the question of whether an object satisfies $Cube(b)$ simply boils down to the question of whether $Cube(b)$ is true. Thus, $\forall x\, Cube(b)$ and $Cube(b)$ are true in exactly the same worlds, and so are logically equivalent. The same holds of $\exists x\, Cube(b)$, which is also equivalent to $Cube(b)$. More generally, if the variable x is not free in wff P, then we have the following equivalences:

$$\forall x\, P \quad \Leftrightarrow \quad P$$

$$\exists x\, P \quad \Leftrightarrow \quad P$$

If null quantification seems counterintuitive to you, take a moment to do Exercise 10.23 now.

What does this have to do with conjunctions and disjunctions? Well, there is a more general observation we can make about null quantification. Suppose we have the sentence $\forall x\, (Cube(b) \vee Small(x))$, where x does not occur free in the first disjunct. This sentence will be true if $Cube(b)$ is true (in which case, every object will satisfy the disjunctive wff trivially), or if every object is small (in which case, they will all satisfy the disjunctive wff by satisfying the second disjunct), or both. That is, this sentence imposes the same conditions on a world as the sentence Cube(b) ∨ ∀x Small(x). Indeed, when the variable x is not free in a wff P, we have both of the following:

$$\forall x\,(P \lor Q(x)) \quad \Leftrightarrow \quad P \lor \forall x\,Q(x)$$
$$\exists x\,(P \land Q(x)) \quad \Leftrightarrow \quad P \land \exists x\,Q(x)$$

Compare these two equivalences to the non-equivalences highlighted a moment ago, and make sure you understand the differences. The equivalences involving null quantification, surprisingly enough, will become very useful later, when we learn how to put sentences into what is called *prenex form,* where all the quantifiers are out in front.

replacing bound variables

The last principles we mention are so basic they are easily overlooked. When you are translating from English to FOL and find a quantified noun phrase, you must pick some variable to use in your translation. But we have given you no guidelines as to which one is "right." This is because it doesn't matter which variable you use, as long as you don't end up with two quantifiers whose scope overlaps but which bind the same variable. In general, the variable itself makes no difference whatsoever. For example, these sentences are logically equivalent:

$$\exists x\,(Dodec(x) \land Larger(x, b)) \quad \Leftrightarrow \quad \exists y\,(Dodec(y) \land Larger(y, b))$$

We codify this by means of the following: For any wff P(x) and variable y that does not occur in P(x)

$$\forall x\,P(x) \quad \Leftrightarrow \quad \forall y\,P(y)$$
$$\exists x\,P(x) \quad \Leftrightarrow \quad \exists y\,P(y)$$

Remember

1. (Pushing quantifiers past ∧ and ∨) For any wffs P(x) and Q(x):

    (a) $\forall x\,(P(x) \wedge Q(x)) \Leftrightarrow \forall x\,P(x) \wedge \forall x\,Q(x)$

    (b) $\exists x\,(P(x) \vee Q(x)) \Leftrightarrow \exists x\,P(x) \vee \exists x\,Q(x)$

2. (Null quantification) For any wff P in which x is not free:

    (a) $\forall x\,P \Leftrightarrow P$

    (b) $\exists x\,P \Leftrightarrow P$

    (c) $\forall x\,(P \vee Q(x)) \Leftrightarrow P \vee \forall x\,Q(x)$

    (d) $\exists x\,(P \wedge Q(x)) \Leftrightarrow P \wedge \exists x\,Q(x)$

3. (Replacing bound variables) For any wff P(x) and variable y that does not occur in P(x):

    (a) $\forall x\,P(x) \Leftrightarrow \forall y\,P(y)$

    (b) $\exists x\,P(x) \Leftrightarrow \exists y\,P(y)$

Exercises

10.23 (Null quantification) Open Null Quantification Sentences. In this file you will find sentences in the odd numbered slots. Notice that each sentence is obtained by putting a quantifier in front of a sentence in which the quantified variable is not free.

    1. Open Godel's World and evaluate the truth of the first sentence. Do you understand why it is false? Repeatedly play the game committed to the truth of this sentence, each time choosing a different block when your turn comes around. Not only do you always lose, but your choice has no impact on the remainder of the game. Frustrating, eh?

    2. Check the truth of the remaining sentences and make sure you understand why they have the truth values they do. Play the game a few times on the second sentence, committed to both true and false. Notice that neither your choice of a block (when committed to false) nor Tarski's World's choice (when committed to true) has any effect on the game.

3. In the even numbered slots, write the sentence from which the one above it was obtained. Check that the even and odd numbered sentences have the same truth value, no matter how you modify the world. This is because they are logically equivalent. Save and submit your sentence file.

Some of the following biconditionals are logical truths (which is the same as saying that the two sides of the biconditional are logically equivalent); some are not. If you think the biconditional is a logical truth, create a file with Fitch, enter the sentence, and check it using **FO Con**. *If the sentence is not a logical truth, create a world in Tarski's World in which it is false. Submit the file you create.*

10.24 $(\forall x\, Cube(x) \lor \forall x\, Dodec(x))$
$\hspace{1.5em} \leftrightarrow \forall x\, (Cube(x) \lor Dodec(x))$

10.25 $\neg\exists z\, Small(z) \leftrightarrow \exists z\, \neg Small(z)$

10.26 $\forall x\, Tet(b) \leftrightarrow \exists w\, Tet(b)$

10.27 $\exists w\, (Dodec(w) \land Large(w))$
$\hspace{1.5em} \leftrightarrow (\exists w\, Dodec(w) \land \exists w\, Large(w))$

10.28 $\exists w\, (Dodec(w) \land Large(b))$
$\hspace{1.5em} \leftrightarrow (\exists w\, Dodec(w) \land Large(b))$

10.29 $\neg\forall x\, (Cube(x) \rightarrow (Small(x) \lor Large(x)))$
$\hspace{1.5em} \leftrightarrow \exists z\, (Cube(z) \land \neg Small(z) \land \neg Large(z))$

SECTION 10.5

The axiomatic method

As we will see in the coming chapters, first-order consequence comes much closer to capturing the logical consequence relation of ordinary language than does tautological consequence. This will be apparent from the kinds of sentences that we can translate into the quantified language and from the kinds of inference that turn out to be first-order valid.

Still, we have already encountered several arguments that are intuitively valid but not first-order valid. Let's look at an example where the replacement method reveals that the conclusion is not a first-order consequence of the premises:

$$\forall x\, (Cube(x) \leftrightarrow SameShape(x, c))$$
$$Cube(c)$$

Using the replacement method, we substitute meaningless predicate symbols, say P and Q, for the predicates Cube and SameShape. The result is

$$\forall x\, (P(x) \leftrightarrow Q(x, c))$$
$$P(c)$$

which is clearly *not* a valid argument. If we wanted the conclusion of our argument to be a first-order consequence of the premises, we would need to add a new premise or premises expressing facts about the predicates involved in the original inference. For the present argument, here is what we might do:

$$\forall x\, (\text{Cube}(x) \leftrightarrow \text{SameShape}(x, c))$$
$$\forall x\, \text{SameShape}(x, x)$$

$$\text{Cube}(c)$$

The premise we've added is clearly justified by the meaning of SameShape. What's more, the replacement method now gives us

$$\forall x\, (\text{P}(x) \leftrightarrow \text{Q}(x, c))$$
$$\forall x\, \text{Q}(x, x)$$

$$\text{P}(c)$$

which is logically valid no matter what the predicates P and Q mean. The conclusion is a first-order consequence of the two premises.

This technique of adding a premise whose truth is justified by the meanings of the predicates is one aspect of what is known as the *axiomatic method*. It is often possible to bridge the gap between the intuitive notion of consequence and the more restricted notion of first-order consequence by systematically expressing facts about the predicates involved in our inferences. The sentences used to express these facts are sometimes called *meaning postulates*, a special type of *axiom*.

axiomatic method

meaning postulates

Suppose we wanted to write out axioms that bridged the gap between first-order consequence and the intuitive notion of consequence that we've been using in connection with the blocks language. That would be a big task, but we can make a start on it by axiomatizing the shape predicates. We might begin by taking four basic shape axioms as a starting point. These axioms express the fact that every block has one and only one shape.

Basic Shape Axioms:

1. $\neg\exists x\, (\text{Cube}(x) \wedge \text{Tet}(x))$

2. $\neg\exists x\, (\text{Tet}(x) \wedge \text{Dodec}(x))$

basic shape axioms

3. $\neg\exists x\, (\text{Dodec}(x) \wedge \text{Cube}(x))$

4. $\forall x\, (\text{Tet}(x) \vee \text{Dodec}(x) \vee \text{Cube}(x))$

The first three axioms stem from the meanings of our three basic shape predicates. Being one of these shapes simply precludes being another. The fourth axiom, however, is presumably not part of the meaning of the three predicates, as there are certainly other possible shapes. Still, if our goal is to capture reasoning about blocks worlds of the sort that can be built in Tarski's World, we will want to include 4 as an axiom.

Any argument that is intuitively valid and involves only the three basic shape predicates is first-order valid if we add these axioms as additional premises. For example, the following intuitively valid argument is not first-order valid:

$$\neg \exists x\, \mathsf{Tet}(x)$$
$$\forall x\, (\mathsf{Cube}(x) \leftrightarrow \neg \mathsf{Dodec}(x))$$

If we add axioms 3 and 4 as premises, however, then the resulting argument is

$$\neg \exists x\, \mathsf{Tet}(x)$$
$$\neg \exists x\, (\mathsf{Dodec}(x) \wedge \mathsf{Cube}(x))$$
$$\forall x\, (\mathsf{Tet}(x) \vee \mathsf{Dodec}(x) \vee \mathsf{Cube}(x))$$
$$\forall x\, (\mathsf{Cube}(x) \leftrightarrow \neg \mathsf{Dodec}(x))$$

This argument is first-order valid, as can be seen by replacing the shape predicates with meaningless predicates, say P, Q, and R:

$$\neg \exists x\, \mathsf{P}(x)$$
$$\neg \exists x\, (\mathsf{Q}(x) \wedge \mathsf{R}(x))$$
$$\forall x\, (\mathsf{P}(x) \vee \mathsf{Q}(x) \vee \mathsf{R}(x))$$
$$\forall x\, (\mathsf{R}(x) \leftrightarrow \neg \mathsf{Q}(x))$$

If the validity of this argument is not entirely obvious, try to construct a first-order counterexample. You'll see that you can't.

Let's look at an example where we can replace some instances of **Ana Con** by the basic shape axioms and the weaker **FO Con** rule.

You try it
. . . .

▶   1. Open the Fitch file Axioms 1. The premises in this file are just the four basic shape axioms. Below the Fitch bar are four sentences. Each is justified by a use of the rule **Ana Con**, without any sentences cited in support. Verify that each of the steps checks out.

2. Now change each of the justifications from **Ana Con** to **FO Con**. Verify ◀
that none of the steps now checks out. See if you can make each of them
check out by finding a single shape axiom to cite in its support.

3. When you are finished, save your proof as Proof Axioms 1. ◀
. . . . *Congratulations*

The basic shape axioms don't express everything there is to say about the   *other shape axioms*
shapes in Tarski's World. We have not yet said anything about the binary
predicate SameShape. At the very least, we would need the following as a fifth
axiom:

$\forall x\, \mathsf{SameShape}(x, x)$

We will not in fact add this as an axiom since, as we will see in Chapter 12, it
leaves out essential facts about the relation between SameShape and the basic
shape predicates. When we add these other facts, it turns out that the above
axiom is unnecessary.

Axiomatization has another interesting use that can be illustrated with our
axioms about shape. Notice that if we systematically replace the predicates
Cube, Tet, and Dodec by the predicates Small, Medium, and Large, the resulting
sentences are true in all of the blocks worlds. It follows from this that if we
take any valid argument involving just the shape predicates and perform the
stated substitution, the result will be another valid argument.

Presupposing a range of circumstances

The intuitive difference between the first three shape axioms and the fourth,
which asserts a general fact about our block worlds but not one that follows
from the meanings of the predicates, highlights an important characteristic
of much everyday reasoning. More often than not, when we reason, we do so
against the background of an assumed range of possibilities. When you reason   *assuming a range*
about Tarski's World, it is natural to presuppose the various constraints that   *of possibilities*
have been built into the program. When you reason about what movie to
go to, you implicitly presuppose that your options are limited to the movies
showing in the vicinity.

The inferences that you make against this background may not, strictly
speaking, be logically valid. That is, they may be correct relative to the presupposed circumstances but not correct relative to some broader range of possibilities. For example, if you reason from ¬Cube(d) and ¬Tet(d) to Dodec(d),
your reasoning is valid within the domain of Tarski's World, but not relative
to worlds where there are spheres and icosohedra. When you decide that the

290 / The Logic of Quantifiers

latest Harrison Ford movie is the best choice, this may be a correct inference in your vicinity, but perhaps not if you were willing to fly to other cities.

In general, background assumptions about the range of relevant circumstances are not made an explicit part of everyday reasoning, and this can give rise to disagreements about the reasoning's validity. People with different assumptions may come up with very different assessments about the validity of some explicit piece of reasoning. In such cases, it is often helpful to articulate general facts about the presupposed circumstances. By making these explicit, we can often identify the source of the disagreement.

The axiomatic method can be thought of as a natural extension of this everyday process. Using this method, it is often possible to transform arguments that are valid only relative to a particular range of circumstances into arguments that are first-order valid. The axioms that result express facts about the meanings of the relevant predicates, but also facts about the presupposed circumstances.

The history of the axiomatic method is closely entwined with the history of logic. You were probably already familiar with axioms from studying Euclidean geometry. In investigating the properties of points, lines, and geometrical shapes, the ancient Greeks discovered the notion of proof which lies at the heart of deductive reasoning. This came about as follows. By the time of the Greeks, an enormous number of interesting and important discoveries about geometrical objects had already been made, some dating back to the time of the ancient Babylonians. For example, ancient clay tablets show that the Babylonians knew what is now called the Pythagorean Theorem. But for the Babylonians, geometry was an empirical science, one whose facts were discovered by observation.

Somewhere lost in the prehistory of mathematics, someone had a brilliant idea. They realized that there are logical relationships among the known facts of geometry. Some follow from others by logic alone. Might it not be possible to choose a few, relatively clear observations as basic, and derive all the others by logic? The starting truths are accepted as *axioms,* while their consequences are called *theorems.* Since the axioms are supposed to be obviously true, and since the methods of proof are logically valid, we can be sure the theorems are true as well.

This general procedure for systematizing a body of knowledge became known as the axiomatic method. A set of axioms is chosen, statements which we are certain hold of the "worlds" or "circumstances" under consideration. Some of these may be meaning postulates, truths that hold simply in virtue of meaning. Others may express obvious facts that hold in the domain in question, facts like our fourth shape axiom. We can be sure that anything

background assumptions

axioms and theorems

that follows from the axioms by valid methods of inference is on just as firm
a footing as the axioms from which we start.

Exercises

10.30 If you skipped the **You try it** section, go back and do it now. Submit the file Proof Axioms 1.
✐

10.31 Suppose we state our four basic shape axioms in the following schematic form:

    1. $\neg \exists x \, (R(x) \land P(x))$
    2. $\neg \exists x \, (P(x) \land Q(x))$
    3. $\neg \exists x \, (Q(x) \land R(x))$
    4. $\forall x \, (P(x) \lor Q(x) \lor R(x))$

We noted that any valid argument involving just the three shape predicates remains valid
when you substitute other predicates, like the Tarski's World size predicates, that satisfy these
axioms. Which of the following triplets of properties satisfy the axioms in the indicated domain
(that is, make them true when you substitute them for P, Q, and R)? If they don't, say which
axioms fail and why.

    1. *Red, yellow,* and *blue* in the domain of automobiles.
    2. *Entirely red, entirely yellow,* and *entirely blue* in the domain of automobiles.
    3. *Small, medium,* and *large* in the domain of Tarski's World blocks.
    4. *Small, medium,* and *large* in the domain of physical objects.

[Note: Your answers in some cases will depend on how you are construing the predicates. The
important thing is that you explain your interpretations clearly, and how the interpretations
lead to the success or failure of the axioms.]

SECTION 10.6

Lemmas

One important feature of the axiomatic method is that as theorems are proved,
they become available as new facts that can be used in later proofs. If S has
been shown to be a consequence of the axioms, then S can be used in any
proof which employs those same axioms. After all, we could just re-prove S
in the new proof, since all of its premises are available in the new proof, and
then cite S in subsequent inference steps.

    For example, suppose that you have proved that

$$\neg \exists x (Cube(x) \land Dodec(x) \land Tet(x))$$

follows from the shape axioms that we presented in the previous section. If you are later asked to prove

$$\neg\exists x(\mathsf{Cube}(x) \land \mathsf{Dodec}(x) \land \mathsf{Tet}(x)) \lor \mathsf{Small}(a)$$

from the same axioms, you *could* simply reproduce the steps of the earlier proof, and then use the ∨-**Intro** rule to reach the desired conclusion. But you have to repeat the work that you already did. That could be a drag if the proof of the original result is long and complex.

Logicians are a lazy lot, and to avoid repeating work that they have already done they usually simply say something like "we already know that $\neg\exists x(\mathsf{Cube}(x) \land \mathsf{Dodec}(x) \land \mathsf{Tet}(x))$ follows from the axioms, and so obviously, $\neg\exists x(\mathsf{Cube}(x) \land \mathsf{Dodec}(x) \land \mathsf{Tet}(x)) \lor \mathsf{Small}(a)$ does too." They know that if a picky person were to ask them to justify this move, they could simply reproduce the proof. But they know better than to do all that work without having been asked.

lemmas

A result that is used in this way is often called a lemma. There's nothing special about a lemma, it's a result that has been proved and that is being used in the course of proving another result. Often a lemma is of little intrinsic interest but is needed for some larger end, or a result that is generally useful and will be used in many later proofs. Lemmas have the same formal status as theorems or propositions, but are usually less important.

Fitch has a rule that takes advantage of the ability to reuse theorems as lemmas in later proofs. The **Lemma** rule allows you to use a result that you have previously proved within a new proof. To use this rule, you must select a proof file that you have already made, and that file must have a single goal which checks out. You cite the formulas in your proof which correspond to the premises in the file, and you are permitted to conclude the goal formula of that lemma in your proof. Lets try it out.

You try it

. . . .

▶ 1. Open the file **Lemma 1**, which contains a proof of the validity of the following argument

> $\neg\exists x(\mathsf{Tet}(x) \land \mathsf{Cube}(x))$
> $\neg\exists x(\mathsf{Tet}(x) \land \mathsf{Dodec}(x))$
> $\neg\exists x(\mathsf{Cube}(x) \land \mathsf{Dodec}(x))$
>
> $\neg\exists x(\mathsf{Tet}(x) \land \mathsf{Cube}(x) \land \mathsf{Dodec}(x))$

▶ 2. Open the file **Lemma Example 1**, which contains the goal of proving

$$\neg\exists x(\text{Tet}(x) \land \text{Cube}(x))$$
$$\neg\exists x(\text{Tet}(x) \land \text{Dodec}(x))$$
$$\neg\exists x(\text{Cube}(x) \land \text{Dodec}(x))$$

$$\neg\exists x(\text{Tet}(x) \land \text{Cube}(x) \land \text{Dodec}(x)) \lor \text{Small}(a)$$

Create a new step and insert the formula $\neg\exists x(\text{Tet}(x) \land \text{Cube}(x) \land \text{Dodec}(x))$, this is the goal formula of **Lemma 1**. Cite all three premises, which are also the premises of **Lemma 1**. Finally, open the **Rule** menu, and look for the **Lemma** item, this is a submenu which initially contains **Add Lemma. . ..** Select this item. A file chooser dialog will appear. Navigate to and select the file **Lemma 1**. Check the step (it will check out.)

3. The **Lemma** rule will check out when the formula at the lemma step is the only goal in the lemma file, the same number of formulas are cited as there are premises in the lemma file, and these cited steps contain all of the premise formulas (in any order). The premises and the cited formulas have to match exactly. If the lemma contains the premises P and Q, for example, citing the formula P $\land$ Q won't work. ◀

4. Complete the proof of this exercise by using a single application of $\lor$- **Intro**. ◀

. . . . *Congratulations*

Situations where you can use a lemma arise frequently when doing reasoning with a collection of presupposed axioms because every proof in the collection will share the axioms as common premises. Consequently, results that are proved only from the axioms will be readily reusable in other proofs. However there is nothing special about the axioms. As we have seen, they are just premises used in proofs. If you complete a proof of a result using some premises, and then you later have exactly the same premises (perhaps in addition to others) in another proof, then you are entitled to us that previous result as a lemma.

The premises of the lemma do not even have to be *premises* of the new proof as long as they appear in the new proof and can be cited from the step that uses the lemma. Thus the premises of the lemma could instead be derived from whatever premises you have in the new proof. For instance, in the next example, you are required to derive the premises of the lemma before you can use it.

You try it
. . . .

1. Open the file **Lemma 2**, which contains a proof of the validity of the following argument: ◀

> Dodec(d) ∨ Cube(c)
> Dodec(d) → Large(d)
> Cube(c) → Small(c)
> ___
> Large(d) ∨ Small(c)

▶ 2. Open the file **Lemma Example 2**, which contains the goal of proving:

> Dodec(d) ∨ Cube(c)
> ∀x(Dodec(x) → Large(x))
> ∀x(Cube(x) → Small(x))
> ___
> Large(d) ∨ Small(c)

▶ 3. Notice that the second and third premises of the lemma are instances of the second and third premises of this proof, respectively. We can complete the proof by deriving those formula, and then applying the lemma.

First, use ∀-**Elim** to derive the formula Dodec(d) → Large(d) in the first non-premise step of the proof. In the next step, again use ∀-**Elim** to derive the formula Cube(c) → Small(c). Finally complete the proof by applying the lemma.

. . . . *Congratulations*

Whenever you prove a result, you want it to be as general as possible just in case you will need to use this result as a lemma in a later proof. If you look closely at the proof in the file **Lemma 2** which you used in the shape axioms, you will notice that only one of the premises of the lemma is used in the proof of the lemma. That means that the number of situations in which the lemma can be used is much smaller than it needs to be, since to use the lemma we have to be able to cite *all* of the premises. We want the number of premises to be as small as possible to ensure that the proof is most general and most useful. In fact, even if the proof is never used as the lemma, we want this to be true, since otherwise we might be fooled into believing that the result is more specific than it actually is.

There is another kind of generality that is very important. Imagine that we had a proof of the following result:

> ¬(Tet(a) ∨ Small(b))
> ___
> ¬Tet(a) ∧ ¬Small(b)

This is, of course, an instance of De Morgans laws where the specific formula Tet(a) and Small(b) are used. We know that this argument is valid regardless of the sentences that are chosen in place of Tet(a) and Small(b), a fact that we usually express by

$$\neg(P \lor Q)$$

$$\neg P \land \neg Q$$

Because the general form of the result is valid we know that any specific instance is valid. Ideally, we could prove the result in the general form, and then use the general form as a lemma any time that a specific instance of the result is needed. The **Lemma** rule in Fitch allows us to do exactly that.

You try it
. . . .

1. Open the file Lemma 3, which contains a proof of the validity of the following argument ◀

   $$A \lor B$$
   $$A \rightarrow C$$
   $$B \rightarrow D$$

   $$C \lor D$$

2. Open the file Lemma Example 3, which asks you to prove the following argument: ◀

   $$\text{LeftOf}(a, b) \lor \text{RightOf}(a, b)$$
   $$\text{LeftOf}(a, b) \rightarrow (\text{Small}(a) \land \text{Tet}(a))$$
   $$\text{RightOf}(a, b) \rightarrow (\text{Large}(b) \land \text{Cube}(b))$$

   $$(\text{Small}(a) \land \text{Tet}(a)) \lor (\text{Large}(b) \land \text{Cube}(b))$$

3. Notice that the two arguments have identical form. Lemma 3 contains a proof that arguments of a general form (known as constructive dilemma) are valid, while Lemma Example 3 contains a particular instance of this result. ◀

4. Add a new step to Lemma Example 3, and insert the goal formula. Cite all three premises, and use the **Lemma** rule, selecting the file Lemma 3. Verify that the step checks out. ◀

   This works because it is possible to substitute specific formulas from the step in Lemma Example 3 for the general formulas in the lemma, so that the citations and conclusion of the step are obtained. In this case we substitute:

A	by	$\text{LeftOf}(a, b)$
B	by	$\text{RightOf}(a, b)$
C	by	$\text{Small}(a) \land \text{Tet}(a)$
D	by	$\text{Large}(b) \land \text{Cube}(b)$

In general, if a substitution like this can be found that associates a general formula in the lemma with a formula in the citations and goal, then the step will check out.

. . . . *Congratulations*

Just as for predicate symbols and formulae, you can use the general constant symbols $n_1, \ldots, n_2$ in your lemmas to say that the lemma does not concern specific constant symbols. A lemma such as $\mathsf{Tet}(a) \rightarrow \neg\mathsf{Cube}(a)$ can be used only to match that exact formula, but $\mathsf{Tet}(n_1) \rightarrow \neg\mathsf{Cube}(n_1)$ could be used to prove $\mathsf{Tet}(a) \rightarrow \neg\mathsf{Cube}(a)$, $\mathsf{Tet}(f) \rightarrow \neg\mathsf{Cube}(f)$, or even $\mathsf{Tet}(n_2) \rightarrow \neg\mathsf{Cube}(n_2)$.

Remember

1. Lemmas are just proofs that you have completed that are being used to justify a step in another proof.

2. Fitch's lemma rule requires that each cited formula matches a premise in the lemma file, and the derived formula matches the goal of the lemma file.

3. The letters P, Q etc; in a lemma file will match any formula.

4. The constant symbols $n_1, \ldots, n_9$ in a lemma file will match any constant symbol.

Exercises

10.32 If you skipped the **You try it** sections, go back and do them now. Submit the files Proof Lemma Example 1, Proof Lemma Example 2 and Proof Lemma Example 3.

10.33 Which of the following arguments can be justified by a single application of **Lemma 3**? For each one that can be justified, turn in the substitution of formulas that is required to justify the application. For each argument that cannot be justified, explain why not.

1.
$$\begin{array}{l} \mathsf{Cube}(a) \vee \mathsf{Cube}(b) \\ \mathsf{Cube}(b) \rightarrow \mathsf{SameSize}(a, b) \\ \mathsf{Cube}(a) \rightarrow \mathsf{SameShape}(a, b) \\ \hline \mathsf{SameSize}(a, b) \vee \mathsf{SameShape}(a, b) \end{array}$$

2. | Tet(a) ∨ Tet(b)
   | Tet(b) → Larger(e, f)
   | Tet(a) → Smaller(f, e)

   | Smaller(f, e) ∨ Larger(e, f)

3. | (Dodec(d) ∧ Small(d)) → Larger(e, f)
   | Large(e) → (Adjoins(a, b) ∨ Adjoins(a, c))
   | Large(e) ∨ (Dodec(d) ∧ Small(d))

   | (Adjoins(a, b) ∨ Adjoins(a, c)) ∨ Larger(e, f)

4. | Tet(e)
   | Tet(e) → Small(e)
   | Tet(e) → SameSize(e, e)

   | Small(e) ∨ SameSize(e, e)

5. | LeftOf(a, b) ∨ Smaller(a, b)
   | Smaller(a, b) → P
   | LeftOf(a, b) → Q

   | Q ∨ P

10.34 Prove a single lemma which can be used to complete each of the proofs in the files Exercise 10.34.1 and Exercise 10.34.2 in one step using the **Lemma** rule. Submit the lemma file as Proof 10.34.

CHAPTER 11

Multiple Quantifiers

So far, we've considered only sentences that contain a single quantifier symbol. This was enough to express the simple quantified forms studied by Aristotle, but hardly shows the expressive power of the modern quantifiers of first-order logic. Where the quantifiers of FOL come into their own is in expressing claims which, in English, involve several quantified noun phrases.

Long, long ago, probably before you were even born, there was an advertising campaign that ended with the tag line: *Everybody doesn't like something, but nobody doesn't like Sara Lee.* Now there's a quantified sentence! It goes without saying that this was every logician's favorite ad campaign. Or consider Lincoln's famous line: *You may fool all of the people some of the time; you can even fool some of the people all of the time; but you can't fool all of the people all of the time.* Why, the mind reels!

To express claims like these, and to reveal their logic, we need to juggle more than one quantifier in a single sentence. But it turns out that, like juggling, this requires a fair bit of preparation and practice.

Multiple uses of a single quantifier

When you learn to juggle, you start by tossing balls in a single hand, not crossing back and forth.[1] We'll start by looking at sentences that have multiple instances of $\forall$, or multiple instances of $\exists$, but no mixing of the two. Here are a couple of sentences that contain multiple quantifiers:

$$\exists x \, \exists y \, [\mathsf{Cube}(x) \land \mathsf{Tet}(y) \land \mathsf{LeftOf}(x, y)]$$
$$\forall x \, \forall y \, [(\mathsf{Cube}(x) \land \mathsf{Tet}(y)) \to \mathsf{LeftOf}(x, y)]$$

Try to guess what these say. You shouldn't have any trouble: The first says that some cube is left of a tetrahedron; the second says that every cube is left of every tetrahedron.

In these examples, all the quantifiers are out in front (in what we'll later call *prenex form*) but there is no need for them to be. In fact the same claims could be expressed, perhaps more clearly, by the following sentences:

---

[1] We thank juggler and Stanford student, Daniel Jacobs, for pointing out that this is not, in fact, how most beginning jugglers are taught. However, we prefer the simile to the facts.

$$\exists x \, [\mathsf{Cube}(x) \land \exists y \, (\mathsf{Tet}(y) \land \mathsf{LeftOf}(x, y))]$$
$$\forall x \, [\mathsf{Cube}(x) \rightarrow \forall y \, (\mathsf{Tet}(y) \rightarrow \mathsf{LeftOf}(x, y))]$$

The reason these may seem clearer is that they show that the claims have an overall Aristotelian structure. The first says that some cube has the property expressed by $\exists y \, (\mathsf{Tet}(y) \land \mathsf{LeftOf}(x, y))$, namely, being left of some tetrahedron. The second says that every cube has the property expressed by $\forall y \, (\mathsf{Tet}(y) \rightarrow \mathsf{LeftOf}(x, y))$, namely, being left of every tetrahedron.

It is easy to see that these make the same claims as the first pair, even though, in the case of the universal claim, the structure of the FOL sentence has changed considerably. The principles studied in Chapter 10 would allow us to prove these equivalences, if we wanted to take the time.

There is one tricky point that arises with the use of multiple existential quantifiers or multiple universal quantifiers. It's a simple one, but there isn't a logician alive who hasn't been caught by it at some time or other. It'll catch you too. We'll illustrate it in the following **Try It**.

You try it
. . . .

1. Suppose you are evaluating the following sentence in a world with four ◀
   cubes lined up in the front row:

   $$\forall x \, \forall y \, [(\mathsf{Cube}(x) \land \mathsf{Cube}(y)) \rightarrow (\mathsf{LeftOf}(x, y) \lor \mathsf{RightOf}(x, y))]$$

   Do you think the sentence is true in such a world?

2. Open Cantor's Sentences and Cantor's World, and evaluate the first sentence ◀
   in the world. If you are surprised by the outcome, play the game committed
   to the truth of the sentence.

3. It is tempting to read this sentence as claiming that if x and y are cubes, ◀
   then either x is left of y or x is right of y. But there is a conversational
   implicature in this way of speaking, one that is very misleading. The use of
   the plural "cube*s*" suggests that x and y are distinct cubes, but this is not
   part of the claim made by the first-order sentence. In fact, our sentence is
   false in this world, as it must be in *any* world that contains even one cube.

4. If we really wanted to express the claim that every cube is to the left or ◀
   right of every *other* cube, then we would have to write

   $$\forall x \, \forall y \, [(\mathsf{Cube}(x) \land \mathsf{Cube}(y) \land x \neq y) \rightarrow (\mathsf{LeftOf}(x, y) \lor \mathsf{RightOf}(x, y))]$$

   Modify the first sentence in this way and check it in the world.

▶ 5. The second sentence in the file looks for all the world like it says there are two cubes. But it doesn't. Delete all but one cube in the world and check to see that it's still true. Play the game committed to FALSE and see what happens.

▶ 6. See if you can modify the second sentence so it is false in a world with only one cube, but true if there are two or more. (Use ≠ like we did above.) Save the modified sentences as Sentences Multiple 1.

. . . . *Congratulations*

identity and variables

In general, to say that every *pair* of distinct objects stands in some relation, you need a sentence of the form $\forall x \, \forall y \, (x \neq y \rightarrow \ldots)$, and to say that there are *two* objects with a certain property, you need a sentence of the form $\exists x \, \exists y \, (x \neq y \wedge \ldots)$. Of course, other parts of the sentence often guarantee the distinctness for you. For example if you say that every tetrahedron is larger than every cube:

$$\forall x \, \forall y \, ((\mathsf{Tet}(x) \wedge \mathsf{Cube}(y)) \rightarrow \mathsf{Larger}(x, y))$$

then the fact that x must be a tetrahedron and y a cube ensures that your claim says what you intended.

> **Remember**
>
> When evaluating a sentence with multiple quantifiers, don't fall into the trap of thinking that distinct variables range over distinct objects. In fact, the sentence $\forall x \, \forall y \, \mathsf{P}(x, y)$ logically implies $\forall x \, \mathsf{P}(x, x)$, and the sentence $\exists x \, \mathsf{P}(x, x)$ logically implies $\exists x \, \exists y \, \mathsf{P}(x, y)$!

Exercises

11.1 If you skipped the **You try it** section, go back and do it now. Submit the file Sentences Multiple 1.

11.2 (Simple multiple quantifier sentences) The file Frege's Sentences contains 14 sentences; the first seven begin with a pair of existential quantifiers, the second seven with a pair of universal quantifiers. Go through the sentences one by one, evaluating them in Peirce's World. Though you probably won't have any trouble understanding these sentences, don't forget to use the game if you do. When you understand all the sentences, modify the size and location of a single block so that the first seven sentences are true and the second seven false. Submit the resulting world.

11.3 (Getting fancier) Open up Peano's World and Peano's Sentences. The sentence file contains 30 assertions that Alex made about this world. Evaluate Alex's claims. If you have trouble with any, play the game (several times if necessary) until you see where you are going wrong. Then change each of Alex's false claims into a true claim. If you can make the sentence true by adding a clause of the form x ≠ y, do so. Otherwise, see if you can turn the false claim into an interesting truth: don't just add a negation sign to the front of the sentence. Submit your corrected list of sentences.

11.4 (Describing a world) Let's try our hand describing a world using multiple quantifiers. Open Finsler's World and start a new sentence file.

1. Notice that all the small blocks are in front of all the large blocks. Use your first sentence to say this.
2. With your second sentence, point out that there's a cube that is larger than a tetrahedron.
3. Next, say that all the cubes are in the same column.
4. Notice, however, that this is not true of the tetrahedra. So write the same sentence about the tetrahedra, but put a negation sign out front.
5. Every cube is also in a different row from every other cube. Say this.
6. Again, this isn't true of the tetrahedra, so say that it's not.
7. Notice there are different tetrahedra that are the same size. Express this fact.
8. But there aren't different cubes of the same size, so say that, too.

Are all your translations true in Finsler's World? If not, try to figure out why. In fact, play around with the world and see if your first-order sentences always have the same truth values as the claims you meant to express. Check them out in Konig's World, where all of the original claims are false. Are your sentences all false? When you think you've got them right, submit your sentence file.

11.5 (Building a world) Open Ramsey's Sentences. Build a world in which sentences 1–10 are all true at once (ignore sentences 11–20 for now). These first ten sentences all make either *particular* claims (that is, they contain no quantifiers) or *existential* claims (that is, they assert that things of a certain sort exist). Consequently, you could make them true by successively adding objects to the world. But part of the exercise is to make them all true *with as few objects as possible*. You should be able to do it with a total of six objects. So rather than adding objects for each new sentence, only add new objects when absolutely necessary. Again, be sure to go back and check that all the sentences are true when you are finished. Submit your world as World 11.5. [Hint: To make all the sentences true with six blocks, you will have to watch out for some intentionally misleading implicatures. For example, one of the objects will have to have two names.]

11.6 (Modifying the world) Sentences 11-20 of **Ramsey's Sentences** all make *universal* claims. That is,
they all say that every object in the world has some property or other. Check to see whether the
world you have built in Exercise 11.5 satisfies the universal claims expressed by these sentences.
If not, modify the world so it makes all 20 sentences true at once. Submit your modified world
as **World 11.6**. (Make sure you submit both **World 11.5** and **World 11.6** to get credit for both
exercises.)

11.7 (Block parties) The interaction of quantifiers and negation gives rise to subtleties that can be
pretty confusing. Open **Löwenheim's Sentences**, which contains eight sentences divided into two
sets. Suppose we imagine a column containing blocks to be a *party* and think of the blocks in
the column as the attendees. We'll say a party is *lonely* if there's only one block attending it,
and say a party is *exclusive* if there's any block who's not there (i.e., who's in another column).

1. Using this terminology, give simple and clear English renditions of each of the sentences.
   For example, sentence 2 says *some of the parties are not lonely*, and sentence 7 says *there's
   only one party*. You'll find sentences 4 and 9 the hardest to understand. Construct a lot
   of worlds to see what they mean.

2. With the exception of 4 and 9, all of the sentences are first-order equivalent to other
   sentences on the list, or to negations of other sentences (or both). Which sentences are 3
   and 5 equivalent to? Which sentences do 3 and 5 negate?

3. Sentences 4 and 9 are logically independent: it's possible for the two to have any pattern
   of truth values. Construct four worlds: one in which both are true (**World 11.7.1**), one
   in which 4 is true and 9 false (**World 11.7.2**), one in which 4 is false and 9 true (**World
   11.7.3**), and one in which both are false (**World 11.7.4**).

Submit the worlds you've constructed and turn the remaining answers in to your instructor.

SECTION 11.2
Mixed quantifiers

Ready to start juggling with both hands? We now turn to the important case
in which universal and existential quantifiers get mixed together. Let's start
with the following sentence:

$$\forall x \, [\mathsf{Cube}(x) \rightarrow \exists y \, (\mathsf{Tet}(y) \wedge \mathsf{LeftOf}(x, y))]$$

This sentence shouldn't throw you. It has the overall Aristotelian form
$\forall x \, [\mathsf{P}(x) \rightarrow \mathsf{Q}(x)]$, which we have seen many times before. It says that every
cube has some property or other. What property? The property expressed

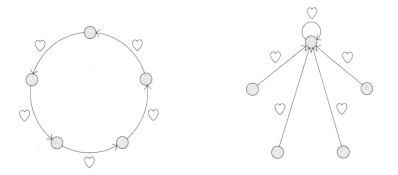

Figure 11.1: A circumstance in which $\forall x\, \exists y\, \mathsf{Likes}(x, y)$ holds versus one in which $\exists y\, \forall x\, \mathsf{Likes}(x, y)$ holds. It makes a big difference to someone!

by $\exists y\, (\mathsf{Tet}(y) \wedge \mathsf{LeftOf}(x, y))$, that is, the property of being left of a tetrahedron. Thus our first-order sentence claims that every cube is to the left of a tetrahedron.

This same claim could also be expressed in a number of other ways. The most important alternative puts the quantifiers all out front, in *prenex* form. Though the prenex form is less natural as a translation of the English, *Every cube is left of some tetrahedron,* it is logically equivalent:

$$\forall x\, \exists y\, [\mathsf{Cube}(x) \rightarrow (\mathsf{Tet}(y) \wedge \mathsf{LeftOf}(x, y))]$$

When we have a sentence with a string of mixed quantifiers, the order of the quantifiers makes a difference. This is something we haven't had to worry about with sentences that contain only universal or only existential quantifiers. Clearly, the sentence $\forall x\, \forall y\, \mathsf{Likes}(x, y)$ is logically equivalent to the sentence where the order of the quantifiers is reversed: $\forall y\, \forall x\, \mathsf{Likes}(x, y)$. They are both true just in case everything in the domain of discourse (say, people) likes everything in the domain of discourse. Similarly, $\exists x\, \exists y\, \mathsf{Likes}(x, y)$ is logically equivalent to $\exists y\, \exists x\, \mathsf{Likes}(x, y)$: both are true if something likes something.

order of quantifiers

This is not the case when the quantifiers are mixed. $\forall x\, \exists y\, \mathsf{Likes}(x, y)$ says that everyone likes someone, which is true in both circumstances shown in Figure 11.1. But $\exists y\, \forall x\, \mathsf{Likes}(x, y)$ says that there is some lucky devil who everyone likes. This is a far stronger claim, and is only true in the second circumstance shown in Figure 11.1. So when dealing with mixed quantifiers, you have to be very sensitive to the order of quantifiers. We'll learn more about getting the order of quantifiers right in the sections that follow.

You try it

. . . .

▶ 1. Open the files Mixed Sentences and Konig's World. If you evaluate the two sentences, you'll see that the first is true and the second false. We're going to play the game to see why they aren't both true.

▶ 2. Play the game on the first sentence, specifying your initial commitment as TRUE. Since this sentence is indeed true, you should find it easy to win. When Tarski's World makes its choice, all you need to do is choose any block in the same row as Tarski's.

▶ 3. Now play the game with the second sentence, again specifying your initial commitment as TRUE. This time Tarski's World is going to beat you because you've got to choose first. As soon as you choose a block, Tarski chooses a block in the other row. Play a couple of times, choosing blocks in different rows. See who's got the advantage now?

▶ 4. Just for fun, delete a row of blocks so that both of the sentences come out true. Now you can win the game. So there, Tarski! She who laughs last laughs best. Save the modified world as World Mixed 1.

. . . . *Congratulations*

order of variables

Have you noticed that switching the order of the quantifiers does something quite different from switching around the variables in the body of the sentence? For example, consider the sentences

$$\forall x\, \exists y\, \text{Likes}(x, y)$$

$$\forall x\, \exists y\, \text{Likes}(y, x)$$

Assuming our domain consists of people, the first of these says that everybody likes somebody or other, while the second says everybody is liked *by* somebody or other. These are both very different claims from either of these:

$$\exists y\, \forall x\, \text{Likes}(x, y)$$

$$\exists y\, \forall x\, \text{Likes}(y, x)$$

Here, the first claims that there is a (very popular) person whom everybody likes, while the second claims that there is a (very indiscriminate?) person who likes absolutely everyone.

In the last section, we saw how using two existential quantifiers and the identity predicate, we can say that there are at least two things with a particular property (say cubes):

$$\exists x\, \exists y\, (x \neq y \wedge \text{Cube}(x) \wedge \text{Cube}(y))$$

With mixed quantifiers and identity, we can say quite a bit more. For example, consider the sentence

$$\exists x \, (\mathsf{Cube}(x) \wedge \forall y \, (\mathsf{Cube}(y) \rightarrow y = x))$$

This says that there is a cube, and furthermore every cube is identical to it. Some cube, in other words, is the *only* cube. Thus, this sentence will be true if and only if there is exactly one cube. There are many ways of saying things like this in FOL; we'll run across others in the exercises. We discuss numerical claims more systematically in Chapter 14.

exactly one

Remember

When you are dealing with mixed quantifiers, the order is very important. $\forall x \, \exists y \, R(x, y)$ is not logically equivalent to $\exists y \, \forall x \, R(x, y)$.

Exercises

___

11.8 If you skipped the **You try it** section, go back and do it now. Submit the file World Mixed 1.
✴

11.9 (Simple mixed quantifier sentences) Open Hilbert's Sentences and Peano's World. Evaluate the
✴ sentences one by one, playing the game if an evaluation surprises you. Once you understand the sentences, modify the false ones by adding a single negation sign so that they come out true. The catch is that you aren't allowed to add the negation sign to the front of the sentence! Add it to an atomic formula, if possible, and try to make the claim nonvacuously true. (This won't always be possible.) Make sure you understand both why the original sentence is false and why your modified sentence is true. When you're done, submit your sentence list with the changes.

11.10 (Mixed quantifier sentences with identity) Open Leibniz's World and use it to evaluate the
✴ sentences in Leibniz's Sentences. Make sure you understand all the sentences and follow any instructions in the file. Submit your modified sentence list.

11.11 (Building a world) Create a world in which all ten sentences in Arnault's Sentences are true.
✴ Submit your world.

11.12 (Name that object) Open Carroll's World and Hercule's Sentences. Try to figure out which objects
✴ have names, and what they are. You should be able to figure this out from the sentences, all of which are true. Once you have come to your conclusion, add the names to the objects and check to see if all the sentences are true. Submit your modified world.

The remaining three exercises all have to do with the sentences in the file Buridan's Sentences *and build on one another.*

11.13 (Building a world) Open Buridan's Sentences. Build a world in which all ten sentences are true.
↗ Submit your world.

11.14 (Consequence) These two English sentences are consequences of the ten sentences in Buridan's
↗ Sentences.
  1. *There are no cubes.*
  2. *There is exactly one large tetrahedron.*

Because of this, they must be true in any world in which Buridan's sentences are all true. So of course they must be true in World 11.13, no matter how you built it.

  ○ Translate the two sentences, adding them to the list in Buridan's Sentences. Name the expanded list Sentences 11.14. Verify that they are all true in World 11.13.

  ○ Modify the world by adding a cube. Try placing it at various locations and giving it various sizes to see what happens to the truth values of the sentences in your file. One or more of the original ten sentences will always be false, though different ones at different times. Find a world in which only one of the original ten sentences is false and name it World 11.14.1.

  ○ Next, get rid of the cube and add a second large tetrahedron. Again, move it around and see what happens to the truth values of the sentences. Find a world in which only one of the original ten sentences is false and name it World 11.14.2.

Submit your sentence file and two world files.

11.15 (Independence) Show that the following sentence is independent of those in Buridan's Sentences,
↗* that is, neither it nor its negation is a consequence of those sentences.

$$\exists x \, \exists y \, (x \neq y \land \text{Tet}(x) \land \text{Tet}(y) \land \text{Medium}(x) \land \text{Medium}(y))$$

You will do this by building two worlds, one in which this sentence is false (call this World 11.15.1) and one in which it is true (World 11.15.2)—but both of which make all of Buridan's sentences true.

The step-by-step method of translation

When an English sentence contains more than one quantified noun phrase, translating it can become quite confusing unless you approach it in a very systematic way. It often helps to go through a few intermediate steps, treating the quantified noun phrases one at a time.

Suppose, for example, we wanted to translate the sentence *Each cube is to the left of a tetrahedron*. Here, there are two quantified noun phrases: *each cube* and *a tetrahedron*. We can start by dealing with the first noun phrase, temporarily treating the complex phrase *is-to-the-left-of-a-tetrahedron* as a single unit. In other words, we can think of the sentence as a single quantifier sentence, on the order of *Each cube is small*. The translation would look like this:

$$\forall x\,(\text{Cube}(x) \to x\ \textit{is-to-the-left-of-a-tetrahedron})$$

Of course, this is not a sentence in our language, so we need to translate the expression x *is-to-the-left-of-a-tetrahedron*. But we can think of this expression as a single quantifier sentence, at least if we pretend that x is a name. It has the same general form as the sentence **b** *is to the left of a tetrahedron*, and would be translated as

$$\exists y\,(\text{Tet}(y) \wedge \text{LeftOf}(x, y))$$

Substituting this in the above, we get the desired translation of the original English sentence:

$$\forall x\,(\text{Cube}(x) \to \exists y\,(\text{Tet}(y) \wedge \text{LeftOf}(x, y)))$$

This is exactly the sentence with which we began our discussion of mixed quantifiers.

This step-by-step process really comes into its own when there are lots of quantifiers in a sentence. It would be very difficult for a beginner to translate a sentence like *No cube to the right of a tetrahedron is to the left of a larger dodecahedron* in a single blow. Using the step-by-step method makes it straightforward. Eventually, though, you will be able to translate quite complex sentences, going through the intermediate steps in your head.

Exercises

11.16 (Using the step-by-step method of translation)
⬧⋆

    ○ Open Montague's Sentences. This file contains expressions that are halfway between English and first-order logic. Our goal is to edit this file until it contains translations of the following English sentences. You should read the English sentence below, make sure you understand how we got to the halfway point, and then complete the translation by replacing the hyphenated expression with a wff of first-order logic.

        1. *Every cube is to the left of every tetrahedron.* [In the Sentence window, you see the halfway completed translation, together with some blanks that need to be replaced by wffs. Commented out below this, you will find an intermediate "sentence." Make sure you understand how we got to this intermediate stage of the translation. Then complete the translation by replacing the blank with

$$\forall y \, (\text{Tet}(y) \to \text{LeftOf}(x, y))$$

Once this is done, check to see if you have a well-formed sentence. Does it look like a proper translation of the original English? It should.]

        2. *Every small cube is in back of a large cube.*
        3. *Some cube is in front of every tetrahedron.*
        4. *A large cube is in front of a small cube.*
        5. *Nothing is larger than everything.*
        6. *Every cube in front of every tetrahedron is large.*
        7. *Everything to the right of a large cube is small.*
        8. *Nothing in back of a cube and in front of a cube is large.*
        9. *Anything with nothing in back of it is a cube.*
       10. *Every dodecahedron is smaller than some tetrahedron.*

    Save your sentences as Sentences 11.16.

    ○ Open Peirce's World. Notice that all the English sentences are true in this world. Check to see that all of your translations are true as well. If they are not, see if you can figure out where you went wrong.

    ○ Open Leibniz's World. Note that the English sentences 5, 6, 8, and 10 are true in this world, while the rest are false. Verify that your translations have the same truth values. If they don't, fix them.

    ○ Open Ron's World. Here, the true sentences are 2, 3, 4, 5, and 8. Check that your translations have the right values, and correct them if they don't.

11.17 (More multiple quantifier sentences) Now, we will try translating some multiple quantifier
↙ sentences completely from scratch. You should try to use the step-by-step procedure.

○ Start a new sentence file and translate the following English sentences.

1. *Every tetrahedron is in front of every dodecahedron.*
2. *No dodecahedron has anything in back of it.*
3. *No tetrahedron is the same size as any cube.*
4. *Every dodecahedron is the same size as some cube.*
5. *Anything between two dodecahedra is a cube.* [Note: This use of *two* really can be paraphrased using *between a dodecahedron and a dodecahedron.*]
6. *Every cube falls between two objects.*
7. *Every cube with something in back of it is small.*
8. *Every dodecahedron with nothing to its right is small.*
9. (⋆) *Every dodecahedron with nothing to its right has something to its left.*
10. *Any dodecahedron to the left of a cube is large.*

○ Open Bolzano's World. All of the above English sentences are true in this world. Verify that all your translations are true as well.

○ Now open Ron's World. The English sentences 4, 5, 8, 9, and 10 are true, but the rest are false. Verify that the same holds of your translations.

○ Open Claire's World. Here you will find that the English sentences 1, 3, 5, 7, 9, and 10 are true, the rest false. Again, check to see that your translations have the appropriate truth value.

○ Finally, open Peano's World. Notice that only sentences 8 and 9 are true. Check to see that your translations have the same truth values.

SECTION 11.4

Paraphrasing English

Some English sentences do not easily lend themselves to direct translation using the step-by-step procedure. With such sentences, however, it is often quite easy to come up with an English paraphrase that is amenable to the procedure. Consider, for example, *If a freshman takes a logic class, then he or she must be smart.* The step-by-step procedure does not work here. If we try to apply the procedure we would get something like

$$\exists x\,(\mathsf{Freshman}(x) \wedge \exists y\,(\mathsf{LogicClass}(y) \wedge \mathsf{Takes}(x,y))) \rightarrow \mathsf{Smart}(x)$$

The problem is that this "translation" is not a sentence, since the last occurrence of x is free. However, we can paraphrase the sentences as *Every freshman who takes a logic class must be smart.* This is easily treated by the procedure, with the result being

$$\forall x\,[(\mathsf{Freshman}(x) \wedge \exists y\,(\mathsf{LogicClass}(y) \wedge \mathsf{Takes}(x,y))) \rightarrow \mathsf{Smart}(x)]$$

donkey sentences

There is one particularly notorious kind of sentence that needs paraphrasing to get an adequate first-order translation. They are known as *donkey sentences,* because the first and most discussed example of this kind is the sentence

Every farmer who owns a donkey beats it.

What makes such a sentence a bit tricky is the existential noun phrase "a donkey" in the noun phrase "every farmer who owns a donkey." The existential noun phrase serves as the antecedent of the pronoun "it" in the verb phrase; its the donkey that gets beaten. Applying the step-by-step method might lead you to translate this as follows:

$$\forall x\,(\mathsf{Farmer}(x) \wedge \exists y\,(\mathsf{Donkey}(y) \wedge \mathsf{Owns}(x,y)) \rightarrow \mathsf{Beats}(x,y))$$

This translation, however, cannot be correct since it's not even a sentence; the occurrence of y in $\mathsf{Beats}(x,y)$ is free, not bound. If we move the parenthesis to capture this free variable, we obtain the following, which means something quite different from our English sentence.

$$\forall x\,(\mathsf{Farmer}(x) \wedge \exists y\,(\mathsf{Donkey}(y) \wedge \mathsf{Owns}(x,y) \wedge \mathsf{Beats}(x,y)))$$

This means that everything in the domain of discourse is a farmer who owns and beats a donkey, something which neither implies nor is implied by the original sentence.

To get a correct first-order translation of the original donkey sentence, it can be paraphrased as

Every donkey owned by any farmer is beaten by them.

This sentence clearly needs two universal quantifiers in its translation:

$$\forall x\,(\mathsf{Donkey}(x) \rightarrow \forall y\,((\mathsf{Farmer}(y) \wedge \mathsf{Owns}(y,x)) \rightarrow \mathsf{Beats}(y,x)))$$

> **Remember**
>
> In translating from English to FOL, the goal is to get a sentence that has the same meaning as the original. This sometimes requires changes in the surface form of the sentence.

Exercises

11.18 (Sentences that need paraphrasing before translation) Translate the following sentences by first
✦* giving a suitable English paraphrase. Some of them are donkey sentences, so be careful.
1. *Only large objects have nothing in front of them.*
2. *If a cube has something in front of it, then it's small.*
3. *Every cube in back of a dodecahedron is also smaller than it.*
4. *If **e** is between two objects, then they are both small.*
5. *If a tetrahedron is between two objects, then they are both small.*

Open Ron's World. Recall that there are lots of hidden things in this world. Each of the above English sentences is true in this world, so the same should hold of your translations. Check to see that it does. Now open Bolzano's World. In this world, only sentence 3 is true. Check that the same holds of your translations. Next open Wittgenstein's World. In this world, only the English sentence 5 is true. Verify that your translations have the same truth values. Submit your sentence file.

11.19 (More sentences that need paraphrasing before translation) Translate the following sentences
✦* by first giving a suitable English paraphrase.
1. *Every dodecahedron is as large as every cube.* [Hint: Since we do not have anything corresponding to *as large as* (by which we mean at least as large as) in our language, you will first need to paraphrase this predicate using *larger than or same size as*.]
2. *If a cube is to the right of a dodecahedron but not in back of it, then it is as large as the dodecahedron.*
3. *No cube with nothing to its left is between two cubes.*
4. *The only large cubes are **b** and **c**.*
5. *At most **b** and **c** are large cubes.* [Note: There is a significant difference between this sentence and the previous one. This one does not imply that b and c are large cubes, while the previous sentence does.]

Open Ron's World. Each of the above English sentences is true in this world, so the same should hold of your translations. Check to see that it does. Now open Bolzano's World. In this world, only sentences 3 and 5 are true. Check that the same holds of your translations. Next open Wittgenstein's World. In this world, only the English sentences 2 and 3 are true. Verify that your translations have the same truth values. Submit your sentence file.

11.20 (More translations) The following English sentences are true in Godel's World. Translate them, and make sure your translations are also true. Then modify the world in various ways, and check that your translations track the truth value of the English sentence.

1. *Nothing to the left of **a** is larger than everything to the left of **b**.*
2. *Nothing to the left of **a** is smaller than anything to the left of **b**.*
3. *The same things are left of **a** as are left of **b**.*
4. *Anything to the left of **a** is smaller than something that is in back of every cube to the right of **b**.*
5. *Every cube is smaller than some dodecahedron but no cube is smaller than every dodecahedron.*
6. *If **a** is larger than some cube then it is smaller than every tetrahedron.*
7. *Only dodecahedra are larger than everything else.*
8. *All objects with nothing in front of them are tetrahedra.*
9. *Nothing is between two objects which are the same shape.*
10. *Nothing but a cube is between two other objects.*
11. ***b** has something behind it which has at least two objects behind it.*
12. *More than one thing is smaller than something larger than **b**.*

Submit your sentence file.

11.21 Using the symbols introduced in Table 1.2, page 30, translate the following into FOL. Do not introduce any additional names or predicates. Comment on any shortcomings in your translations. When you are done, submit your sentence file and turn in your comments to your instructor.

1. *Every student gave a pet to some other student sometime or other.*
2. *Claire is not a student unless she owned a pet (at some time or other).*
3. *No one ever owned both Folly and Scruffy at the same time.*
4. *No student fed every pet.*
5. *No one who owned a pet at 2:00 was angry.*
6. *No one gave Claire a pet this morning.* (Assume that "this morning" simply means before 12:00.)
7. *If Max ever gave Claire a pet, she owned it then and he didn't.*
8. *You can't give someone something you don't own.*
9. *Max fed all of his pets before Claire fed any of her pets.* (Assume that "Max's pets" are the pets he owned at 2:00, and the same for Claire.)
10. *Max gave Claire a pet between 2:00 and 3:00. It was hungry.*

11.22 Using the symbols introduced in Table 1.2, page 30, translate the following into colloquial English. Assume that each of the sentences is asserted at 2 p.m. on January 2, 2011, and use this fact to make your translations more natural. For example, you could translate Owned(max, folly, 2:00) as *Max owns Folly.*

1. $\forall x\,[\text{Student}(x) \rightarrow \exists z\,(\text{Pet}(z) \wedge \text{Owned}(x, z, 2:00))]$
2. $\exists x\,[\text{Student}(x) \wedge \forall z\,(\text{Pet}(z) \rightarrow \text{Owned}(x, z, 2:00))]$
3. $\forall x\,\forall t\,[\text{Gave}(\text{max}, x, \text{claire}, t) \rightarrow \exists y\,\exists t'\,\text{Gave}(\text{claire}, x, y, t')]$
4. $\exists x\,[\text{Owned}(\text{claire}, x, 2:00) \wedge \exists t\,(t < 2:00 \wedge \text{Gave}(\text{max}, x, \text{claire}, t))]$
5. $\exists x\,\exists t\,(1:55 < t \wedge t < 2:00 \wedge \text{Gave}(\text{max}, x, \text{claire}, t))$
6. $\forall y\,[\text{Person}(y) \rightarrow \exists x\,\exists t\,(1:55 < t \wedge t < 2:00 \wedge \text{Gave}(\text{max}, x, y, t))]$
7. $\exists z\,\{\text{Student}(z) \wedge \forall y\,[\text{Person}(y) \rightarrow \exists x\,\exists t\,(1:55 < t \wedge t < 2:00 \wedge \text{Gave}(z, x, y, t))]\}$

11.23 Translate the following into FOL. As usual, explain the meanings of the names, predicates, and function symbols you use, and comment on any shortcomings in your translations.

1. *There's a sucker born every minute.*
2. *Whither thou goest, I will go.*
3. *Soothsayers make a better living in the world than truthsayers.*
4. *To whom nothing is given, nothing can be required.*
5. *If you always do right, you will gratify some people and astonish the rest.*

Ambiguity and context sensitivity

There are a couple of things that make the task of translating between English and first-order logic difficult. One is the sparseness of primitive concepts in FOL. While this sparseness makes the language easy to learn, it also means that there are frequently no very natural ways of saying what you want to say. You have to try to find circumlocutions available with the resources at hand. While this is often possible in mathematical discourse, it is frequently impossible for ordinary English. (We will return to this matter later.)

The other thing that makes it difficult is that English is rife with ambi-   *ambiguity*
guities, whereas the expressions of first-order logic are unambiguous (at least if the predicates used are unambiguous). Thus, confronted with a sentence of English, we often have to choose one among many possible interpretations in deciding on an appropriate translation. Just which is appropriate usually depends on context.

The ambiguities become especially vexing with quantified noun phrases. Consider, for example, the following joke, taken from *Saturday Night Live*:

> *Every minute a man is mugged in New York City. We are going to interview him tonight.*

What makes this joke possible is the ambiguity in the first sentence. The most natural reading would be translated by

$$\forall x\, (\text{Minute}(x) \rightarrow \exists y\, (\text{Man}(y) \wedge \text{MuggedDuring}(y, x)))$$

But the second sentence forces us to go back and reinterpret the first in a rather unlikely way, one that would be translated by

$$\exists y\, (\text{Man}(y) \wedge \forall x\, (\text{Minute}(x) \rightarrow \text{MuggedDuring}(y, x)))$$

This is often called the *strong* reading, the first the *weak* reading, since this one entails the first but not vice versa.

context sensitivity

Notice that the reason the strong translation is less likely is not determined by the form of the original sentence. You can find examples of the same form where the strong reading is more natural. For example, suppose you have been out all day and, upon returning to your room, your roommate says, "Every ten minutes some guy from the registrar's office has called trying to reach you." Here it is the strong reading where the existential "some guy" is given wide scope that is most likely the one intended.

There is another important way in which context often helps us disambiguate an ambiguous utterance or claim. We often speak about situations that we can see, and say something about it in a way that makes perfectly clear, given that what we see. Someone looking at the same scene typically finds it clear and unambiguous, while someone to whom the scene is not visible may find our utterance quite unclear. Let's look at an example.

You try it

▶ 1. It is hard to get too many blocks to adjoin a single block in Tarski's World, because many of the blocks overflow their squares and so do not leave room for similar sized blocks on adjacent squares. How many medium dodecahedra do you think it is possible to have adjacent to a single medium cube?

▶ 2. Open Anderson's First World. Notice that this world has four medium dodecahedra surrounding a single medium cube.

▶ 3. Imagine that Max makes the following claim about this situation:

At least four medium dodecahedra are adjacent to a medium cube.

The most natural understanding of Max's claim in this context is as the claim that there is a single cube to which at least four dodecahedra are adjacent.

4. There is, however, another reading of Max's sentence. Imagine that a tyrant tetrahedron is determined to assassinate any medium dodecahedron with the effrontery to be adjacent to a medium cube. Open Anderson's Second World and assume that Max makes a claim about this world with the above sentence. Here a weaker reading of his claim would be the more reasonable, one where Max is asserting that at least four medium dodecahedra are each adjacent to some medium cube or other.

5. We would ask you to translate these two readings of the one sentence into FOL, but unfortunately you have not yet learned how translate "at least four" into FOL yet; this will come in Chapter 14 (see Exercise 14.5 in particular). Instead consider the following sentence:

Every medium dodecahedron is adjacent to a medium cube.

Write the stronger and weaker translations in a file, in that order. Check that the stronger reading is only true in the first of Anderson's worlds, while the weaker reading is true in both. Save your file as Sentences Max 1.

. . . . *Congratulations*

The problems of translation are much more difficult when we look at extended discourse, where more than one sentence comes in. To get a feeling for the difficulty, we start of with a couple of problems about extended discourse.

extended discourse

> **Remember**
>
> A important source of ambiguity in English stems from the order in which quantifiers are interpreted. To translate such a sentence into FOL, you must know which order the speaker of the sentence had in mind. This can often be determined by looking at the context in which the sentence was used.

Exercises

11.24 If you skipped the **You try it** section, go back and do it now. Save your sentence file as Sentences Max 1.

11.25 (Translating extended discourse)

○ Open Reichenbach's World 1 and examine it. Check to see that all of the sentences in the following discourse are true in this world.

> *There are (at least) two cubes. There is something between them. It is a medium dodecahedron. It is in front of a large dodecahedron. These two are left of a small dodecahedron. There are two tetrahedra.*

Translate this discourse into a single first-order sentence. Check to see that your translation is true. Now check to see that your translation is false in Reichenbach's World 2.

○ Open Reichenbach's World 2. Check to see that all of the sentences in the following discourse are true in this world.

> *There are two tetrahedra. There is something between them. It is a medium dodecahedron. It is in front of a large dodecahedron. There are two cubes. These two are left of a small dodecahedron.*

Translate this into a single first-order sentence. Check to see that your translation is true. Now check to see that your translation is false in Reichenbach's World 1. However, note that the English sentences in the two discourses are in fact exactly the same; they have just been rearranged! The moral of this exercise is that the correct translation of a sentence into first-order logic (or any other language) can be very dependent on context. Submit your sentence file.

11.26 (Ambiguity) Use Tarski's World to create a new sentence file and use it to translate the following sentences into FOL. Each of these sentences is ambiguous, so you should have two different translations of each. Put the two translations of sentence 1 in slots 1 and 2, the two translations of sentence 3 in slots 3 and 4, and so forth.

1. *Every cube is between a pair of dodecahedra.*
3. *Every cube to the right of a dodecahedron is smaller than it is.*
5. *Cube **a** is not larger than every dodecahedron.*

7. *No cube is to the left of some dodecahedron.*

9. *(At least) two cubes are between (at least) two dodecahedra.*

Now open Carroll's World. Which of your sentences are true in this world? You should find that exactly one translation of each sentence is true. If not, you should correct one or both of your translations. Notice that if you had had the world in front of you when you did the translations, it would have been harder to see the ambiguity in the English sentences. The world would have provided a context that made one interpretation the natural one. Submit your sentence file.

(Ambiguity and inference) Whether or not an argument is valid often hinges on how some ambiguous claim is taken. Here are two arguments, each of whose first premise is ambiguous. Translate each argument into FOL twice, corresponding to the ambiguity in the first premise. (In 11.27, ignore the reading where "someone" means "everyone.") Under one translation the conclusion follows: prove it. Under the other, it does not: describe a situation in which the premises are true but the conclusion false.

11.27
✎

Everyone admires someone who has red hair.

Anyone who admires himself is conceited.

─────

Someone with red hair is conceited.

11.28
✎

All that glitters is not gold.

This ring glitters.

─────

This ring is not gold.

Translations using function symbols

Intuitively, functions are a kind of relation. One's mother is one's mother because of a certain relationship you and she bear to one another. Similarly, $2 + 3 = 5$ because of a certain relationship between two, three, and five. Building on this intuition, it is not hard to see that anything that can be expressed in FOL with function symbols can also be expressed in a version of FOL where the function symbols have been replaced by relation symbols.

relations and functions

The basic idea can be illustrated easily. Let us use mother as a unary function symbol, but MotherOf as a *binary* relation symbol. Thus, for example, mother(max) = nancy and MotherOf(nancy, max) both state that Nancy is the mother of Max.

The basic claim is that anything we can say with the function symbol we can say in some other way using the relation symbol. As an example, here is a simple sentence using the function symbol:

$$\forall x\ \text{OlderThan}(\text{mother}(x), x)$$

It expresses the claim that a person's mother is always older than the person. To express the same thing with the relation symbol, we might write

$$\forall x \, \exists y \, [\mathsf{MotherOf}(y, x) \wedge \mathsf{OlderThan}(y, x)]$$

Actually, one might wonder whether the second sentence quite manages to express the claim made by the first, since all it says is that everyone has at least one mother who is older than they are. One might prefer something like

$$\forall x \, \forall y \, [\mathsf{MotherOf}(y, x) \rightarrow \mathsf{OlderThan}(y, x)]$$

This says that every mother of everyone is older than they are. But this too seems somewhat deficient. A still better translation would be to conjoin one of the above sentences with the following two sentences which, together, assert that the relation of being the mother of someone is functional. Everyone has at least one, and everyone has at most one.

$$\forall x \, \exists y \, \mathsf{MotherOf}(y, x)$$

and

$$\forall x \, \forall y \, \forall z \, [(\mathsf{MotherOf}(y, x) \wedge \mathsf{MotherOf}(z, x)) \rightarrow y = z]$$

We will study this sort of thing much more in Chapter 14, where we will see that these two sentences can jointly be expressed by one rather opaque sentence:

$$\forall x \, \exists y \, [\mathsf{MotherOf}(y, x) \wedge \forall z \, [\mathsf{MotherOf}(z, x) \rightarrow y = z]]$$

And, if we wanted to, we could then incorporate our earlier sentence and express the first claim by means of the horrendous looking:

$$\forall x \, \exists y \, [\mathsf{MotherOf}(y, x) \wedge \mathsf{OlderThan}(y, x) \wedge \forall z \, [\mathsf{MotherOf}(z, x) \rightarrow y = z]]$$

By now it should be clearer why function symbols are so useful. Look at all the connectives and additional quantifiers that have come into translating our very simple sentence

$$\forall x \, \mathsf{OlderThan}(\mathsf{mother}(x), x)$$

We present some exercises below that will give you practice translating sentences from English into FOL, sentences that show why it is nice to have function symbols around.

Remember

Anything you can express using an n-ary function symbol can also be expressed using an $n + 1$-ary relation symbol, plus the identity predicate, but at a cost in terms of the complexity of the sentences used.

11.29   Translate the following sentences into FOL twice, once using the function symbol mother, once using the relation symbol MotherOf.
1. *Claire's mother is older than Max's mother.*
2. *Everyone's mother's mother is older than Melanie.*
3. *Someone's mother's mother is younger than Mary.*

11.30   Translate the following into a version of FOL that has function symbols height, mother, and father, the predicate >, and names for the people mentioned.
1. *Mary's father is taller than Mary but not taller than Claire's father.*
2. *Someone is taller than Claire's father.*
3. *Someone's mother is taller than their father.*
4. *Everyone is taller than someone else.*
5. *No one is taller than himself.*
6. *Everyone but J.R. who is taller than Claire is taller than J.R.*
7. *Everyone who is shorter than Claire is shorter than someone who is shorter than Melanie's father.*
8. *Someone is taller than Jon's paternal grandmother but shorter than his maternal grandfather.*

Say which sentences are true, referring to the table in Figure 9.1 (p. 256). Take the domain of quantification to be the people mentioned in the table. Turn in your answers.

11.31   Translate the following sentences into the blocks language augmented with the four function symbols lm, rm, fm, and bm discussed in Section 1.5 (page 33) and further discussed in connection with quantifiers in Section 9.7 (page 254). Tell which of these sentences are true in Malcev's World.
1. *Every cube is to the right of the leftmost block in the same row.*
2. *Every block is in the same row as the leftmost block in the same row.*
3. *Some block is in the same row as the backmost block in the same column.*
4. *Given any two blocks, the first is the leftmost block in the same row as the second if and only if there is nothing to the left of the second.*
5. *Given any two blocks, the first is the leftmost block in the same row as the second if and only if there is nothing to the left of the second and the the two blocks are in the same row.*

Turn in your answers.

11.32 Using the first-order language of arithmetic described earlier, express each of the following in FOL.

1. Every number is either 0 or greater than 0.
2. The sum of any two numbers greater than 1 is smaller than the product of the same two numbers.
3. Every number is even. [This is false, of course.]
4. If $x^2 = 1$ then $x = 1$. [Hint: Don't forget the implicit quantifier.]
5. ** For any number x, if $ax^2 + bx + c = 0$ then either $x = \frac{-b + \sqrt{b^2 - 4ac}}{2a}$ or $x = \frac{-b - \sqrt{b^2 - 4ac}}{2a}$. In this problem treat a, b, c as constants but x as a variable, as usual in algebra.

Section 11.7

Prenex form

When we translate complex sentences of English into FOL, it is common to end up with sentences where the quantifiers and connectives are all scrambled together. This is usually due to the way in which the translations of complex noun phrases of English use both quantifiers and connectives:

$$\forall x\,(P(x) \rightarrow \dots)$$

$$\exists x\,(P(x) \wedge \dots)$$

As a result, the translation of (the most likely reading of) a sentence like *Every cube to the left of a tetrahedron is in back of a dodecahedron* ends up looking like

$$\forall x\,[(\text{Cube}(x) \wedge \exists y\,(\text{Tet}(y) \wedge \text{LeftOf}(x,y))) \rightarrow \exists y\,(\text{Dodec}(y) \wedge \text{BackOf}(x,y))]$$

While this is the most natural translation of our sentence, there are situations where it is not the most convenient one. It is sometimes important that we be able to rearrange sentences like this so that all the quantifiers are *prenex form* out in front and all the connectives in back. Such a sentence is said to be in *prenex form*, since all the quantifiers come first.

Stated more precisely, a wff is in *prenex normal form* if either it contains no quantifiers at all, or else is of the form

$$Q_1 v_1 Q_2 v_2 \dots Q_n v_n P$$

where each Q_i is either $\forall$ or $\exists$, each v_i is some variable, and the wff P is quantifier-free.

There are several reasons one might want to put sentences into prenex form. One is that it gives you a nice measure of the logical complexity of the sentences. What turns out to matter is not so much the number of quantifiers, as the number of times you get a flip from ∀ to ∃ or the other way round. The more of these so-called *alternations,* the more complex the sentence is, logically speaking. Another reason is that this prenex form is quite analogous to the conjunctive normal form for quantifier-free wffs we studied earlier. And like that normal form, it is used extensively in automated theorem proving.

quantifier alternations

It turns out that every sentence is logically equivalent to one (in fact many) in prenex form. In this section we will present some rules for carrying out this transformation. When we apply the rules to our earlier example, we will get

$$\forall x \, \forall y \, \exists z \, [(\text{Cube}(x) \land \text{Tet}(y) \land \text{LeftOf}(x, y)) \to (\text{Dodec}(z) \land \text{BackOf}(x, z))]$$

To arrive at this sentence, we did not just blindly pull quantifiers out in front. If we had, it would have come out all wrong. There are two problems. One is that the first ∃y in the original sentence is, logically speaking, inside a ¬. (To see why, replace → by its definition in terms of ¬ and ∨.) The DeMorgan laws for quantifiers tell us that it will end up being a universal quantifier. Another problem is that the original sentence has two quantifiers that bind the variable y. There is no problem with this, but if we pull the quantifiers out front, there is suddenly a clash. So we must first change one of the ys to some other variable, say z.

converting to prenex form

We have already seen the logical equivalences that are needed for putting sentences in prenex form. They were summarized in a box on page 285. They allowed us to move negations inside quantifiers by switching quantifiers, to distribute ∀ over ∧, ∃ over ∨, to replace bound variables by other variables, and to move quantifiers past formulas in which the variable being quantified is not free. In order to apply these maneuvers to sentences with → or ↔, one needs to either replace these symbols with equivalent versions using ¬, ∨ and ∧ (or else derive some similar rules for these symbols).

The basic strategy for putting sentences into prenex form is to work from the inside out, working on parts, then putting them together. By way of example, here is a chain of equivalences where we start with a sentence not in prenex normal form and turn it into a logically equivalent one that is prenex normal form, explaining why we do each step as we go.

$$\exists x \, P(x) \to \exists y \, Q(y)$$

In getting a formula into prenex form, it's a good idea to get rid of conditionals in favor of Boolean connectives, since these interact more straightforwardly with the quantifiers in our principles. So our first step is to obtain

$$\neg\exists x\, P(x) \vee \exists y\, Q(y)$$

Now we have a disjunction, but the first disjunct is no longer in prenex form. That can be fixed using DeMorgan's law:

$$\forall x\, \neg P(x) \vee \exists y\, Q(y)$$

Now we can use the Null Quantification Principle to move either of the quantifiers. We chose to move $\exists y$ first, for no particular reason.

$$\exists y\, [\forall x\, \neg P(x) \vee Q(y)]$$

Finally, we move $\forall x$

$$\exists y\, \forall x\, (\neg P(x) \vee Q(y))$$

If we had done it in the other order, we would have obtained the superficially different

$$\forall x\, \exists y\, (\neg P(x) \vee Q(y))$$

While the order of mixed quantifiers is usually quite important, in this case it does not matter because of the pattern of variables within the matrix of the wff.

Here is another example:

$$(\exists x\, P(x) \vee R(b)) \to \forall x\, (P(x) \wedge \forall x\, Q(x))$$

Again we have a conditional, but this time neither the antecedent nor the consequent is in prenex normal form. Following the basic strategy of working from the inside out, let's first put the antecedent and then the consequent each into prenex form and then worry about what to do about the conditional. Using the principle of null quantification on the antecedent we obtain

$$\exists x\, (P(x) \vee R(b)) \to \forall x\, (P(x) \wedge \forall x\, Q(x))$$

Next we use the principle involving the distribution of $\forall$ and $\wedge$ on the consequent:

$$\exists x\, (P(x) \vee R(b)) \to \forall x\, (P(x) \wedge Q(x))$$

Now both the antecedent and consequent are in prenex form. Recall that it's a good idea to get rid of conditionals in favor of Boolean connectives. Hence, we replace $\to$ by its equivalent using $\neg$ and $\vee$:

$$\neg\exists x\, (P(x) \vee R(b)) \vee \forall x\, (P(x) \wedge Q(x))$$

Now we have a disjunction, but one of the disjuncts is not in prenex form. Again, that can be fixed using DeMorgan's law:

$$\forall x \,\neg(P(x) \vee R(b)) \vee \forall x \,(P(x) \wedge Q(x))$$

Now both disjuncts are in prenex form. We need to pull the $\forall$'s out in front. (If they were both $\exists$'s, we could do this easily, but they aren't.) Here is probably the least obvious step in the process: In order to get ready to pull the $\forall$'s out in front, we replace the x in the second disjunct by a variable (say z) not in the first disjunct:

$$\forall x \,\neg(P(x) \vee R(b)) \vee \forall z \,(P(z) \wedge Q(z))$$

We now use the principle of null quantification twice, first on $\forall x$:

$$\forall x \,[\neg(P(x) \vee R(b)) \vee \forall z \,(P(z) \wedge Q(z))]$$

Finally, we use the same principle on $\forall z$, giving a wff in prenex form:

$$\forall x \,\forall z \,[\neg(P(x) \vee R(b)) \vee (P(z) \wedge Q(z))]$$

It is at this step that things would have gone wrong if we had not first changed the second x to a z. Do you see why? The wrong quantifiers would have bound the variables in the second disjunct.

If we wanted to, for some reason, we could now go on and put the inner part, the part following all the quantifiers, into one of our propositional normal forms, CNF or DNF.

With these examples behind us, here is a step-by-step transformation of our original sentence into the one in prenex form given above. We have abbreviated the predicates in order to make it easier to read.

$$
\begin{aligned}
&\forall x \,[(C(x) \wedge \exists y \,(T(y) \wedge L(x,y))) \rightarrow \exists y \,(D(y) \wedge B(x,y))] && \Leftrightarrow \\
&\forall x \,[\neg(C(x) \wedge \exists y \,(T(y) \wedge L(x,y))) \vee \exists y \,(D(y) \wedge B(x,y))] && \Leftrightarrow \\
&\forall x \,[\neg\exists y \,(C(x) \wedge T(y) \wedge L(x,y)) \vee \exists y \,(D(y) \wedge B(x,y))] && \Leftrightarrow \\
&\forall x \,[\forall y \,\neg(C(x) \wedge T(y) \wedge L(x,y)) \vee \exists y \,(D(y) \wedge B(x,y))] && \Leftrightarrow \\
&\forall x \,[\forall y \,\neg(C(x) \wedge T(y) \wedge L(x,y)) \vee \exists z \,(D(z) \wedge B(x,z))] && \Leftrightarrow \\
&\forall x \,\forall y \,[\neg(C(x) \wedge T(y) \wedge L(x,y)) \vee \exists z \,(D(z) \wedge B(x,z))] && \Leftrightarrow \\
&\forall x \,\forall y \,[\exists z \,\neg(C(x) \wedge T(y) \wedge L(x,y)) \vee \exists z \,(D(z) \wedge B(x,z))] && \Leftrightarrow \\
&\forall x \,\forall y \,\exists z \,[\neg(C(x) \wedge T(y) \wedge L(x,y)) \vee (D(z) \wedge B(x,z))] && \Leftrightarrow \\
&\forall x \,\forall y \,\exists z \,[(C(x) \wedge T(y) \wedge L(x,y)) \rightarrow (D(z) \wedge B(x,z))]
\end{aligned}
$$

Remember

A sentence is in prenex form if any quantifiers contained in it are out in front. Any sentence is logically equivalent to one in prenex form.

Exercises

Derive the following from the principles given earlier, by replacing → by its definition in terms of ∨ and ¬.

11.33 $\forall x\, P \to Q \quad \Leftrightarrow \quad \exists x\,[P \to Q] \quad$ if x not free in Q
✎

11.34 $\exists x\, P \to Q \quad \Leftrightarrow \quad \forall x\,[P \to Q] \quad$ if x not free in Q
✎

11.35 $P \to \forall x\, Q \quad \Leftrightarrow \quad \forall x\,[P \to Q] \quad$ if x not free in P
✎

11.36 $P \to \exists x\, Q \quad \Leftrightarrow \quad \exists x\,[P \to Q] \quad$ if x not free in P
✎

11.37   (Putting sentences in Prenex form) Open Jon Russell's Sentences. You will find ten sentences, at the odd numbered positions. Write a prenex form of each sentence in the space below it. Save your sentences. Open a few worlds, and make sure that your prenex form has the same truth value as the sentence above it.

11.38   (Some invalid quantifier manipulations) We remarked above on the invalidity of some quantifier manipulations that are superficially similar to the valid ones. In fact, in both cases one side is a logical consequence of the other side, but not vice versa. We will illustrate this. Build a world in which (1) and (3) below are true, but (2) and (4) are false.
1. $\forall x\,[\mathsf{Cube}(x) \lor \mathsf{Tet}(x)]$
2. $\forall x\, \mathsf{Cube}(x) \lor \forall x\, \mathsf{Tet}(x)$
3. $\exists x\, \mathsf{Cube}(x) \land \exists x\, \mathsf{Small}(x)$
4. $\exists x\,[\mathsf{Cube}(x) \land \mathsf{Small}(x)]$

Section 11.8

Some extra translation problems

Some instructors concentrate more on translation than others. For those who like to emphasize this skill, we present some additional challenging exercises here.

Exercises

11.39 (Translation) Open Peirce's World. Look at it in 2-D to remind yourself of the hidden objects. Start a new sentence file where you will translate the following English sentences. Again, be sure to check each of your translations to see that it is indeed a true sentence.

1. *Everything is either a cube or a tetrahedron.*
2. *Every cube is to the left of every tetrahedron.*
3. *There are at least three tetrahedra.*
4. *Every small cube is in back of a particular large cube.*
5. *Every tetrahedron is small.*
6. *Every dodecahedron is smaller than some tetrahedron.* [Note: This is vacuously true in this world.]

Now let's change the world so that none of the English sentences are true. (We can do this by changing the large cube in front to a dodecahedron, the large cube in back to a tetrahedron, and deleting the two small tetrahedra in the far right column.) If your answers to 1–5 are correct, all of your translations should be false as well. If not, you have made a mistake in translation. Make further changes, and check to see that the truth values of your translations track those of the English sentences. Submit your sentence file.

11.40 (More translations for practice) This exercise is just to give you more practice translating sentences of various sorts. They are all true in Skolem's World, in case you want to look while translating.

○ Translate the following sentences.

1. *Not every cube is smaller than every tetrahedron.*
2. *No cube is to the right of anything.*
3. *There is a dodecahedron unless there are at least two large objects.*
4. *No cube with nothing in back of it is smaller than another cube.*
5. *If any dodecahedra are small, then they are between two cubes.*
6. *If a cube is medium or is in back of something medium, then it has nothing to its right except for tetrahedra.*
7. *The further back a thing is, the larger it is.*
8. *Everything is the same size as something else.*
9. *Every cube has a tetrahedron of the same size to its right.*
10. *Nothing is the same size as two (or more) other things.*
11. *Nothing is between objects of shapes other than its own.*

○ Open **Skolem's World**. Notice that all of the above English sentences are true. Verify that the same holds of your translations.

○ This time, rather than open other worlds, make changes to **Skolem's World** and see that the truth value of your translations track that of the English sentence. For example, consider sentence 5. Add a small dodecahedron between the front two cubes. The English sentence is still true. Is your translation? Now move the dodecahedron over between two tetrahedra. The English sentence is false. Is your translation? Now make the dodecahedron medium. The English sentence is again true. How about your translation?

Submit your sentence file.

11.41 Using the symbols introduced in Table 1.2, page 30, translate the following into FOL. Do not introduce any additional names or predicates. Comment on any shortcomings in your translations.

1. *No student owned two pets at a time.*
2. *No student owned two pets until Claire did.*
3. *Anyone who owns a pet feeds it sometime.*
4. *Anyone who owns a pet feeds it sometime while they own it.*
5. *Only pets that are hungry are fed.*

11.42 Translate the following into FOL. As usual, explain the meanings of the names, predicates, and function symbols you use, and comment on any shortcomings in your translations.

1. *You should always except the present company.*
2. *There was a jolly miller once*
   Lived on the River Dee;
   He worked and sang from morn till night
   No lark more blithe than he.
3. *Man is the only animal that blushes. Or needs to.*
4. *You can fool all of the people some of the time, and some of the people all of the time, but you can't fool all of the people all of the time.*
5. *Everybody loves a lover.*

11.43 Give two translations of each of the following and discuss which is the most plausible reading, and why.

1. *Every senior in the class likes his or her computer, and so does the professor.* [Treat "the professor" as a name here and in the next sentence.]
2. *Every senior in the class likes his or her advisor, and so does the professor.*
3. *In some countries, every student must take an exam before going to college.*
4. *In some countries, every student learns a foreign language before going to college.*

11.44 (Using DeMorgan's Laws in mathematics) The DeMorgan Laws for quantifiers are quite helpful in mathematics. A function f on real numbers is said to be continuous at 0 if, intuitively, $f(x)$ can be kept close to $f(0)$ by keeping x close enough to 0. If you have had calculus then you will probably recognize the following a way to make this definition precise:

$$\forall \epsilon > 0 \, \exists \delta > 0 \, \forall x \, (\mid x \mid < \delta \to \mid f(x) - f(0) \mid < \epsilon)$$

Here "$\forall \epsilon > 0 (\ldots)$" is shorthand for "$\forall \epsilon (\epsilon > 0 \to \ldots)$". Similarly, "$\exists \delta > 0 (\ldots)$" is shorthand for "$\exists \delta (\delta > 0 \wedge \ldots)$". Use DeMorgan's Laws to express the claim that f is not continuous at 0 in prenex form. You may use the same kind of shorthand we have used. Turn in your solution.

11.45 Translate the following two sentences into FOL:

1. *If everyone comes to the party, I will have to buy more food.*
2. *There is someone such that if that person comes to the party, I will have to buy more food.*

The natural translations of these turn out to have forms that are equivalent, according to the equivalence in Problem 11.33. But clearly the English sentences do not mean the same thing. Explain what is going on here. Are the natural translations really correct?

CHAPTER 12

Methods of Proof for Quantifiers

In earlier chapters we discussed valid patterns of reasoning that arise from the various truth-functional connectives of FOL. This investigation of valid inference patterns becomes more interesting and more important now that we've added the quantifiers ∀ and ∃ to our language.

Our aim in this chapter and the next is to discover methods of proof that allow us to prove all and only the first-order validities, and all and only the first-order consequences of a given set of premises. In other words, our aim is to devise methods of proof sufficient to prove everything that follows in virtue of the meanings of the quantifiers, identity, and the truth-functional connectives. The resulting deductive system does indeed accomplish this goal, but our proof of that fact will have to wait until the final chapter of this book. That chapter will also discuss the issue of logical consequence when we take into account the meanings of other predicates in a first-order language.

Again, we begin looking at informal patterns of inference and then present their formal counterparts. As with the connectives, there are both simple proof steps and more substantive methods of proof. We will start by discussing the simple proof steps that are most often used with ∀ and ∃. We first discuss proofs involving single quantifier sentences and then explore what happens when we have multiple and mixed quantifier sentences.

Valid quantifier steps

There are two very simple valid quantifier steps, one for each quantifier. They work in opposite directions, however.

Universal elimination

Suppose we are given as a premise (or have otherwise established) that everything in the domain of discourse is either a cube or a tetrahedron. And suppose we also know that c is in the domain of discourse. It follows, of course, that c is either a cube or a tetrahedron, since everything is.

More generally, suppose we have established $\forall x\, S(x)$, and we know that c names an object in the domain of discourse. We may legitimately infer $S(c)$.

After all, there is no way the universal claim could be true without the specific claim also being true. This inference step is called *universal instantiation* or *universal elimination*. Notice that it allows you to move from a known result that begins with a quantifier $\forall x\, (\ldots x \ldots)$ to one $(\ldots c \ldots)$ where the quantifier has been eliminated.

universal elimination
(instantiation)

Existential introduction

There is also a simple proof step for $\exists$, but it allows you to *introduce* the quantifier. Suppose you have established that c is a small tetrahedron. It follows, of course, that there is a small tetrahedron. There is no way for the specific claim about c to be true without the existential claim also being true. More generally, if we have established a claim of the form $S(c)$ then we may infer $\exists x\, S(x)$. This step is called *existential generalization* or *existential introduction*.

existential introduction
(generalization)

In mathematical proofs, the preferred way to demonstrate the truth of an existential claim is to find (or construct) a specific instance that satisfies the requirement, and then apply existential generalization. For example, if we wanted to prove that there are natural numbers x, y, and z for which $x^2 + y^2 = z^2$, we could simply note that $3^2 + 4^2 = 5^2$ and apply existential generalization (thrice over).

The validity of both of these inference steps is not unconditional in English. They are valid as long as any name used denotes some object in the domain of discourse. This holds for FOL by convention, as we have already stressed, but English is a bit more subtle here. Consider, for example, the name *Santa*. The sentence

presuppositions
of these rules

Santa does not exist

might be true in circumstances where one would be reluctant to conclude

There is something that does not exist.

The trouble, of course, is that the name *Santa* does not denote anything. So we have to be careful applying this rule in ordinary arguments where there might be names in use that do not refer to actually existing objects.

Let's give an informal proof that uses both steps, as well as some other things we have learned. We will show that the following argument is valid:

$\forall x\, [\mathsf{Cube}(x) \rightarrow \mathsf{Large}(x)]$
$\forall x\, [\mathsf{Large}(x) \rightarrow \mathsf{LeftOf}(x, b)]$
$\mathsf{Cube}(d)$

$\exists x\, [\mathsf{Large}(x) \wedge \mathsf{LeftOf}(x, b)]$

This is a rather obvious result, which is all the better for illustrating the obviousness of these steps.

Proof: Using universal instantiation, we get

$$\mathsf{Cube(d)} \rightarrow \mathsf{Large(d)}$$

and

$$\mathsf{Large(d)} \rightarrow \mathsf{LeftOf(d, b)}$$

Applying modus ponens to $\mathsf{Cube(d)}$ and the first of these conditional claims gives us $\mathsf{Large(d)}$. Another application of modus ponens gives us $\mathsf{LeftOf(d, b)}$. But then we have

$$\mathsf{Large(d)} \wedge \mathsf{LeftOf(d, b)}$$

Finally, applying existential introduction gives us our desired conclusion:

$$\exists x\, [\mathsf{Large(x)} \wedge \mathsf{LeftOf(x, b)}]$$

Before leaving this section, we should point out that there are ways to prove existential statements other than by existential generalization. In particular, to prove $\exists x\, \mathsf{P(x)}$ we could use proof by contradiction, assuming $\neg \exists x\, \mathsf{P(x)}$ and deriving a contradiction. This method of proceeding is somewhat less satisfying, since it does not actually tell you which object it is that satisfies the condition $\mathsf{P(x)}$. Still, it does show that there is some such object, which is all that is claimed. This was in fact the method we used back on page 132 to prove that there are irrational numbers x and y such that x^y is rational.

Remember

1. Universal instantiation: From $\forall x\, \mathsf{S(x)}$, infer $\mathsf{S(c)}$, so long as c denotes an object in the domain of discourse.

2. Existential generalization: From $\mathsf{S(c)}$, infer $\exists x\, \mathsf{S(x)}$, so long as c denotes an object in the domain of discourse.

SECTION 12.2

The method of existential instantiation

Existential instantiation is one of the more interesting and subtle methods of proof. It allows you to prove results when you are given an existential statement. Suppose our domain of discourse consists of all children, and you are told that some boy is at home. If you want to use this fact in your reasoning, you are of course not entitled to infer that Max is at home. Neither are you allowed to infer that John is at home. In fact, there is no particular boy about whom you can safely conclude that he is at home, at least if this is all you know. So how should we proceed? What we could do is give a temporary name to one of the boys who is at home, and refer to him using that name, as long as we are careful not to use a name already used in the premises or the desired conclusion.

temporary names

This sort or reasoning is used in everyday life when we know that someone (or something) satisfies a certain condition, but do not know who (or what) satisfies it. For example, when Scotland Yard found out there was a serial killer at large, they dubbed him "Jack the Ripper," and used this name in reasoning about him. No one thought that this meant they knew who the killer was; rather, they simply introduced the name to refer to whoever was doing the killing. Note that if the town tailor were already called Jack the Ripper, then the detectives' use of this name would (probably) have been a gross injustice.

This is a basic strategy used when giving proofs in FOL. If we have correctly proven that $\exists x\, S(x)$, then we can give a name, say c, to one of the objects satisfying $S(x)$, as long as the name is not one that is already in use. We may then assume $S(c)$ and use it in our proof. This is the rule known as *existential instantiation* or *existential elimination*.

existential elimination (instantiation)

Generally, when existential instantiation is used in a mathematical proof, this will be marked by an explicit introduction of a new name. For example, the author of the proof might say, "So we have shown that there is a prime number between n and m. Call it p." Another phrase that serves the same function is: "Let p be such a prime number."

Let's give an example of how this rule might be used, by modifying our preceding example. The desired conclusion is the same but one of the premises is changed.

$$\begin{array}{|l} \forall x\,[\mathsf{Cube}(x) \rightarrow \mathsf{Large}(x)] \\ \forall x\,[\mathsf{Large}(x) \rightarrow \mathsf{LeftOf}(x, b)] \\ \exists x\,\mathsf{Cube}(x) \\ \hline \exists x\,[\mathsf{Large}(x) \wedge \mathsf{LeftOf}(x, b)] \end{array}$$

The first two premises are the same but the third is weaker, since it does not tell us which block is a cube, only that there is one. We would like to eliminate the $\exists$ in our third premise, since then we would be back to the case we have already examined. How then should we proceed? The proof would take the following form:

> **Proof:** We first note that the third premise assures us that there is at least one cube. Let "e" name one of these cubes. We can now proceed just as in our earlier reasoning. Applying the first premise, we see that e must be large. (What steps are we using here?) Applying the second premise, we see that e must also be left of b. Thus, we have shown that e is both large and left of b. Our desired conclusion follows (by what inference step?) from this claim.

an important condition   In applying existential instantiation, it is very important to make sure you use a new name, not one that is already in use or that appears in the conclusion you wish to prove. Looking at the above example shows why. Suppose we had thoughtlessly used the name "b" for the cube e. Then we would have been able to prove $\exists x\,\mathsf{LeftOf}(x, x)$, which is impossible. But our original premises are obviously satisfiable: they are true in many different worlds. So if we do not observe this condition, we can be led from true premises to false (even impossible) conclusions.

The method of general conditional proof

One of the most important methods of proof involves reasoning about an arbitrary object of a particular kind in order to prove a universal claim about all such objects. This is known as the method of *general* conditional proof. It is a more powerful version of conditional proof, and similar in spirit to the method of existential instantiation just discussed.

Let's start out with an example. This time let us assume that the domain of discourse consists of students at a particular college. We suppose that we are given a bunch of information about these students in the form of premises.

Finally, let us suppose we are able to prove from these premises that Sandy, a math major, is smart. Under what conditions would we be entitled to infer that every math major at the school is smart?

At first sight, it seems that we could never draw such a conclusion, unless there were only one math major at the school. After all, it does not follow from the fact that one math major is smart that all math majors are. But what if our proof that Sandy is smart uses *nothing at all* that is particular to Sandy? What if the proof would apply equally well to any math major? Then it seems that we should be able to conclude that every math major is smart.

How might one use this in a real example? Let us suppose that our argument took the following form:

> Anyone who passes Logic 101 with an A is smart.
> Every math major has passed Logic 101 with an A.
>
> Every math major is smart.

Our reasoning proceeds as follows.

> **Proof:** Let "Sandy" refer to any one of the math majors. By the second premise, Sandy passed Logic 101 with an A. By the first premise, then, Sandy is smart. But since Sandy is an arbitrarily chosen math major, it follows that every math major is smart.

This method of reasoning is used at every turn in doing mathematics. The general form is the following: Suppose we want to prove $\forall x\,[P(x) \to Q(x)]$ from some premises. The most straightforward way to proceed is to choose a name that is not in use, say c, assume $P(c)$, and prove $Q(c)$. If you are able to do this, then you are entitled to infer the desired result.

general conditional proof

Let's look at another example. Suppose we wanted to prove that every prime number has an irrational square root. To apply general conditional proof, we begin by assuming that p is an arbitrary prime number. Our goal is to show that $\sqrt{p}$ is irrational. If we can do this, we will have established the general claim. We have already proven that this holds if $p = 2$. But our proof relied on specific facts about 2, and so the general claim certainly doesn't follow from our proof. The proof, however, can be generalized to show what we want. Here is how the generalization goes.

> **Proof:** Let p be an arbitrary prime number. (That is, let "p" refer to any prime number.) Since p is prime, it follows that if p divides a square, say k^2, then it divides k. Hence, if p divides k^2, p^2 also divides k^2. Now assume, for proof by contradiction, that $\sqrt{p}$ is rational. Write it in lowest terms as $\sqrt{p} = n/m$. In particular, we can make sure that

p does not divide both n and m without remainder. Now, squaring both sides, we see that

$$p = \frac{n^2}{m^2}$$

and hence

$$pm^2 = n^2$$

But then it follows that p divides n^2, and so, as we have seen, p divides n and p^2 divides n^2. But from the latter of these it follows that p^2 divides pm^2 so p divides m^2. But then p divides m. So we have shown that p divides both n and m, contradicting our choice of n and m. This contradiction shows that $\sqrt{p}$ is indeed irrational.

It is perhaps worth mentioning an aspect of mathematical discourse illustrated in this proof that often confuses newcomers to mathematics. In mathematics lectures and textbooks, one often hears or reads "Let r (or n, or f, etc.) be an arbitrary real number (natural number, function, etc.)." What is confusing about this way of talking is that while you might say "Let Sandy take the exam next week," no one would ever say "Let Sandy be an arbitrary student in the class." That is not the way "let" normally works with names in English. If you want to say something like that, you should say "Let's use 'Sandy' to stand for any student in the class," or "Let 'Sandy' denote any student in the class." In mathematics, though, this "let" locution is a standard way of speaking. What is meant by "Let r be an arbitrary real number" is "Let 'r' denote any real number." In the above proof we paraphrased the first sentence to make it clearer. In the future, we will not be so pedantic.

Universal generalization

In formal systems of deduction, the method of general conditional proof is usually broken down into two parts, conditional proof and a method for proving completely general claims, claims of the form $\forall x\, S(x)$. The latter method is called *universal generalization* or *universal introduction*. It tells us that if we are able to introduce a new name c to stand for a completely arbitrary member of the domain of discourse and go on to prove the sentence $S(c)$, then we can conclude $\forall x\, S(x)$.

universal introduction (generalization)

Here is a very simple example. Suppose we give an informal proof that the following argument is valid.

$\forall x\, (\text{Cube}(x) \rightarrow \text{Small}(x))$
$\forall x\, \text{Cube}(x)$

$\forall x\, \text{Small}(x)$

In fact, it was the first example we looked at back in Chapter 10. Let's give a proof of this argument.

> **Proof:** We begin by taking a new name d, and think of it as standing for any member of the domain of discourse. Applying universal instantiation twice, once to each premise, gives us
>
> 1. Cube(d) → Small(d)
> 2. Cube(d)
>
> By modus ponens, we conclude Small(d). But d denotes an arbitrary object in the domain, so our conclusion, ∀x Small(x), follows by universal generalization.

Any proof using general conditional proof can be converted into a proof using universal generalization, together with the method of conditional proof. Suppose we have managed to prove ∀x [P(x) → Q(x)] using general conditional proof. Here is how we would go about proving it with universal generalization instead. First we would introduce a new name c, and think of it as standing for an arbitrary member of the domain of discourse. We know we can then prove P(c) → Q(c) using ordinary conditional proof, since that is what we did in our original proof. But then, since c stands for an arbitrary member of the domain, we can use universal generalization to get ∀x [P(x) → Q(x)].

universal generalization and general conditional proof

This is how formal systems of deduction can get by without having an explicit rule of general conditional proof. One could in a sense think of universal generalization as a special case of general conditional proof. After all, if we wanted to prove ∀x S(x) we could apply general conditional proof to the logically equivalent sentence ∀x [x = x → S(x)]. Or, if our language has the predicate Thing(x) that holds of everything in the domain of discourse, we could use general conditional proof to obtain ∀x [Thing(x) → S(x)]. But since general conditional proof may not allow us to prove ∀x S(x) alone, universal generalization is, well, more general.[1] (The relation between general conditional proof and universal generalization will become clearer when we get to the topic of generalized quantifiers in Section 14.4.)

We have chosen to emphasize general conditional proof since it is the method most often used in giving rigorous informal proofs. The division of this method into conditional proof and universal generalization is a clever trick, but it does not correspond well to actual reasoning. This is at least in part due to the fact that universal noun phrases of English are always restricted by some common noun, if only the noun *thing*. The natural counterparts of such statements in FOL have the form ∀x [P(x) → Q(x)], which is why we typically prove them by general conditional proof.

---

[1] We would like to thank S. Marc Cohen for his observations on the relationship between universal generalization and general conditional proof.

We began the discussion of the logic of quantified sentences in Chapter 10 by looking at the following arguments:

1.
> $\forall x\,(\mathsf{Cube}(x) \to \mathsf{Small}(x))$
> $\forall x\,\mathsf{Cube}(x)$
>
> $\forall x\,\mathsf{Small}(x)$

2.
> $\forall x\,\mathsf{Cube}(x)$
> $\forall x\,\mathsf{Small}(x)$
>
> $\forall x\,(\mathsf{Cube}(x) \land \mathsf{Small}(x))$

We saw there that the truth functional rules did not suffice to establish these arguments. In this chapter we have seen (on page 335) how to establish the first using valid methods that apply to the quantifiers. Let's conclude this discussion by giving an informal proof of the second.

> **Proof:** Let d be any object in the domain of discourse. By the first premise, we obtain (by universal elimination) $\mathsf{Cube}(d)$. By the second premise, we obtain $\mathsf{Small}(d)$. Hence we have $(\mathsf{Cube}(d) \land \mathsf{Small}(d))$. But since d is an arbitrary object in the domain, we can conclude $\forall x\,(\mathsf{Cube}(x) \land \mathsf{Small}(x))$, by universal generalization.

Exercises

The following exercises each contain a formal argument and something that purports to be an informal proof of it. Some of these proofs are correct while others are not. Give a logical critique of the purported proof. Your critique should take the form of a short essay that makes explicit each proof step or method of proof used, indicating whether it is valid or not. If there is a mistake, see if can you patch it up by giving a correct proof of the conclusion from the premises. If the argument in question is valid, you should be able to fix up the proof. If the argument is invalid, then of course you will not be able to fix the proof.

12.1

> $\forall x\,[(\mathsf{Brillig}(x) \lor \mathsf{Tove}(x)) \to (\mathsf{Mimsy}(x) \land \mathsf{Gyre}(x))]$
> $\forall y\,[(\mathsf{Slithy}(y) \lor \mathsf{Mimsy}(y)) \to \mathsf{Tove}(y)]$
> $\exists x\,\mathsf{Slithy}(x)$
>
> $\exists x\,[\mathsf{Slithy}(x) \land \mathsf{Mimsy}(x)]$

> **Purported proof:** By the third premise, we know that something in the domain of discourse is slithy. Let b be one of these slithy things. By the second premise, we know that b is a tove. By the first premise, we see that b is mimsy. Thus, b is both slithy and mimsy. Hence, something is both slithy and mimsy.

12.2

$\forall x\,[\text{Brillig}(x) \rightarrow (\text{Mimsy}(x) \land \text{Slithy}(x))]$
$\forall y\,[(\text{Slithy}(y) \lor \text{Mimsy}(y)) \rightarrow \text{Tove}(y)]$
$\forall x\,[\text{Tove}(x) \rightarrow (\text{Outgrabe}(x, b) \land \text{Brillig}(x))]$

$\forall z\,[\text{Brillig}(z) \leftrightarrow \text{Mimsy}(z)]$

Purported proof: In order to prove the conclusion, it suffices to prove the logically equivalent sentence obtained by conjoining the following two sentences:

(1) $\forall x\,[\text{Brillig}(x) \rightarrow \text{Mimsy}(x)]$

(2) $\forall x\,[\text{Mimsy}(x) \rightarrow \text{Brillig}(x)]$

We prove these by the method of general conditional proof, in turn. To prove (1), let b be anything that is brillig. Then by the first premise it is both mimsy and slithy. Hence it is mimsy, as desired. Thus we have established (1).

To prove (2), let b be anything that is mimsy. By the second premise, b is also tove. But then by the final premise, b is brillig, as desired. This concludes the proof.

12.3

$\forall x\,[(\text{Brillig}(x) \land \text{Tove}(x)) \rightarrow \text{Mimsy}(x)]$
$\forall y\,[(\text{Tove}(y) \lor \text{Mimsy}(y)) \rightarrow \text{Slithy}(y)]$
$\exists x\,\text{Brillig}(x) \land \exists x\,\text{Tove}(x)$

$\exists z\,\text{Slithy}(z)$

Purported proof: By the third premise, we know that there are brillig toves. Let b be one of them. By the first premise, we know that b is mimsy. By the second premise, we know that b is slithy. Hence, there is something that is slithy.

The following exercises each contains an argument; some are valid, some not. If the argument is valid, give an informal proof. If it is not valid, use Tarski's World to construct a counterexample.

12.4

$\forall y\,[\text{Cube}(y) \lor \text{Dodec}(y)]$
$\forall x\,[\text{Cube}(x) \rightarrow \text{Large}(x)]$
$\exists x\,\neg\text{Large}(x)$

$\exists x\,\text{Dodec}(x)$

12.5

$\forall y\,[\text{Cube}(y) \lor \text{Dodec}(y)]$
$\forall x\,[\text{Cube}(x) \rightarrow \text{Large}(x)]$
$\exists x\,\neg\text{Large}(x)$

$\exists x\,[\text{Dodec}(x) \land \text{Small}(x)]$

12.6

$\forall x\,[\text{Cube}(x) \lor \text{Dodec}(x)]$
$\forall x\,[\neg\text{Small}(x) \rightarrow \text{Tet}(x)]$

$\neg\exists x\,\text{Small}(x)$

12.7

$\forall x\,[\text{Cube}(x) \lor \text{Dodec}(x)]$
$\forall x\,[\text{Cube}(x) \rightarrow (\text{Large}(x) \land \text{LeftOf}(c, x))]$
$\forall x\,[\neg\text{Small}(x) \rightarrow \text{Tet}(x)]$

$\exists z\,\text{Dodec}(z)$

12.8
✦|✎

> $\forall x\,[\mathsf{Cube}(x) \lor (\mathsf{Tet}(x) \land \mathsf{Small}(x))]$
> $\exists x\,[\mathsf{Large}(x) \land \mathsf{BackOf}(x, c)]$
>
> $\exists x\,[\mathsf{FrontOf}(c, x) \land \mathsf{Cube}(x)]$

12.9
✦|✎

> $\forall x\,[(\mathsf{Cube}(x) \land \mathsf{Large}(x)) \lor (\mathsf{Tet}(x) \land \mathsf{Small}(x))]$
> $\forall x\,[\mathsf{Tet}(x) \rightarrow \mathsf{BackOf}(x, c)]$
>
> $\forall x\,[\mathsf{Small}(x) \rightarrow \mathsf{BackOf}(x, c)]$

12.10
✦|✎

> $\forall x\,[\mathsf{Cube}(x) \lor (\mathsf{Tet}(x) \land \mathsf{Small}(x))]$
> $\exists x\,[\mathsf{Large}(x) \land \mathsf{BackOf}(x, c)]$
>
> $\forall x\,[\mathsf{Small}(x) \rightarrow \neg\mathsf{BackOf}(x, c)]$

Section 12.4

Proofs involving mixed quantifiers

There are no new methods of proof that apply specifically to sentences with mixed quantifiers, but the introduction of mixed quantifiers forces us to be more explicit about some subtleties having to do with the interaction of methods that introduce new names into a proof: existential instantiation, general conditional proof, and universal generalization. It turns out that problems can arise from the interaction of these methods of proof.

Let us begin by illustrating the problem. Consider the following argument:

> $\exists y\,[\mathsf{Girl}(y) \land \forall x\,(\mathsf{Boy}(x) \rightarrow \mathsf{Likes}(x, y))]$
>
> $\forall x\,[\mathsf{Boy}(x) \rightarrow \exists y\,(\mathsf{Girl}(y) \land \mathsf{Likes}(x, y))]$

If the domain of discourse were the set of children in a kindergarten class, the conclusion would say every boy in the class likes some girl or other, while the premise would say that there is some girl who is liked by every boy. Since this is valid, let's start by giving a proof of it.

> **Proof:** Assume the premise. Thus, at least one girl is liked by every boy. Let c be one of these popular girls. To prove the conclusion we will use general conditional proof. Assume that d is any boy in the class. We want to prove that d likes some girl. But every boy likes c, so d likes c. Thus d likes some girl, by existential generalization. Since d was an arbitrarily chosen boy, the conclusion follows.

This is a perfectly legitimate proof. The problem we want to illustrate, however, is the superficial similarity between the above proof and the following incorrect "proof" of the argument that reverses the order of the premise and conclusion:

$$\forall x\,[\mathsf{Boy}(x) \to \exists y\,(\mathsf{Girl}(y) \land \mathsf{Likes}(x,y))]$$

$$\exists y\,[\mathsf{Girl}(y) \land \forall x\,(\mathsf{Boy}(x) \to \mathsf{Likes}(x,y))]$$

This is obviously invalid. The fact that every boy likes some girl or other doesn't imply that some girl is liked by every boy. So we can't really prove that the conclusion follows from the premise. But the following pseudo-proof might appear to do just that.

> **Pseudo-proof:** Assume the premise, that is, that every boy likes some girl or other. Let e be any boy in the domain. By our premise, e likes some girl. Let us introduce the new name "f" for some girl that e likes. Since the boy e was chosen arbitrarily, we conclude that every boy likes f, by general conditional proof. But then, by existential generalization, we have the desired result, namely, that some girl is liked by every boy.

This reasoning is fallacious. Seeing why it is fallacious is extremely important, if we are to avoid missteps in reasoning. The problem centers on our conclusion that every boy likes f. Recall how the name "f" came into the proof. We knew that e, being one of the boys, liked some girl, and we chose one of those girls and dubbed her with the name "f". This choice of a girl depends crucially on which boy e we are talking about. If e was Matt or Alex, we could have picked Zoe and dubbed her f. But if e was Eric, we couldn't pick Zoe. Eric likes one of the girls, but certainly not Zoe.

hidden dependencies

The problem is this. Recall that in order to conclude a universal claim based on reasoning about a single individual, it is imperative that we not appeal to anything specific about that individual. But after we give the name "f" to one of the girls that e likes, any conclusion we come to about e and f may well violate this imperative. We can't be positive that it would apply equally to all the boys.

Stepping back from this particular example, the upshot is this. Suppose we assume $P(c)$, where c is a new name, and prove $Q(c)$. We cannot conclude $\forall x\,[P(x) \to Q(x)]$ if $Q(c)$ mentions a specific individual whose choice depended on the individual denoted by c. In practice, the best way to insure that no such individual is specifically mentioned is to insist that $Q(c)$ not contain *any* name that was introduced by existential instantiation under the assumption that $P(c)$.

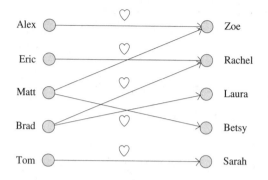

Figure 12.1: A circumstance in which $\forall x\,[\mathsf{Boy}(x) \to \exists y\,(\mathsf{Girl}(y) \land \mathsf{Likes}(x, y))]$.

a new restriction

A similar restriction must be placed on the use of universal generalization. Recall that universal generalization involves the introduction of a new constant, say c, standing for an arbitrary member c of the domain of discourse. We said that if we could prove a sentence $\mathsf{S}(\mathsf{c})$, we could then conclude $\forall x\,\mathsf{S}(x)$. However, we must now add the restriction that $\mathsf{S}(\mathsf{c})$ not contain any constant introduced by existential instantiation after the introduction of the constant c. This restriction prevents invalid proofs like the following.

> **Pseudo-proof:** Assume $\forall x\,\exists y\,\mathsf{Adjoins}(x, y)$. We will show that, ignoring the above restriction, we can "prove" $\exists y\,\forall x\,\mathsf{Adjoins}(x, y)$. We begin by taking c as a name for an arbitrary member of the domain. By universal instantiation, we get $\exists y\,\mathsf{Adjoins}(\mathsf{c}, y)$. Let d be such that $\mathsf{Adjoins}(\mathsf{c}, \mathsf{d})$. Since c stands for an arbitrary object, we have $\forall x\,\mathsf{Adjoins}(x, \mathsf{d})$. Hence, by existential generalization, we get $\exists y\,\forall x\,\mathsf{Adjoins}(x, y)$.

Can you spot the fallacious step in this proof? The problem is that we generalized from $\mathsf{Adjoins}(\mathsf{c}, \mathsf{d})$ to $\forall x\,\mathsf{Adjoins}(x, \mathsf{d})$. But the constant d was introduced by existential instantiation (though we did not say so explicitly) after the constant c was introduced. Hence, the choice of the object d depends on which object c we are talking about. The subsequent universal generalization is just what our restriction rules out.

Let us now give a summary statement of the main methods of proof involving the first-order quantifiers.

> ### Remember
>
> Let $S(x)$, $P(x)$, and $Q(x)$ be wffs.
>
> 1. Existential Instantiation: If you have proven $\exists x\, S(x)$ then you may choose a *new* constant symbol c to stand for any object satisfying $S(x)$ and so you may assume $S(c)$.
>
> 2. General Conditional Proof: If you want to prove $\forall x\, [P(x) \rightarrow Q(x)]$ then you may choose a *new* constant symbol c, assume $P(c)$, and prove $Q(c)$, making sure that Q does not contain any names introduced by existential instantiation after the assumption of $P(c)$.
>
> 3. Universal Generalization: If you want to prove $\forall x\, S(x)$ then you may choose a new constant symbol c and prove $S(c)$, making sure that $S(c)$ does not contain any names introduced by existential instantiation after the introduction of c.

Two famous proofs

There are, of course, endless applications of the methods we have discussed above. We illustrate the correct uses of these methods with two famous examples. One of the examples goes back to the ancient Greeks. The other, about a hundred years old, is known as the Barber Paradox and is due to the English logician Bertrand Russell. The Barber Paradox may seem rather frivolous, but the result is actually closely connected to Russell's Paradox, a result that had a very significant impact on the history of mathematics and logic. It is also connected with the famous result known as Gödel's Theorem. (We'll discuss Russell's Paradox in Chapter 15, and Gödel's Theorem in the final section of the book.)

Euclid's Theorem

Recall that a prime number is a whole number greater than 1 that is not divisible by any whole numbers other than 1 and itself. The first ten primes are $2, 3, 5, 7, 11, 13, 17, 19, 23$ and 29. The prime numbers become increasingly scarce as the numbers get larger. The question arises as to whether there is a largest one, or whether the primes go on forever. Euclid's Theorem is the statement that they go on forever, that there is no largest prime. In FOL, we might put it this way:

Euclid's Theorem

$$\forall x\, \exists y\, [y \geq x \wedge \mathsf{Prime}(y)]$$

Here the intended domain of discourse is the natural numbers, of course.

> **Proof:** We see that this sentence is a mixed quantifier sentence of just the sort we have been examining. To prove it, we let n be an arbitrary natural number and try to prove that there exists a prime number at least as large as n. To prove this, let k be the product of all the prime numbers less than n. Thus each prime less than n divides k without remainder. So now let $m = k + 1$. Each prime less than n divides m with remainder 1. But we know that m can be factored into primes. Let p be one of these primes. Clearly, by the earlier observation, p must be greater than or equal to n. Hence, by existential generalization, we see that there does indeed exist a prime number greater than or equal to n. But n was arbitrary, so we have established our result.

Notice the order of the last two steps. Had we violated the new condition on the application of general conditional proof to conclude that p is a prime number greater than or equal to every natural number, we would have obtained a patently false result.

Twin Prime Conjecture   Here, by the way, is a closely related conjecture, called the Twin Prime Conjecture. No one knows whether it is true or not.

$$\forall x\, \exists y\, [y > x \land \mathsf{Prime}(y) \land \mathsf{Prime}(y + 2)]$$

The Barber Paradox

There was once a small town in Indiana where there was a barber who shaved all and only the men of the town who did not shave themselves. We might formalize this in FOL as follows:

$$\exists z\, \exists x\, [\mathsf{BarberOf}(x, z) \land \forall y\, (\mathsf{ManOf}(y, z) \to (\mathsf{Shave}(x, y) \leftrightarrow \neg\mathsf{Shave}(y, y)))]$$

Now there does not on the face of it seem to be anything logically incoherent about the existence of such a town. But here is a proof that there can be no such town.

> **Purported proof:** Suppose there is such a town. Let's call it Hoosierville, and let's call Hoosierville's barber Fred. By assumption, Fred shaves all and only those men of Hoosierville who do not shave themselves.
>
> Now either Fred shaves himself, or he doesn't. But either possibility leads to a contradiction, as we now show. As to the first possibility, if

Fred does shave himself, then he doesn't, since by the assumption he does not shave any man of the town who shaves himself. So now assume the other possibility, namely, that Fred doesn't shave himself. But then since Fred shaves every man of the town who doesn't shave himself, he must shave himself. We have shown that a contradiction follows from each possibility. By proof by cases, then, we have established a contradiction from our original assumption. This contradiction shows that our assumption is wrong, so there is no such town.

The conflict between our intuition that there could be such a town on the one hand, and the proof that there can be none on the other hand, has caused this result to be known as the Barber Paradox.

Barber Paradox

Actually, though, there is a subtle sexist flaw in this proof. Did you spot it? It came in our use of the name "Fred." By naming the barber Fred, we implicitly assumed the barber was a man, an assumption that was needed to complete the proof. After all, it is only about *men* that we know the barber shaves those who do not shave themselves. Nothing is said about women, children, or other inhabitants of the town.

The proof, though flawed, is not worthless. What it really shows is that if there is a town with such a barber, then that barber is not a man of the town. The barber might be a woman, or maybe a man from some other town. In other words, the proof works fine to show that the following is a first-order validity:

$$\neg \exists z \, \exists x \, [\mathsf{ManOf}(x, z) \land \forall y \, (\mathsf{ManOf}(y, z) \to (\mathsf{Shave}(x, y) \leftrightarrow \neg \mathsf{Shave}(y, y)))]$$

There are many variations on this example that you can use to amaze, amuse, or annoy your family with when you go home for the holidays. We give a couple examples in the exercises (see Exercises 12.13 and 12.28).

Exercises

These exercises each contain a purported proof. If it is correct, say so. If it is incorrect, explain what goes wrong using the notions presented above.

12.11

There is a number greater than every other number.

> **Purported proof:** Let n be an arbitrary number. Then n is less than some other number, $n + 1$ for example. Let m be any such number. Thus $n \leq m$. But n is an arbitrary number, so every number is less or equal m. Hence there is a number that is greater than every other number.

12.12

✎

$$\forall x\, [\text{Person}(x) \to \exists y\, \forall z\, [\text{Person}(z) \to \text{GivesTo}(x, y, z)]]$$

$$\forall x\, [\text{Person}(x) \to \forall z\, (\text{Person}(z) \to \exists y\, \text{GivesTo}(x, y, z))]$$

Purported proof: Let us assume the premise and prove the conclusion. Let **b** be an arbitrary person in the domain of discourse. We need to prove

$$\forall z\, (\text{Person}(z) \to \exists y\, \text{GivesTo}(b, y, z))$$

Let c be an arbitrary person in the domain of discourse. We need to prove

$$\exists y\, \text{GivesTo}(b, y, c)$$

But this follows directly from our premise, since there is something that **b** gives to everyone.

12.13

✎

Harrison admires only great actors who do not admire themselves
Harrison admires all great actors who do not admire themselves.

Harrison is not a great actor.

Purported proof: Toward a proof by contradiction, suppose that Harrison is a great actor. Either Harrison admires himself or he doesn't. We will show that either case leads to a contradiction, so that our assumption that Harrison is a great actor must be wrong. First, assume that Harrison does admire himself. By the first premise and our assumption that Harrison is a great actor, Harrison does not admire himself, which is a contradiction. For the other case, assume that Harrison does not admire himself. But then by the second premise and our assumption that Harrison is a great actor, Harrison does admire himself after all. Thus, under either alternative, we have our contradiction.

12.14

✎

There is at most one object.

Purported proof: Toward a proof by contradiction, suppose that there is more than one object in the domain of discourse. Let c be any one of these objects. Then there is some other object d, so that $d \neq c$. But since c was arbitrary, $\forall x\, (d \neq x)$. But then, by universal instantiation, $d \neq d$. But $d = d$, so we have our contradiction. Hence there can be at most one object in the domain of discourse.

12.15

$\forall x \, \forall y \, \forall z \, [(\text{Outgrabe}(x, y) \wedge \text{Outgrabe}(y, z)) \rightarrow \text{Outgrabe}(x, z)]$

$\forall x \, \forall y \, [\text{Outgrabe}(x, y) \rightarrow \text{Outgrabe}(y, x)]$

$\exists x \, \exists y \, \text{Outgrabe}(x, y)$

$\forall x \, \text{Outgrabe}(x, x)$

> **Purported proof:** Applying existential instantiation to the third premise, let b and c be arbitrary objects in the domain of discourse such that b outgrabes c. By the second premise, we also know that c outgrabes b. Applying the first premise (with $x = z = b$ and $y = c$ we see that b outgrabes itself. But b was arbitrary. Thus by universal generalization, $\forall x \, \text{Outgrabe}(x, x)$.

The next three exercises contain arguments from a single set of premises. In each case decide whether or not the argument is valid. If it is, give an informal proof. If it isn't, use Tarski's World to construct a counterexample.

12.16

$\forall x \, \forall y \, [\text{LeftOf}(x, y) \rightarrow \text{Larger}(x, y)]$

$\forall x \, [\text{Cube}(x) \rightarrow \text{Small}(x)]$

$\forall x \, [\text{Tet}(x) \rightarrow \text{Large}(x)]$

$\forall x \, \forall y \, [(\text{Small}(x) \wedge \text{Small}(y)) \rightarrow \neg\text{Larger}(x, y)]$

$\neg \exists x \, \exists y \, [\text{Cube}(x) \wedge \text{Cube}(y) \wedge \text{RightOf}(x, y)]$

12.17

$\forall x \, \forall y \, [\text{LeftOf}(x, y) \rightarrow \text{Larger}(x, y)]$

$\forall x \, [\text{Cube}(x) \rightarrow \text{Small}(x)]$

$\forall x \, [\text{Tet}(x) \rightarrow \text{Large}(x)]$

$\forall x \, \forall y \, [(\text{Small}(x) \wedge \text{Small}(y)) \rightarrow \neg\text{Larger}(x, y)]$

$\forall z \, [\text{Medium}(z) \rightarrow \text{Tet}(z)]$

12.18

$\forall x \, \forall y \, [\text{LeftOf}(x, y) \rightarrow \text{Larger}(x, y)]$

$\forall x \, [\text{Cube}(x) \rightarrow \text{Small}(x)]$

$\forall x \, [\text{Tet}(x) \rightarrow \text{Large}(x)]$

$\forall x \, \forall y \, [(\text{Small}(x) \wedge \text{Small}(y)) \rightarrow \neg\text{Larger}(x, y)]$

$\forall z \, \forall w \, [(\text{Tet}(z) \wedge \text{Cube}(w)) \rightarrow \text{LeftOf}(z, w)]$

The next three exercises contain arguments from a single set of premises. In each, decide whether the argument is valid. If it is, give an informal proof. If it isn't valid, use Tarski's World to build a counterexample.

12.19
✐|✎

> $\forall x\,[\text{Cube}(x) \rightarrow \exists y\,\text{LeftOf}(x, y)]$
> $\neg \exists x\,\exists z\,[\text{Cube}(x) \wedge \text{Cube}(z) \wedge \text{LeftOf}(x, z)]$
> $\exists x\,\exists y\,[\text{Cube}(x) \wedge \text{Cube}(y) \wedge x \neq y]$
>
> $\exists x\,\exists y\,\exists z\,[\text{BackOf}(y, z) \wedge \text{LeftOf}(x, z)]$

12.20
✐|✎

> $\forall x\,[\text{Cube}(x) \rightarrow \exists y\,\text{LeftOf}(x, y)]$
> $\neg \exists x\,\exists z\,[\text{Cube}(x) \wedge \text{Cube}(z) \wedge \text{LeftOf}(x, z)]$
> $\exists x\,\exists y\,[\text{Cube}(x) \wedge \text{Cube}(y) \wedge x \neq y]$
>
> $\exists x\,\neg\text{Cube}(x)$

12.21
✐|✎

> $\forall x\,[\text{Cube}(x) \rightarrow \exists y\,\text{LeftOf}(x, y)]$
> $\neg \exists x\,\exists z\,[\text{Cube}(x) \wedge \text{Cube}(z) \wedge \text{LeftOf}(x, z)]$
> $\exists x\,\exists y\,[\text{Cube}(x) \wedge \text{Cube}(y) \wedge x \neq y]$
>
> $\exists x\,\exists y\,(x \neq y \wedge \neg\text{Cube}(x) \wedge \neg\text{Cube}(y))$

12.22
✐|✎⋆

Is the following logically true?

$$\exists x\,[\text{Cube}(x) \rightarrow \forall y\,\text{Cube}(y)]$$

If so, given an informal proof. If not, build a world where it is false.

12.23
✎

Translate the following argument into FOL and determine whether or not the conclusion follows from the premises. If it does, give a proof.

> Every child is either right-handed or intelligent.
> No intelligent child eats liver.
> There is a child who eats liver and onions.
>
> There is a right-handed child who eats onions.

In the next three exercises, we work in the first-order language of arithmetic with the added predicates Even(x), Prime(x), *and* DivisibleBy(x, y), *where these have the obvious meanings (the last means that the natural number* y *divides the number* x *without remainder.) Prove the result stated in the exercise. In some cases, you have already done all the hard work in earlier problems.*

12.24
✎

$\exists y\,[\text{Prime}(y) \wedge \text{Even}(y)]$

12.25
✎

$\forall x\,[\text{Even}(x) \leftrightarrow \text{Even}(x^2)]$

12.26
✎

$\forall x\,[\text{DivisibleBy}(x^2, 3) \rightarrow \text{DivisibleBy}(x^2, 9)]$

12.27
✎

Are sentences (1) and (2) in Exercise 9.19 on page 252 logically equivalent? If so, give a proof. If not, explain why not.

12.28
✎

Show that it would be impossible to construct a reference book that lists all and only those reference books that do not list themselves.

12.29 Call a natural number a *near prime* if its prime factorization contains at most two distinct
✎* primes. The first number which is not a near prime is $2 \times 3 \times 5 = 30$. Prove

$$\forall x \, \exists y \, [y > x \land \neg \mathsf{NearPrime}(y)]$$

You may appeal to our earlier result that there is no largest prime.

SECTION 12.5

Axiomatizing shape

Let's return to the project of giving axioms for the shape properties in Tarski's
World. In Section 10.5, we gave axioms that described basic facts about the
three shapes, but we stopped short of giving axioms for the binary relation
SameShape. The reason we stopped was that the needed axioms require multiple quantifiers, which we had not covered at the time.

How do we choose which sentences to take as axioms? The main consideration is *correctness*: the axioms must be true in all relevant circumstances, either in virtue of the meanings of the predicates involved, or because we have restricted our attention to a specific type of circumstance.

correctness of axioms

The two possibilities are reflected in our first four axioms about shape,
which we repeat here for ease of reference:

Basic Shape Axioms:

1. $\neg \exists x \, (\mathsf{Cube}(x) \land \mathsf{Tet}(x))$

2. $\neg \exists x \, (\mathsf{Tet}(x) \land \mathsf{Dodec}(x))$

3. $\neg \exists x \, (\mathsf{Dodec}(x) \land \mathsf{Cube}(x))$

4. $\forall x \, (\mathsf{Tet}(x) \lor \mathsf{Dodec}(x) \lor \mathsf{Cube}(x))$

The first three of these are correct in virtue of the meanings of the predicates;
the fourth expresses a truth about all worlds of the sort that can be built in
Tarski's World.

Of second importance, just behind correctness, is *completeness*. We say
that a set of axioms is complete if, whenever an argument is intuitively
valid (given the meanings of the predicates and the intended range of circumstances), its conclusion is a first-order consequence of its premises taken
together with the axioms in question.

completeness of axioms

The notion of completeness, like that of correctness, is not precise, depending as it does on the vague notions of meaning and "intended circumstances."

For example, in axiomatizing the basic shape predicates of Tarski's World, there is an issue about what kinds of worlds are included among the intended circumstances. Do we admit only those having at most twelve objects, or do we also consider those with more objects, so long as they have one of the three shapes? If we make this latter assumption, then the basic shape axioms are complete. This is not totally obvious, but we will justify the claim in Section 18.4. Here we will simply illustrate it.

Consider the following argument:

$$\exists x\, \exists y\, (\mathsf{Tet}(x) \land \mathsf{Dodec}(y) \land \forall z\, (z = x \lor z = y))$$
$$\neg \exists x\, \mathsf{Cube}(x)$$

This argument is clearly valid, in the sense that in any world in which the premise is true, the conclusion will be true as well. But the conclusion is certainly not a first-order consequence of the premise. (Why?) If we treat the four basic shape axioms as additional premises, though, we can prove the conclusion using just the first-order methods of proof available to us.

> **Proof:** By the explicit premise, we know there are blocks e and f such that e is a tetrahedron, f is a dodecahedron, and everything is one of these two blocks. Toward a contradiction, suppose there were a cube, say c. Then either $c = e$ or $c = f$. If $c = e$ then by the indiscernibility of identicals, c is both a cube and a tetrahedron, contradicting axiom 1. Similarly, if $c = f$ then c is both a cube and a dodecahedron, contradicting axiom 3. So we have a contradiction in either case, showing that our assumption cannot be true.

While our four axioms are complete if we restrict attention to the three shape predicates, they are clearly not complete when we consider sentences involving SameShape. If we were to give inference rules for this predicate, it would be natural to state them in the form of introduction and elimination rules: the former specifying when we can conclude that two blocks are the same shape; the latter specifying what we can infer from such a fact. This suggests the following axioms:

SameShape Introduction Axioms:

5. $\forall x\, \forall y\, ((\mathsf{Cube}(x) \land \mathsf{Cube}(y)) \rightarrow \mathsf{SameShape}(x, y))$

6. $\forall x\, \forall y\, ((\mathsf{Dodec}(x) \land \mathsf{Dodec}(y)) \rightarrow \mathsf{SameShape}(x, y))$

7. $\forall x\, \forall y\, ((\mathsf{Tet}(x) \land \mathsf{Tet}(y)) \rightarrow \mathsf{SameShape}(x, y))$

SameShape **Elimination Axioms:**

8. $\forall x \forall y ((\text{SameShape}(x, y) \wedge \text{Cube}(x)) \rightarrow \text{Cube}(y))$

9. $\forall x \forall y ((\text{SameShape}(x, y) \wedge \text{Dodec}(x)) \rightarrow \text{Dodec}(y))$

10. $\forall x \forall y ((\text{SameShape}(x, y) \wedge \text{Tet}(x)) \rightarrow \text{Tet}(y))$

As it happens, these ten axioms give us a complete axiomatization of the shape predicates in our blocks language. This means that any argument that is valid and which uses only these predicates can be turned into one where the conclusion is a first-order consequence of the premises plus these ten axioms. It also means that any sentence that is true simply in virtue of the meanings of the four shape predicates is a first-order consequence of these axioms. For example, the sentence $\forall x \, \text{SameShape}(x, x)$, which we considered as a possible axiom in Chapter 10, follows from our ten axioms:

> **Proof:** Let b be an arbitrary block. By axiom 4, we know that b is a tetrahedron, a dodecahedron, or a cube. If b is a tetrahedron, then axiom 7 guarantees that b is the same shape as b. If b is a dodecahedron or a cube, this same conclusion follows from axioms 6 or 5, respectively. Consequently, we know that b is the same shape as b. Since b was arbitrary, we can conclude that every block is the same shape as itself.

During the course of this book, we have proven many claims about natural numbers, and asked you to prove some as well. You may have noticed that these proofs did not appeal to explicit premises. Rather, the proofs freely cited any obvious facts about the natural numbers. However, we could (and will) make the needed assumptions explicit by means of the axiomatic method. In Section 16.4, we will discuss the standard *Peano axioms* that are used to axiomatize the obvious truths of arithmetic. While we will not do so, it would be possible to take each of our proofs about natural numbers and turn it into a proof that used only these Peano Axioms as premises.

Peano axioms

We will later show, however, that the Peano Axioms are not complete, and that it is in fact impossible to present first-order axioms that are complete for arithmetic. This is the famous Gödel Incompleteness Theorem and is discussed in the final section of this book.

Gödel's Incompleteness Theorem

Exercises

Give informal proofs of the following arguments, if they are valid, making use of any of the ten shape axioms as needed, so that your proof uses only first-order methods of proof. Be very explicit about which axioms you are using at various steps. If the argument is not valid, use Tarski's World to provide a counterexample.

12.30

$\exists x\,(\neg Cube(x) \wedge \neg Dodec(x))$
$\exists x\,\forall y\,SameShape(x, y)$

$\forall x\,Tet(x)$

12.31

$\forall x\,(Cube(x) \rightarrow SameShape(x, c))$

$Cube(c)$

12.32

$\forall x\,Cube(x) \vee \forall x\,Tet(x) \vee \forall x\,Dodec(x)$

$\forall x\,\forall y\,SameShape(x, y)$

12.33

$\forall x\,\forall y\,SameShape(x, y)$

$\forall x\,Cube(x) \vee \forall x\,Tet(x) \vee \forall x\,Dodec(x)$

12.34

$SameShape(b, c)$

$SameShape(c, b)$

12.35

$SameShape(b, c)$
$SameShape(c, d)$

$SameShape(b, d)$

12.36 The last six shape axioms are quite intuitive and easy to remember, but we could have gotten by with fewer. In fact, there is a single sentence that completely captures the meaning of SameShape, given the first four axioms. This is the sentence that says that two things are the same shape if and only if they are both cubes, both tetrahedra, or both dodecahedra:

$$\forall x\,\forall y\,(SameShape(x, y) \quad \leftrightarrow \quad ((Cube(x) \wedge Cube(y))$$
$$\vee (Tet(x) \wedge Tet(y))$$
$$\vee (Dodec(x) \wedge Dodec(y)))$$

Use this axiom and and the basic shape axioms (1)-(4) to give informal proofs of axioms (5) and (8).

12.37 Let us imagine adding as new atomic sentences involving a binary predicate MoreSides. We assume that MoreSides(b, c) holds if block b has more sides than block c. See if you can come up with axioms that completely capture the meaning of this predicate. The natural way to do this involves two or three introduction axioms and three or four elimination axioms. Turn in your axioms to your instructor.

12.38 Find first-order axioms for the six size predicates of the blocks language. [Hint: use the axiomatization of shape to guide you.]

Chapter 13

Formal Proofs and Quantifiers

Now that we have learned the basic informal methods of proof for quantifiers, we turn to the task of giving formal rules that correspond to them. Again, we can to do this by having two rules for each quantifier.

Before getting down to the rules, though, we should emphasize that formal proofs in the system $\mathcal{F}$ contain only sentences, never wffs with free variables. This is because we want every line of a proof to make a definite claim. Wffs with free variables do not make claims, as we have noted. Some deductive systems do allow proofs containing formulas with free variables, where such variables are interpreted universally, but that is not how the system $\mathcal{F}$ works.

Universal quantifier rules

The valid inference step of universal instantiation or elimination is easily formalized. Here is the schematic version of the rule:

Universal Elimination ($\forall$ Elim):

$$
\begin{array}{c|l}
 & \forall x\, S(x) \\
 & \quad \vdots \\
\triangleright & S(c) \\
\end{array}
$$

Here x stands for any variable, c stands for any individual constant (whether or not it has been used elsewhere in the proof), and $S(c)$ stands for the result of replacing free occurrences of x in $S(x)$ with c. If the language contains function symbols, c can also be any complex term that contains no variables.

Next, let us formalize the more interesting methods of general conditional proof and universal generalization. This requires that we decide how to represent the fact that a constant symbol, say c, has been introduced to stand for an arbitrary object satisfying some condition, say $P(c)$. We indicate this by means of a subproof with assumption $P(c)$, insisting that the constant c in question occur only within that subproof. This will guarantee, for example, that the constant does not appear in the premises of the overall proof.

boxed constant

To remind ourselves of this crucial restriction, we will introduce a new graphical device, boxing the constant symbol in question and putting it in front of the assumption. We will think of the boxed constant as the formal analog of the English phrase "Let c denote an arbitrary object satisfying P(c)."

General Conditional Proof (∀ Intro):

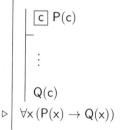

Where c does not occur outside the subproof where it is introduced.

When we give the justification for universal introduction, we will cite the subproof, as we do in the case of conditional introduction. The requirement that c not occur outside the subproof in which it is introduced does not preclude it occurring within subproofs of that subproof. A sentence in a subproof of a subproof still counts as a sentence of the larger subproof.

As a special case of ∀ **Intro** we allow a subproof where there is no sentential assumption at all, just the boxed constant on its own. This corresponds to the method of universal generalization discussed earlier, where one assumes that the constant in question stands for an arbitrary object in the domain of discourse.

Universal Introduction (∀ Intro):

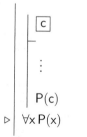

Where c does not occur outside the subproof where it is introduced.

As we have indicated, we don't really need both forms of ∀ **Intro**. We use both because the first is more natural while the second is more general.

Let's illustrate how to use these rules by giving a formal proof mirroring the informal proof given on page 335. We prove that the following argument is valid:

$$\forall x\,(P(x) \rightarrow Q(x))$$
$$\forall z\,(Q(z) \rightarrow R(z))$$

$$\forall x\,(P(x) \rightarrow R(x))$$

(This is a general form of the argument about all math majors being smart given earlier.) Here is a completed proof:

1. $\forall x\,(P(x) \rightarrow Q(x))$
2. $\forall z\,(Q(z) \rightarrow R(z))$

   3. $\boxed{d}\; P(d)$

   4. $P(d) \rightarrow Q(d)$ $\forall$ **Elim**: 1
   5. $Q(d)$ $\rightarrow$ **Elim**: 3, 4
   6. $Q(d) \rightarrow R(d)$ $\forall$ **Elim**: 2
   7. $R(d)$ $\rightarrow$ **Elim**: 5, 6
8. $\forall x\,(P(x) \rightarrow R(x))$ $\forall$ **Intro**: 3-7

Notice that the constant symbol d does not appear outside the subproof. It is newly introduced at the beginning of that subproof, and occurs nowhere else outside it. That is what allows the introduction of the universal quantifier in the final step.

You try it
. . . .

1. Open the file Universal 1. This file contains the argument proven above. ◄ We'll show you how to construct this proof in Fitch.

2. Start a new subproof immediately after the premises. Before typing any- ◄ thing in, notice that there is a blue, downward pointing triangle to the left of the blinking cursor. It looks just like the focus slider, but sort of stand- ing on its head. Use your mouse to click down on this triangle. A menu will pop up, allowing you to choose the constant(s) you want "boxed" in this subproof. Choose d from the menu. (If you choose the wrong one, say c, then choose it again to "unbox" it.)

3. After you have d in the constant box, enter the sentence P(d) as your ◄ assumption. Then add a step and continue the subproof.

▶ 4. You should now be able to complete the proof on your own. When you're done, save it as Proof Universal 1.

. . . . *Congratulations*

Default and generous uses of the ∀ rules

default uses of ∀ rules Both of the universal quantifier rules have default uses. If you cite a universal sentence and apply ∀ **Elim** without entering a sentence, Fitch will replace the universally quantified variable with its best guess of the name you intended. It will either choose the alphabetically first name that does not already appear in the sentence, or the first name that appears as a boxed constant in the current subproof. For example, in steps 4 and 6 of the above proof, the default mechanism would choose d, and so generate the correct instances.

indicating substitutions If you know you want a different name substituted for the universally quantified variable, you can indicate this by typing a colon (:), followed by the variable, followed by the greater-than sign (>), followed by the name you want. In other words, if instead of a sentence you enter ": x > c", Fitch will instantiate $\forall x\, P(x)$ as $P(c)$, rather than picking its own default instance. (Think of ": x > c" as saying "substitute c for x.")

If you apply ∀ **Intro** to a subproof that starts with a boxed constant on its own, without entering a sentence, Fitch will take the last sentence in the cited subproof and universally quantify the name introduced at the beginning of the subproof. If the cited subproof starts with a boxed constant and a sentence, then Fitch will write the corresponding universal conditional, using the first sentence and the last sentence of the proof to create the conditional.

You try it
. . . .

▶ 1. Open the file Universal 2. Look at the goal to see what sentence we are trying to prove. Then focus on each step in succession and check the step. Before moving to the next step, make sure you understand why the step checks out and, more important, why we are doing what we are doing at that step. At the empty steps, try to predict which sentence Fitch will provide as a default before you check the step.

▶ 2. When you are done, make sure you understand the completed proof. Save the file as Proof Universal 2.

. . . . *Congratulations*

Fitch has generous uses of both ∀ rules. ∀ **Elim** will allow you to remove several universal quantifiers from the front a sentence simultaneously. For example, if you have proven ∀x ∀y SameCol(x, y) you could infer SameCol(f, c) in one step in Fitch. If you want to use the default mechanism to generate this step, you can enter the substitutions ": x > f : y > c" before checking the step.

In a like manner, you can often prove a sentence starting with more than one universal quantifier by means of a single application of ∀ **Intro**. You do this by starting a subproof with the appropriate number of boxed constants. If you then prove a sentence containing these constants you may end the subproof and infer the result of universally quantifying each of these constants using ∀ **Intro**. The default mechanism allows you to specify the variables to be used in the generated sentence by indicating the desired substitutions, for example ": a > z : b > w" will generate ∀z ∀w R(w, z) when applied to R(b, a). Notice the order used to specify substitutions: for ∀ **Elim** it will always be ": variable > name," while for ∀ **Intro** it must be ": name > variable."

Add Support Steps can be used with the ∀ **Intro** rule. If the focus step contains a universally quantified formula then the support will be a subproof with a new constant at the assumption step. The last step of the subproof will contain the appropriate instance of the universal formula. With the ∀ **Elim** rule a single support step containing a universal formula will be added. The support formula will be a universal generalization of the formula at the focus step, with the first constant replaced by the variable of quantification.

Remember

The formal rule of ∀ **Intro** corresponds to the informal method of general conditional proof, including the special case of universal generalization.

Exercises

13.1 If you skipped the **You try it** sections, go back and do them now. Submit the files Proof Universal 1 and Proof Universal 2.

For each of the following arguments, decide whether or not it is valid. If it is, use Fitch to give a formal proof. If it isn't, use Tarski's World to give a counterexample. In this chapter you are free to use **Taut Con** *to justify proof steps involving only propositional connectives.*

13.2

> $\forall x\,(Cube(x) \leftrightarrow Small(x))$
> $\forall x\,Cube(x)$
>
> $\forall x\,Small(x)$

13.3

> $\forall x\,Cube(x)$
> $\forall x\,Small(x)$
>
> $\forall x\,(Cube(x) \wedge Small(x))$

13.4

> $\neg\forall x\,Cube(x)$
>
> $\neg\forall x\,(Cube(x) \wedge Small(x))$

13.5

> $\forall x\,\forall y\,((Cube(x) \wedge Dodec(y))$
> $\qquad \rightarrow Larger(y, x))$
> $\forall x\,\forall y\,(Larger(x, y) \leftrightarrow LeftOf(x, y))$
>
> $\forall x\,\forall y\,((Cube(x) \wedge Dodec(y))$
> $\qquad \rightarrow LeftOf(y, x))$

13.6

> $\forall x\,((Cube(x) \wedge Large(x))$
> $\qquad \vee (Tet(x) \wedge Small(x)))$
> $\forall x\,(Tet(x) \rightarrow BackOf(x, c))$
> $\forall x\,\neg(Small(x) \wedge Large(x))$
>
> $\forall x\,(Small(x) \rightarrow BackOf(x, c))$

(See Exercise 12.9. Notice that we have included a logical truth as an additional premise here.)

13.7

> $\forall x\,\forall y\,((Cube(x) \wedge Dodec(y))$
> $\qquad \rightarrow FrontOf(x, y))$
>
> $\forall x\,(Cube(x) \rightarrow \forall y\,(Dodec(y)$
> $\qquad \rightarrow FrontOf(x, y)))$

13.8

> $\forall x\,(Cube(x) \rightarrow \forall y\,(Dodec(y)$
> $\qquad \rightarrow FrontOf(x, y)))$
>
> $\forall x\,\forall y\,((Cube(x) \wedge Dodec(y))$
> $\qquad \rightarrow FrontOf(x, y))$

13.9

> $\forall x\,\forall y\,((Cube(x) \wedge Dodec(y))$
> $\qquad \rightarrow Larger(x, y))$
> $\forall x\,\forall y\,((Dodec(x) \wedge Tet(y))$
> $\qquad \rightarrow Larger(x, y))$
>
> $\forall x\,\forall y\,((Cube(x) \wedge Tet(y))$
> $\qquad \rightarrow Larger(x, y))$

Section 13.2

Existential quantifier rules

Recall that in our discussion of informal proofs, existential introduction was a simple proof step, whereas the elimination of $\exists$ was a subtle method of proof. Thus, in presenting our formal system, we begin with the introduction rule.

Existential Introduction (∃ Intro):

$$\begin{array}{l} \mathsf{S(c)} \\ \vdots \\ \rhd \quad \exists\mathsf{x}\, \mathsf{S(x)} \end{array}$$

Here too x stands for any variable, c stands for any individual constant (or complex term without variables), and S(c) stands for the result of replacing free occurrences of x in S(x) with c. Note that there may be other occurrences of c in S(x) as well.

When we turn to the rule of existential elimination, we employ the same, "boxed constant" device as with universal introduction. If we have proven ∃x S(x), then we introduce a new constant symbol, say c, along with the assumption that the object denoted by c satisfies the formula S(x). If, from this assumption, we can derive some sentence Q not containing the constant c, then we can conclude that Q follows from the original premises.

Existential Elimination (∃ Elim):

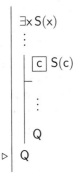

Where c does not occur outside the subproof where it is introduced.

Again we think of the notation at the beginning of the subproof as the formal counterpart of the English "Let c be an arbitrary individual such that S(c)."

comparison with
∨ Elim

The rule of existential elimination is quite analogous to the rule of disjunction elimination, both formally and intuitively. With disjunction elimination, we have a disjunction and break into cases, one for each disjunct, and establish the same result in each case. With existential elimination, we can think of having one case for each object in the domain of discourse. We are required to show that, whichever object it is that satisfies the condition S(x), the same result Q can be obtained. If we can do this, we may conclude Q.

To illustrate the two existential rules, we will give a formal counterpart to the proof given on page 332.

1. $\forall x\,[\mathsf{Cube}(x) \to \mathsf{Large}(x)]$
2. $\forall x\,[\mathsf{Large}(x) \to \mathsf{LeftOf}(x, b)]$
3. $\exists x\,\mathsf{Cube}(x)$

    4. $\boxed{e}$ $\mathsf{Cube}(e)$

    5. $\mathsf{Cube}(e) \to \mathsf{Large}(e)$ $\forall$ **Elim**: 1
    6. $\mathsf{Large}(e)$ $\to$ **Elim**: 5, 4
    7. $\mathsf{Large}(e) \to \mathsf{LeftOf}(e, b)$ $\forall$ **Elim**: 2
    8. $\mathsf{LeftOf}(e, b)$ $\to$ **Elim**: 7, 6
    9. $\mathsf{Large}(e) \land \mathsf{LeftOf}(e, b)$ $\land$ **Intro**: 6, 8
    10. $\exists x\,(\mathsf{Large}(x) \land \mathsf{LeftOf}(x, b))$ $\exists$ **Intro**: 9
11. $\exists x\,(\mathsf{Large}(x) \land \mathsf{LeftOf}(x, b))$ $\exists$ **Elim**: 3, 4-10

Default and generous uses of the $\exists$ rules

default uses of $\exists$ rules

Defaults for the existential quantifier rules work similarly to those for the universal quantifier. If you cite a sentence and apply $\exists$ **Intro** without typing a sentence, Fitch will supply a sentence that existentially quantifies the alphabetically first name appearing in the cited sentence. When replacing the name with a variable, Fitch will choose the first variable in the list of variables that does not already appear in the cited sentence. If this isn't the name or variable you want used, you can specify the substitution yourself; for example ": $max > z$" will replace max with z and add $\exists z$ to the front of the result.

In a default application of $\exists$ **Elim**, Fitch will supply the last sentence in the cited subproof, providing that sentence does not contain the temporary name introduced at the beginning of the subproof.

You try it
. . . .

▶ 1. Open the file Existential 1. Look at the goal to see the sentence we are trying to prove. Then focus on each step in succession and check the step. Before moving to the next step, make sure you understand why the step checks out and, more important, why we are doing what we are doing at that step.

▶ 2. At any empty steps, you should try to predict which sentence Fitch will provide as a default before you check the step. Notice in particular step eight, the one that contains ": $a > y$". Can you guess what sentence would have been supplied by Fitch had we not specified this substitution? You could try it if you like.

3. When you are done, make sure you understand the completed proof. Save the file as Proof Existential 1.   ◀

. . . . *Congratulations*

As with ∀, Fitch has generous uses of both ∃ rules. ∃ **Intro** will allow you to add several existential quantifiers to the front a sentence. For example, if you have proved SameCol(b, a) you could infer ∃y ∃z SameCol(y, z) in one step in Fitch. In a like manner, you can use a sentence beginning with more than one existential quantifier in a single application of ∃ **Elim**. You do this by starting a subproof with the appropriate number of boxed constants. If you then prove a sentence not containing these constants, you may end the subproof and infer the result using ∃ **Elim**.

The **Add Support Steps** command cannot be used with either of the ∃ rules.

> **Remember**
>
> The formal rule of ∃ **Elim** corresponds to the informal method of existential instantiation.

Exercises

13.10 If you skipped the **You try it** section, go back and do it now. Submit the file Proof Existential 1.
✎

For each of the following arguments, decide whether or not it is valid. If it is, use Fitch to give a formal proof. If it isn't, use Tarski's World to give a counterexample. Remember that in this chapter you are free to use **Taut Con** *to justify proof steps involving only propositional connectives.*

13.11
✎
| ∀x (Cube(x) ∨ Tet(x))
| ∃x ¬Cube(x)
|
| ∃x ¬Tet(x)

13.12
✎
| ∀x (Cube(x) ∨ Tet(x))
| ∃x ¬Cube(x)
|
| ∃x Tet(x)

13.13
✎
| ∀y [Cube(y) ∨ Dodec(y)]
| ∀x [Cube(x) → Large(x)]
| ∃x ¬Large(x)
|
| ∃x Dodec(x)

13.14
✎
| ∀x (Cube(x) ↔ Small(x))
| ∃x ¬Cube(x)
|
| ∃x ¬Small(x)

13.15
↗

> $\exists x\,(\text{Cube}(x) \rightarrow \text{Small}(x))$
> $\forall x\,\text{Cube}(x)$
>
> $\exists x\,\text{Small}(x)$

13.16
↗

> $\exists x\,\exists y\,\text{Adjoins}(x, y)$
> $\forall x\,\forall y\,(\text{Adjoins}(x, y)$
> $\qquad\qquad \rightarrow \neg\text{SameSize}(x, y))$
>
> $\exists x\,\exists y\,\neg\text{SameSize}(y, x)$

In our discussion of the informal methods, we observed that the method that introduces new constants can interact to give defective proofs, if not used with care. The formal system $\mathcal{F}$ automatically prevents these misapplications of the quantifier rules. The next two exercises are designed to show you how the formal rules prevent these invalid steps by formalizing one of the fallacious informal proofs we gave earlier.

13.17 Here is a formalization of the pseudo-proof given on page 340:
↗

> 1. $\forall x\,\exists y\,\text{SameCol}(x, y)$
>
>> 2. $\boxed{c}$
>>
>> 3. $\exists y\,\text{SameCol}(c, y)$ $\forall$ **Elim**: 1
>>> 4. $\boxed{d}\ \text{SameCol}(c, d)$
>>
>> 5. $\text{SameCol}(c, d)$ **Reit**: 4
>> 6. $\text{SameCol}(c, d)$ $\exists$ **Elim**: 3, 4–5
> 7. $\forall x\,\text{SameCol}(x, d)$ $\forall$ **Intro**: 2–6
> 8. $\exists y\,\forall x\,\text{SameCol}(x, y)$ $\exists$ **Intro**: 7

1. Write this "proof" in a file using Fitch and check it out. You will discover that step 6 is incorrect; it violates the restriction on existential elimination that requires the constant d to appear only in the subproof where it is introduced. Notice that the other steps all check out, so if we *could* make that move, then the rest of the proof would be fine.
2. Construct a counterexample to the argument to show that no proof is possible.

Submit both files.

13.18 Let's contrast the faulty proof from the preceding exercise with a genuine proof
↗   that $\forall x\,\exists y\,R(x, y)$ follows from $\exists y\,\forall x\,R(x, y)$. Use Fitch to create the following proof.

1. $\exists y \, \forall x \, \mathsf{SameCol}(x, y)$

  2. $\boxed{d} \; \forall x \, \mathsf{SameCol}(x, d)$

    3. $\boxed{c}$

    4. $\mathsf{SameCol}(c, d)$ $\forall$ **Elim**: 2
    5. $\exists y \, \mathsf{SameCol}(c, y)$ $\exists$ **Intro**: 4
  6. $\forall x \, \exists y \, \mathsf{SameCol}(x, y)$ $\forall$ **Intro**: 3–5
7. $\forall x \, \exists y \, \mathsf{SameCol}(x, y)$ $\exists$ **Elim**: 1, 2–6

Notice that in this proof, unlike the one in the previous exercise, both constant symbols c and d are properly sequestered within the subproofs where they are introduced. Therefore the quantifier rules have been applied properly. Submit your proof.

SECTION 13.3

Strategy and tactics

We have seen some rather simple examples of proofs using the new rules. In more interesting examples, however, the job of finding a proof can get pretty challenging. So a few words on how to approach these proofs will be helpful.

We have given you a general maxim and two strategies for finding sentential proofs. The maxim—to consider what the various sentences mean—is even more important with the quantifiers. Only if you remember what they mean and how the formal methods mirror common-sense informal methods will you be able to do any but the most boring of exercises.

consider meaning

Our first strategy was to try to come up with an informal proof of the goal sentence from the premises, and use it to try to figure out how your formal proof will proceed. This strategy, too, is even more important in proofs involving quantifiers, but it is a bit harder to apply. The key skill in applying the strategy is the ability to identify the formal rules implicit in your informal reasoning. This takes a bit of practice. Let's work through an example, to see some of the things you should be looking out for.

informal proof as guide

Suppose we want to prove that the following argument is valid:

$\exists x \, (\mathsf{Tet}(x) \wedge \mathsf{Small}(x))$
$\forall x \, (\mathsf{Small}(x) \rightarrow \mathsf{LeftOf}(x, b))$

$\exists x \, \mathsf{LeftOf}(x, b)$

Obviously, the conclusion follows from the given sentences. But ask yourself how you would prove it, say, to your stubborn roommate, the one who likes to play devil's advocate. You might argue as follows:

> Look, Bozo, we're told that there is a small tetrahedron. So we know that it is small, right? But we're also told that anything that's small is left of b. So if it's small, it's got to be left of b, too. So something's left of b, namely the small tetrahedron.

Now we don't recommend calling your roommate "Bozo," so ignore that bit. The important thing to notice here is the implicit use of three of our quantifier rules: ∃ **Elim**, ∀ **Elim**, and ∃ **Intro**. Do you see them?

What indicates the use of ∃ **Elim** is the "it" appearing in the second sentence. What we are doing there is introducing a temporary name (in this case, the pronoun "it") and using it to refer to a small tetrahedron. That corresponds to starting the subproof needed for an application of ∃ **Elim**. So after the second sentence of our informal proof, we can already see the following steps in our reasoning (using "c" for "it"):

1. $\exists x\,(\mathsf{Tet}(x) \wedge \mathsf{Small}(x))$
2. $\forall x\,(\mathsf{Small}(x) \rightarrow \mathsf{LeftOf}(x, b))$

> 3. $\boxed{\mathsf{c}}\ \mathsf{Tet}(c) \wedge \mathsf{Small}(c)$

> 4. $\mathsf{Small}(c)$ ∧ **Elim**: 3
> $\vdots$
> 5. $\exists x\, \mathsf{LeftOf}(x, b)$??

6. $\exists x\, \mathsf{LeftOf}(x, b)$ ∃ **Elim**: 1, 3-5

In general, the key to recognizing ∃ **Elim** is to watch out for any reference to an object whose existence is guaranteed by an existential claim. The reference might use a pronoun (*it, he, she*), as in our example, or it might use a definite noun phrase (*the small tetrahedron*), or finally it might use an actual name (*let n be a small tetrahedron*). Any of these are signs that the reasoning is proceeding via existential elimination.

The third and fourth sentences of our informal argument are where the implicit use of ∀ **Elim** shows up. There we apply the claim about all small things to the small tetrahedron we are calling "it." This gives us a couple more steps in our formal proof:

1. $\exists x\,(\text{Tet}(x) \land \text{Small}(x))$
2. $\forall x\,(\text{Small}(x) \to \text{LeftOf}(x, b))$

    3. $\boxed{c}\;\text{Tet}(c) \land \text{Small}(c)$

    4. $\text{Small}(c)$ $\land$ **Elim**: 3
    5. $\text{Small}(c) \to \text{LeftOf}(c, b)$ $\forall$ **Elim**: 2
    6. $\text{LeftOf}(c, b)$ $\to$ **Elim**: 4, 5
    $\vdots$
    8. $\exists x\,\text{LeftOf}(x, b)$?

9. $\exists x\,\text{LeftOf}(x, b)$ $\exists$ **Elim**: 1, 3-8

The distinctive mark of universal elimination is just the application of a general claim to a specific individual. For example, we might also have said at this point: "So the small tetrahedron [there's the specific individual] must be left of b."

The implicit use of $\exists$ **Intro** appears in the last sentence of the informal reasoning, where we conclude that something is left of b, based on the fact that "it," the small tetrahedron, is left of b. In our formal proof, this application of $\exists$ **Intro** will be done within the subproof, giving us a sentence that we can export out of the subproof since it doesn't contain the temporary name c.

1. $\exists x\,(\text{Tet}(x) \land \text{Small}(x))$
2. $\forall x\,(\text{Small}(x) \to \text{LeftOf}(x, b))$

    3. $\boxed{c}\;\text{Tet}(c) \land \text{Small}(c)$

    4. $\text{Small}(c)$ $\land$ **Elim**: 3
    5. $\text{Small}(c) \to \text{LeftOf}(c, b)$ $\forall$ **Elim**: 2
    6. $\text{LeftOf}(c, b)$ $\to$ **Elim**: 4, 5
    7. $\exists x\,\text{LeftOf}(x, b)$ $\exists$ **Intro**: 6

8. $\exists x\,\text{LeftOf}(x, b)$ $\exists$ **Elim**: 1, 3-7

One thing that's a bit tricky is that in informal reasoning we often leave out simple steps like $\exists$ **Intro**, since they are so obvious. Thus in our example, we might have left out the last sentence completely. After all, once we conclude that the small tetrahedron is left of b, it hardly seems necessary to point out that *something* is left of b. So you've got to watch out for these omitted steps.

This completes our formal proof. To a trained eye, the proof matches the informal reasoning exactly. But you shouldn't feel discouraged if you would have missed it on your own. It takes a lot of practice to recognize the steps implicit in our own reasoning, but it is practice that in the end makes us more careful and able reasoners.

working backward   The second strategy that we stressed is that of working backwards: starting from the goal sentence and inserting steps or subproofs that would enable us to infer that goal. It turns out that of the four new quantifier rules, only ∀ **Intro** really lends itself to this technique.

Suppose your goal sentence is of the form $\forall x\,(P(x) \to Q(x))$. After surveying your given sentences to see whether there is any immediate way to infer this conclusion, it is almost always a good idea to start a subproof in which you introduce an arbitrary name, say c, and assume $P(c)$. Then add a step to the subproof and enter the sentence $Q(c)$, leaving the rule unspecified. Next, end the subproof and infer $\forall x\,(P(x) \to Q(x))$ by ∀ **Intro**, citing the subproof in support. When you check this partial proof, an X will appear next to the sentence $Q(c)$, indicating that your new goal is to prove this sentence.

Remember

1. Always be clear about the meaning of the sentences you are using.

2. A good strategy is to find an informal proof and then try to formalize it.

3. Working backwards can be very useful in proving universal claims, especially those of the form $\forall x\,(P(x) \to Q(x))$.

4. Working backwards is not useful in proving an existential claim $\exists x\,S(x)$ unless you can think of a particular instance $S(c)$ of the claim that follows from the premises.

5. If you get stuck, consider using proof by contradiction.

A worked example

We are going to work through a moderately difficult proof, step by step, using what we have learned in this section. Consider the the following argument:

$$\neg \forall x\, P(x)$$
$$\exists x\, \neg P(x)$$

This is one of four such inferences associated with the DeMorgan rules relating quantifiers and negation. The fact that this inference can be validated in $\mathcal{F}$ is one we will need in our proof of the Completeness Theorem for the system $\mathcal{F}$ in the final chapter. (The other three DeMorgan rules will be given in the review exercises at the end of this chapter. Understanding this example will be a big help in doing those exercises.)

Before embarking on the proof, we mention that this inference is one of the hallmarks of first-order logic. Notice that it allows us to assert the existence of something having a property from a negative fact: that not everything has the opposite property.

In the late nineteenth and early twentieth century, the validity of this sort of inference was hotly debated in mathematical circles. While it seems obvious to us now, it is because we have come to understand existence claims in a somewhat different way than some (the so-called "intuitionists") understood them. While the first-order understanding of $\exists x\, Q(x)$ is as asserting that some Q exists, the intuitionist took it as asserting something far stronger: that the asserter had actually *found* a Q and proven it to be a Q. Under this stronger reading, the DeMorgan principle under discussion would not be valid. This point will be relevant to our proof.

intuitionists

Let us now turn to the proof. Following our strategy, we begin with an informal proof, and then formalize it.

> **Proof:** Since we are trying to prove an existential sentence, our first thought would be to use existential introduction, say by proving $\neg P(c)$ for some c. But if we think a bit about what our premise means, we see there is no hope of proving of any particular thing that it satisfies $\neg P(x)$. From the fact that not everything satisfies $P(x)$, we aren't going to prove of some specific c that $\neg P(c)$. So this is surely a dead end. (It is also why the intuitionist would not accept the argument as valid, given his or her understanding of $\exists$.)
>
> This leaves only one possible route to our desired conclusion: proof by contradiction. Thus we will negate our desired conclusion and try to obtain a contradiction. Thus, we assume $\neg\exists x\, \neg P(x)$. How can we hope to obtain a contradiction? Since our only premise is $\neg\forall x\, P(x)$, the most promising line of attack would be to try for a proof of $\forall x\, P(x)$ using universal generalization. Thus, let c be an arbitrary individual in our domain of discourse. Our goal is to prove $P(c)$. How can we do this? Another proof by contradiction, for if $P(c)$ were not the case, then we would have $\neg P(c)$, and hence $\exists x\, \neg P(x)$. But this contradicts our assumption. Hence $P(c)$ is the case. Since

c was arbitrary, we get $\forall x\, P(x)$. But this contradicts our premise. Hence, we get our desired conclusion using the method of proof by contradiction.

We now begin the process of turning our informal proof into a formal proof.

You try it

. . . .

▶ 1. Open Quantifier Strategy 1. This contains the skeleton of our proof:

> 1. $\neg\forall x\, P(x)$
> $\vdots$
> 2. $\exists x\, \neg P(x)$

▶ 2. The first step in our informal proof was to decide to try to give a proof by contradiction. Formalize this idea by filling in the following:

> 1. $\neg\forall x\, P(x)$
> > 2. $\neg\exists x\, \neg P(x)$
> > $\vdots$
> > 4. $\bot$ $\bot$ **Intro**: ?, ?
> 5. $\exists x\, \neg P(x)$ $\neg$ **Intro**: 2–4

This step will check out because of the generous nature of Fitch's $\neg$ **Intro** rule, which lets us strip off as well as add a negation.

▶ 3. We next decided to try to contradict $\neg\forall x\, P(x)$ by proving $\forall x\, P(x)$ using universal generalization. Formalize this as follows:

> 1. $\neg\forall x\, P(x)$
> > 2. $\neg\exists x\, \neg P(x)$
> > > 3. $\boxed{c}$
> > > $\vdots$
> > > 5. $P(c)$?
> > 6. $\forall x\, P(x)$ $\forall$ **Intro**: 3–5
> > 7. $\bot$ $\bot$ **Intro**: 6, 1
> 8. $\exists x\, \neg P(x)$ $\neg$ **Intro**: 2–7

4. Recall how we proved P(c). We said that if P(c) were not the case, then ◀
   we would have ¬P(c), and hence ∃x ¬P(x). But this contradicted the assumption at step 2. Formalize this reasoning by filling in the rest of the
   proof.

> 1. ¬∀x P(x)
>> 2. ¬∃x ¬P(x)
>>> 3. $\boxed{c}$
>>>> 4. ¬P(c)
>>>> 5. ∃x ¬P(x)   ∃ **Intro**: 4
>>>> 6. ⊥   ⊥ **Intro**: 5, 2
>>> 7. ¬¬P(c)   ¬ **Intro**: 4–6
>>> 8. P(c)   ¬ **Elim**: 7
>> 9. ∀x P(x)   ∀ **Intro**: 3–8
>> 10. ⊥   ⊥ **Intro**: 9, 1
> 11. ∃x ¬P(x)   ¬ **Intro**: 2–10

5. This completes our formal proof of ∃x ¬P(x) from the premise ¬∀x P(x). ◀
   Verify your proof and save it as Proof Quantifier Strategy 1.
   . . . . *Congratulations*

Exercises

13.20 If you skipped the **You try it** section, go back and do it now. Submit the file Proof Quantifier
⤻   Strategy 1.

Recall that in Exercises 12.1–12.3 on page 336, you were asked to give logical analyses of purported
proofs of some arguments involving nonsense predicates. In the following exercises, we return to these
arguments. If the argument is valid, submit a formal proof. If it is invalid, turn in an informal coun-
terexample. If you submit a formal proof, be sure to use the Exercise file supplied with Fitch. In order
to keep your hand in at using the propositional rules, we ask you not to use **Taut Con** *in these proofs.*

13.20
⤻|✎
> ∀x [(Brillig(x) ∨ Tove(x)) → (Mimsy(x) ∧ Gyre(x))]
> ∀y [(Slithy(y) ∨ Mimsy(y)) → Tove(y)]
> ∃x Slithy(x)
>
> ∃x [Slithy(x) ∧ Mimsy(x)]

(See Exercise 12.1 on p. 336.)

13.21
✒️|✏️

$\forall x \, [\text{Brillig}(x) \rightarrow (\text{Mimsy}(x) \wedge \text{Slithy}(x))]$
$\forall y \, [(\text{Slithy}(y) \vee \text{Mimsy}(y)) \rightarrow \text{Tove}(y)]$
$\forall x \, [\text{Tove}(x) \rightarrow (\text{Outgrabe}(x, b) \wedge \text{Brillig}(x))]$

$\forall z \, [\text{Brillig}(z) \leftrightarrow \text{Mimsy}(z)]$

(See Exercise 12.2 on p. 337.)

13.22
✒️|✏️

$\forall x \, [(\text{Brillig}(x) \wedge \text{Tove}(x)) \rightarrow \text{Mimsy}(x)]$
$\forall y \, [(\text{Tove}(y) \vee \text{Mimsy}(y)) \rightarrow \text{Slithy}(y)]$
$\exists x \, \text{Brillig}(x) \wedge \exists x \, \text{Tove}(x)$

$\exists z \, \text{Slithy}(z)$

(See Exercise 12.3, p. 337)

Some of the following arguments are valid, some are not. For each, either use Fitch to give a formal proof or use Tarski's World to construct a counterexample. In giving proofs, feel free to use **Taut Con** *if it helps.*

13.23
✒️

$\forall y \, [\text{Cube}(y) \vee \text{Dodec}(y)]$
$\forall x \, [\text{Cube}(x) \rightarrow \text{Large}(x)]$
$\exists x \, \neg \text{Large}(x)$

$\exists x \, \text{Dodec}(x)$

13.24
✒️

$\exists x \, (\text{Cube}(x) \wedge \text{Small}(x))$

$\exists x \, \text{Cube}(x) \wedge \exists x \, \text{Small}(x)$

13.25
✒️

$\exists x \, \text{Cube}(x) \wedge \exists x \, \text{Small}(x)$

$\exists x \, (\text{Cube}(x) \wedge \text{Small}(x))$

13.26
✒️

$\forall x \, (\text{Cube}(x) \rightarrow \text{Small}(x))$
$\forall x \, (\text{Adjoins}(x, b) \rightarrow \text{Small}(x))$

$\forall x \, ((\text{Cube}(x) \vee \text{Small}(x))$
$\quad \rightarrow \text{Adjoins}(x, b))$

13.27
✒️

$\forall x \, (\text{Cube}(x) \rightarrow \text{Small}(x))$
$\forall x \, (\neg \text{Adjoins}(x, b) \rightarrow \neg \text{Small}(x))$

$\forall x \, ((\text{Cube}(x) \vee \text{Small}(x)) \rightarrow \text{Adjoins}(x, b))$

For each of the following, use Fitch to give a formal proof of the argument. These look simple but some of them are a bit tricky. Don't forget to first figure out an informal proof. Use **Taut Con** *whenever it is convenient but do not use* **FO Con**.

13.28
✒️⋆

$\forall x \, \forall y \, \text{Likes}(x, y)$

$\forall x \, \exists y \, \text{Likes}(x, y)$

13.29
✒️⋆

$\forall x \, (\text{Small}(x) \rightarrow \text{Cube}(x))$
$\exists x \, \neg \text{Cube}(x) \rightarrow \exists x \, \text{Small}(x)$

$\exists x \, \text{Cube}(x)$

13.30
✒️⋆

$\text{Likes}(\text{carl}, \text{max})$
$\forall x \, [\exists y \, (\text{Likes}(y, x) \vee \text{Likes}(x, y))$
$\quad \rightarrow \text{Likes}(x, x)]$

$\exists x \, \text{Likes}(x, \text{carl})$

13.31
✒️⋆

$\forall x \, \forall y \, [\text{Likes}(x, y) \rightarrow \text{Likes}(y, x)]$
$\exists x \, \forall y \, \text{Likes}(x, y)$

$\forall x \, \exists y \, \text{Likes}(x, y)$

*The following valid arguments come in pairs. The validity of the first of the pair makes crucial use of the meanings of the blocks language predicates, whereas the second adds one or more premises, making the result a first-order valid argument. For the latter, give a proof that does not make use of **Ana Con**. For the former, give a proof that uses **Ana Con** but only where the premises and conclusions of the citation are literals (including ⊥). You may use **Taut Con** but do not use **FO Con** in any of the proofs.*

13.32

> ¬∃x (Tet(x) ∧ Small(x))
>
> ∀x [Tet(x) → (Large(x) ∨ Medium(x))]

13.33

> ¬∃x (Tet(x) ∧ Small(x))
> ∀y (Small(y) ∨ Medium(y) ∨ Large(y))
>
> ∀x [Tet(x) → (Large(x) ∨ Medium(x))]

13.34

> ∀x (Dodec(x) → SameCol(x, a))
> SameCol(a, c)
>
> ∀x (Dodec(x) → SameCol(x, c))

13.35

> ∀x (Dodec(x) → SameCol(x, a))
> SameCol(a, c)
> ∀x ∀y ∀z ((SameCol(x, y) ∧ SameCol(y, z))
>   → SameCol(x, z))
>
> ∀x (Dodec(x) → SameCol(x, c))

13.36

> ∀x (Dodec(x) → LeftOf(x, a))
> ∀x (Tet(x) → RightOf(x, a))
>
> ∀x (SameCol(x, a) → Cube(x))

13.37

> ∀x (Dodec(x) → LeftOf(x, a))
> ∀x (Tet(x) → RightOf(x, a))
> ∀x ∀y (LeftOf(x, y) → ¬SameCol(x, y))
> ∀x ∀y (RightOf(x, y) → ¬SameCol(x, y))
> ∀x (Cube(x) ∨ Dodec(x) ∨ Tet(x))
>
> ∀x (SameCol(x, a) → Cube(x))

13.38

> ∀x (Cube(x) → ∀y (Dodec(y)
>   → Larger(x, y)))
> ∀x (Dodec(x) → ∀y (Tet(y) → Larger(x, y)))
> ∃x Dodec(x)
>
> ∀x (Cube(x) → ∀y (Tet(y) → Larger(x, y)))

(Compare this with Exercise 13.9. The crucial difference is the presence of the third premise, not the difference in form of the first two premises.)

13.39

> ∀x (Cube(x) → ∀y (Dodec(y)
>   → Larger(x, y)))
> ∀x (Dodec(x) → ∀y (Tet(y)
>   → Larger(x, y)))
> ∃x Dodec(x)
> ∀x ∀y ∀z ((Larger(x, y) ∧ Larger(y, z))
>   → Larger(x, z))
>
> ∀x (Cube(x) → ∀y (Tet(y)
>   → Larger(x, y)))

Section 13.4

Soundness and completeness

In Chapter 8 we raised the question of whether the deductive system $\mathcal{F}_T$ was sound and complete with respect to tautological consequence. The same issues arise with the full system $\mathcal{F}$, which contains the rules for the quantifiers and identity, in addition to the rules for the truth-functional connectives. Here, the target consequence relation is the notion of first-order consequence, rather than tautological consequence.

soundness of $\mathcal{F}$

The *soundness* question asks whether anything we can prove in $\mathcal{F}$ from premises $P_1, \ldots, P_n$ is indeed a first-order consequence of the premises. The

completeness of $\mathcal{F}$

completeness question asks the converse: whether every first-order consequence of a set of sentences can be proven from that set using the rules of $\mathcal{F}$.

It turns out that both of these questions can be answered in the affirmative. Before actually proving this, however, we need to add more precision to the notion of first-order consequence, and this presupposes tools from set theory that we will introduce in Chapters 15 and 16. We state and prove the soundness theorem for first-order logic in Chapter 18. The completeness theorem for first-order logic is the main topic of Chapter 19.

Section 13.5

Some review exercises

In this section we present more problems to help you solidify your understanding of the methods of reasoning involving quantifiers. We also present some more interesting problems from a theoretical point of view.

Exercises

Some of the following arguments are valid, some are not. For each, either use Fitch to give a formal proof or use Tarski's World to construct a counterexample. In giving proofs, feel free to use **Taut Con** *if it helps.*

13.40
$$\exists x\, \text{Cube}(x) \land \text{Small}(d)$$

$$\exists x\, (\text{Cube}(x) \land \text{Small}(d))$$

13.41
$$\forall x\, (\text{Cube}(x) \lor \text{Small}(x))$$

$$\forall x\, \text{Cube}(x) \lor \forall x\, \text{Small}(x)$$

13.42

$\forall x\, \text{Cube}(x) \lor \forall x\, \text{Small}(x)$

$\forall x\, (\text{Cube}(x) \lor \text{Small}(x))$

Each of the following is a valid argument of a type discussed in Section 10.3. Use Fitch to give a proof of its validity. You may use **Taut Con** *freely in these proofs.*

13.43

$\neg \forall x\, \text{Cube}(x)$

$\exists x\, \neg\text{Cube}(x)$

13.44

$\neg \exists x\, \text{Cube}(x)$

$\forall x\, \neg\text{Cube}(x)$

13.45

$\forall x\, \neg\text{Cube}(x)$

$\neg \exists x\, \text{Cube}(x)$

13.46 (Change of bound variables)

$\forall x\, \text{Cube}(x)$

$\forall y\, \text{Cube}(y)$

13.47 (Change of bound variables)

$\exists x\, \text{Tet}(x)$

$\exists y\, \text{Tet}(y)$

13.48 (Null quantification)

$\text{Cube}(b) \leftrightarrow \forall x\, \text{Cube}(b)$

13.49

$\exists x\, P(x)$

$\forall x\, \forall y\, ((P(x) \land P(y)) \rightarrow x = y)$

$\exists x\, (P(x) \land \forall y\, (P(y) \rightarrow y = x))$

13.50

$\exists x\, (P(x) \land \forall y\, (P(y) \rightarrow y = x))$

$\forall x\, \forall y\, ((P(x) \land P(y)) \rightarrow x = y)$

13.51

$\exists x\, (P(x) \rightarrow \forall y\, P(y))$

[Hint: Review your answer to Exercise 12.22 where you should have given an informal proof of something of this form.]

13.52

$\neg \exists x\, \forall y\, [E(x, y) \leftrightarrow \neg E(y, y)]$

This result might be called Russell's Theorem. It is connected with the famous result known as Russell's Paradox, which is discussed in Section 15.9. In fact, it was upon discovering this that Russell invented the Barber Paradox, to explain his result to a general public.

13.53 Is $\exists x\, \exists y\, \neg\text{LeftOf}(x, y)$ a first-order consequence of $\exists x\, \neg\text{LeftOf}(x, x)$? If so, give a formal proof. If not, give a reinterpretation of LeftOf and an example where the premise is true and the conclusion is false.

The next exercises are intended to help you review the difference between first-order satisfiability and true logical possibility. All involve the four sentences in the file Padoa's Sentences. *Open that file now.*

13.54
↯* Any three of the sentences in Padoa's Sentences form a satisfiable set. There are four sets of three sentences, so to show this, build four worlds, World 13.54.123, World 13.54.124, World 13.54.134, and World 13.54.234, where the four sets are true. (Thus, for example, sentences 1, 2 and 4 should be true in World 13.54.124.)

13.55
✎* Give an informal proof that the four sentences in Padoa's Sentences taken together are inconsistent.

13.56
✎* Is the set of sentences in Padoa's Sentences first-order satisfiable, that is, satisfiable with some reinterpretation of the predicates other than identity? [Hint: Imagine a world where one of the blocks is a sphere.]

13.57
✎* Reinterpret the predicates Tet and Dodec in such a way that sentence 3 of Padoa's Sentences comes out true in World 13.54.124. Since this is the only sentence that uses these predicates, it follows that all four sentences would, with this reinterpretation, be true in this world. (This shows that the set is first-order satisfiable.)

13.58
↯|✎** (Logical truth *versus* non-logical truth in all worlds) A distinction Tarski's World helps us to understand is the difference between sentences that are logically true and sentences that are, for reasons that have nothing to do with logic, true in all worlds. The notion of logical truth has to do with a sentence being true simply in virtue of the *meaning* of the sentence, and so no matter how the world is. However, some sentences are true in all worlds, not because of the meaning of the sentence or its parts, but because of, say, laws governing the world. We can think of the constraints imposed by the innards of Tarski's World as analogues of physical laws governing how the world can be. For example, the sentence which asserts that there are at most 12 objects happens to hold in all the worlds that we can construct with Tarski's World. However, it is not a logical truth.

Open Post's Sentences. Classify each sentence in one of the following ways: (A) a logical truth, (B) true in all worlds that can be depicted using Tarski's World, but not a logical truth, or (C) falsifiable in some world that can be depicted by Tarski's World. For each sentence of type (C), build a world in which it is false, and save it as World 13.58.x, where x is the number of the sentence. For each sentence of type (B), use a pencil and paper to depict a world in which it is false. (In doing this exercise, assume that Medium simply means *neither small nor large*, which seems plausible. However, it is not plausible to assume that Cube means *neither a dodecahedron nor tetrahedron*, so you should not assume anything like this.)

CHAPTER 14

More about Quantification

Many English sentences take the form

$$Q \, A \, B$$

where Q is a *determiner* expression like *every, some, the, more than half the, at least three, no, many, Max's*, etc.; A is a common noun phrase like *cube, student of logic, thing*, etc.; and B is a verb phrase like *sits in the corner* or *is small*.

Such sentences are used to express quantitative relationships between the set of objects satisfying the common noun phrase and the set of objects satisfying the verb phrase. Here are some examples, with the determiner in bold:

> **Every** *cube is small.*
> **Some** *cube is small.*
> **More than half the** *cubes are small.*
> **At least three** *cubes are small.*
> **No** *cube is small.*
> **Many** *cubes are small.*
> **Max's** *cube is small.*

These sentences say of the set A of cubes in the domain of discourse and the set B of small things in the domain of discourse that

> every A is a B,
> some A is a B,
> more than half the A's are B's,
> at least three A's are B's,
> no A is a B,
> many A's are B's, and
> Max's A is a B.

Each of these can be thought of as expressing a kind of binary relation between A and B.

Linguistically, these words and phrases are known as *determiners*. The relation expressed by a determiner is usually, though not always, a *quantitative* relation between A and B. Sometimes this quantitative relation can be captured using the FOL quantifiers ∀ and ∃, though sometimes it can't. For

determiners and quantifiers

example, to express *more than half the A's are B's*, it turns out that we need to supplement FOL to include new expressions that behave something like ∀ and ∃. When we add such expressions to the formal language, we call them *generalized quantifiers*, since they extend the kinds of quantification we can express in the language.

generalized quantifiers

In this chapter, we will look at the logic of some English determiners beyond *some* and *all*. We will consider not only determiners that can be expressed using the usual quantifiers of FOL, but also determiners whose meanings can only be captured by adding new quantifiers to FOL.

In English, there are ways of expressing quantification other than determiners. For example, the sentences

> Max **always** eats pizza.
> Max **usually** eats pizza.
> Max **often** eats pizza.
> Max **seldom** eats pizza.
> Max **sometimes** eats pizza.
> Max **never** eats pizza.

each express a quantitative relation between the set of times when Max eats and the set of times when he eats pizza. But in these sentences it is the adverb that is expressing quantification, not a determiner. While we are going to discuss the logic only of determiners, much of what we say can be extended to other forms of quantification, including this kind of adverbial quantification.

adverbial quantification

In a sentence of the form $Q\,A\,B$, different determiners express very different relations between A and B and so have very different logical properties. A valid argument typically becomes invalid if we change any of the determiners. For instance, while

> No cube is small
> d is a cube
> ───────────
> d is not small

is a valid argument, it would become invalid if we replaced *no* by any of the other determiners listed above. On the other hand, the valid argument

> Many cubes are small
> Every small block is left of d
> ───────────
> Many cubes are left of d

remains valid if *many* is replaced by any of the above determiners other than *no*. These are clearly logical facts, things we'd like to understand at a more theoretical level. For example, we'll soon see that the determiners that can replace *Many* in the second argument and still yield a valid argument are the *monotone increasing* determiners.

There are two rather different approaches to studying quantification. One approach studies determiners that can be expressed using the existing resources of FOL. In the first three sections, we look at several important English determiners that can be defined in terms of $\forall$, $\exists$, $=$, and the truth-functional connectives, and then analyze their logical properties by means of these definitions. The second approach is to strengthen FOL by allowing a wider range of quantifiers, capturing kinds of quantification not already expressible in FOL. In the final three sections, we look briefly at this second approach and its resulting logic.

approaches to quantification

Numerical quantification

We have already seen that many complex noun phrases can be expressed in terms of $\forall$ (which really means "everything", not just "every") and $\exists$ (which means "something," not "some"). For example, *Every cube left of **b** is small* can be paraphrased as *Everything that is a cube and left of **b** is small*, a sentence that can easily be translated into FOL using $\forall$, $\wedge$ and $\rightarrow$. Similarly, *No cube is small* can be paraphrased as *Everything is such that if it is a cube then it is not small*, which can again be easily translated into FOL.

Other important examples of quantification that can be indirectly expressed in FOL are numerical claims. By a "numerical claim" we mean a one that explicitly uses the numbers $1, 2, 3, \ldots$ to say something about the relation between the A's and the B's. Here are three different kinds of numerical claims:

numerical claims

> **At least two** *books arrived this week.*
> **At most two** *books are missing from the shelf.*
> **Exactly two** *books are on the table.*

First-order languages do not in general allow us to talk directly about numbers, only about elements of our domain of discourse. The blocks language, for example, only talks about blocks, not about numbers. Still, it is possible to express these three kinds of numerical claims in FOL.

Recall that in FOL, distinct names do not necessarily refer to distinct objects. Similarly, distinct variables need not vary over distinct objects. For example, both of the following sentences can be made true in a world with one object:

$$Cube(a) \wedge Small(a) \wedge Cube(b)$$

$$\exists x \, \exists y \, [Cube(x) \wedge Small(x) \wedge Cube(y)]$$

at least two

In order to say that there are at least two cubes, you must find a way to guarantee that they are different. For example, either of the following would do:

$$Cube(a) \wedge Small(a) \wedge Cube(b) \wedge Large(b)$$

$$\exists x \, \exists y \, [Cube(x) \wedge Small(x) \wedge Cube(y) \wedge LeftOf(x, y)]$$

The most direct way, though, is simply to say that they are different:

$$\exists x \, \exists y \, [Cube(x) \wedge Cube(y) \wedge x \neq y]$$

This sentence asserts that there are at least two cubes. To say that there are at least three cubes we need to add another $\exists$ and some more inequalities:

$$\exists x \, \exists y \, \exists z \, [Cube(x) \wedge Cube(y) \wedge Cube(z) \wedge x \neq y \wedge x \neq z \wedge y \neq z]$$

You will see in the **You try it** section below that all three of these inequalities are really needed. To say that there are at least four objects takes four $\exists$'s and six $(= 3 + 2 + 1)$ inequalities; to say there are at least five takes five $\exists$'s and 10 $(= 4 + 3 + 2 + 1)$ inequalities, and so forth.

at most two

Turning to the second kind of numerical quantification, how can we say that there are *at most* two cubes? Well, one way to do it is by saying that there are not at least three cubes:

$$\neg \exists x \, \exists y \, \exists z \, [Cube(x) \wedge Cube(y) \wedge Cube(z) \wedge x \neq y \wedge x \neq z \wedge y \neq z]$$

Applying some (by now familiar) quantifier equivalences, starting with De-Morgan's Law, gives us the following equivalent sentence:

$$\forall x \, \forall y \, \forall z \, [(Cube(x) \wedge Cube(y) \wedge Cube(z)) \rightarrow (x = y \vee x = z \vee y = z)]$$

We will take this as our official way of expressing *at most two*.

Notice that while it took two existential quantifiers to express *there are at least two cubes*, it took three universal quantifiers to say that there are at most two cubes. More generally, to translate the determiner *at least n* into FOL, we need n existential quantifiers, while to translate *at most n* we need $n + 1$ universal quantifiers.

exactly two

To express the sentence *there are exactly two cubes*, we could paraphrase it as follows: *There are at least two cubes and there are at most two cubes.* Translating each conjunct gives us a rather long sentence using five quantifiers:

$$\exists x \, \exists y \, [\text{Cube}(x) \wedge \text{Cube}(y) \wedge x \neq y] \wedge$$

$$\forall x \, \forall y \, \forall z \, [(\text{Cube}(x) \wedge \text{Cube}(y) \wedge \text{Cube}(z)) \rightarrow (x = y \vee x = z \vee y = z)]$$

The same claim can be expressed more succinctly, however, as follows:

$$\exists x \, \exists y \, [\text{Cube}(x) \wedge \text{Cube}(y) \wedge x \neq y \wedge \forall z \, (\text{Cube}(z) \rightarrow (z = x \vee z = y))]$$

If we translate this into English, we see that it says there are two distinct objects, both cubes, and that any cube is one of these. This is a different way of saying that there are exactly two cubes. (We ask you to give formal proofs of their equivalence in Exercises 14.12 and 14.13.) Notice that this sentence uses two existential quantifiers and one universal quantifier. An equivalent way of saying this is as follows:

$$\exists x \, \exists y \, [x \neq y \wedge \forall z \, (\text{Cube}(z) \leftrightarrow (z = x \vee z = y))]$$

Put in prenex form, this becomes:

$$\exists x \, \exists y \, \forall z \, [x \neq y \wedge (\text{Cube}(z) \leftrightarrow (z = x \vee z = y))]$$

All three expressions consist of two existential quantifiers followed by a single universal quantifier. More generally, to say that there are exactly n objects satisfying some condition requires $n + 1$ quantifiers, n existential followed by one universal.

You try it
. . . .

1. In this **Try It**, you will get to examine some of the claims made above in more detail. Open Whitehead's Sentences.   ◄

2. The first sentence says that there are at least two objects and the second sentence says that there are at most two objects. (Do you see how they manage to say these things?) Build a model where the first two sentences are both true.   ◄

3. Sentence 3 is the conjunction of the first two. Hence, it asserts, in one sentence, that there are exactly two objects. Check to see that it is true in the world you have just built.   ◄

▶ 4. The fourth sentence is in fact equivalent to the third sentence. It is a shorter way of saying that there are exactly two objects. Play the game three times with this sentence, committed to true each time. First play it in a world with one object, then in a world with two objects, then in a world with three objects. You will be able to win only in the second world.

▶ 5. Sentence 5 appears, at first sight, to assert that there are at least three objects, so it should be false in a world with two objects. Check to see if it is indeed false in such a world. Why isn't it? Play the game to confirm your suspicions.

▶ 6. The sixth sentence actually manages to express the claim that there are at least three objects. Do you see how it's different from the fifth sentence? Check to see that it is false in the current world, but is true if you add a third object to the world.

▶ 7. The seventh sentence says that there are exactly three objects in the world. Check to see that it is true in the world with three objects, but false if you either delete an object or add another object.

▶ 8. Sentence 8 asserts that a is a large object, and in fact the *only* large object. To see just how the sentence manages to say this, start with a world with three small objects and name one of them "a." Play the game committed to true to see why the sentence is false. You can quit the game as soon as you understand why the sentence is false. Now make a large. Again play the game committed to true and see why you can now win (does it matter which block Tarski picks?). Finally, make one of the other objects large as well, and play the game committed to true to see why it is false.

▶ 9. Sentence 8 asserted that a was the only large object. How might we say that there is exactly one large object, without using a name for the object? Compare sentence 8 with sentence 9. The latter asserts that there is something that is the only large object. Check to see that it is true only in worlds in which there is exactly one large object.

▶ 10. If you have understood sentence 9, you should also be able to understand sentence 10. Construct a world in which sentence 10 is true. Save this world as World Numerical 1.

▶ 11. Sentence 11 says there is exactly one medium dodecahedron, while sentence 12 says there are at least two dodecahedra. There is nothing incompatible about these claims. Make sentences 11 and 12 true in a single world. Save the world as World Numerical 2.

12. Sentence 13 is another way to assert that there is a unique dodecahedron. ◀
    That is, sentence 13 is equivalent to sentence 10. Can you see why? Check
    three worlds to see that the two sentences are true in the same worlds,
    those in which there is a single dodecahedron.

13. Sentence 14 says that there are exactly two tetrahedra. Check that it is ◀
    true in such worlds, but false if there are fewer or more than two.

. . . . *Congratulations*

Numerical quantification, when written out in full in FOL, is hard to read   *abbreviations for*
because of all the inequalities, especially when the numbers get to be more   *numerical claims*
than 3 or 4, so a special notation has become fairly common:

  o $\exists^{\geq n} x\, P(x)$ for the FOL sentence asserting "There are at least n objects
    satisfying $P(x)$."

  o $\exists^{\leq n} x\, P(x)$ for the FOL sentence asserting "There are at most n objects
    satisfying $P(x)$."

  o $\exists^{!n} x\, P(x)$ for the FOL sentence asserting "There are exactly n objects
    satisfying $P(x)$."

It is important to remember that this notation is *not* part of the official
language of FOL, but an abbreviation for a much longer FOL expression.

The special case of $n = 1$ is important enough to warrant special comment.   *exactly one*
The assertion that there is exactly one object satisfying some condition $P(x)$
can be expressed in FOL as follows:

$$\exists x\, [P(x) \wedge \forall y\, (P(y) \rightarrow y = x)]$$

as long as y does not occur already in the wff $P(x)$. According to the conventions we have just established, this should be abbreviated as $\exists^{!1} x\, P(x)$. In
practice, though, this is used so often that it is further shortened to $\exists! x\, P(x)$.
It is read "there is a unique x such that $P(x)$." Again, this is not a new quantifier; wffs in which it occurs are just abbreviations for longer wffs involving
the old quantifiers.

We started out with the goal of learning how to express claims of the form
$Q\, A\, B$ where Q is a numerical determiner and A is any common noun. But all
we have seen so far is how to express claims of the form *there are at least/at
most/exactly n **things** satisfying P*. Having learned how to do this, however,
it's easy to express claims of the desired form. For example, to say *At least
n cubes are small*, we say *There are at least n things that are small cubes.*

Similarly, to say *There are at most n cubes that are small*, we say *There are at most n things that are small cubes*. Finally, to say *There are exactly n cubes that are small*, we say *There are exactly n things that are small cubes*. These observations probably seem so obvious that they don't require mentioning. But we will soon see that nothing like this holds for some determiners, and that the consequences are rather important for the general theory of quantification.

> **Remember**
>
> The notations $\exists^{\geq n}$, $\exists^{\leq n}$, and $\exists^{!n}$ are abbreviations for complex FOL expressions meaning "there are at least/at most/exactly n things such that . . . ."

Exercises

14.1 If you skipped the **You try it** section, go back and do it now. Submit the files World Numerical 1 and World Numerical 2.

14.2 Give clear English translations of the following sentences of FOL. Which of the following are logically equivalent and which are not? Explain your answers.
1. $\exists! x\, \text{Tove}(x)$ [Remember that the notation $\exists!$ is an abbreviation, as explained above.]
2. $\exists x\, \forall y\, [\text{Tove}(y) \rightarrow y = x]$
3. $\exists x\, \forall y\, [\text{Tove}(y) \leftrightarrow y = x]$
4. $\forall x\, \forall y\, [(\text{Tove}(x) \land \text{Tove}(y)) \rightarrow x = y]$
5. $\forall x\, \forall y\, [(\text{Tove}(x) \land \text{Tove}(y)) \leftrightarrow x = y]$

14.3 (Translating numerical claims) In this exercise we will try our hand at translating English sentences involving numerical claims.

○ Using Tarski's World, translate the following English sentences.

1. *There are at least two dodecahedra.*
2. *There are at most two tetrahedra.*
3. *There are exactly two cubes.*
4. *There are only three things that are not small.*
5. *There is a single large cube. No dodecahedron is in back of it.*

○ Open Peano's World. Note that all of the English sentences are true in this world. Check to see that your translations are as well.

○ Open **Bolzano's World**. Here sentences 1, 3, and 5 are the only true ones. Verify that your translations have the right truth values in this world.

○ Open **Skolem's World**. Only sentence 5 is true in this world. Check your translations.

○ Finally, open **Montague's World**. In this world, sentences 2, 3, and 5 are the only true ones. Check your translations.

14.4 (Saying more complicated things) Open **Skolem's World**. Create a file called **Sentences 14.4** and describe the following features of **Skolem's World**.

1. Use your first sentence to say that there are only cubes and tetrahedra.
2. Next say that there are exactly three cubes.
3. Express the fact that every cube has a tetrahedron that is to its right but is neither in front of or in back of it.
4. Express the fact that at least one of the tetrahedra is between two other tetrahedra.
5. Notice that the further back something is, the larger it is. Say this.
6. Note that none of the cubes is to the right of any of the other cubes. Try to say this.
7. Observe that there is a single small tetrahedron and that it is in front of but to neither side of all the other tetrahedra. State this.

If you have expressed yourself correctly, there is very little you can do to **Skolem's World** without making at least one of your sentences false. Basically, all you can do is "stretch" things out, that is, move things apart while keeping them aligned. To see this, try making the following changes. (There's no need to turn in your answers, but try the changes.)

1. Add a new tetrahedron to the world. Find one of your sentences that comes out false. Move the new tetrahedron so that a different sentence comes out false.
2. Change the size of one of the objects. What sentence now comes out false?
3. Change the shape of one of the objects. What sentence comes out false?
4. Slide one of the cubes to the left. What sentence comes out false?
5. Rearrange the three cubes. What goes wrong now?

14.5 (Ambiguity and numerical quantification) In the **Try It** on page 314, we saw that the sentence

At least four medium dodecahedra are adjacent to a medium cube.

is ambiguous, having both a strong and a weak reading. Using Tarski's World, open a new sentence file and translate the strong and weak readings of this sentence into FOL as sentences (1) and (2). Remember that Tarski's World does not understand our abbreviation for "at least four" so you will need to write this out in full. Check that the first sentence is true in **Anderson's First World** but not in **Anderson's Second World**, while the second sentence is true in both worlds. Make some changes to the worlds to help you check that your translations express what you intend. Submit your sentence file.

14.6 (Games of incomplete information) As you recall, you can sometimes know that a sentence is true in a world without knowing how to play the game and win. Open Mostowski's World. Translate the following into first-order logic. Save your sentences as Sentences 14.6. Now, without using the 2-D view, make as good a guess as you can about whether the sentences are true or not in the world. Once you have assessed a given sentence, use **Verify** to see if you are right. Then, with the correct truth value checked, see how far you can go in playing the game. Quit whenever you get stuck, and play again. Can you predict in advance when you will be able to win? Do not look at the 2-D view until you have finished the whole exercise.

1. *There are at least two tetrahedra.*
2. *There are at least three tetrahedra.*
3. *There are at least two dodecahedra.*
4. *There are at least three dodecahedra.*
5. *Either there is a small tetrahedron behind a small cube or there isn't.*
6. *Every large cube is in front of something.*
7. *Every tetrahedron is in back of something.*
8. *Every small cube is in back of something.*
9. *Every cube has something behind it.*
10. *Every dodecahedron is small, medium, or large.*
11. *If e is to the left of every dodecahedron, then it is not a dodecahedron.*

Now modify the world so that the true sentences are still true, but so that it will be clear how to play the game and win. When you are done, just submit your sentence file.

14.7 (Satisfiability) Recall that a set of sentences is satisfiable if there is world in which it is true. Determine whether the following set of sentences is satisfiable. If it is, build a world. If it is not, use informal methods of proof to derive a contradiction from the set.

1. *Every cube is to the left of every tetrahedron.*
2. *There are no dodecahedra.*
3. *There are exactly four cubes.*
4. *There are exactly four tetrahedra.*
5. *No tetrahedron is large.*
6. *Nothing is larger than anything to its right.*
7. *One thing is to the left of another just in case the latter is behind the former.*

14.8 (Numbers of variables) Tarski's World only allows you to use six variables. Let's explore what kind of limitation this imposes on our language.

1. Translate the sentence *There are at least two objects*, using only the predicate =. How many variables do you need?
2. Translate *There are at least three objects*. How many variables do you need?

3. It is impossible express the sentence *There are at least seven objects* using only = and the six variables available in Tarski's World, no matter how many quantifiers you use. Try to prove this. [Warning: This is true, but it is very challenging to prove. Contrast this problem with the one below.] Submit your two sentences and turn in your proof.

14.9 (Reusing variables) In spite of the above exercise, there are in fact sentences we can express
✐⋆ using just the six available variables that can only be true in worlds with at least seven objects. For example, in Robinson's Sentences, we give such a sentence, one that only uses the variables x and y.

1. Open this file. Build a world where there are six small cubes arranged on the front row and test the sentence's truth. Now add one more small cube to the front row, and test the sentence's truth again. Then play the game committed (incorrectly) to false. Can you see the pattern in Tarski's World's choice of objects? When it needs to pick an object for the variable x, it picks the leftmost object to the right of all the previous choices. Then, when it needs to pick an object for the variable y, it picks the last object chosen. Can you now see how the reused variables are working?

2. Now delete one of the cubes, and play the game committed (incorrectly) to true. Do you see why you can't win?

3. Now write a sentence that says there are at least four objects, one in front of the next. Use only variables x and y. Build some worlds to check whether your sentence is true under the right conditions. Submit your sentence file.

Proving numerical claims

Since numerical claims can be expressed in FOL, we can use the methods of proof developed in previous chapters to prove numerical claims. However, as you may have noticed in doing the exercises, numerical claims are not always terribly perspicuous when expressed in FOL notation. Indeed, expressing a numerical claim in FOL and then trying to prove the result is a recipe for disaster. It is all too easy to lose one's grasp on what needs to be proved.

Suppose, for example, that you are told there are exactly two logic classrooms and that each classroom contains exactly three computers. Suppose you also know that every computer is in some logic classroom. From these assumptions it is of course quite easy to prove that there are exactly six computers. How would the proof go?

Proof: To prove there are exactly six computers it suffices to prove that there are at least six, and at most six. To prove that there are at most six, we simply note that every computer must be in one of the two classrooms, and that each classroom contains at most three, so there can be at most six altogether, since $2 \times 3 = 6$. To prove that there are at least six, we note that each classroom contains at least three. But now we need another assumption that was not explicitly stated in the exercise. Namely, we need to know that no computer can be in two classrooms. Given that, we see that there must be at least six computers, and so exactly six.

This may seem like making pretty heavy weather of an obvious fact, but it illustrates two things. First, to prove a numerical claim of the form *there exist exactly n objects x such that $P(x)$*, which we agreed to abbreviate as $\exists^{!n} x\, P(x)$, you need to prove two things: that there are *at least n* such objects, and that there are *at most n* such objects.

formal proofs of numerical claims

The proof also illustrates a point about FOL. If we were to translate our premises and desired conclusion into FOL, things would get quite complicated. If we then tried to prove our FOL conclusion from the FOL premises using the rules we have presented earlier, we would completely lose track of the basic fact that makes the proof work, namely, that $2 \times 3 = 6$. Rather than explicitly state and use this fact, as we did above, we would have to rely on it in a hidden way in the combinatorial details of the proof. While it would be possible to give such a proof, no one would really do it that way.

The problem has to do with a syntactic shortcoming of FOL. Not having quantifiers that directly express numerical claims in terms of numbers, such claims must be translated using just $\forall$ and $\exists$. If we were to add numerical quantifiers to FOL, we would be able to give proofs that correspond much more closely to the intuitive proofs. Still, the theoretical expressive power of the language would remain the same.

a new method of proof

We can think of the above proof as illustrating a new method of proof. When trying to prove $\exists^{!n} x\, P(x)$, prove two things: that there are at least n objects satisfying $P(x)$, and that there are at most n such objects.

A particularly important special case of this method is with uniqueness claims, those of the form $\exists! x\, P(x)$, which say there is exactly one object with some property. To prove such a claim, we must prove two things, existence and uniqueness. In proving existence, we prove that there is at least one object satisfying $P(x)$. Given that, we can then show uniqueness by showing that there is at most one such object. To give an example, let us prove $\exists! x\, [\mathsf{Even}(x) \land \mathsf{Prime}(x)]$.

Proof: We first prove existence, that is, that there is an even prime. This we do simply by noting that 2 is even and a prime. Thus, by existential generalization, there is an even prime. Next we prove uniqueness. That is, we prove that for any number x, if x is an even prime, then $x = 2$, by general conditional proof. Suppose x is an even prime. Since it is even, it must be divisible by 2. But being prime, it is divisible only by itself and 1. So $x = 2$. This concludes our proof of uniqueness.

With one significant exception (induction, which we take up in Chapter 16), we have now introduced all the major methods of proof. When these are used in mathematical proofs, it is common to suppress a lot of detail, including explicit mention of the methods being used. To a certain extent, we have already been doing this in our proofs. From now on, though, we will present proofs in a more abbreviated fashion, and expect you to be able to fill in the details. For example, here is a more abbreviated version of our proof that there is exactly one even prime. You should check to see that you could have filled in the details on your own.

> **Proof:** We first prove existence, that is, that there is an even prime. This we do simply by noting that 2 is even and a prime. We then prove uniqueness, by proving that any even prime must be 2. First, since it is even, it must be divisible by 2. But being prime, if it is divisible by 2, it is 2.

Since the numerical quantifiers are really shorthand for more complicated expressions in our language, there is no real need to introduce rules that specifically apply to them. Of course the same could have been said for $\rightarrow$, but we saw that it was much more convenient to have rules of proof for $\rightarrow$ than to reduce things to $\vee$ and $\neg$ and use their rules of proof. But the situation is different with numerical quantifiers. In practice, people rarely give formal proofs of numerical claims expressed in FOL, since they quickly become too complex, with or without special rules for these quantifiers. With numerical claims, informal proofs are the order of the day.

Remember

To prove $\exists^{!n} x\, P(x)$, prove two things:

o that there are at least n objects satisfying $P(x)$, and

o that there are at most n objects satisfying $P(x)$.

Exercises

---

*Use Fitch to give formal proofs of the following arguments. You may use **Taut Con** where it is convenient. We urge you to work backwards, especially with the last problem, whose proof is simple in conception but complex in execution.*

14.10
✐

$$\exists x\, (\text{Cube}(x) \land \forall y\, (\text{Cube}(y) \to y = x))$$

$$\exists x\, \forall y\, (\text{Cube}(y) \leftrightarrow y = x)$$

14.11
✐

$$\exists x\, \forall y\, (\text{Cube}(y) \leftrightarrow y = x)$$

$$\exists x\, (\text{Cube}(x) \land \forall y\, (\text{Cube}(y) \to y = x))$$

14.12
✐

$$\exists x\, \exists y\, (\text{Cube}(x) \land \text{Cube}(y) \land x \neq y)$$
$$\forall x\, \forall y\, \forall z\, ((\text{Cube}(x) \land \text{Cube}(y) \land \text{Cube}(z)) \to (x = y \lor x = z \lor y = z))$$

$$\exists x\, \exists y\, (\text{Cube}(x) \land \text{Cube}(y) \land x \neq y \land \forall z\, (\text{Cube}(z) \to (z = x \lor z = y)))$$

14.13
✐★

$$\exists x\, \exists y\, (\text{Cube}(x) \land \text{Cube}(y) \land x \neq y \land \forall z\, (\text{Cube}(z) \to (z = x \lor z = y)))$$

$$\exists x\, \exists y\, (\text{Cube}(x) \land \text{Cube}(y) \land x \neq y) \land \forall x\, \forall y\, \forall z\, ((\text{Cube}(x) \land \text{Cube}(y) \land \text{Cube}(z))$$
$$\to (x = y \lor x = z \lor y = z))$$

The next two exercises contain arguments with similar premises and the same conclusion. If the argument is valid, turn in an informal proof. If it is not, submit a world in which the premises are true but the conclusion is false.

14.14
✐|✎

There are exactly four cubes.
Any column that contains a cube contains a tetrahedron, and vice versa.
No tetrahedron is in back of any other tetrahedron.

There are exactly four tetrahedra.

14.15
✐|✎

There are exactly four cubes.
Any column that contains a cube contains a tetrahedron, and vice versa.
No column contains two objects of the same shape.

There are exactly four tetrahedra.

The following exercises state some logical truths or valid arguments involving numerical quantifiers. Give informal proofs of each. Contemplate what it would be like to give a formal proof (for specific values of n and m) and be thankful we didn't ask you to give one!

14.16

$\vdash$

$\exists^{\leq 0}x\, S(x) \leftrightarrow \forall x\, \neg S(x)$

[The only hard part about this is figuring out what $\exists^{\leq 0}x\, S(x)$ abbreviates.]

14.17

$\vdash$

$\neg\exists^{\geq n+1}x\, S(x) \leftrightarrow \exists^{\leq n}x\, S(x)$

14.18

$\exists^{\leq n}x\, A(x)$
$\exists^{\leq m}x\, B(x)$

$\exists^{\leq n+m}x\, (A(x) \vee B(x))$

14.19

$\exists^{\geq n}x\, A(x)$
$\exists^{\geq m}x\, B(x)$
$\neg\exists x\, (A(x) \wedge B(x))$

$\exists^{\geq n+m}x\, (A(x) \vee B(x))$

14.20

$\forall x\, [A(x) \rightarrow \exists! y\, R(x,y)]$
$\exists^{\leq n}x\, A(x)$

$\exists^{\leq n}y\, \exists x\, [A(x) \wedge R(x,y)]$

14.21 We have seen that $\exists x\, \exists y\, R(x,y)$ is logically equivalent to $\exists y\, \exists x\, R(x,y)$, and similarly for $\forall$. What happens if we replace both of these quantifiers by some numerical quantifier? In particular, is the following argument valid?

$\exists! x\, \exists! y\, R(x,y)$

$\exists! y\, \exists! x\, R(x,y)$

If so, give an informal proof. If not, describe a counterexample.

The following exercises contain true statements about the domain of natural numbers 0, 1, ... Give informal proofs of these statements.

14.22 $\exists! x\, [x^2 - 2x + 1 = 0]$

14.23 $\exists^{!2}y\, [y + y = y \times y]$

14.24 $\exists^{!2}x\, [x^2 - 4x + 3 = 0]$

14.25 $\exists! x\, [(x^2 - 5x + 6 = 0) \wedge (x > 2)]$

Section 14.3

The, both, and *neither*

The English determiners *the, both,* and *neither* are extremely common. Indeed, *the* is one of the most frequently used words in the English language. (We used it twice in that one sentence.) In spite of their familiarity, their logical properties are subtle and, for that matter, still a matter of some dispute.

To see why, suppose I say "The elephant in my closet is not wrinkling my clothes." What would you make of this, given that, as you probably guessed, there is no elephant in my closet? Is it simply false? Or is there something else wrong with it? If it is false, then it seems like its negation should be true. But the negation seems to be the claim that the elephant in my closet *is* wrinkling my clothes. Similar puzzles arise with *both* and *neither*:

> *Both elephants in my closet are wrinkling my clothes.*
> *Neither elephant in my closet is wrinkling my clothes.*

What are you to make of these if there are no elephants in my closet, or if there are three?

the

Early in the twentieth century, the logician Bertrand Russell proposed an analysis of such sentences. He proposed that a sentence like *The cube is small* should be analyzed as asserting that there is exactly one cube, and that it is small. According to his analysis, the sentence will be false if there is no cube, or if there is more than one, or if there is exactly one, but it's not small. If Russell's analysis is correct, then such sentences can easily be expressed in first-order logic as follows:

$$\exists x\, [\mathsf{Cube}(x) \wedge \forall y\, (\mathsf{Cube}(y) \to y = x) \wedge \mathsf{Small}(x)]$$

More generally, a sentence of the form *The A is a B*, on the Russellian analysis, would be translated as:

$$\exists x\, [\mathsf{A}(x) \wedge \forall y\, (\mathsf{A}(y) \to x = y) \wedge \mathsf{B}(x)]$$

definite descriptions

Noun phrases of the form *the A* are called *definite descriptions* and the above analysis is called the *Russellian analysis of definite descriptions.*

both, neither

While Russell did not explicitly consider *both* or *neither*, the spirit of his analysis extends naturally to these determiners. We could analyze *Both cubes are small* as saying that there are exactly two cubes and each of them is small:

$$\exists^{!2} x\, \mathsf{Cube}(x) \wedge \forall x\, [\mathsf{Cube}(x) \to \mathsf{Small}(x)]$$

Similarly, *Neither cube is small* would be construed as saying that there are exactly two cubes and each of them is not small:

$$\exists^{!2}x\, \mathsf{Cube}(x) \land \forall x\, [\mathsf{Cube}(x) \rightarrow \neg\mathsf{Small}(x)]$$

More generally, *Both A's are B's* would be translated as:

$$\exists^{!2}x\, \mathsf{A}(x) \land \forall x\, [\mathsf{A}(x) \rightarrow \mathsf{B}(x)]$$

and *Neither A is a B* would be translated as:

$$\exists^{!2}x\, \mathsf{A}(x) \land \forall x\, [\mathsf{A}(x) \rightarrow \neg\mathsf{B}(x)]$$

Notice that on Russell's analysis of definite descriptions, the sentence *The cube is not small* would be translated as:

$$\exists x\, [\mathsf{Cube}(x) \land \forall y\, (\mathsf{Cube}(y) \rightarrow y = x) \land \neg\mathsf{Small}(x)]$$

This is not, logically speaking, the negation of *The cube is small.* Indeed both sentences could be false if there are no cubes or if there are too many. The superficial form of the English sentences makes them look like negations of one another, but according to Russell, the negation of *The cube is small* is something like *Either there is not exactly one cube or it is not small.* Or perhaps more clearly, *If there is exactly one cube then it is not small.* Similarly, the negation of *Both cubes are small* would not be *Both cubes are not small* but *If there are exactly two cubes then they are not both small.*

definite descriptions and negation

Russell's analysis is not without its detractors. The philosopher P. F. Strawson, for example, argued that Russell's analysis misses an important feature of our use of the determiner *the.* Return to our example of the elephant. Consider these three sentences:

> *The elephant in my closet is wrinkling my clothes.*
> *The elephant in my closet is not wrinkling my clothes.*
> *It is not the case that the elephant in my closet is wrinkling my clothes.*

It seems as if none of these sentences is appropriate if there is no elephant in my closet. That is to say, they all seem to presuppose that there is a unique elephant in my closet. According to Strawson, they all do presuppose this, but they do not claim it.

Strawson's general picture is this. Some sentences carry certain *presuppositions.* They can be used to make a claim only when those presuppositions are fulfilled. Just as you can't drive a car unless there is a car present, you

presuppositions

cannot make a successful claim unless the presuppositions of your claim are satisfied. With our elephant example, the sentence can only be used to make a claim in case there is one, and only one, elephant in the speaker's closet. Otherwise the sentence simply misfires, and so does not have a truth value at all. It is much like using an FOL sentence containing a name b to describe a world where no object is named b. Similarly, on Strawson's approach, if we use *both elephants in my closet* or *neither elephant in my closet*, our statement simply misfires unless there are exactly two elephants in my closet.

If Strawson's objection is right, then there will be no general way of translating *the, both,* or *neither* into FOL, since FOL sentences (at least those without names in them) always have truth values. There is nothing to stop us from enriching FOL to have expressions that work this way. Indeed, this has been proposed and studied, but that is a different, richer language than FOL.

conversational implicature

On the other hand, there have been rejoinders to Strawson's objection. For example, it has been suggested that when we say *The elephant in my closet is not wrinkling my clothes*, the suggestion that there is an elephant in my closet is simply a conversational implicature. To see if this is plausible, we try the cancellability test. Does the following seem coherent or not? "The elephant in my closet is not wrinkling my clothes. In fact, there is no elephant in my closet." Some people think that, read with the right intonation, this makes perfectly good sense. Others disagree.

As we said at the start of this section, these are subtle matters and there is still no universally accepted theory of how these determiners work in English. What we can say is that the Russellian analysis is as close as we can come in FOL, that it is important, and that it captures at least some uses of these determiners. It is the one we will treat in the exercises that follow.

Remember

1. The Russellian analysis of *The A is a B* is the FOL translation of *There is exactly one A and it is a B.*

2. The Russellian analysis of *Both A's are B's* is the FOL translation of *There are exactly two A's and each of them is a B.*

3. The Russellian analysis of *Neither A is a B* is the FOL translation of *There are exactly two A's and each of them is not a B.*

4. The competing Strawsonian analysis of these determiners treats them as having presuppositions, and so as only making claims when these presuppositions are met. On Strawson's analysis, these determiners cannot be adequately translated in FOL.

Exercises

14.26 (The Russellian analysis of definite descriptions)

1. Open Russell's Sentences. Sentence 1 is the second of the two ways we saw in the **You try it** section on page 377 for saying that there is a single cube. Compare sentence 1 with sentence 2. Sentence 2 is the Russellian analysis of our sentence *The cube is small.* Construct a world in which sentence 2 is true.

2. Construct a world in which sentences 2-7 are all true. (Sentence 7 contains the Russellian analysis of *The small dodecahedron is to the left of the medium dodecahedron.*)

Submit your world.

14.27 (The Strawsonian analysis of definite descriptions) Using Tarski's World, open a sentence file and write the Russellian analysis of the following two sentences:

1. *b is left of the cube.*
2. *b is not left of the cube.*

Build a world containing a dodec named b and one other block in which neither of your translations is true. To do so, you will need to violate what Strawson would call the common presupposition of these two sentences. Submit both the sentence and world files.

14.28 (The Russellian analysis of *both* and *neither*) Open Russell's World. Notice that the following sentences are all true:

1. *Both cubes are medium.*
2. *Neither dodec is small.*
3. *Both cubes are in front of the tetrahedron.*
4. *Both cubes are left of both dodecahedra.*
5. *Neither cube is in back of either dodecahedron.*

Start a new sentence file and write the Russellian analysis of these five sentences. Since Tarski's World doesn't let you use the notation $\exists^{!2}$, you may find it easier to write the sentences on paper first, using this abbreviation, and then translate them into proper FOL. Check that your translations are true in Russell's World. Then make some changes to the sizes and positions of the blocks and again check that your translations have the same truth values as the English sentences.

14.29 Discuss the meaning of the determiner *Max's.* Notice that you can say *Max's pet is happy,* but also *Max's pets are happy.* Give a Russellian and a Strawsonian analysis of this determiner. Which do you think is better?

Section 14.4

Adding other determiners to FOL

most, more than half

We have seen that many English determiners can be captured in FOL, though by somewhat convoluted circumlocutions. But there are also many determiners that simply aren't expressible in FOL. A simple example is the determiner *Most*, as in *Most cubes are large*. There are two difficulties. One is that the meaning of *most* is a bit indeterminate. *Most cubes are large* clearly implies *More than half the cubes are large*, but does the latter imply the former? Intuitions differ. But even if we take it to mean the same as *More than half*, it cannot be expressed in FOL, since the determiner *More than half* is not expressible in FOL.

It is possible to give a mathematical proof of this fact. For example, consider the sentence:

More than half the dodecahedra are small.

To see the problem, notice that the English sentence makes a claim about the relative sizes of the set A of small dodecahedra and the set B of dodecahedra that are not small. It says that the set A is larger than the set B and it does so without claiming anything about how many objects there are in these sets or in the domain of discourse. To express the desired sentence, we might try something like the following (where we use $A(x)$ as shorthand for $Dodec(x) \land Small(x)$, and $B(x)$ as shorthand for $Dodec(x) \land \neg Small(x)$):

$$[\exists x\, A(x) \land \forall x\, \neg B(x)] \lor [\exists^{\geq 2} x\, A(x) \land \exists^{\leq 1} x\, B(x)] \lor [\exists^{\geq 3} x\, A(x) \land \exists^{\leq 2} x\, B(x)] \lor \ldots$$

The trouble is, there is no place to stop this disjunction! Without some fixed finite upper bound on the total number of objects in the domain, we need all of the disjuncts, and so the translation of the English sentence would be an infinitely long sentence, which FOL does not allow. If we knew there were a maximum of twelve objects in the world, as in Tarski's World, then we could write a sentence that said what we needed; but without this constraint, the sentence would have to be infinite.

unexpressible in FOL

This is not in itself a proof that the English sentence cannot be expressed in FOL. But it does pinpoint the problem and, using this idea, one can actually give such a proof. In particular, it is possible to show that for any first-order sentence S of the blocks language, if S is true in every world where more than half the dodecahedra are small, then it is also true in some world where less than half the dodecahedra are small. Unfortunately, the proof of this would take us beyond the scope of this book.

The fact that we cannot express *more than half* in FOL doesn't mean there is anything suspect about this determiner. It just means that it does not fall within the expressive resources of the invented language FOL. Nothing stops us from enriching FOL by adding a new quantifier symbol, say Most. Let's explore this idea for a moment, since it will shed light on some topics from earlier in the book.

How not to add a determiner

We'll begin by telling you how *not* to add the determiner Most to the language. Following the lead from ∀ and ∃, we might start by adding the following clause to our grammatical rules on page 233:

If S is a wff and ν is a variable, then Most ν S is a wff, and any occurrence of ν in Most ν S is said to be bound.

We might then say that the sentence Most x S(x) is true in a world just in case more objects in the domain satisfy S(x) than don't.[1] Thus the sentence Most x Cube(x) says that most things are cubes.

How can we use our new language to express our sentence *Most dodecahedra are small*? The answer is, we can't. If we look back at ∀, ∃, and the numerical determiners, we note something interesting. It so happens that we can paraphrase *every cube is small* and *some cube is small* using *everything* and *something*; namely, *Everything is such that if it is a cube then it is small* and *Something is a cube and it is small*. At the end of the section on numerical quantification, we made a similar observation. There is, however, simply no way to paraphrase *Most dodecahedra are small* using *Most things* and expressions that can be translated into FOL. After all, it may be that most cubes are small, even when there are only three or four cubes and millions of dodecahedra and tetrahedra in our domain. Talking about most things is not going to let us say much of interest about the lonely cubes.

These observations point to something interesting about quantification and the way it is represented in FOL. For any determiner Q, let us mean by its *general form* any use of the form $Q \; A \; B$ as described at the beginning of this chapter. In contrast, by its *special form* we'll mean a use of the form Q *thing(s) B*. The following table of examples makes this clearer.

---

[1]For the set-theoretically sophisticated, we note that this definition make sense even if the domain of discourse is infinite.

Determiner	Special form	General form
every	*everything*	*every cube, every student of logic, ...*
some	*something*	*some cube, some student of logic, ...*
no	*nothing*	*no cube, no student of logic, ...*
exactly two	*exactly two things*	*exactly two cubes, exactly two students of logic, ...*
most	*most things*	*most cubes, most students of logic, ...*

reducibility

Many determiners have the property that the general form can be reduced to the special form by the suitable use of truth-functional connectives. Let's call such a determiner *reducible*. We have seen that *every, some, no,* and the various numerical determiners are reducible in this sense. Here are a couple of the reductions:

> *Every A B ⇔ Everything is such that if it is an A then it is a B*
> *Exactly two A B ⇔ Exactly two things satisfy A and B*

But some determiners, including *most, many, few,* and *the,* are not reducible. For non-reducible determiners Q, we cannot add Q to FOL by simply adding the special form in the way we attempted here. We will see how we can add such determiners in a moment.

There was some good fortune involved when logicians added ∀ and ∃ as they did. Since *every* and *some* are reducible, the definition of FOL can get away with just the special forms, which makes the language particularly simple. On the other hand, the fact that FOL takes the special form as basic also results in many of the difficulties in translating from English to FOL that we have noted. In particular, the fact that the reduction of *Every A* uses →, while that of *Some A* uses ∧, causes a lot of confusion among beginning students.

How *to* add a determiner

new grammatical form

The observations made above show that if we are going to add a quantifier like Most to our language, we must add the general form, not just the special form. Thus, the formation rule should take *two* wffs and a variable and create a new wff:

> If A and B are wffs and ν is a variable, then Most ν (A, B) is a wff, and any occurrence of ν in Most ν (A, B) is said to be bound.

The wff Most x (A, B) is read "most x satisfying A satisfy B." Notice that the syntactic form of this wff exhibits the fact that Most x (A, B) expresses a binary relation between the set A of things satisfying A and the set B of things that satisfy B. We could use the abbreviation Most x (S) for Most x (x = x, S); this is read "most things x satisfy S." This, of course, is the special form of the determiner, whereas the general form takes two wffs.

We need to make sure that our new symbol Most means what we want it to. Toward this end, let us agree that the sentence Most x (A, B) is true in a world just in case most objects that satisfy A(x) satisfy B(x) (where by this we mean more objects satisfy A(x) and B(x) than satisfy A(x) and ¬B(x)). With these conventions, we can translate our English sentence faithfully as:

$$\text{Most}\,x\,(\text{Dodec}(x), \text{Small}(x))$$

The order here is very important. While the above sentences says that most dodecahedra are small, the sentence

$$\text{Most}\,x\,(\text{Small}(x), \text{Dodec}(x))$$

says that most small things are dodecahedra. These sentences are true under very different conditions. We will look more closely at the logical properties of Most and some other determiners in the next section.

Once we see the general pattern, we see that any meaningful determiner Q of English can be added to FOL in a similar manner.

> If A and B are wffs and ν is a variable, then $Q\,\nu\,(A, B)$ is a wff, and any occurrence of ν in $Q\,\nu\,(A, B)$ is said to be bound.

The wff $Q\,x\,(A, B)$ is read "Q x satisfying A satisfy B," or more simply, "Q A's are B's." Thus, for example,

$$\text{Few}\,x\,(\text{Cube}(x), \text{Small}(x))$$

is read "Few cubes are small."

As for the special form, again we use the abbreviation

$$Q\,x\,(S)$$

for $Q\,x\,(x = x, S)$; this is read "Q things x satisfy S." For instance, the wff Many x (Cube(x)) is shorthand for Many x (x = x, Cube(x)), and is read "Many things are cubes."

What about the truth conditions of such wffs? Our reading of them suggests how we might define their truth conditions. We say that the sentence $Q\,x\,(A, B)$ is true in a world just in case Q of the objects that satisfy A(x) also satisfy B(x). Here are some instances of this definition:

new semantic rules

1. At least a quarter x (Cube(x), Small(x)) is true in a world iff at least a quarter of the cubes in that world are small.

2. At least two x (Cube(x), Small(x)) is true in a world iff at least two cubes in that world are small.

3. Finitely many x (Cube(x), Small(x)) is true in a world iff finitely many of the cubes in that world are small.

4. Many x (Cube(x), Small(x)) is true in a world iff many of the cubes in that world are small.

The first of these examples illustrates a kind of determiner we have not even mentioned before. The second shows that we could treat the numerical determiners of the preceding section in a different manner, by adding them as new primitives to an expansion of FOL. The third example is of another determiner that cannot be expressed in FOL.

But wait a minute. There is something rather unsettling about the fourth example. The problem is that the English determiner *many*, unlike the other examples, is context dependent; just what counts as many varies from one context to another. If we are talking about a class of twenty, 18 or 19 would count as many. If we are talking about atoms in the universe, this would count as very few, not many.

This context dependence infects our definition of the truth conditions for Many x (Cube(x), Small(x)). What might count as many cubes for one purpose, or one speaker, might not count as many for another purpose or another speaker. Logic is supposed to be the *science* of reasoning. But if we are trying to be scientific, the incursion of context dependence into the theory is most unwelcome.

dealing with context dependence

There are two things to say about this context dependence. The first is that even with context dependent determiners, there are certain clear logical principles that can be uncovered and explained. We will take some of these up in the next section. The second point is that the context dependence problem has a solution. It is possible to model the meaning of context dependent determiners in a perfectly precise, mathematically rigorous manner. Unfortunately, the modeling requires ideas from set theory that we have not yet covered. The basic idea, which will only make sense after the next chapter, is to model the meaning of any determiner as a binary relation on subsets of the domain of discourse. Just what relation is the best model in the case of a determiner like Many will depend on the intentions of the person using the determiner. But all such relations will have certain features in common, features that help explain the logical properties of the determiners just alluded to. For further details, see Exercises 18.5 and 18.6.

Remember

Given any English determiner Q, we can add a corresponding quantifier Q to FOL. In this extended language, the sentence $Q x (A, B)$ is true in a world just in case Q of the objects that satisfy $A(x)$ also satisfy $B(x)$.

Exercises

14.30 Some of the following English determiners are reducible, some are not. If they are reducible, explain how the general form can be reduced to the special form. If they do not seem to be reducible, simply say so.

1. *At least three*
2. *Both*
3. *Finitely many*
4. *At least a third*
5. *All but one*

14.31 Open Cooper's World. Suppose we have expanded FOL by adding the following expressions:

$\forall^{b1}$, meaning all but one,

Few, interpreted as meaning at most 10%, and

Most, interpreted as meaning more than half.

Translate the following sentences into this extended language. Then say which are true in Cooper's World. (You will have to use paper to write out your translations, since Tarski's World does not understand these quantifiers. If the sentence is ambiguous—for example, sentence 5— give both translations and say whether each is true.)

1. *Few cubes are small.*
2. *Few cubes are large.*
3. *All but one cube is not large.*
4. *Few blocks are in the same column as **b**.*
5. *Most things are adjacent to some cube.*
6. *A cube is adjacent to most tetrahedra.*
7. *Nothing is adjacent to most things.*
8. *Something is adjacent to something, but only to a few things.*
9. *All but one tetrahedron is adjacent to a cube.*

14.32 Once again open Cooper's World. This time translate the following sentences into English and say which are true in Cooper's World. Make sure your English translations are clear and unambiguous.

1. Most y (Tet(y), Small(y))
2. Most z (Cube(z), LeftOf(z, b))
3. Most y Cube(y)
4. Most x (Tet(x), ∃y Adjoins(x, y))
5. ∃y Most x (Tet(x), Adjoins(x, y))
6. Most x (Cube(x), ∃y Adjoins(x, y))
7. ∃y Most x (Cube(x), Adjoins(x, y))
8. Most y (y ≠ b)
9. ∀x (Most y (y ≠ x))
10. Most x (Cube(x), Most y (Tet(y), FrontOf(x, y)))

Section 14.5
The logic of generalized quantification

In this section we look briefly at some of the logical properties of determiners. Since different determiners typically have different meanings, we expect them to have different logical properties. In particular, we expect the logical truths and valid arguments involving determiners to be highly sensitive to the particular determiners involved. Some of the logical properties of determiners fall into nice clusters, though, and this allows us to classify determiners in logically significant ways.

We will assume that Q is some determiner of English and that we have introduced a formal counterpart Q into FOL in the manner described at the end of the last section.

Conservativity

As it happens, there is one logical property that holds of virtually all single-word determiners in every natural language. Namely, for any predicates A and B, the following are logically equivalent:

$$Q \, x \, (A(x), B(x)) \; \Leftrightarrow \; Q \, x \, (A(x), (A(x) \wedge B(x)))$$

conservativity property

This is called the *conservativity property* of determiners. Here are two instances of the ⇐ half of conservativity, followed by two instances of the ⇒ half:

> *If **no** doctor is a doctor and a lawyer, then **no** doctor is a lawyer.*
> *If **exactly three** cubes are small cubes, then **exactly three** cubes*

are small.

If *few* actors are rich, then *few* actors are rich and actors.
If *all* good actors are rich, then *all* good actors are rich and good
actors.

It is interesting to speculate why this principle holds of single word determiners in human languages. There is no logical reason why there could not be determiners that did not satisfy it. (See, for example, Exercise 14.52.) It might have something to do with the difficulty of understanding quantification that does not satisfy the condition, but if so, exactly why remains a puzzle.

There is one word which has the superficial appearance of a determiner that is not conservative, namely the word *only*. For example, it is true that only actors are rich actors but it does not follow that only actors are rich, as it would if *only* were conservative. There are independent linguistic grounds for thinking that *only* is not really a determiner. One piece of evidence is the fact that determiners can't be attached to complete noun phrases. You can't say *Many some books are on the table* or *Few Claire eats pizza*. But you can say *Only some books are on the table* and *Only Claire eats pizza*, suggesting that it is not functioning as a determiner. In addition, *only* is much more versatile than determiners, as is shown by the sentences *Claire only eats pizza* and *Claire eats only pizza*. You can't replace *only* in these sentences with a determiner and get a grammatical sentence. If *only* is not a determiner, it is not a counterexample to the conservativity principle.

Monotonicity

The *monotonicity* of a determiner has to do with what happens when we increase or decrease the set B of things satisfying the verb phrase in a sentence of the form $Q\,A\,B$. The determiner Q is said to be *monotone increasing* provided for all A, B, and B', the following argument is valid:

monotone increasing

$$
\begin{array}{|l}
\mathsf{Q\,x}\,(\mathsf{A(x)}, \mathsf{B(x)}) \\
\forall \mathsf{x}\,(\mathsf{B(x)} \to \mathsf{B'(x)}) \\
\hline
\mathsf{Q\,x}\,(\mathsf{A(x)}, \mathsf{B'(x)})
\end{array}
$$

In words, if $Q(A, B)$ and you increase B to a larger set B', then $Q(A, B')$. There is a simple test to determine whether a determiner is monotone increasing:

Test for monotone increasing determiners: Q is monotone increasing if and only if the following argument is valid:

Table 14.1: Monotonically increasing and decreasing determiners.

Monotone increasing	Monotone decreasing	Neither
every	*no*	*all but one*
some		
the		
both	*neither*	
many	*few*	
several		
most		
at least two	*at most two*	*exactly two*
infinitely many	*finitely many*	
Max's		

Q cube(s) is (are) small and in the same row as c.

Q cube(s) is (are) small.

The reason this test works is that the second premise in the definition of monotone increasing, $\forall x\,(B(x) \to B'(x))$, is automatically true. If we try out the test with a few determiners, we see, for example, that *some, every,* and *most* are monotone increasing, but *few* is not.

monotone decreasing   On the other hand, Q is said to be *monotone decreasing* if things work in the opposite direction, moving from the larger set B' to a smaller set B:

$Q x\,(A(x), B'(x))$
$\forall x\,(B(x) \to B'(x))$

$Q x\,(A(x), B(x))$

The test for monotone decreasing determiners is just the opposite as for monotone increasing determiners:

Test for monotone decreasing determiners: Q is monotone decreasing if and only if the following argument is valid:

Q cube(s) is (are) small.

Q cube(s) is (are) small and in the same row as c.

Many determiners are monotone increasing, several are monotone decreasing, but some are neither. Using our tests, you can easily verify for yourself the classifications shown in Table 14.1. To apply our test to the first column of the table, note that the following argument is valid, and remains so even if *most* is replaced by any determiner in this column:

> **Most** cubes are small and in the same row as *c*.
>
> **Most** cubes are small.

On the other hand, if we replace *most* by any of the determiners in the other columns, the resulting argument is clearly invalid.

To apply the test to the list of monotone decreasing determiners we observe that the following argument is valid, and remains so if *no* is replaced by any of the other determiners in the second column:

> **No** cubes are small.
>
> **No** cubes are small and in the same row as *c*.

On the other hand, if we replace *no* by the determiners in the other columns, the resulting argument is no longer valid.

If you examine Table 14.1, you might notice that there are no simple one-word determiners in the third column. This is because there aren't any. It so happens that all the one-word determiners are either monotone increasing or monotone decreasing, and only a few fall into the decreasing category. Again, this may have to do with the relative simplicity of monotonic *versus* non-monotonic quantification.

Persistence

Persistence is a property of determiners very similar to monotonicity, but persistence has to do which what happens if we increase or decrease the set of things satisfying the common noun: the A in a sentence of the form $Q\,A\,B$. The determiner Q is said to be *persistent* provided for all A, A', and B, the following argument is valid:[2]

persistence

> $Q\,x\,(A(x), B(x))$
> $\forall x\,(A(x) \rightarrow A'(x))$
>
> $Q\,x\,(A'(x), B(x))$

In words, if $Q\,A\,B$ and you increase A to a larger A', then $Q\,A'\,B$. On the other hand, Q is said to be *anti-persistent* if things work in the opposite direction:

anti-persistence

---

[2]Some authors refer to persistence as *left monotonicity*, and what we have been calling monotonicity as *right monotonicity*, since they have to do with the left and right arguments, respectively, when we look at $Q\,A\,B$ as a binary relation $Q(A, B)$.

Table 14.2: Persistent and anti-persistent determiners.

Persistent	Anti-persistent	Neither
some	*every*	*all but one*
several	*few*	*most*
at least two	*at most two*	*exactly two*
infinitely many	*finitely many*	*many*
	no	*the*
		both
		neither

$$Q\,x\,(A'(x), B(x))$$
$$\forall x\,(A(x) \rightarrow A'(x))$$

$$Q\,x\,(A(x), B(x))$$

To test a determiner for persistence or anti-persistence, try out the two argument forms given below and see whether the result is valid:

Test for Persistence: The determiner Q is persistent if and only if the following argument is valid:

Q small cube(s) is (are) left of b.

Q cube(s) is (are) left of b.

Test for Anti-persistence: The determiner Q is anti-persistent if and only if the following argument is valid:

Q cube(s) is (are) left of b.

Q small cube(s) is (are) left of b.

Applying these tests gives us the results shown in Table 14.2. Make sure you try out some or all of the entries to make sure you understand how the tests work. You will want to refer to this table in doing the exercises.

The properties of monotonicity and persistence play a large role in ordinary reasoning with determiners. Suppose, by way of example, that your father is trying to convince you to stay on the family farm rather than become an actor. He might argue as follows:

You want to be rich, right? Well, according to this report, few actors have incomes above the federal poverty level. Hence, few actors are rich.

Your father's argument depends on the fact that *few* is monotone decreasing. The set of rich people is a subset of those with incomes above the poverty level, so if few actors are in the second set, few are in the first. Notice that we immediately recognize the validity of this inference without even thinking twice about it.

Suppose you were to continue the discussion by pointing out that the actor Brad Pitt is extraordinarily rich. Your father might go on this way:

Several organic farmers I know are richer than Brad Pitt. So even some farmers are extraordinarily rich.

This may seem like an implausible premise, but you know fathers. In any case, the argument is valid, though perhaps unsound. Its validity rests on the fact that *Several* is both persistent and monotone increasing. By persistence, we can conclude that several farmers are richer than Brad Pitt (since the organic farmers are a subset of the farmers), and by monotonicity that several farmers are extraordinarily rich (since everyone richer than Brad Pitt is). Finally, from the fact that several farmers are extraordinarily rich it obviously follows that some farmers are (see Exercise 14.51).

There are many other interesting topics related to the study of determiners, but this introduction should give you a feel for the kinds of things we can discover about determiners, and the large role they play in everyday reasoning.

Remember

1. There are three properties of determiners that are critical to their logical behavior: *conservativity*, *monotonicity*, and *persistence*.

2. All English determiners are conservative (with the exception of *only*, which is not usually considered a determiner).

3. Monotonicity has to do with the behavior of the second argument of the determiner. All *basic* determiners in English are monotone increasing or decreasing, with most being monotone increasing.

4. Persistence has to do with the behavior of the first argument of the determiner. It is less common than monotonicity.

Exercises

For each of the following arguments, decide whether it is valid. If it is, explain why. This explanation could consist in referring to one of the determiner properties mentioned in this section or it could consist in an informal proof. If the argument is not valid, carefully describe a counterexample.

14.33

Few cubes are large.

Few cubes are large cubes.

14.34

Few cubes are large.

Few large things are cubes.

14.35

Many cubes are large.

Many cubes are not small.

14.36

Few cubes are large.

Few cubes are not small.

14.37

Few cubes are not small.

Few cubes are large.

14.38

Most cubes are left of b.

Most small cubes are left of b.

14.39

At most three cubes are left of b.

At most three small cubes are left of b.

14.40

Most cubes are not small.

Most cubes are large.

14.41

$\exists x \, [\text{Dodec}(x) \wedge \text{Most} \, y \, (\text{Dodec}(y), y = x))]$

$\exists ! x \, \text{Dodec}(x)$

14.42

At least three small cubes are left of b.

At least three cubes are left of b.

14.43

Most small cubes are left of b.

Most cubes are left of b.

14.44

Most tetrahedra are left of b.

a is a tetrahedron in the same column as b.

a is not right of anything in the same row as b.

Most tetrahedra are not in the same row as b.

14.45

Only cubes are large.

Only cubes are large cubes.

14.46

Only tetrahedra are large tetrahedra.

Only tetrahedra are large.

14.47

Most of the students brought a snack to class.
Most of the students were late to class.

Most of the students were late to class and brought a snack.

14.48

Most of the students brought a snack to class.
Most of the students were late to class.

At least one student was late to class and brought a snack.

14.49

Most former British colonies are democracies.
All English speaking countries were formerly British colonies.

Most English speaking countries are democracies.

14.50

Many are called.
Few are chosen.

Most are rejected.

14.51 In one of our example arguments, we noted that *Several A B* implies *Some A B*. In general, a determiner Q is said to have *existential import* if *Q A B* logically implies *Some A B*. Classify each of the determiners listed in Table 14.2 as to whether it has existential import. For those that don't, give informal counterexamples. Discuss any cases that seem problematic.

14.52 Consider a hypothetical English determiner "allbut." For example, we might say *Allbut cubes are small* to mean that all the blocks except the cubes are small. Give an example to show that "allbut" is not conservative. Is it monotone increasing or decreasing? Persistent or antipersistent? Illustrate with arguments expressed in English augmented with "allbut."

14.53 (Only) Whether or not *only* is a determiner, it could still be added to FOL, allowing expressions of the form $\mathsf{Only}\,x\,(\mathsf{A}, \mathsf{B})$, which would be true if and only if only A's are B's.
   1. While *Only* is not conservative, it does satisfy a very similar property. What is it?
   2. Discuss monotonicity and persistence for *Only*. Illustrate your discussion with arguments expressed in English.

14.54 (Adverbs of temporal quantification) It is interesting to extend the above discussion of quantification from determiners to so-called adverbs of temporal quantification, like *always, often, usually, seldom, sometimes,* and *never*. To get a hint how this might go, let's explore the ambiguities in the English sentence *Max usually feeds Carl at 2:00 p.m.*

Earlier, we treated expressions like 2:00 as names of times on a particular day. To interpret this sentence in a reasonable way, however, we need to treat such expressions as predicates of times. So we need to add to our language a predicate $\mathsf{2pm(t)}$ that holds of those times t (in the domain of discourse) that occur at 2 p.m., no matter on what day they occur. Let us suppose that *Usually* means most times. Thus,

$$\text{Usually t } (\mathsf{A}(\mathsf{t}), \mathsf{B}(\mathsf{t}))$$

means that most times satisfying $\mathsf{A}(\mathsf{t})$ also satisfy $\mathsf{B}(\mathsf{t})$.

1. One interpretation of *Max usually feeds Carl at 2:00 p.m.* is expressed by

$$\text{Usually t } (\mathsf{2pm}(\mathsf{t}), \mathsf{Feeds}(\mathsf{max}, \mathsf{carl}, \mathsf{t}))$$

   Express this claim using an unambiguous English sentence.

2. A different interpretation of the sentence is expressed by

$$\text{Usually t } (\mathsf{Feeds}(\mathsf{max}, \mathsf{carl}, \mathsf{t}), \mathsf{2pm}(\mathsf{t}))$$

   Express this claim using an unambiguous English sentence. Then elucidate the difference between this claim and the first by describing situations in which each is true while the other isn't.

3. Are the same ambiguities present in the sentence *Claire seldom feeds Folly at 2:00 p.m.*? How about with the other adverbs listed above?

4. Can you think of yet a third interpretation of *Max usually feeds Carl at 2:00 p.m.*, one that is not captured by either of these translations? If so, try to express it in our language or some expansion of it.

Section 14.6
Other expressive limitations of first-order logic

The study of generalized quantification is a response to one expressive limitation of FOL, and so to its inability to illuminate the full logic inherent in natural languages like English. The determiners studied in the preceding sections are actually just some of the ways of expressing quantification that we find in natural languages. Consider, for example, the sentences

> **More** cubes **than** tetrahedra are on the same row as e.
> **Twice as many** cubes **as** tetrahedra are in the same column as f.
> **Not as many** tetrahedra **as** dodecahedra are large.

The expressions in bold take two common noun expressions and a verb expression to make a sentence. The techniques used to study generalized quantification in earlier sections can be extended to study these determiners, but we have to think of them as expressing *three place* relations on sets, not just two place relations. Thus, if we added these determiners to the language, they would have the general form $\mathsf{Q} \times (\mathsf{A}(\mathsf{x}), \mathsf{B}(\mathsf{x}), \mathsf{C}(\mathsf{x}))$.

*three place
quantification*

A related difference in expressive power between FOL and English comes in the ability of English to use both singular and plural noun phrases. There is a difference between saying *The boys argued with the teacher* and saying *Every boy argued with the teacher*. The first describes a single argument between a teacher and a group of boys, while the second may describe a sequence of distinct arguments. FOL does not allow us to capture this difference.

plurals

Quantification is just the tip of an iceberg, however. There are many expressions of natural languages that go beyond first-order logic in various ways. Some of these we have already discussed at various points, both with examples and exercises. As one example, we saw that there are many uses of the natural language conditional *if...then...* that are not truth functional, and so not captured by the truth-functional connective →.

Another dimension in which FOL is limited, in contrast to English, comes in the latter's flexible use of tense. FOL assumes a timeless domain of unchanging relationships, whereas in English, we can exploit our location in time and space to say things about the present, the past, and locations around us. For example, in FOL we cannot easily say that it is hot here today but it was cool yesterday. To say something similar in FOL, we need to allow quantifiers over times and locations, and add corresponding argument positions to our atomic predicates.

tense

Similarly, languages like English have a rich modal structure, allowing us not only to say how things are, but how they must be, how they might be, how they can't be, how they should be, how they would be if we had our way, and so forth. So, for example, we can say *All the king's horses couldn't put Humpty Dumpty together again*. Or *Humpty shouldn't have climbed on the wall*. Or *Humpty might be dead*. Such statements lie outside the realm of FOL.

modality

All of these expressions have their own logic, and we can explore and try to understand just which claims involving these expressions follow logically from others. Building on the great success of FOL, logicians have studied (and are continuing to study) extensions of FOL in which these and similar expressive deficiencies are addressed. But as of now there is no single such extension of FOL that has gained anything like its currency.

Exercises

14.55 Try to translate the nursery rhyme about Humpty Dumpty into FOL. Point out the various linguistic mechanisms that go beyond FOL. Discuss this in class.

14.56 Consider the following two claims. Does either follow logically from the other? Are they logically
equivalent? Explain your answers.

1. I can eat every apple in the bowl.
2. I can eat any apple in the bowl.

14.57 Recall the first-order language introduced in Table 1.2, page 30. Some of the following can be
given first-order translations using that language, some cannot. Translate those that can be.
For the others, explain why they cannot be faithfully translated, and discuss whether they
could be translated with additional names, predicates, function symbols, and quantifiers, or if
the shortcoming in the language is more serious.

1. *Claire gave Max at least two pets at 2:00 pm.*
2. *Claire gave Max at most two pets at 2:00 pm.*
3. *Claire gave Max several pets at 2:00 pm.*
4. *Claire was a student before Max was.*
5. *The pet Max gave Claire at 2:00 pm was hungry.*
6. *Most pets were hungry at noon.*
7. *All but two pets were hungry at noon.*
8. *There is at least one student who made Max angry every time he (or she) gave Max a
pet.*
9. *Max was angry whenever a particular student gave him a pet.*
10. *If someone gave Max a pet, it must have been Claire.*
11. *No pet fed by Max between 2:00 and 2:05 belonged to Claire.*
12. *If Claire fed one of Max's pets before 2:00 pm, then Max was angry at 2:00 pm.*
13. *Folly's owner was a student.*
14. *Before 3:00, no one gave anyone a pet unless it was hungry.*
15. *No one should give anyone a pet unless it is hungry.*
16. *A pet that is not hungry always belongs to someone or other.*
17. *A pet that is not hungry must belong to someone or other.*
18. *Max was angry at 2:00 pm because Claire had fed one of his pets.*
19. *When Max gave Folly to Claire, Folly was hungry, but Folly was not hungry five minutes
later.*
20. *No student could possibly be a pet.*

14.58 Here is a famous puzzle. There was a Roman who went by two names, "Cicero" and "Tully."
Discuss the validity or invalidity of the following argument.

> Bill claims Cicero was a great orator.
> Cicero is Tully.
>
> Bill claims Tully was a great orator.

What is at stake here is nothing more or less than the principle that if $(\ldots\, a\, \ldots)$ is true, and $a = b$, then $(\ldots\, b\, \ldots)$ is true. [Hint: Does the argument sound more reasonable if we replace "claims" by "claims that"? By the way, the puzzle is usually stated with "believes" rather than "claims."]

The following more difficult exercises are not specifically relevant to this section, but to the general topic of truth of quantified sentences. They can be considered as research projects in certain types of classes.

14.59 (Persistence through expansion) As we saw in Exercise 11.5, page 301, some sentences simply ✎★★ can't be made false by adding objects of various sorts to the world. Once they are true, they stay true. For example, the sentence *There is at least one cube and one tetrahedron*, if true, cannot be made false by adding objects to the world. This exercise delves into the analysis of this phenomenon in a bit more depth.

Let's say that a sentence A is *persistent through expansion* if, whenever it is true, it remains true no matter how many objects are added to the world. (In logic books, this is usually called just persistence, or persistence under extensions.) Notice that this is a semantic notion. That is, it's defined in terms of truth in worlds. But there is a corresponding syntactic notion. Call a sentence *existential* if it is logically equivalent to a prenex sentence containing only existential quantifiers.

  ○ Show that $\mathsf{Cube(a)} \rightarrow \exists x\, \mathsf{FrontOf(x, a)}$ is an existential sentence.

  ○ Is $\exists x\, \mathsf{FrontOf(x, a)} \rightarrow \mathsf{Cube(a)}$ an existential sentence?

  ○ Show that every existential sentence is persistent through expansion. [Hint: You will have to prove something slightly stronger, by induction on wffs. If you are not familiar with induction on wffs, just try to understand why this is the case. If you are familiar with induction, try to give a rigorous proof.] Conclude that every sentence equivalent to an existential sentence is persistent through expansion.

It is a theorem, due to Tarski and Łoś (a Polish logician whose name is pronounced more like "wash" than like "loss"), that any sentence that is persistent through expansion is existential. Since this is the converse of what you were asked to prove, we can conclude that a sentence is persistent through expansion if and only if it is existential. This is a classic example of a theorem that gives a syntactic characterization of some semantic notion. For a proof of the theorem, see any textbook in model theory.

14.60 (Invariance under motion, part 1) The real world does not hold still, the way the world of ✗ mathematical objects does. Things move around. The truth values of some sentences change with such motion, while the truth values of other sentences don't. Open **Ockham's World** and **Ockham's Sentences**. Verify that all the sentences are true in the given world. Make as many of Ockham's Sentences false as you can by just moving objects around. Don't add or remove any objects from the world, or change their size or shape. You should be able to make false (in a single world) all of the sentences containing any spatial predicates, that is, containing LeftOf, RightOf, FrontOf, BackOf, or Between. (However, this is a quirk of this list of sentences, as we will see in the next exercise.) Save the world as World 14.60.

14.61 (Invariance under motion, part 2) Call a sentence *invariant under motion* if, for every world, the truth value of the sentence (whether true *or* false) does not vary as objects move around in that world.

1. Prove that if a sentence does not contain any spatial predicates, then it is invariant under motion.

2. Give an example of a sentence containing a spatial predicate that is nonetheless invariant under motion.

3. Give another such example. But this time, make sure your sentence is not first-order equivalent to any sentence that doesn't contain spatial predicates.

14.62 (Persistence under growth, part 1) In the real world, things not only move around, they also grow larger. (Some things also shrink, but ignore that for now.) Starting with Ockham's World, make the following sentences true by allowing some of the objects to grow:

1. $\forall x \, \neg Small(x)$
2. $\exists x \, \exists y \, (Cube(x) \land Dodec(y) \land Larger(y, x))$
3. $\forall y \, (Cube(y) \rightarrow \forall v \, (v \neq y \rightarrow Larger(v, y)))$
4. $\neg \exists x \, \exists y \, (\neg Large(x) \land \neg Large(y) \land x \neq y)$

How many of Ockham's Sentences are false in this world? Save your world as World 14.62.

14.63 (Persistence under growth, part 2) Say that a sentence S is *persistent under growth* if, for every world in which S is true, S remains true if some or all of the objects in that world get larger. Thus, Large(a) and ¬Small(a) are persistent under growth, but Smaller(a, b) isn't. Give a syntactic definition of as large a set of sentences as you can for which every sentence in the set is persistent under growth. Can you prove that all of these sentences are persistent under growth?

PART III

Applications and Metatheory

Chapter 15

First-order Set Theory

Over the past hundred years, set theory has become an important and useful part of mathematics. It is used both in mathematics itself, as a sort of universal framework for describing other mathematical theories, and also in applications outside of mathematics, especially in computer science, linguistics, and the other symbolic sciences. The reason set theory is so useful is that it provides us with tools for modeling an extraordinary variety of structures.

Personally, we think of sets as being a lot like Tinkertoys or Lego blocks: basic kits out of which we can construct models of practically anything. If you go on to study mathematics, you will no doubt take courses in which natural numbers are modeled by sets of a particular kind, and real numbers are modeled by sets of another kind. In the study of rational decision making, economists use sets to model situations in which rational agents choose among competing alternatives. Later in this chapter, we'll do a little of this, modeling properties, relations, and functions as sets. These models are used extensively in philosophy, computer science, and mathematics. In Chapter 18 we will use these same tools to make rigorous our notions of first-order consequence and first-order validity.

modeling in set theory

In this chapter, though, we will start the other way around, applying what we have learned about first-order logic to the study of set theory. Since set theory is generally presented as an axiomatized theory within a first-order language, this gives us an opportunity to apply just about everything we've learned so far. We will be expressing various set-theoretic claims in FOL, figuring out consequences of these claims, and giving informal proofs of these claims. The one thing we won't be doing very much is constructing formal proofs of set-theoretic claims. This may disappoint you. Many students are initially intimidated by formal proofs, but come to prefer them over informal proofs because the rules of the game are so clear-cut. For better or worse, however, formal proofs of substantive set-theoretic claims can be hundreds or even thousands of steps long. In cases where the formal proof is manageable, we will ask you to formalize it in the exercises. If you want to prove more results in set theory, we recommend building a library of lemma files so you can avoid repeating work.

logic and set theory

Set theory has a rather complicated and interesting history. Another objective of this chapter is to give you a feeling for this history. We will start out with an untutored, or "naive" notion of set, the one that you were no doubt

naive set theory

exposed to in elementary school. We begin by isolating two basic principles that seem, at first sight, to be clearly true of this intuitive notion of set. The principles are called the Axiom of Extensionality and the Axiom of Comprehension. We'll state the axioms in the first-order language of set theory and draw out some of their consequences.

We don't have to go too far, however, before we discover that we can prove a contradiction from these axioms. This contradiction will demonstrate that the axioms are inconsistent. There simply can't be a domain of discourse satisfying our axioms; which is to say that the intuitive notion of a set is just plain inconsistent. The inconsistency, in its simplest incarnation, is known as Russell's Paradox.

Russell's Paradox

Russell's Paradox has had a profound impact on modern set theory and logic. It forced the founders of set theory to go back and think more critically about the intuitive notion of a set. The aim of much early work in set theory was to refine the conception in a way that avoids inconsistency, but retains the power of the intuitive notion. Examining this refined conception of set leads to a modification of the axioms. We end the chapter by stating the revised axioms that make up the most widely used set theory, known as Zermelo-Frankel set theory, or ZFC. Most of the proofs given from the naive theory carry over to ZFC, but not any of the known proofs of inconsistency. ZFC is believed by almost every mathematician to be not only consistent, but true of our intuitive notion of sets.

This may seem like a rather tortured route to the modern theory, but it is very hard to understand and appreciate ZFC without first being exposed to naive set theory, understanding what is wrong with it, and seeing how the modern theory derives from this understanding.

Section 15.1

Naive set theory

sets and membership

The first person to study sets extensively and to appreciate the inconsistencies lurking in the naive conception was the nineteenth century German mathematician Georg Cantor. According to the naive conception, a set is just a collection of things, like a set of chairs, a set of dominoes, or a set of numbers. The things in the collection are said to be *members* of the set. We write $a \in b$, and read "a is a member (or an element) of b," if a is one of the objects that makes up the set b.

There is only one constant symbol in set theory and this is denoted by a special symbol. We therefore don't need to use the letters at the beginning of the alphabet as constants, so instead we will use them as a new kind of variable in the language of set theory. Specifically, we will use the variables $a, b, c, \ldots$, with and without subscripts, as variables which range over sets, and the variables $x, y, z, \ldots$ to range over everything—ordinary objects as well as sets. Thus, for example, if we wanted to say that everything is a member of some set or other, we would write

$$\forall x \, \exists a \, (x \in a)$$

To say the same thing using only one kind of variable we would need a predicate, $Set(x)$, true of only sets, and we would have to write

$$\forall x \, \exists y \, [Set(y) \wedge x \in y]$$

This is a common way to extend FOL, and produces what is known as a *many-sorted* language. We've in fact seen something like this before, when we translated sentences like *Max gave something to Claire between 2:00 and 3:00*. In translating such sentences, we often use one sort of variable for quantifying over ordinary objects ($\exists x$) and another to quantify over times ($\exists t$).[1]

many-sorted language

There are two principles that together completely describe naive set theory. We will discuss each in turn.

The Axiom of Comprehension

The first principle tells us exactly which sets exist. Ideally we would like to be able to form any set in which all of the members share a property, for example, the set of tetrahedra, the set of small cubes, or the set of blocks that are to the left of a. However, if we were to try to make this precise, we would have to begin by defining what we mean by a property, and this is at least as difficult as defining what we mean by a set. We don't want to get into the business of having to axiomatize properties as well as sets. To get around this, we use formulas of first-order logic instead of properties. While we would like to be able to form the set of all objects that share any property, we will have to satisfy ourselves with the set of all objects that satisfy some formula. We can approximate this by saying that for each formula $P(x)$ of FOL, we have the axiom:

$$\exists a \, \forall x \, [x \in a \leftrightarrow P(x)]$$

This says that there is a set a whose members are all and only those things that satisfy the formula $P(x)$. (To make sure it says this, we demand that the variable a not occur in the wff $P(x)$.)

---

[1]If you read the section on generalized quantifiers in Chapter 14, you will recognize these as the quantifiers formed from *some* and the nouns *thing* and *time* respectively.

Notice that this is not just one axiom, but an infinite collection of axioms, one for each wff $P(x)$. For this reason, it is called an *axiom scheme*. When we replace $P(x)$ by some specific wff then we call the result an *instance* of the axiom scheme. We will see later that some instances of this axiom scheme are inconsistent, so we will have to modify the scheme. But for now we assume all of its instances as axioms in our theory of sets.

Actually, the Axiom of Comprehension is a bit more general than our notation suggests, since the wff $P(x)$ can contain variables other than x, say

$z_1, \ldots, z_n$. What we really want is the *universal closure* of the displayed formula, where all the other variables are universally quantified:

Axiom 1. Axiom of Unrestricted Comprehension

$$\forall z_1 \ldots \forall z_n \, \exists a \, \forall x \, [x \in a \leftrightarrow P(x)]$$

Most applications of the axiom will in fact make use of these additional variables. For example, the claim that for any objects z_1 and z_2, there is a set containing z_1 and z_2 as its only members, is an instance of this axiom scheme:

$$\forall z_1 \, \forall z_2 \, \exists a \, \forall x \, [x \in a \leftrightarrow (x = z_1 \vee x = z_2)]$$

Here the formula $P(x)$ that we are using as an instance is $x = z_1 \vee x = z_2$, which contains the free variables z_1 and z_2. These must be universally quantified to form an instance of comprehension.

The Axiom of Comprehension, as we have stated it, is weaker than the intuitive principle that motivated it. After all, we have already seen that there are many determinate properties expressible in English that cannot be expressed in any particular version of FOL. For example, we can't express the English connective *because* or the quantifier *many* in FOL. Since there is no formula of FOL capable of expressing properties requiring these connectives, the axiom does not guarantee that there are sets containing just those objects that satisfy such properties. These sets are getting left out of our axiomatization. Still, the axiom as stated is quite strong. In fact, it is too strong, as we will soon see.

The Axiom of Extensionality

As we said, there are two principles that capture the naive conception of a set. The second principle is that a set is completely determined by its members. If you know the members of a set b, then you know everything there is to

know about the identity of the set. This principle is captured by the *Axiom of Extensionality*. Stated precisely, the Axiom of Extensionality says that if sets a and b have the same elements, then $a = b$. We can express this in FOL as follows:

Axiom 2. (Axiom of Extensionality)

$$\forall a \, \forall b \, [\forall x \, (x \in a \leftrightarrow x \in b) \rightarrow a = b]$$

In particular, the identity of a set does not depend on how it is described. For example, suppose we have the set containing just the two numbers 7 and 11. It can be described as the set of prime numbers between 6 and 12, or as the set of solutions to the equation $x^2 - 18x + 77 = 0$. It might even be the set of Max's favorite numbers, who knows? The important point is that the Axiom of Extensionality tells us that all of these descriptions pick out the same set.

Notice that if we were developing a theory of properties rather than sets, we would not take extensionality as an axiom. It is perfectly reasonable to have two distinct properties that apply to exactly the same things. For example, the property of being a prime number between 6 and 12 is a different property from that of being a solution to the equation $x^2 - 18x + 77 = 0$, and both of these are different properties from the property of being one of Max's favorite numbers. It happens that these properties hold of exactly the same numbers, but the properties themselves are still different.

sets vs. properties

We can use the Axioms of Extensionality and Comprehension together to prove an important claim about sets, namely that each formula gives rise to a unique set.

Proposition 1. *For each wff $P(x)$ we can prove that there is a unique set of objects that satisfy $P(x)$. Using the notation introduced in Section 14.1:*

uniqueness theorem

$$\forall z_1 \ldots \forall z_n \, \exists! a \, \forall x \, [x \in a \leftrightarrow P(x)]$$

This is our first chance to apply our techniques of informal proof to a claim in set theory. Our proof might look like this:

Proof: We will prove the claim using universal generalization. Let $z_1, \ldots, z_n$ be arbitrary objects. The Axiom of Comprehension assures us that there is at least one set of objects that satisfy $P(x)$. So we need only prove that there is *at most* one such set. Suppose a and b are both sets that have as members exactly those things that satisfy $P(x)$. That is, a and b satisfy:

$$\forall x \, [x \in a \leftrightarrow P(x)]$$
$$\forall x \, [x \in b \leftrightarrow P(x)]$$

But then it follows that a and b satisfy:

$$\forall x\,[x \in a \leftrightarrow x \in b]$$

(This rather obvious step actually uses a variety of the methods of proof we have discussed, and would be rather lengthy if we wrote it out in complete, formal detail. You are asked to give a formal proof in Exercise 15.2.) Applying the Axiom of Extensionality to this last claim gives us $a = b$. This is what we needed to prove.

Proposition 1 shows that given any first-order wff $P(x)$, our axioms allow us to deduce the existence of *the* set of objects that satisfy that wff. The set of all objects x that satisfy $P(x)$ is often written informally as follows:

$$\{x \mid P(x)\}$$

This is read: "the set of x such that $P(x)$." Note that if we had used a different variable, say "y" rather than "x," we would have had different notation for the very same set:

$$\{y \mid P(y)\}$$

brace notation

This *brace notation* for sets is convenient but inessential. It is not part of the official first-order language of set theory, since it doesn't fit the format of first-order languages. We will only use it in informal contexts. In any event, anything that can be said using brace notation can be said in the official language. For example, $b \in \{x \mid P(x)\}$ could be written:

$$\exists a\,[\forall x\,(x \in a \leftrightarrow P(x)) \wedge b \in a]$$

Remember

Naive set theory has the Axiom of Extensionality and the Axiom Scheme of Comprehension. Comprehension asserts that every first-order formula determines a set. Extensionality says that sets with the same members are identical.

Exercises

15.1   List three members of the sets defined by the following properties:
1. Being a prime number larger than 15.
2. Being one of your ancestors.
3. Being a grammatical sentence of English.
4. Being a prefix of English.
5. Being a palindrome of English, that is, a phrase whose reverse is the very same phrase, as with "Madam, I'm Adam".

15.2   In the Fitch file Exercise 15.2, you are asked to give a formal proof of the main step in our proof of Proposition 1. You should give a complete proof, without using any of the **Con** rules. (You will find the symbol ∈ on the Fitch toolbar in the tab named **Set**.)

15.3   Consider the following true statement:

The set whose only members are the prime numbers between 6 and 12 is the same as the set whose only members are the solutions to the equation $x^2 - 18x + 77 = 0$.

Write this statement using brace notation. Then write it out in the first-order language of set theory, without using the brace notation. In both cases you may allow yourself natural predicates like NatNum, Prime, and Set.

SECTION 15.2

The empty set, singletons and pairs

We have seen that the axioms of Comprehension and Extensionality allow us to form the set of objects that satisfy any formula $P(x)$. There are two special kinds of sets that sometimes cause confusion. One is when there is no object at all which satisfies the formula $P(x)$, and the other is when only one object satisfies that formula. Let's take these up in turn.

Suppose that no object satisfies the formula $P(x)$, for example when the formula is $x \neq x$. What are we to make of the set

$$\{x \mid x \neq x\}$$

There is no object that satisfies this property, and so the set is said to be *empty*, since it contains nothing. It is easy to prove that there can be at most one such set, so it is called *the* empty set, and we use the symbol ∅   *empty set (∅)*

to represent it. Some authors use 0 to denote the empty set. It can also be informally denoted by {}.

When there is one and only one object x satisfying $P(x)$ the Axiom of Comprehension guarantees there is a set whose only member is that object. We call this the *singleton set containing* x, and denote it by $\{x\}$.

singleton set

Some students are tempted to confuse an object with the singleton set containing that object. But in that direction lies, if not madness, at least dreadful confusion. After all, a singleton set is a set (an abstract object) and its member might have been any object at all, say the Washington Monument. The Washington Monument is a physical object, not a set. So we must not confuse an object x with the set $\{x\}$. Even if x is a set, we must not confuse it with its own singleton. For example, a set x might have any number of elements in it, but $\{x\}$ has exactly one element: x.

The notation for singleton sets is an example of *list notation* for sets, which allows us to just list the elements of the set in braces. We write $\{3\}$ for the set that contains only the number 3, $\{2,3\}$ for the set that contains just the numbers 2 and 3, and $\{\text{Washington Monument}, \text{White House}, \text{Lincoln Memorial}\}$ for the set that contains those three landmarks. Like brace notation, list notation is a convenient but dispensable part of the notation that we use for discussing set theory. The description of a set in list notation is just an abbreviation for a longer expression in the official language.

Introducing list notation suggests that there are sets with two elements, three elements, four elements and so on, but before we can be sure of this we have to justify the existence of any such set using the Axiom of Comprehension. The next proposition proves the existence of a set containing any two elements that we choose.

Proposition 2. (Unordered Pairs) *For any objects x and y there is a (unique) set $a = \{x, y\}$. In symbols:*

unordered pairs

$$\forall x \, \forall y \, \exists! a \, \forall w \, (w \in a \leftrightarrow (w = x \vee w = y))$$

Proof: Let x and y be arbitrary objects, and let

$$a = \{w \mid w = x \vee w = y\}$$

The existence of a is guaranteed by Comprehension, and its uniqueness follows from the Axiom of Extensionality. Clearly a has x and y and nothing else as elements.

We did not previously prove the existence of singleton sets, but we now have two ways to show that singleton sets exist. One is to use an instance of the Axiom of Comprehension with a single equality as the instance of $P(x)$, like this:

$$\forall y\, \exists a\, \forall x\, (x \in a \leftrightarrow x = y)$$

which says that for any y there is a set whose elements are those that are equal to y, *i.e.* $\{y\}$.

Another is to apply the previous result about pairs, with the same object playing the role of both x and y, thus:

Proposition 3. (Singletons) *For any object x there is a singleton set $\{x\}$.*

singleton sets

> **Proof:** To prove this, apply the previous proposition in the case where $x = y$.

A word of caution about list notation. For any set, every object is either member of the set or it is not. An expression like this: $\{3, 3\}$ is at best redundant as far as set theory is concerned. Think about the brace notation that this abbreviates: $\{x \mid x = 3 \vee x = 3\}$. The formula $x = 3 \vee x = 3$ is satisfied by exactly one object, namely 3. If an object satisfies the formula, then it is in the set, if it doesn't then it is not. There is no notion of repeated occurrences of an element in a set. This tells us that $\{x \mid x = 3 \vee x = 3\}$, whose existence is justified by Proposition 2, has exactly one element, namely the number 3. In general, you should avoid confusion and be certain that you never repeat elements of the set when using list notation.

Exercises

15.4   Are the following true or false? Prove your claims.
  1. $\{7, 8, 9\} = \{7, 8, 10\}$
  2. $\{7, 8, 9, 10\} = \{7, 8, 10, 9\}$
  3. $\{7, 8, 9, 9\} = \{7, 8, 9\}$

15.5   Give a list description of the following sets.
  1. The set of all prime numbers between 5 and 15.
  2. $\{\, x \mid x$ is a member of your family $\}$
  3. The set of letters of the English alphabet.
  4. The set of words of English with three successive double letters.

15.6   Are the following true or false?
  1. $y \in \{x \mid x$ is a prime less than 10$\}$ if and only if y is one of 2, 3, 5 and 7.
  2. $\{x \mid x$ is a prime less than 10$\} = \{2, 3, 5, 7\}$.
  3. Ronald Reagan $\in \{x \mid x$ was President of the US$\}$.
  4. "Ronald Reagan" $\in \{x \mid x$ was President of the US$\}$.

15.7 Suppose that a_1 and a_2 are sets, each of which has only the Washington Monument as a member. Prove (informally) that $a_1 = a_2$.

15.8 Give an informal proof that there is only one empty set. (Hint: Use the Axiom of Extensionality.)

15.9 Give an informal proof that the set of even primes greater than 10 is equal to the set of even primes greater than 100.

15.10 How many elements do the following sets contain?
1. $\{7, 8, 9\}$
2. $\{3 + 4, 4 + 4, 5 + 4\}$
3. $\{7, 8, 9, 3 + 4\}$
4. $\{7, 3 + 4, 5 + 2, 6 + 1\}$

SECTION 15.3

Subsets

The next notion is closely related to the membership relation, but fundamentally different. It is the subset relation, and is defined as follows:

subset ($\subseteq$)

Definition Given sets a and b, we say that a is a *subset* of b, written $a \subseteq b$, provided every member of a is also a member of b.

For example, the set of vowels, $\{\mathsf{a, e, i, o, u}\}$, is a subset of the set of letters of the alphabet, $\{\mathsf{a, b, c}, \ldots, \mathsf{z}\}$, but not vice versa. Similarly, the singleton set $\{\text{Washington Monument}\}$ is a subset of the set $\{x \mid x \text{ is taller than 100 feet}\}$.

subset vs. membership

It is very important to read the sentences "$a \in b$" and "$a \subseteq b$" carefully. The first is read "a is a member of b" or "a is an element of b." The latter is read "a is a subset of b." Sometimes it is tempting to read one or the other of these as "a is included in b." However, this is a very bad idea, since the term "included" is ambiguous between membership and subset. (If you can't resist using "included," use it only for the subset relation.)

From the point of view of FOL, there are two ways to think of our definition of "subset." One is to think of it as saying that the formula "$a \subseteq b$" is an abbreviation of the following wff:

$$\forall x \, [x \in a \rightarrow x \in b]$$

Another way is to think of $\subseteq$ as an additional binary relation symbol in our language, and to construe the definition as an axiom:

$$\forall a \, \forall b \, [a \subseteq b \leftrightarrow \forall x \, (x \in a \rightarrow x \in b)]$$

It doesn't make much difference which way you think of it. Different people prefer different understandings. The first is probably the most common, since it keeps the official language of set theory pretty sparse.

Let's prove a proposition involving the subset relation that is very obvious, but worth noting.

Proposition 4. *For any set a, $a \subseteq a$.*

> **Proof:** Let a be an arbitrary set. For purposes of general conditional proof, assume that c is an arbitrary member of a. Then trivially (by reiteration), c is a member of a. So $\forall x(x \in a \rightarrow x \in a)$. But then we can apply our definition of subset to conclude that $a \subseteq a$. Hence, $\forall a(a \subseteq a)$. (You are asked to formalize this proof in Exercise 15.14.)

The following proposition is very easy to prove, but it is also extremely useful. You will have many opportunities to apply it in what follows.

Proposition 5. *For all sets a and b, $a = b$ if and only if $a \subseteq b$ and $b \subseteq a$. In symbols:*

$$\forall a \, \forall b \, (a = b \leftrightarrow (a \subseteq b \wedge b \subseteq a))$$

> **Proof:** Again, we use the method of universal generalization. Let a and b be arbitrary sets. To prove the biconditional, we first prove that if $a = b$ then $a \subseteq b$ and $b \subseteq a$. So, assume that $a = b$. We need to prove that $a \subseteq b$ and $b \subseteq a$. But this follows from Proposition 4 and two uses of the indiscernibility of identicals.
>
> To prove the other direction of the biconditional, we assume that $a \subseteq b$ and $b \subseteq a$, and show that $a = b$. To prove this, we use the Axiom of Extensionality. By that axiom, it suffices to prove that a and b have the same members. But this follows from our assumptions, which tell us that every member of a is a member of b and vice versa.
>
> Since a and b were arbitrary sets, our proof is complete. (You are asked to formalize this proof in Exercise 15.15.)

Remember

Let a and b be sets.

1. $a \subseteq b$ iff every element of a is an element of b.

2. $a = b$ iff $a \subseteq b$ and $b \subseteq a$.

Exercises

15.11 Which of the following are true?

1. The set of all US senators $\subseteq$ the set of US citizens.
2. The set of all students at your school $\subseteq$ the set of US citizens.
3. The set of all male students at your school $\subseteq$ the set of all males.
4. The set of all John's brothers $\subseteq$ the set of all John's relatives.
5. The set of all John's relatives $\subseteq$ the set of all John's brothers.
6. $\{2, 3, 4\} \subseteq \{1 + 1, 1 + 2, 1 + 3, 1 + 4\}$
7. $\{\text{"2"}, \text{"3"}, \text{"4"}\} \subseteq \{\text{"1 + 1"}, \text{"1 + 2"}, \text{"1 + 3"}, \text{"1 + 4"}\}$

15.12 Give an informal proof of the following simple theorem: *For every set a,* $\emptyset \subseteq a$.

15.13 Give an formal proof of the following simple theorem: *For every set a,* $\emptyset \subseteq a$.

15.14 In the file Exercise 15.14, you are asked to give a formal proof of Proposition 4 from the definition of the subset relation. The proof is very easy, so you should not use any of the **Con** rules. (You will find the symbol $\subseteq$ if you choose the Set tab in the Fitch toolbar.)

15.15 In the file Exercise 15.15, you are asked to give a formal proof of Proposition 5 from the Axiom of Extensionality, the definition of subset, and Proposition 4. The proof is a bit more complex, so you may use **Taut Con** if you like.

15.16 Give a formal proof that the subset relation is transitive, that is, that

$$\forall x \, \forall y \, \forall z \, ((x \subseteq y \land y \subseteq z) \rightarrow x \subseteq z)$$

15.17 Proposition 4 tells us that every set is a subset of itself. Sometimes we are interested in just those subsets of a set that are not equal to the set. We can define these *proper* subsets using the formula $\forall x \, \forall y \, (x \subset y \leftrightarrow (x \subseteq y \land x \neq y))$. Give an informal proof of

$$\forall x \, \forall y \, (x \subset y \rightarrow \neg y \subset x)$$

Section 15.4

Intersection and union

There are two important operations on sets that you have probably seen before: intersection and union. These operations take two sets and form a third.

Definition Let a and b be sets.

1. The *intersection* of a and b is the set whose members are just those objects in both a and b. This set is generally written $a \cap b$. ("$a \cap b$" is a complex term built up using a binary function symbol $\cap$ placed in infix notation.[2]) In symbols:

$$\forall a \, \forall b \, \forall z \, (z \in a \cap b \leftrightarrow (z \in a \wedge z \in b))$$

intersection ($\cap$)

2. The *union* of a and b is the set whose members are just those objects in either a or b or both. This set is generally written $a \cup b$. In symbols:

$$\forall a \, \forall b \, \forall z \, (z \in a \cup b \leftrightarrow (z \in a \vee z \in b))$$

union ($\cup$)

At first sight, these definitions seem no more problematic than the definition of the subset relation. But if you think about it, you will see that there is actually something a bit fishy about them as they stand. For how do we know that there *are* sets of the kind described? For example, even if we know that a and b are sets, how do we know that there is a set whose members are the objects in both a and b? And how do we know that there is exactly one such set? Remember the rules of the road. We have to prove everything from explicitly given axioms. Can we prove, based on our axioms, that there is such a unique set?

It turns out that we can, at least with the naive axioms. But later, we will have to modify the Axiom of Comprehension to avoid inconsistencies. The modified form of this axiom will allow us to justify only one of these two operations. To justify the union operation, we will need a new axiom. But we will get to that in good time.

Proposition 6. (Intersection) *For any pair of sets a and b there is one and only one set c whose members are the objects in both a and b. In symbols:*

existence and uniqueness of $a \cap b$

$$\forall a \, \forall b \, \exists! c \, \forall x \, (x \in c \leftrightarrow (x \in a \wedge x \in b))$$

This proposition is actually just an instance of Proposition 1 on page 417. Look back at the formula displayed for that proposition, and consider the special case where z_1 is a, z_2 is b, and $P(x)$ is the wff $x \in a \wedge x \in b$. So Proposition 6 is really just a corollary (that is, an immediate consequence) of Proposition 1.

---

[2]Function symbols are discussed in the optional Section 1.5. You should read this section now if you skipped over it.

We can make this same point using our brace notation. Proposition 1 guarantees a unique set $\{x \mid P(x)\}$ for any formula $P(x)$, and we are simply noting that the intersection of sets a and b is the set $c = \{x \mid x \in a \wedge x \in b\}$.

The union operation is very similar to intersection, except that it forms the set of objects that are in *either* of its argument sets, not both as for intersection.

existence and uniqueness of $a \cup b$

Proposition 7. (Union) *For any pair of sets a and b there is one and only one set c whose members are the objects in either a or b or both. In symbols:*

$$\forall a \, \forall b \, \exists! c \, \forall x \, (x \in c \leftrightarrow (x \in a \vee x \in b))$$

Again, this is a corollary of Proposition 1, since $c = \{x \mid x \in a \vee x \in b\}$. This set clearly has the desired members.

The definition of union is superficially similar to the definition of pair sets that we encountered on page 420. Can you spot the difference? The pair set axiom says that there is a set containing objects that are *equal* to one of two objects, while the union axiom says that there is a set containing objects that are *members of* one of two objects. A pair set contains at most two objects, while there is no limit to the number of objects that a union could contain.

Here are several theorems we can prove using the above definitions and results.

Proposition 8. *Let a, b, and c be any sets.*

1. $a \cap b = b \cap a$

2. $a \cup b = b \cup a$

3. $a \cap b = b$ *if and only if $b \subseteq a$*

4. $a \cup b = b$ *if and only if $a \subseteq b$*

5. $a \cap (b \cup c) = (a \cap b) \cup (a \cap c)$

6. $a \cup (b \cap c) = (a \cup b) \cap (a \cup c)$

We prove two of these and leave the rest as exercises.

Proof of 8.1: This follows quite easily from the definition of intersection and the Axiom of Extensionality. To show that $a \cap b = b \cap a$, we need only show that $a \cap b$ and $b \cap a$ have the same members. By the definition of intersection, the members of $a \cap b$ are the things that are in both a and b, whereas the members of $b \cap a$ are the things that are in both b and a. These are clearly the same things. We will look at a formal proof of this in the next **You try it** section.

Proof of 8.3: Since (8.3) is the most interesting, we prove it. Let a and b be arbitrary sets. We need to prove $a \cap b = b$ iff $b \subseteq a$. To prove this, we give two conditional proofs. First, assume $a \cap b = b$. We need to prove that $b \subseteq a$. But this means $\forall x(x \in b \rightarrow x \in a)$, so we will use the method of general conditional proof. Let x be an arbitrary member of b. We need to show that $x \in a$. But since $b = a \cap b$, we see that $x \in a \cap b$. Thus $x \in a \wedge x \in b$ by the definition of intersection. Then it follows, of course, that $x \in a$, as desired.

Now let's prove the other half of the biconditional. Thus, assume that $b \subseteq a$ and let us prove that $a \cap b = b$. By Proposition 5, it suffices to prove $a \cap b \subseteq b$ and $b \subseteq a \cap b$. The first of these is easy, and does not even use our assumption. So let's prove the second, that $b \subseteq a \cap b$. That is, we must prove that $\forall x(x \in b \rightarrow x \in (a \cap b))$. This is proven by general conditional proof. Thus, let x be an arbitrary member of b. We need to prove that $x \in a \cap b$. But by our assumption, $b \subseteq a$, so $x \in a$. Hence, $x \in a \cap b$, as desired.

You try it
. . . .

1. Open the Fitch file Intersection 1. Here we have given a complete formal proof of Proposition 8.1 from the definition of intersection and the Axiom of Extensionality. (We have written "int(x, y)" for "$x \cap y$.") We haven't specified the rules or support steps in the proof, so this is what you need to do. This is the first formal proof we've given using function symbols. The appearance of complex terms makes it a little harder to spot the instances of the quantifier rules. ◀

2. Specify the rules and support steps for each step except the next to last (i.e., step 22). The heart of the proof is really the steps in which $c \in a \wedge c \in b$ is commuted to $c \in b \wedge c \in a$, and vice versa. ◀

3. Although it doesn't look like it, the formula in step 22 is actually an instance of the Axiom of Extensionality. Cite the axiom, which is one of your premises, and see that this sentence follows using $\forall$ **Elim**. ◀

4. When you have a completed proof specifying all rules and supports, save it as Proof Intersection 1. ◀

. . . . *Congratulations*

The following reminder shows us that ∩ is the set-theoretic counterpart of ∧ while ∪ is the counterpart of ∨.

Remember
Let b and c be sets.
1. $x \in b \cap c$ if and only if $x \in b \land x \in c$
2. $x \in b \cup c$ if and only if $x \in b \lor x \in c$

Exercises

15.18 If you skipped the **You try it** section, go back and do it now. Submit the file Proof Intersection 1.

15.19 Let $a = \{2, 3, 4, 5\}$, $b = \{2, 4, 6, 8\}$, and $c = \{3, 5, 7, 9\}$. Compute the following and express your answer in list notation.
1. $a \cap b$
2. $b \cap a$
3. $a \cup b$
4. $b \cap c$
5. $b \cup c$
6. $(a \cap b) \cup c$
7. $a \cap (b \cup c)$

15.20 Give an informal proof of Proposition 8.2.

15.21 Use Fitch to give a formal proof of Proposition 8.2. You will find the problem set up in the file Exercise 15.21. You may use **Taut Con**, since a completely formal proof would be quite tedious.

15.22 Give an informal proof of Proposition 8.4.

15.23 Use Fitch to give a formal proof of Proposition 8.4. You will find the problem set up in the file Exercise 15.23. You may use **Taut Con** in your proof.

15.24 Give an informal proof of Proposition 8.5.

15.25 Give an informal proof of Proposition 8.6.

15.26 Give an informal proof that for every set a there is a unique set c such that for all x, $x \in c$ iff $x \notin a$. This set c is called the *absolute complement* of a, and is denoted by $\bar{a}$. (This result will not follow from the axioms we eventually adopt. In fact, it will follow that *no* set has an absolute complement.) If you were to formalize this proof, what instance of the Axiom of Comprehension would you need? Write it out explicitly.

SECTION 15.5

Ordered Pairs

In order for set theory to be a useful framework for modeling structures of various sorts, it is important to find a way to represent order. For example, in high school you learned about the representation of lines and curves as sets of "ordered pairs" of real numbers. A circle of radius one, centered at the origin, is represented as the following set of ordered pairs:

modeling order

$$\{\langle x, y \rangle \mid x^2 + y^2 = 1\}$$

But sets themselves are unordered. For example $\{1, 0\} = \{0, 1\}$ by Extensionality. So how are we to represent ordered pairs and other ordered objects?

What we need is some way of modeling ordered pairs that allows us to prove the following:

$$\langle x, y \rangle = \langle u, v \rangle \leftrightarrow (x = u \wedge y = v)$$

If we can prove that this holds of our representation of ordered pairs, then we know that the representation allows us to determine which is the first element of the ordered pair and which is the second.

It turns out that there are many ways to do this. The simplest and most widely used is to model the ordered pair $\langle x, y \rangle$ by means of the unlikely set $\{\{x\}, \{x, y\}\}$.

Definition For any objects x and y, we take the ordered pair $\langle x, y \rangle$ to be the set $\{\{x\}, \{x, y\}\}$. In symbols:

ordered pair

$$\forall x \, \forall y \, \langle x, y \rangle = \{\{x\}, \{x, y\}\}$$

Later, we will ask you to prove that the fundamental property of ordered pairs displayed above holds when we represent them this way. Here we simply point out that the set $\{\{x\}, \{x, y\}\}$ exists and is unique, using Propositions 2 and 3.

Notice that there is something about this set that we have not encountered explicitly before, namely that an ordered pair is a set whose members are themselves sets: it is a *set of sets*. There is nothing mysterious about this fact; it follows directly from the fact that the axiom of comprehension is stated very generally. The range of the quantifier x in the axiom is the entire domain of naive set theory, which certainly contains all of the sets, in addition to any other objects that might exist.

ordered n-tuples

Once we have figured out how to represent ordered pairs, the way is open for us to represent ordered triples, quadruples, etc. For example, we will represent the ordered triple $\langle x, y, z \rangle$ as $\langle x, \langle y, z \rangle \rangle$. More generally, we will represent ordered n-tuples as $\langle x_1, \langle x_2, \dots x_n \rangle \rangle$.

By the way, as with brace notation for sets, the ordered pair notation $\langle x, y \rangle$ is not part of the official language of set theory. It can be eliminated from formulas without difficulty, though the formulas get rather long.

Exercises

15.27 Using propositions 2 and 3, let $a = \{2, 3\}$ and let $b = \{a\}$. How many members does a have? How many members does b have? Does $a = b$? That is, is $\{2, 3\} = \{\{2, 3\}\}$?

15.28 How many sets are members of the set described below?

$$\{\{\}, \{\{\}, 3, \{\}\}, \{\}\}$$

[Hint: First rewrite this using "$\emptyset$" as a notation for the empty set. Then delete from each description of a set any redundancies.]

15.29 Apply the Unordered Pair theorem (Proposition 2) to $x = y = \emptyset$. What set is obtained? Call this set c. Now apply the theorem to $x = \emptyset, y = c$. Do you obtain the same set or a different set?

15.30 This exercise and the one to follow lead you through the basic properties of ordered pairs.
   1. How many members does the set $\{\{x\}, \{x, y\}\}$ contain if $x \neq y$? How many if $x = y$?
   2. Recall that we defined $\langle x, y \rangle = \{\{x\}, \{x, y\}\}$. How do we know that for any x and y there is a unique set $\langle x, y \rangle$?
   3. Give an informal proof that the easy half of the fundamental property of ordered pairs holds with this definition:

$$(x = u \land y = v) \to \langle x, y \rangle = \langle u, v \rangle$$

4. (★★) Finally, prove the harder half of the fundamental property:

$$\langle x, y \rangle = \langle u, v \rangle \to (x = u \wedge y = v)$$

[Hint: Break into two cases, depending on whether or not $x = y$.]

15.31 Building on Problem 15.30, prove that for any two sets a and b, there is a set of all ordered ✎★ pairs $\langle x, y \rangle$ such that $x \in a$ and $y \in b$. This set is called the *Cartesian Product* of a and b, and is denoted by $a \times b$.

15.32 Suppose that a has three elements and b has five. What can you say about the size of $a \cup b$, ✎★ $a \cap b$, and $a \times b$? ($a \times b$ is defined in Exercise 15.31.) [Hint: in some of these cases, all you can do is give upper and lower bounds on the size of the resulting set. In other words, you'll have to say the set contains at least such and such members and at most so and so.]

SECTION 15.6

Modeling relations in set theory

Suppose we are talking about some domain D of objects, for instance a set of blocks from Tarski's World. Intuitively, we use unary predicates like Tet and Small to express properties of these blocks. How might we model the properties expressed by these predicates? The subsets of D defined by the predicates,

{x | x ∈ D and x is a tetrahedron}

{x | x ∈ D and x is small}

are not themselves properties, but they are the closest representatives that we have in set theory. We call these sets the *extension* (in domain D) of the corresponding property, and use them to model the properties expressed by the unary predicates in that domain.

Similarly, a binary predicate like Larger expresses a binary relation between objects in domain D. In set theory, we model this relation by means of a set of ordered pairs, specifically the set

{⟨x, y⟩ | x ∈ D, y ∈ D, and x is larger than y}

This set is called the *extension* (in domain D) of the predicate or relation. More generally, given some set D, we call any set of pairs $\langle x, y \rangle$, where x and y are in D, a binary *relation on* D. We model ternary relations similarly, as sets of ordered triples, and so forth for higher arities.

extension

relation in set theory

properties of relations

It is important to remember that the extension of a predicate can depend on the circumstances that hold in the domain of discourse. For example, if we rotate a world 90 degrees clockwise in Tarski's World, the domain of objects remains unchanged but the extension of *left of* becomes the new extension of *back of*. Similarly, if someone in the domain of discourse sits down, then the extension of *is sitting* changes. The binary predicates themselves do not change, nor does what they express, but the things that stand in these relations do, that is, their extensions change.

There are a few special kinds of binary relations that it is useful to have names for. In fact, we have already talked about some of these informally in Chapter 2. A relation R is said to be *transitive* if it satisfies the following:

Transitivity: $\forall x \, \forall y \, \forall z \, [(R(x,y) \wedge R(y,z)) \rightarrow R(x,z)]$

As examples, we mention that the relation *larger than* is transitive, whereas the relation *adjoins* is not. Since we are modeling relations by sets of ordered pairs, this condition becomes the following condition on a set R of ordered pairs: if $\langle x,y \rangle \in R$ and $\langle y,z \rangle \in R$ then $\langle x,z \rangle \in R$.

Here are several more special properties of binary relations:

Reflexivity:	$\forall x \, R(x,x)$
Irreflexivity:	$\forall x \, \neg R(x,x)$
Symmetry:	$\forall x \, \forall y \, (R(x,y) \rightarrow R(y,x))$
Asymmetry:	$\forall x \, \forall y \, (R(x,y) \rightarrow \neg R(y,x))$
Antisymmetry:	$\forall x \, \forall y \, [(R(x,y) \wedge R(y,x)) \rightarrow x = y]$

Each of these conditions can be expressed as conditions on the extension of the relation. The first, for example, says that for every $x \in D$, $\langle x,x \rangle \in R$.

To check whether you understand these properties, see if you agree with the following claims: The *larger than* relation is irreflexive and asymmetric. The *adjoins* relation is irreflexive but symmetric. The relation of *being the same shape as* is reflexive, symmetric, and transitive. The relation of $\leq$ on natural numbers is reflexive, antisymmetric, and transitive.

These properties of relations are intimately connected with the logic of atomic sentences discussed in Chapter 2. For example, to say that the following argument is valid is equivalent to saying that the predicate in question (Larger, for example) has a transitive extension under all logically possible circumstances. In that case the following *inference scheme* is valid:

inference scheme

$$
\begin{array}{|l}
R(a,b) \\
R(b,c) \\
\hline
R(a,c)
\end{array}
$$

Similarly, to say of some binary predicate R that

$$\forall x \, R(x, x)$$

is logically true is to say that the extension of R is reflexive in all logically possible circumstances. Identity is an example of this.

In connection with the logic of atomic sentences, let's look at two particularly important topics, inverse relations and equivalence relations, in a bit more detail.

Inverse relations

In our discussion of the logic of atomic sentences in Section 2.2, we noted that some of the logical relations between atomic sentences stem from the fact that one relation is the "inverse" of another (page 52). Examples were *right of* and *left of, larger* and *smaller*, and *less than* and *greater than*. We can now see what being inverses of one another says about the extensions of such pairs of predicates.

Given any set-theoretic binary relation R on a set D, the *inverse* (sometimes called the *converse*) of that relation is the relation R^{-1} defined by

inverse or converse

$$R^{-1} \;=\; \{\langle x, y \rangle \mid \langle y, x \rangle \in R\}$$

Thus, for example, the extension of *smaller* in some domain is always the inverse of the extension of *larger*. In an exercise, we ask you to prove some simple properties of inverse relations, including one showing that if S is the inverse of R, then R is the inverse of S.

Equivalence relations and equivalence classes

Many relations have the properties of reflexivity, symmetry, and transitivity. We have seen one example: *being the same shape as*. Such relations are called *equivalence relations*, since they each express some kind of equivalence among objects. Some other equivalence relations expressible in the blocks language include *being the same size as, being in the same row as*, and *being in the same column as*. Other equivalence relations include *has the same birthday as, has the same parents as*, and *wears the same size shoes as*. The identity relation is also an equivalence relation, even though it never classifies distinct objects as equivalent, the way others do.

equivalence relations

As these examples illustrate, equivalence relations group together objects that are the same in some dimension or other. This fact makes it natural to talk about the collections of objects that are the same as one another along the given dimension. For example, if we are talking about the *same size* relation, say among shirts in a store, we can talk about all the shirts of a particular size, say small, medium, and large, and even group them onto three appropriate racks.

We can model this grouping process very nicely in set theory with an important construction known as equivalence classes. This construction is widely used in mathematics and will be needed in our proof of the Completeness Theorem for the formal proof system $\mathcal{F}$.

Given any equivalence relation R on a set D, we can group together the objects that are deemed equivalent by means of R. Specifically, for each $x \in D$, let $[x]_R$ be the set

$$\{y \in D \mid \langle x, y \rangle \in R\}$$

equivalence classes

In words, $[x]_R$ is the set of things equivalent to x with respect to the relation R. It is called the *equivalence class* of x. (If x is a small shirt, then think of $[x]_{SameSize}$ as the store's small rack.) The fact that this grouping operation behaves the way we would hope and expect is captured by the following proposition. (We typically omit writing the subscript R from $[x]_R$ when it is clear from context, as in the following proposition.)

Proposition 9. Let R be an equivalence relation on a set D.

1. For each x, $x \in [x]$.

2. For all x, y, $[x] = [y]$ if and only if $\langle x, y \rangle \in R$.

3. For all x, y, $[x] = [y]$ if and only if $[x] \cap [y] \neq \emptyset$.

> **Proof:** (1) follows from the fact that R is reflexive on D. (2) is more substantive. Suppose that $[x] = [y]$. By (1), $y \in [y]$, so $y \in [x]$. But then by the definition of $[x]$, $\langle x, y \rangle \in R$. For the converse, suppose that $\langle x, y \rangle \in R$. We need to show that $[x] = [y]$. To do this, it suffices to prove that $[x] \subseteq [y]$ and $[y] \subseteq [x]$. We prove the first, the second being entirely similar. Let $z \in [x]$. We need to show that $z \in [y]$. Since $z \in [x]$, $\langle x, z \rangle \in R$. From the fact that $\langle x, y \rangle \in R$, using symmetry, we obtain $\langle y, x \rangle \in R$. By transitivity, from $\langle y, x \rangle \in R$ and $\langle x, z \rangle \in R$ we obtain $\langle y, z \rangle \in R$. But then $z \in [y]$, as desired. The proof of (3) is similar and is left as an exercise.

Exercises

15.33 Open the Fitch file Exercise 15.33. This file contains as goals the sentences expressing that the *same shape* relation is reflexive, symmetric, and transitive (and hence an equivalence relation). You can check that each of these sentences can be proven outright with a single application of **Ana Con**. However, in this exercise we ask you to prove this applying **Ana Con** only to atomic sentences. Thus, the exercise is to show how these sentences follow from the meaning of the basic predicate, using just the quantifier rules and propositional logic.

*For the next six exercises, we define relations R and S so that $R(a, b)$ holds if either a or b is a tetrahedron, and a is in the same row as b, whereas $S(a, b)$ holds if both a and b are tetrahedra, and in the same row. The exercises ask you to decide whether R or S has various of the properties we have been studying. If it does, open the appropriate Fitch exercise file and submit a proof. If it does not, submit a world that provides a counterexample. Thus, for example, when we ask whether R is reflexive, you should create a world in which there is an object that does not bear R to itself, since R is not in fact reflexive. In cases where you give a proof, you may use **Ana Con** applied to literals.*

15.34 Is R reflexive?

15.35 Is R symmetric?

15.36 Is R transitive?

15.37 Is S reflexive?

15.38 Is S symmetric?

15.39 Is S transitive?

15.40 Fill in the following table, putting *yes* or *no* to indicate whether the relation expressed by the predicate at the top of the column has the property indicated at the left.

	Smaller	SameCol	Adjoins	LeftOf
Transitive				
Reflexive				
Irreflexive				
Symmetric				
Asymmetric				
Antisymmetric				

15.41 Use Tarski's World to open the file Venn's World. Write out the extension of the *same column* relation in this world. (It contains eight ordered pairs.) Then write out the extension of the *between* relation in this world. (This will be a set of ordered triples.) Finally, what is the extension of the *adjoins* relation in this world? Turn in your answers.

15.42 Describe a valid inference scheme (similar to the one displayed on page 432) that goes with each of the following properties of binary relations: symmetry, antisymmetry, asymmetry, and irreflexivity.

15.43 What are the inverses of the following binary relations: *older than, as tall as, sibling of, father of,* and *ancestor of*?

15.44 Give informal proofs of the following simple facts about inverse relations.
1. R is symmetric iff $R = R^{-1}$.
2. For any relation R, $(R^{-1})^{-1} = R$.

15.45 Use Tarski's World to open the file Venn's World. Write out equivalence classes that go with each of the following equivalence relations: *same shape, same size, same row,* and identity. You can write the equivalence classes using list notation. For example, one of the *same shape* equivalence classes is $\{a, e\}$.

15.46 Given an equivalence relation R on a set D, we defined, for any $x \in D$:

$$[x]_R = \{y \in D \mid \langle x, y \rangle \in R\}$$

Explain how Proposition 1 can be used to show that the set displayed on the right side of this equation exists.

15.47 (Partitions and equivalence relations) Let D be some set and let $\mathcal{P}$ be some set of non-empty subsets of D with the property that every element of D is in exactly one member of $\mathcal{P}$. Such a set is said to be a *partition* of D. Define a relation E on D by: $\langle a, b \rangle \in E$ iff there is an $X \in \mathcal{P}$ such that $a \in X$ and $b \in X$. Show that E is an equivalence relation and that $\mathcal{P}$ is the set of its equivalence classes.

15.48 If a and b are subsets of D, then the Cartesian product (defined in Exercise 15.31) $a \times b$ is a binary relation on D. Which of the properties of relations discussed in this section does this relation have? (As an example, you will discover that $a \times b$ is irreflexive if and only if $a \cap b = \emptyset$.) Your answer should show that in the case where $a = b = D$, $a \times b$ is an equivalence relation. How many equivalence classes does it have?

15.49 Prove part (3) of Proposition 9.

Section 15.7

Functions

The notion of a function is one of the most important in mathematics. We have already discussed functions to some extent in Section 1.5.

Intuitively, a function is simply a way of doing things to things or assigning things to things: assigning license numbers to cars, assigning grades to students, assigning a temperature to a pair consisting of a place and a time, and so forth. We've already talked about the *father* function, which assigns to each person that person's father, and the *addition* function, which assigns to every pair of numbers another number, their sum.

Like relations, functions are modeled in set theory using sets of ordered pairs. This is possible because we can think of any unary function as a special type of binary relation: the relation that holds between the input of the function and its output. Thus, a relation R on a set D is said to be a *function* if it satisfies the following condition:

Functional: $\quad \forall x\, \exists^{\leq 1} y\, R(x, y)$

In other words, a relation is a function if for any "input" there is at most one "output." If the function also has the following property, then it is called a *total* function on D:

functions as special kind of relation

Totality: $\quad \forall x\, \exists y\, R(x, y)$

Total functions give "answers" for every object in the domain. If a function is not total on D, it is called a *partial* function on D.[3]

total functions

partial functions

Whether or not a function is total or partial depends very much on just what the domain D of discourse is. If D is the set of all people living or dead, then intuitively the *father of* function is total, though admittedly things get a bit hazy at the dawn of humankind. But if D is the set of living people, then this function is definitely partial. It only assigns a value to a person whose father is still living.

There are some standard notational conventions used with functions. First, it is standard to use letters like f, g, h and so forth to range over functions. Second, it is common practice to write $f(x) = y$ rather than $\langle x, y \rangle \in f$ when f is a function.

$f(x)$

The *domain* of a function f is the set

domain of function

$$\{x \mid \exists y(f(x) = y)\}$$

while its *range* is

$$\{y \mid \exists x(f(x) = y)\}$$

It is common to say that a function f is *defined on* x if x is in the domain of f. Thus the *father of* function is defined on the individual Max, but not on the number 5. In the latter case, we say the function is *undefined*. The domain of f will be the entire domain D of discourse if the function f is total, but will be a subset of D if f is partial.

defined vs. undefined

---

[3]Actually, usage varies. Some authors use "partial function" to include total functions, that is, to be synonymous with "function."

When dealing with partial functions, the notion of an *extension* of a function is important. A function f' is an *extension* of function f if the domain of f is a subset of the domain of f' and $f(a) = f'(a)$ for any a in the domain of f. In other words, f and f' agree on their assignments wherever they are both defined, but the extension f' may make additional assignments not made by f. Notice that this is a different meaning of *extension* from the one that we mentioned earlier in the section on modeling relations. Sorry for the confusion, but there you have it.

Notice that the identity relation on D, $\{\langle x, x \rangle \mid x \in D\}$, is a total function on D: it assigns each object to itself. When we think of this relation as a function we usually write it as *id*. Thus, $id(x) = x$ for all $x \in D$.

Later in the book we will be using functions to model rows of truth tables, naming functions for individual constants, and most importantly in defining the notion of a first-order structure, the notion needed to make the concept of first-order consequence mathematically rigorous.

Exercises

15.50 Use Tarski's World to open the file Venn's World. List the ordered pairs in the *frontmost* (fm) function described in Section 1.5 (page 33). Is the function total or partial? What is its range?

15.51 Which of the following sets represent functions on the set $D = \{1, 2, 3, 4\}$? For those which are functions, pick out their domain and range.
1. $\{\langle 1, 3 \rangle, \langle 2, 4 \rangle, \langle 3, 3 \rangle\}$
2. $\{\langle 1, 2 \rangle, \langle 2, 3 \rangle, \langle 3, 4 \rangle, \langle 4, 1 \rangle\}$
3. $\{\langle 1, 2 \rangle, \langle 1, 3 \rangle, \langle 3, 4 \rangle, \langle 4, 1 \rangle\}$
4. $\{\langle 1, 1 \rangle, \langle 2, 2 \rangle, \langle 3, 3 \rangle, \langle 4, 4 \rangle\}$
5. $\emptyset$

15.52 What is the domain and range of the square root function on the set $N = \{0, 1, 2, \ldots\}$ of all natural numbers?

15.53 Open the Fitch file Exercise 15.53. The premise here defines R to be the *frontmost* relation. The goal of the exercise is to prove that this relation is functional. You may use **Taut Con** as well as **Ana Con** applied to literals.

A function f is said to be injective *or one-to-one if it always assigns different values to different objects in its domain. In symbols, if $f(x) = f(y)$ then $x = y$ for all x, y in the domain of f.*

15.54 Which of the following functions are one-to-one: *father of, student id number of, frontmost,* and *fingerprint of*? (You may need to decide just what the domain of the function should be before deciding whether the function is injective. For *frontmost*, take the domain to be Venn's World.)

15.55 Let $f(x) = 2x$ for any natural number x. What is the domain of this function? What is its range? Is the function one-to-one?

15.56 Let $f(x) = x^2$ for any natural number x. What is the domain of this function? What is its range? Is the function one-to-one? How does your answer change if we take the domain to consist of all the integers, both positive and negative?

15.57 Let E be an equivalence relation on a set D. Consider the relation R that holds between any x in D and its equivalence class $[x]_E$. Is this a function? If so, what is its domain? What is its range? Under what conditions is it an one-to-one function?

SECTION 15.8

The powerset of a set

Once we get used to the idea that sets can be members of other sets, it is natural to form the set of all subsets of any given set b. The following theorem, which is easy to prove, shows that there is one and only one such set. This set is called the *powerset* of b and denoted $\wp b$ or $\wp(b)$.

powersets ($\wp$)

Proposition 10. (Powersets) *For any set b there is a unique set whose members are just the subsets of b. In symbols:*

$$\forall b \, \exists c \, \forall x \, (x \in c \leftrightarrow x \subseteq b)$$

Proof: By the Axiom of Comprehension, we may form the set $c = \{x \mid x \subseteq b\}$. This is the desired set. By the Axiom of Extensionality, there can be only one such set.

By way of example, let us form the powerset of the set $b = \{2, 3\}$. Thus, we need a set whose members are all the subsets of b. There are four of these. The most obvious two are the singletons $\{2\}$ and $\{3\}$. The other two are the empty set, which is a subset of every set, as we saw in Problem 15.12, and the set b itself, since every set is a subset of itself. Thus:

$$\wp b = \{\emptyset, \{2\}, \{3\}, \{2, 3\}\}$$

Here are some facts about the powerset operation. We will ask you to prove them in the exercises.

Proposition 11. *Let a and b be any sets.*

1. $b \in \wp b$

SECTION 15.8

2. $\emptyset \in \wp b$

3. $a \subseteq b$ iff $\wp a \subseteq \wp b$

It is possible for a set to have some of its own subsets as elements. For example, any set that has the empty set as an element has a subset as an element, since the empty set is a subset of every set. To take another example, the set

$$\{\text{Washington Monument}\}$$

is both a subset and an element of the set

$$\{\text{Washington Monument}, \{\text{Washington Monument}\}\}$$

However, it turns out that no set can have *all* of its subsets as elements.

Proposition 12. *For any set b, it is not the case that $\wp b \subseteq b$.*

Proof: Let b be any set. We want to prove that $\wp b \not\subseteq b$, or equivalently that there is a member of $\wp b$ that is not a member of b. To prove this, we construct a particular subset of b that is not an element of b. Let

$$c = \{x \mid x \in b \wedge x \notin x\}$$

by the Axiom of Comprehension. This set c is clearly a subset of b since it was defined to consist of those members of b satisfying some additional condition. It follows from the definition of the powerset operation that c is an element of $\wp b$. We will show that $c \notin b$.

Toward a proof by contradiction, suppose that $c \in b$. Then either $c \in c$ or $c \notin c$. But which? It is not hard to see that neither can be the case. First, suppose that $c \in c$. Then by our definition of c, c is one of those members of b that is left out of c. So $c \notin c$. Next consider the possibility that $c \notin c$. But then c is one of those members of b that satisfies the defining condition for c. Thus $c \in c$. Thus we have proven that $c \in c \leftrightarrow c \notin c$, which is a contradiction. So our assumption that $c \in b$ must be false, so $\wp b \not\subseteq b$.

This theorem applies to both finite and infinite sets. The proof shows how to take any set b and find a set c which is a subset of b but not a member of b, namely the set $c = \{x \mid x \in b$ and $x \notin x\}$. This is sometimes called the *Russell set for b*, after Bertrand Russell. So what we have proved in the preceding can be restated as:

Russell set for b

Proposition 13. *For any set b, the Russell set for b, the set*

$$\{x \mid x \in b \land x \notin x\},$$

is a subset of b but not a member of b.

This result is, as we will see, a very important result, one that immediately implies Proposition 12.

Let's compute the Russell set for a few sets. If $b = \{0, 1\}$, then the Russell set for b is just b itself. If $b = \{0, \{0, \{0, \dots\}\}\}$ then the Russell set for b is just $\{0\}$ since $b \in b$. Finally, if $b = \{\text{Washington Monument}\}$, then the Russell set for b is just b itself.

Remember

The powerset of a set b is the set of all its subsets:

$$\wp b = \{a \mid a \subseteq b\}$$

Exercises

15.58 Compute $\wp\{2, 3, 4\}$. Your answer should have eight distinct elements.

15.59 Compute $\wp\{2, 3, 4, 5\}$.

15.60 Compute $\wp\{2\}$.

15.61 Compute $\wp\emptyset$.

15.62 Compute $\wp\wp\{2, 3\}$.

15.63 Prove the results stated in Proposition 11.

15.64 Here are a number of conjectures you might make. Some are true, but some are false. Prove the true ones, and find examples to show that the others are false.

    1. For any set b, $\emptyset \subseteq \wp b$.
    2. For any set b, $b \subseteq \wp b$.
    3. For any sets a and b, $\wp(a \cup b) = \wp a \cup \wp b$.
    4. For any sets a and b, $\wp(a \cap b) = \wp a \cap \wp b$.

15.65 What is the Russell set for each of the following sets?

    1. $\{\emptyset\}$
    2. A set a satisfying $a = \{a\}$
    3. A set $\{1, a\}$ where $a = \{a\}$
    4. The set of all sets

Section 15.9

Russell's Paradox

We are now in a position to show that something is seriously amiss with the theory we have been developing. Namely, we can prove the negation of Proposition 12. In fact, we can prove the following which directly contradicts Proposition 12.

Proposition 14. *There is a set c such that $\wp c \subseteq c$.*

> **Proof:** Using the Axiom of Comprehension, there is a universal set, a set that contains everything. This is the set $c = \{x \mid x = x\}$. But then every subset of c is a member of c, so $\wp c$ is a subset of c.

universal set (V)

The set c used in the above proof is called the *universal set* and is usually denoted by "V." It is called that because it contains everything as a member, including itself. What we have in fact shown is that the powerset of the universal set both is and is not a subset of the universal set.

Let us look at our contradiction a bit more closely. Our proof of Proposition 12, applied to the special case of the universal set, gives rise to the set

$$Z = \{x \mid x \in V \wedge x \notin x\}$$

This is just the Russell set for the universal set. But proposition 13 tells us that for any set a, the Russell set for a is not a member of a. So the Russell set for V is not a member of V, which is a contradiction because *everything* is in V.

Russell's Paradox

This set Z is called the (absolute) Russell set, and the contradiction we have just established is called Russell's Paradox.

It would be hard to overdramatize the impact the discovery of Russell's Paradox had on set theory at the turn of the century. Simple as it is, it shook the subject to its foundations. It is just as if in arithmetic we discovered a proof that 23+27=50 and 23+27≠ 50. Or as if in geometry we could prove that the area of a square both is and is not the square of the side. But here we are in just that position. This shows that there is something wrong with our starting assumptions of the whole theory, the two axioms with which we began. There simply is no domain of sets which satisfies these assumptions. This discovery was regarded as a paradox just because it had earlier seemed to most mathematicians that the intuitive universe of sets did satisfy the axioms.

reactions to the paradox

Russell's Paradox is just the tip of an iceberg of problematic results in naive set theory. These paradoxes resulted in a wide-ranging attempt to clarify the notion of a set, so that a consistent conception could be found to use in mathematics. There is no one single conception which has completely won out in this effort, but all do seem to agree on one thing. The problem with the naive theory is that it is too uncritical in its acceptance of collections such as V used in the last proof. What the result shows is that there is no such set. So our axioms must be wrong. We must not be able to use just any old property in forming a set.

The father of set theory was the German mathematician Georg Cantor. His work in set theory, in the late nineteenth century, preceded Russell's discovery of Russell's paradox in the earlier twentieth century. It is thus natural to imagine that he was working with the naive, hence inconsistent view of sets. However, there is clear evidence in Cantor's writings that he was aware that unrestricted set formation was inconsistent. He discussed consistent versus inconsistent "multiplicities," and only claimed that consistent multiplicities could be treated as objects in their own right, that is, as sets. Cantor was not working within an axiomatic framework and was not at all explicit about just what properties or concepts give rise to inconsistent multiplicities. People following his lead were less aware of the pitfalls in set formation prior to Russell's discovery.

Remember

Russell found a paradox in naive set theory by considering

$$Z = \{x \mid x \notin x\}$$

and showing that the assumption $Z \in Z$ and its negation each entails the other.

Exercises

15.66
✎

In the file Exercise 15.66, you are asked to give a formal proof of $\bot$ from the single premise $\exists y \, \forall x \, (x \in y \leftrightarrow \neg x \in x)$. The problem is, this is a straightforward instance of the Axiom of Comprehension (using $\neg x \in x$ for the formula $P(x)$). Your formal proof will thus show that this axiom is inconsistent. (We could also have asked you to prove, from no premises, the negation of this instance of Comprehension, showing that the negation is a logical truth.)

Section 15.10

Zermelo Frankel set theory (ZFC)

diagnosing the problem

The paradoxes of naive set theory show us that our intuitive notion of set is simply inconsistent. We must go back and rethink the assumptions on which the theory rests. However, in doing this rethinking, we do not want to throw out the baby with the bath water. Set theory has proved so valuable as a useful toolkit for mathematicians that we would like to develop a new set theory in which all of the previous results, except for the proof of inconsistency, are provable.

If we examine the Russell Paradox closely, we see that it is actually a straightforward refutation of the Axiom of Comprehension. It shows that there is no set determined by the property of not belonging to itself. That is, the following is, on the one hand, a logical truth, but also the negation of an instance of Comprehension:

$$\neg \exists c \, \forall x \, (x \in c \leftrightarrow x \notin x)$$

The Axiom of Extensionality is not needed in the derivation of this fact. So it is the Comprehension Axiom that is the problem, as you proved in exercise 15.66. In fact, back in Chapter 13, Exercise 13.52, we asked you to give a formal proof of

$$\neg \exists y \, \forall x \, [\mathsf{E}(x, y) \leftrightarrow \neg \mathsf{E}(x, x)]$$

This is just the above sentence with "$\mathsf{E}(x, y)$" used instead of "$x \in y$". The proof shows that the sentence is actually a first-order validity; its validity does not depend on anything about the meaning of "$\in$." It follows that no coherent conception of set can countenance the Russell set.

But why is there no such set? It is not enough to say that the set leads us to a contradiction. We would like to understand why this is so. Various answers have been proposed to this question.

Cumulative sets

One popular view is the "cumulation" metaphor due to the logician Ernst Zermelo. Zermelo's idea is that sets should be thought of as formed by abstract acts of collecting together previously given objects. We start with some basic objects. We collect sets of these objects. Then we collet sets whose members are the objects and earlier sets, and so on and on. Before one can form a set by this abstract act of collecting, one must already have all of its members, Zermelo suggested.

On this conception, sets come in distinct, discrete "stages," each set arising at the first stage after the stages where all of its members arise. For example, if set x arises as stage 17 and set y at stage 37, then $a = \{x, y\}$ would arise at stage 38. If b is constructed at some stage, then its powerset $\wp b$ will be constructed at the next stage. Each stage gives rise to sets that can be used to form new sets at a later stage. Once a set has been formed, we can always create a new stage by forming its power set and so there is no last stage of this process. On Zermelo's conception, the reason there can never be a universal set is that as any set b arises, there is always its powerset to be formed later, and so there is no stage at which the universal set V can appear. In this sense, V is "too big" to be a set.

The most common form of modern set theory is Zermelo-Frankel set the-ory, also known as zfc. The axioms of zfc capture this cumulative idea of sets. The axioms allow us to prove the existence of a basic collection of sets, those that exist at stage 1, and additional axioms permit us to collect together sets of objects from one stage to form the members of the next stage.

Zermelo-Frankel set theory zfc

In zfc, it is generally assumed that we are dealing with "pure" sets, that is, there is nothing but sets in the domain of discourse. The only basic object with which we start our "collection" operations is the empty set, and its existence must be justified by an axiom. If we want to speak about numbers or any other objects in zfc, we must build models of them within the theory. For example, in zfc, we could model 0 by the empty set, 1 by $\{\emptyset\}$, 2 by $\{\{\emptyset\}\}$, and so on.

Here is a list of the axioms of zfc.

axioms of zfc

1. **Axiom of Extensionality**:

$$\forall a\, \forall b\, (\forall x\, (x \in a \leftrightarrow x \in b) \rightarrow a = b)$$

Since the Axiom of Extensionality is not implicated in any set theoretic paradox, and since it does not even assert the existence of any sets, we are safe using it in the same form as before. If we are developing a theory of pure sets, however, it is more common to state it in a single-sorted language:

$$\forall y\, \forall z\, (\forall x\, (x \in y \leftrightarrow x \in z) \rightarrow y = z)$$

The difference between these two is that the first only addresses the identity of sets, while the second addresses the identity of everything in the domain of discourse. Note in particular that the second version implies that there is at most one thing in the domain that has no members (which would be the empty set). We will continue to state the remaining axioms in our many-sorted language, since it makes them easier to understand.

2. **Axiom of Separation**:

$$\forall z_1 \ldots \forall z_n \, \forall a \, \exists b \, \forall x \, [x \in b \leftrightarrow (x \in a \land P(x))]$$

The Axiom of Separation is a weakened version of the Axiom of Unrestricted Comprehension that led us into contradiction in naive set theory. To see how it differs, compare it to the full version of Comprehension stated on page 416.

Where the Axiom of Comprehension allowed us to prove the existence of any set of objects satisfying a formula, the Axiom of Separation only allows us to use a set that already exists, and form the subset of its members that satisfy some formula. We can *separate* the elements satisfying the property out from some larger set. The idea here is that if the set a has already been collected from previous stages of the cumulation, then any subset must also be available at the same stage.

As promised, the existence of the intersection $a \cap c$ of sets a and c follows directly from the Axiom of Separation. Using the property $x \in c$ as $P(x)$ in the Axiom of Separation yields the same formula as using $x \in a \land x \in c$ in the Axiom of Comprehension. However, there are some sets that the Axiom of Comprehension would permit that cannot be proved to exist on the basis of Separation. This is good, since in particular we cannot prove the existence of the universal set. (In fact we can show that it does not exist.) It is easy to show that the resulting theory is consistent (see Exercise 15.71). Unfortunately, we also cannot prove the existence of sets that we *do* want in our set theory. In particular it blocks the proofs that the empty set, unions and powersets exist. None of the instances of the Axiom of Comprehension that we used to prove the existence of such sets are also instances of the Axiom of Separation. We need to add extra axioms to ensure that these unproblematic operations are still available to us.

3. **Unordered Pair Axiom**: For any two objects there is a set that has both as elements (and no others).

$$\forall x \, \forall y \, \exists a \, \forall z \, (z \in a \leftrightarrow (z = x \lor z = y))$$

It is a requirement of first-order logic that the domain of discourse is non-empty, and so the Unordered Pair Axiom implies the existence of a set that contains this object, whatever it is. From this set, we can use the Axiom of Separation to derive the existence of the empty set as the subset of this set whose elements satisfy $x \neq x$.

4. **Union Axiom**: Given any set a of sets, the union of all the members of a is also a set. That is:

$$\forall a \, \exists b \, \forall x \, [x \in b \leftrightarrow \exists c \, (c \in a \land x \in c)]$$

This is a generalization of the union axiom that we presented earlier. The idea here is that a is a set of sets, and the union set whose existence is guaranteed by this axiom is the set of all objects that are in at least one member of the set a.

Using this axiom, we can show that for any sets c and d, $c \cup d$ exists. $c \cup d$ is the set of objects that are in either c or d, which means that they are members of some member of $\{c, d\}$. The Union axiom generalizes this idea, because the set a may have infinitely many members and so we can form the union of all of the members of this set. There is not necessarily a finite sequence of binary union operations that could be used to form the same set.

5. **Powerset Axiom**: Every set has a powerset.

$$\forall a \, \exists b \, \forall x \, (x \in b \leftrightarrow x \subseteq a)$$

The preceding four axioms are important instances of the Axiom of Unrestricted Comprehension that we used in the development of naive set theory. In zfc we are not allowed to use every instance of this axiom, since this leads to inconsistency. But we can use these instances which we can show lead to a consistent theory.

The next axiom guarantees the existence of an infinite set.

6. **Axiom of Infinity**: There is a set containing the empty set and which contains $\{a\}$ for every a that it contains.

$$\exists a \, (\emptyset \in a \land \forall x \, (x \in a \rightarrow \{x\} \in a))$$

The set whose existence is guaranteed by this axiom has as many elements as there are natural numbers. We can think of the set $\emptyset$ as representing the number 0, $\{\emptyset\}$ as representing 1, $\{\{\emptyset\}\}$ as representing 2 and so on. In general. the singleton set containing the representation of a number is the representation of the next number.

The next two axioms use functions to construct new sets out of existing sets.

7. **Axiom of Replacement**: Suppose that you have a formula $P(x,y)$ which holds of exactly one y for each x in a set a, that is:

$$\forall x\ (x \in a \rightarrow \exists! y\ P(x,y))$$

Then, there is a set

$$\{y \mid \exists x\ (x \in a \wedge P(x,y))\}$$

The Axiom of Replacement tells us that if we have a set and a formula which relates each element of the set to a unique object, then the collection of those objects is also a set. This axiom is actually a schema, since $P(x,y)$ can be any formula with the required uniqueness property, and each such formula defines a set.

8. **Axiom of Choice**: Suppose a is a set whose members are all non-empty sets. Then there is a function f whose domain is a and which satisfies the following:

$$\forall x\ (x \in a \rightarrow f(x) \in x)$$

The idea here is that the function f looks at each set in a and chooses exactly one member of that set — sort of like Max picking his favorite book from each shelf in the bookcase. The function f is called a *choice function* for a. The Axiom of Choice guarantees the existence of f and, thanks to Replacement, of the set that forms the range of f:

$$\{f(x) \mid x \in a\}$$

Axiom of Choice

Unlike the previous axioms, the Axiom of Choice is not a straightforward logical consequence of naive set theory.[4] The Axiom of Choice has a long and somewhat convoluted history. There are many, many equivalent ways of stating it; in fact there is a whole book of statements equivalent to the axiom of choice. In the early days of set theory some authors took it for granted, others saw no reason to suppose it to be true. Nowadays it is taken for granted as being obviously true by most mathematicians. The attitude is that while there may be no way to *define* a choice function for a given set a, and so no way to prove that one exists by means of Separation, such functions exist nonetheless, and so are asserted to exist by this axiom. It is extremely widely used in modern mathematics.

---

[4]Technically speaking, it is a consequence, though, since the naive theory is inconsistent. After all, everything is a consequence of inconsistent premises.

9. **Axiom of Regularity**: No nonempty set has a nonempty intersection with each of its own elements. That is:

$$\forall b \left[b \neq \emptyset \rightarrow \exists y \left(y \in b \wedge y \cap b = \emptyset \right) \right]$$

All of the preceding axioms tell us what sets must exist in our set theoretic universe. The Axiom of Regularity is different because it tells us that certain sets cannot exist. Specifically, this axiom rules out the possibility of "irregular" sets, such as a set which contains itself as its only member. We might write this set as $S = \{S\}$, or as $\{\{\{\ldots\}\}\}$. Your mind might recoil at the thought of such sets, and that is because such sets cannot exist on the cumulative conception of sets. For any set to exist its members must already be formed at a previous stage, and since this set's only member is itself, it couldn't possibly arise at any stage in the cumulative hierarchy.

More generally, let us see why, on the cumulative conception, the Axiom of Regularity is true. That is, let us prove that on this conception, no set has a nonempty intersection with each of its own elements.

regularity and cumulation

> **Proof:** Let a be any nonempty set. We need to show that one of the elements of a has an empty intersection with a. Among a's elements, pick any $b \in a$ that occurs earliest in the cumulation process. That is, for any other $c \in a$, b is constructed at least as early as c. We claim that $b \cap a = \emptyset$. If we can prove this, we will be done. The proof is by contradiction. Suppose that $b \cap a \neq \emptyset$ and let $c \in b \cap a$. Since $c \in b$, c has to occur earlier in the construction process than b. On the other hand, $c \in a$ and b was chosen so that there was no $c \in a$ constructed earlier than b. This contradiction concludes the proof.

The axioms of zfc that we have so far discussed do not imply the existence, or the non-existence, of irregular sets. There are set theories in which all of the preceding axioms of zfc are true and irregular sets exist, and others in which the preceding axioms of zfc are true and in which no irregular sets exist. If we want to explicitly enforce Zermelo's cumulative conception of sets, then we must adopt the Axiom of Regularity.

One of the reasons the Axiom of Regularity is assumed is that it gives us a powerful method for proving theorems about sets "by induction." We discuss various forms of proof by induction in the next chapter. The Axiom of Regularity is sometimes called the Axiom of Foundation because of its relationship to a particular kind of induction called "well-founded induction." For the relation with the Axiom of Regularity, see Exercise 16.10.

You should examine the axioms of ZFC in turn to see if you think they hold on Zermelo's conception of set.

Sizes of infinite sets

One view of the problem caused by considering the collection of all objects to be a set is that this collection is just "too big" to be a completed totality. As we have seen, the universal set V cannot appear at any stage of Zermelo's cumulative hierarchy of sets. But this is a slightly different objection to considering V to be a set, an objection that is due to John von Neumann.

If V is too big, then exactly how big does a collection have to get to be too big to be considered a set? Some philosophers have suggested that the powerset of an infinite set might be too large to be considered as a completed totality, and this has led to concern that the Powerset Axiom is not justified on this conception of set. To see why, let us start by thinking about the size of the powerset of finite sets.

sizes of powersets

We have seen that if we start with a set b of size n, then its powerset $\wp b$ has 2^n members. For example, if b has five members, then its power set has $2^5 = 32$ members. But if b has 1000 members, then its power set has 2^{1000} members, an incredibly large number indeed; larger, they say, than the number of atoms in the universe. And then we could form the powerset of that, and the powerset of that — gargantuan sets indeed.

sizes of infinite sets

But what happens if b is infinite? To address this question, we first have to figure out what exactly we mean by the size of an infinite set. Cantor answered this question by giving a rigorous analysis of size that applies to all sets, finite and infinite. For any set b, the Cantorian size of b is denoted $|b|$. Informally,

$|b|$

$|b| = |c|$ just in case the members of b and the members of c can be associated with one another in a unique fashion. More precisely, what is required is that there be a one-to-one function with domain b and range c. (The notion of a one-to-one function was defined in Exercise 15.54.)

For finite sets, $|b|$ behaves just as one would expect. This notion of size is somewhat subtle when it comes to infinite sets, though. It turns out that for infinite sets, a set can have the same size as some of its proper subsets.[5] The set N of all natural numbers, for example, has the same size as the set E of even numbers; that is $|N| = |E|$. The main idea of the proof is contained in

---

[5]Recall from Exercise 15.17 that a <u>proper</u> subset is a subset which is not equal to the set, in other words it leaves out at least one member of the set.

the following picture:

$$
\begin{array}{cccccc}
0 & 1 & 2 & \ldots & n & \ldots \\
\updownarrow & \updownarrow & \updownarrow & & \updownarrow & \\
0 & 2 & 4 & \ldots & 2n & \ldots
\end{array}
$$

This picture shows the sense in which there are as many even integers as there are integers. (This was really the point of Exercise 15.55.) Indeed, it turns out that many sets have the same size as the set of natural numbers, including the set of all rational numbers. The set of real numbers, however, is strictly larger than the set of natural numbers, as Cantor proved.

Cantor also showed that that for any set b whatsoever,

$$|\wp b| > |b|$$

This result is not surprising, given what we have seen for finite sets. (The proof of Proposition 12 was really extracted from Cantor's proof of this fact.) The two together do raise the question as to whether an infinite set b could be "small" but its powerset "too large" to be a set. Thus the Powerset Axiom is not as unproblematic as the other axioms in terms of Von Neumann's size metaphor. Still, it is almost universally assumed that if b can be coherently regarded as a fixed totality, so can $\wp b$. Thus the Powerset Axiom is a full-fledged part of modern set theory.

questions about powerset axiom

The modern conception of set really combines these two ideas, von Neumann's and Zermelo's. This conception views a set is as a "small" collection that is formed at some stage of the cumulation process.

Remember

1. Modern set theory replaces the naive concept of set, which is inconsistent, with a concept of set as a collection that is not too large.

2. These collections are seen as arising in stages, where a set arises only after all its members are present.

3. The Axiom of Comprehension of set theory is replaced by the Axiom of Separation and some of the intuitively correct consequences of the Axiom of Comprehension.

4. Modern set theory also contains the Axiom of Regularity, which is justified on the basis of (2).

5. All the propositions stated in this chapter—with the exception of Propositions 1 and 14—are theorems of zfc.

Exercises

15.67 Try to derive the existence of the absolute Russell set from the Axiom of Separation. Where does the proof break down?

15.68⋆ Verify our claim that all of Propositions 2–13 are provable using the axioms of ZFC. (Some of the proofs are trivial in that the theorems were thrown in as axioms. Others are not trivial.)

15.69⋆ (Cantor's Theorem) Show that for any set b whatsoever, $|\wp b| \neq |b|$. [Hint: Suppose that f is a function mapping $\wp b$ one-to-one into b and then modify the proof of Proposition 12.]

15.70 (There is no universal set)
1. Verify that our proof of Proposition 12 can be carried out using the axioms of ZFC.
2. Use (1) to prove there is no universal set.

15.71 Prove that the Axiom of Separation and Extensionality are consistent. That is, find a universe of discourse in which both are clearly true. [Hint: consider the domain whose only element is the empty set.]

15.72⋆ Show that the theorem about the existence of $a \cap b$ can be proven using the Axiom of Separation, but that the theorem about the existence of $a \cup b$ cannot be so proven. [Come up with a domain of sets in which the separation axiom is true but the theorem in question is false.]

15.73 (The Union Axiom and ∪) Exercise 15.72 shows us that we cannot prove the existence of $a \cup b$ from the Axiom of Separation. However, the Union Axiom of ZFC is stronger than this. It says not just that $a \cup b$ exists, but that the union of any set of sets exists.
1. Show how to prove the existence of $a \cup b$ from the Union Axiom. What other axioms of ZFC do you need to use?
2. Apply the Union Axiom to show that there is no set of all singletons. [Hint: Use proof by contradiction and the fact that there is no universal set.]

15.74⋆ Prove in ZFC that for any two sets a and b, the Cartesian product $a \times b$ exists. The proof you gave in an earlier exercise will probably not work here, but the result is provable.

15.75 While ∧ and ∨ have set-theoretic counterparts in ∩ and ∪, there is no absolute counterpart to ¬.
1. Use the axioms of ZFC to prove that no set has an absolute complement.
2. In practice, when using set theory, this negative result is not a serious problem. We usually work relative to some domain of discourse, and form relative complements. Justify this by showing, within ZFC, that for any sets a and b, there is a set $c = \{x \mid x \in a \wedge x \notin b\}$. This is called the *relative complement of* b with respect to a.

15.76 Assume the Axiom of Regularity. Show that no set is a member of itself. Conclude that, if we
✎★ assume Regularity, then for any set b, the Russell set for b is simply b itself.

15.77 (Consequences of the Axiom of Regularity)
✎★

1. Show that if there is a sequence of sets with the following property, then the Axiom of
   Regularity is false:
   $$\ldots \in a_{n+1} \in a_n \in \ldots \in a_2 \in a_1$$

2. Show that in ZFC we can prove that there are no sets $b_1, b_2, \ldots, b_n, \ldots$, where $b_n = \{n, b_{n+1}\}$.

3. In computer science, a *stream* is defined to be an ordered pair $\langle x, y \rangle$ whose first element
   is an "atom" and whose second element is a stream. Show that if we work in ZFC and
   define ordered pairs as usual, then there are no streams.

There are alternatives to the Axiom of Regularity which have been explored in recent years.
We mention our own favorite, the axiom AFA, due to Peter Aczel and others. The name "AFA"
stands for "anti-foundation axiom." Using AFA you can prove that a great many sets exist with
properties that contradict the Axiom of Regularity. We wrote a book, *The Liar*, in which we
used AFA to model and analyze the so-called Liar's Paradox (see Exercise 19.32, page 571).

CHAPTER 16

Mathematical Induction

In the first two parts of this book, we covered most of the important methods of proof used in rigorous reasoning. But we left out one extremely important method: proof by *mathematical induction*.

By and large, the methods of proof discussed earlier line up fairly nicely with various connectives and quantifiers, in the sense that you can often tell from the syntactic form of your premises or conclusion what methods you will be using. The most obvious exception is proof by contradiction, or its formal counterpart ¬ **Intro**. This method can in principle be used to prove any form of statement, no matter what its main connective or quantifier. This is because any sentence S is logically equivalent to one that begins with a negation symbol, namely, ¬¬S.

form of statements proved by induction

In terms of syntactic form, mathematical induction is typically used to prove statements of the form

$$\forall x\,[P(x) \rightarrow Q(x)]$$

This is also the form of statements proved using general conditional proof. In fact, proof by induction is really a pumped-up version of this method: general conditional proof on steroids, you might say. It works when these statements involve a predicate P(x) defined in a special way. Specifically, proof by induction is available when the predicate P(x) is defined by what is called an

inductive definition

inductive definition. For this reason, we need to discuss proof by induction and inductive definitions side by side. We will see that whenever a predicate P(x) is defined by means of an inductive definition, proof by induction provides a much more powerful method of proof than ordinary general conditional proof.

induction in science

Before we can discuss either of these, though, we should distinguish both from yet a third process that is also known as *induction*. In science, we use the term "induction" whenever we draw a general conclusion on the basis of a finite number of observations. For example, every day we observe that the sun comes up, that dropped things fall down, and that people smile more when it is sunny. We come to infer that this is always the case: that the sun comes up every morning, that dropped things always fall, that people are always happier when the sun is out.

Of course there is no strict logical justification for such inferences. We may have correctly inferred some general law of nature, or we may have simply observed a bunch of facts without any law that backs them up. Some time in the future, people may be happier if it rains, for example after a long drought. Induction, in this sense, does not guarantee that the conclusion follows necessarily from the premises. It is not a deductively valid form of inference, since it is logically possible for the premises to be true and the conclusion false.

This is all by way of contrast with mathematical induction, where we *can* justify a general conclusion, with infinitely many instances, on the basis of a finite proof. How is this possible? The key lies in the inductive definitions that underwrite this method of proof. Induction, in our sense, is a logically valid method of proof, as certain as any we have studied so far.

vs. mathematical induction

Usually, discussions of mathematical induction start (and end) with induction on the natural numbers, to prove statements of the form

$$\forall x\, [\text{NatNum}(x) \rightarrow Q(x)]$$

We will start with other examples, examples which show that mathematical induction applies much more widely than just to natural numbers. The reason it applies to natural numbers is simply that the natural numbers can be specified by means of an inductive definition. But so can many other things.

Inductive definitions and inductive proofs

Inductive definitions involve setting things up in a certain methodical, step-by-step manner. Proofs by induction take advantage of the structure that results from such inductive definitions. We begin with a simple analogy.

Dominoes

When they were younger, Claire and Max liked to build long chains of dominoes, all around the house. Then they would knock down the first and, if things were set up right, the rest would all fall down. Little did they know that in so doing they were practicing induction. Setting up the dominoes is like giving an inductive definition. Knocking them all down is like proving a theorem by induction.

There are two things required to make all the dominoes fall over. They must be close enough together that when any one domino falls, it knocks down the next. And then, of course, you need to knock down the first. In a proof by induction, these two steps correspond to what are called the inductive step (getting from one to the next) and the basis step (getting the whole thing started).

Notice that there is no need to have just one domino following each domino. You can have two, as long as the one in front will knock down both of its successors. In this way you can build quite elaborate designs, branching out here and there, and, when the time is right, you can knock them all down with a single flick of the finger. The same is true, as we'll see, with induction.

Inductive definitions

inductive definitions

Inductive definitions are used a great deal in logic. In fact, we have been using them implicitly throughout this book. For example, our definitions of the wffs of FOL were really inductive definitions. So was our definition of the set of terms of first-order arithmetic. Both of these definitions started by specifying the simplest members of the defined collection, and then gave rules that told us how to generate "new" members of the collection from "old" ones. This is how inductive definitions work.

Let's look at another example, just to make things more explicit. Suppose that for some reason we wanted to study an ambiguous variant of propositional logic, maybe as a mathematical model of English that builds in some ambiguity. Let's take some primitive symbols, say $A_1, \ldots, A_n$, and call these propositional letters. Next, we will build up "wffs" from these using our old friends $\neg, \wedge, \vee, \rightarrow$, and $\leftrightarrow$. But we are going to let the language be ambiguous, unlike FOL, by leaving out all parentheses. How will we do this? To distinguish these strings from wffs, let us call them *ambig-wffs*. Intuitively, what we want to say is the following:

ambig-wffs

1. Each propositional letter is an ambig-wff.

2. If p is any ambig-wff, so is the string $\neg p$.

3. If p and q are ambig-wffs, so are $p \wedge q$, $p \vee q$, $p \rightarrow q$, and $p \leftrightarrow q$.

4. Nothing is an ambig-wff unless it is generated by repeated applications of (1), (2), and (3).

base clause
inductive clause
final clause

In this definition, clause (1) specifies the basic ambig-wffs. It is called the *base clause* of the definition. Clauses (2) and (3) tell us how to form new ambig-wffs from old ones. They are called *inductive* clauses. The final clause just informs us that all ambig-wffs are generated by the earlier clauses, in case we thought that the World Trade Center or the actor Brad Pitt or the set $\{2\}$ might be an ambig-wff.

> **Remember**
>
> An inductive definition consists of
>
> o a *base clause*, which specifies the basic elements of the defined set,
>
> o one or more *inductive clauses*, which tell us how to generate additional elements, and
>
> o a *final clause*, which tells us that all the elements are either basic or generated by the inductive clauses.

Inductive proofs

Having set up an inductive definition of the set of ambig-wffs, we are in a position to prove things about this set. For example, assuming the clauses of our inductive definition as premises, we can easily prove that $A_1 \vee A_2 \wedge \neg A_3$ is an ambig-wff.

> **Proof:** First, A_1, A_2, and A_3 are ambig-wffs by clause (1). $\neg A_3$ is thus an ambig-wff by clause (2). Then $A_2 \wedge \neg A_3$ is an ambig-wff by clause (3). Another use of clause (3) gives us the desired ambig-wff $A_1 \vee A_2 \wedge \neg A_3$. (Can you give a different derivation of this ambig-wff, one that applies $\wedge$ before $\vee$?)

This proof shows us how the inductive definition of the ambig-wffs is supposed to work, but it is *not* an inductive proof. So let's try to prove something about ambig-wffs using the method of inductive proof. Indeed, let's prove a few things that will help us identify strings that are *not* ambig-wffs.

Consider the string $\neg \vee \to$. Obviously, this is not an ambig-wff. But how do we know? Well, clause (4) says it has to be formed by repeated applications of clauses (1)–(3). Examining these clauses, it seems obvious that anything you get from them will have to contain at least one propositional letter. But what kind of proof is that? What method are we applying when we say "examining these clauses, it seems obvious that . . . "? What we need is a way to prove the following simple fact:

Proposition 1. *Every ambig-wff contains at least one propositional letter.*

Notice that this claim has the form of a general conditional, where the antecedent involves an inductively defined predicate:

$$\forall p\,[(p \text{ is an ambig-wff}) \to Q(p)]$$

proof by induction

basis step

inductive step

Here, Q is the property of containing at least one propositional letter. What the method of induction allows us to do is prove just such a claim. The way we do it is by showing two things. First, we show that all the basic ambig-wffs, those specified by clause (1), have the property Q. We call this the *basis step* of our inductive proof. Second, we show that if some "old" members of ambig-wff have the property Q, then so will the "new" members generated from them by the inductive clauses (2) and (3). We call this the *inductive step* of the proof. This is just like knocking down dominoes, albeit in reverse: the inductive step shows that if a domino falls down, so will the next one; the basis step tips over the initial domino. Here is the actual inductive proof:

an inductive proof

Proof: We will prove this proposition by induction on the ambig-wffs.

Basis: For our basis case, we need to show that all the propositional letters are strings that contain at least one propositional letter. But they do, since they in fact consist of exactly one such letter.

Induction: Suppose p and q are ambig-wffs that each contain at least one propositional letter. We want to show that the new ambig-wffs generated from these by clauses (2) and (3) will also contain at least one propositional letter. This is clearly true, since $\neg p$ contains all the propositional letters contained in p, and so contains at least one propositional letter; and $p \wedge q$, $p \vee q$, $p \to q$, and $p \leftrightarrow q$ contain all the propositional letters contained in p and q, and so contain at least one (indeed at least two) propositional letters.

By induction, we can thus conclude that all ambig-wffs contain at least one propositional letter.

As far as substance goes, this is a pretty trivial proof. But it is important to have a good grasp of the form of the proof, and particularly the form of the inductive step. The inductive step is always a subproof whose assumption is that the property in question, Q, holds of some arbitrarily selected members of the inductively defined set. In the above example, we assumed that the ambig-wffs p and q each had Q, that is, each contained at least one propositional letter. This assumption is called the *inductive hypothesis*. The goal of the step is to show that it follows from the inductive hypothesis that any new members generated from these—new ambig-wffs in our example—must have the property Q as well.

inductive hypothesis

What ultimately justifies the conclusion of an inductive proof is the last clause of the inductive definition. In our example, since nothing is an ambig-wff except the basic elements and things that can be generated from them by repeated applications of our two rules, we can be sure that all the ambig-wffs have the property in question.

Let's try another example. Suppose we want to prove that the string $A_1 \neg \rightarrow A_2$ is not an ambig-wff. Again, this is pretty obvious, but to prove it we need to prove a general fact about the ambig-wffs, one that will allow us to conclude that this particular string does not qualify. The following fact would suffice:

Proposition 2. *No ambig-wff has the symbol $\neg$ occurring immediately before one of the binary connectives: $\wedge, \vee, \rightarrow, \leftrightarrow$.*

Once again, note that the desired result has the form of a general conditional claim, where the antecedent is our inductively defined predicate:

$$\forall p \left[(p \text{ is an ambig-wff}) \rightarrow Q(p) \right]$$

This time, Q is the property of not having $\neg$ occurring immediately in front of a binary connective. To prove this, we need a basis step and an inductive step. The basis step must show that $Q(p)$ holds for those expressions p that are ambig-wffs in virtue of clause (1), that is, the propositional letters. The inductive step involves two cases, one corresponding to premise (2), the other to premise (3). For (2), we must show that if an ambig-wff p has property Q, so does $\neg p$. For (3), we need to prove that if p and q are ambig-wffs with property Q then so are $p \wedge q$, $p \vee q$, $p \rightarrow q$, and $p \leftrightarrow q$. If we can do this, the proof will be completed by induction, thanks to clause (4). After all, since every ambig-wff has to be obtained by repeated applications of (1), (2), and (3), every ambig-wff will have been shown to have the property in question.

But there is a problem when we try to carry out the details of this proof. Do you see what it is? Think about trying to do either part of the inductive step, either (2) or (3). For example, in case (2), how do we know that just because p has property Q so does $\neg p$? Well, we don't. For example, $\rightarrow A_1$ has property Q but $\neg \rightarrow A_1$ does not. (Find a similar problem with case (3).)

This is an example of the so-called *Inventor's Paradox*. It is not a real paradox, as in the case of Russell's Paradox, but it is a bit counterintuitive. It turns out that proofs by induction often get stuck, not because you are trying to prove something false, but because you are not aiming high enough. You need to prove more. In this case, what we have to prove to keep the induction from getting stuck is this stronger claim: no ambig-wff either begins with a binary connective, or ends with a negation sign, or has a negation sign immediately preceding a binary connective. So let Q' be this stronger property. It is clear that $\forall p \left[Q'(p) \rightarrow Q(p) \right]$. Thus, what we need to prove by induction is that

inventor's paradox

$$\forall p \left[(p \text{ is an ambig-wff}) \rightarrow Q'(p) \right]$$

another inductive definition

This turns out to be easy, and is left as an exercise.

Let's quickly look at another example of an inductively defined set and a proof by induction based on this definition. Suppose we defined the set *pal* as follows:

1. Each letter in the alphabet $(a, b, c, \ldots, z)$ is a pal.

2. If a string α is a pal, so is the result of putting any letter of the alphabet both in front of and in back of α (e.g., $a\alpha a$, $b\alpha b$, $c\alpha c$, etc.).

3. Nothing is a pal unless it is generated by repeated applications of (1) and (2).

Make sure you understand how this definition works. For example, come up with a string of seven letters that is a pal.

palindromes

Now let's prove that every pal reads the same way back to front and front to back, in other words, every pal is a palindrome. Here is our inductive proof of this fact:

Proof: We prove by induction that every pal reads the same forwards and backwards, that is, when the order of letters in the string is reversed.

Basis: The basic elements of pal are single letters of the alphabet. Clearly, any single letter reads the same way forwards or backwards.

Induction: Suppose that the pal α reads the same way forwards or backwards. (This is our inductive hypothesis.) Then we must show that if you add a letter, say l, to the beginning and end of α, then the result, $l\alpha l$, reads the same way forwards and backwards. When you reverse the string $l\alpha l$, you get $l\alpha' l$, where α' is the result of reversing the string α. But by the inductive hypothesis, $\alpha = \alpha'$, and so the result of reversing $l\alpha l$ is $l\alpha l$, i.e., it reads the same forwards and backwards.

We conclude by induction that every pal is a palindrome.

> **Remember**
>
> Given an inductive definition of a set, an inductive proof requires
>
> o a *basis step,* which shows that the property holds of the basic elements, and
>
> o an *inductive step,* which shows that *if* the property holds of some elements, then it holds of any elements generated from them by the inductive clauses.
>
> The assumption that begins the inductive step is called the *inductive hypothesis.*

Exercises

16.1   In the state of Euphoria, the following two principles hold:
1. If it is sunny on one day, it is sunny the next day.
2. It is sunny today.

Prove that it is going to be sunny from now on.

16.2   Raymond Smullyan, a famous logician/magician, gives the following good advice: (1) always speak the truth, and (2) each day, say "I will repeat this sentence tomorrow." Prove that anyone who did these two things would live forever. Then explain why it won't work.

16.3   Give at least two distinct derivations which show that the following is an ambig-wff: $A_1 \rightarrow A_2 \leftrightarrow \neg A_2$.

16.4   Prove by induction that no ambig-wff begins with a binary connective, ends with a negation sign, or has a negation sign immediately preceding a binary connective. Conclude that the string $A_1 \neg \rightarrow A_2$ is not an ambig-wff.

16.5   Prove that no ambig-wff ever has two binary connectives next to one another. Conclude that $A_1 \rightarrow \vee A_2$ is not an ambig-wff.

16.6   Modify the inductive definition of ambig-wff as follows, to define the set of semi-wffs:
1. Each propositional letter is a semi-wff.
2. If p is any semi-wff, so is the string $\neg p)$.
3. If p and q are semi-wffs, so are $(p \wedge q), (p \vee q), (p \rightarrow q), (p \leftrightarrow q)$.
4. Nothing is a semi-wff except in virtue of repeated applications of (1), (2), and (3).

Prove by induction that every semi-wff has the following property: the number of right parentheses is equal to the number of left parentheses plus the number of negation signs.

16.7 In the text, we proved that every pal is a palindrome, a string of letters that reads the same back to front and front to back. Is the converse true, that is, is every palindrome a pal? If so, prove it. If not, fix up the definition so that it becomes true.

16.8 (Existential wffs) In this problem we return to a topic raised in Problem 14.59. In that problem we defined an existential sentence as one whose prenex form contains only existential quantifiers. A more satisfactory definition can be given by means of the following inductive definition. The existential wffs are defined inductively by the following clauses:

1. Every atomic or negated atomic wff is existential.
2. If $P_1, \ldots, P_n$ are existential, so are $(P_1 \vee \ldots \vee P_n)$ and $(P_1 \wedge \ldots \wedge P_n)$.
3. If P is an existential wff, so is $\exists \nu P$, for any variable ν.
4. Nothing is an existential wff except in virtue of (1)–(3).

Prove the following facts by induction:

○ If P is an existential wff, then it is logically equivalent to a prenex wff with no universal quantifiers.

○ Suppose P is an existential sentence of the blocks language. Prove that if P is true in some world, then it will remain true if new objects are added to the world. [You will need to prove something a bit stronger to keep the induction going.]

Is our new definition equivalent to our old one? If not, how could it be modified to make it equivalent?

16.9 Give a definition of universal wff, just like that of existential wff in the previous problem, but with universal quantifiers instead of existential. State and prove results analogous to the results you proved there. Then show that every universal wff is logically equivalent to the negation of an existential wff.

16.10 Define the class of *wellfounded sets* by means of the following inductive definition:

1. If C is any set of objects, each of which is either not a set or is itself a wellfounded set, then C is a wellfounded set.

2. Nothing is a wellfounded set except as justified by (1).

This exercise explores the relationship between the wellfounded sets and the cumulative conception set discussed in the preceding chapter.

1. Which of the following sets are wellfounded?

$$\emptyset, \{\emptyset\}, \{\text{Washington Monument}\}, \{\{\{\ldots\}\}\}$$

2. Assume that a is wellfounded. Show that $\wp a$ is wellfounded.

3. Assume that a and b are wellfounded. Is the ordered pair $\langle a, b \rangle$ (as defined in the preceding chapter) wellfounded?

4. Assume that $a = \{1, b\}$ and $b = \{2, a\}$. Are a and b wellfounded?

5. Show that the Axiom of Regularity implies that every set is wellfounded.

6. When using set theory, one often wants to be able to prove statements of the form:

$$\forall x \, [\mathsf{Set}(x) \to \mathsf{Q}(x)]$$

One of the advantages of the cumulative conception of set discussed in the preceding chapter is that it allows one to prove such statements "by induction on sets." How?

7. Use mathematical induction to show that there is no infinite sequence of wellfounded sets $a_1, a_2, a_3, \ldots$ such that $a_{n+1} \in a_n$ for each natural number n.

Inductive definitions in set theory

The way we have been stating inductive definitions seems reasonably rigorous. Still, you might wonder about the status of clauses like

4. Nothing is an ambig-wff unless it can be generated by repeated applications of (1), (2), and (3).

This clause is quite different in character from the others, since it mentions not just the objects we are defining, but the other clauses of the definition itself. You might also wonder just what is getting packed into the phrase "repeated applications."

One way to see that there is something different about clause (4) is to note that the other clauses are obviously expressible using first-order formulas. For example, if concat is a symbol for the concatenation function (that is, the function that takes two expressions and places the first immediately to the left of the second), then one could express (2) as

$$\forall p \, [\mathsf{ambig\text{-}wff}(p) \to \mathsf{ambig\text{-}wff}(\mathsf{concat}(\neg, p))]$$

In contrast, clause (4) is not the sort of thing that can be expressed in FOL.

However, it turns out that if we work within set theory, then we can express inductive definitions with first-order sentences. Here, for example, is a definition of the set of ambig-wffs that uses sets. It turns out that this definition can be transcribed into the language of set theory in a straightforward way. The English version of the definition is as follows:

making the final clause more precise

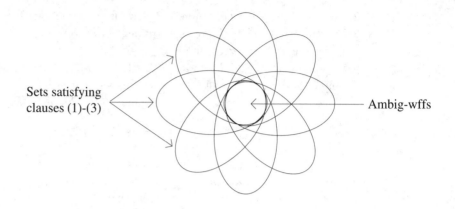

Figure 16.1: The set of ambig-wffs is the intersection of all sets satisfying (1)–(3).

Definition The set S of ambig-wffs is the smallest set satisfying the following clauses:

1. Each propositional letter is in S.

2. If p is in S, then so is $\neg p$.

3. If p and q are in S, then so are $p \wedge q$, $p \vee q$, $p \rightarrow q$, and $p \leftrightarrow q$.

"smallest" set

What we have done here is replace the puzzling clause (4) by one that refers to the smallest set satisfying (1)–(3). How does that help? First of all, what do we mean by *smallest?* We mean smallest in the sense of subset: we want a set that satisfies (1)–(3), but one that is a subset of any *other* set satisfying (1)–(3). How do we know that there is such a smallest set? We need to prove a lemma, to show that our definition makes sense.

Lemma 3. *If S is the intersection of a collection $\mathcal{X}$ of sets, each of which satisfies (1)–(3), S will also satisfy (1)–(3).*

We leave the proof of the lemma as Exercise 16.11.

As a result of this lemma, we know that if we define the set of ambig-wffs to be the intersection of *all* sets that satisfy (1)–(3), then we will have a set that satisfies (1)–(3). Further, it must be the smallest such set, since when you take the intersection of a bunch of sets, the result is always a subset of all of the original sets.

The situation is illustrated in Figure 16.1. There are lots of sets that satisfy clauses (1)–(3) of our definition, most of which contain many elements that are not ambig-wffs. For example, the set of all finite strings of propositional letters and connectives satisfies (1)–(3), but it contains strings like $A_1 \neg \to A_2$ that aren't ambig-wffs. Our set theoretic definition takes the set S of ambig-wffs to be the *smallest*, that is, the *intersection* of all these sets.

Notice that we can now explain exactly why proof by induction is a valid form of reasoning. When we give an inductive proof, say that all ambig-wffs have property Q, what we are really doing is showing that the set $\{x \mid Q(x)\}$ satisfies clauses (1)–(3). We show that the basic elements all have property Q and that if you apply the generation rules to things that have Q, you will get other things that have Q. But if Q satisfies clauses (1)–(3), and S is the intersection of all the sets that satisfy these clauses, then $S \subseteq Q$. Which is to say: all ambig-wffs have property Q.

justifying induction

Exercises

16.11   Prove Lemma 3.

✎

16.12   Give an inductive definition of the set of wffs of propositional logic, similar to the above definition, but putting in the parentheses in clause (3). That is, the set of wffs should be defined as the smallest set satisfying various clauses. Be sure to verify that there is such a smallest set.

✎

16.13   Based on your answer to Exercise 16.12, prove that every wff has the same number of left parentheses as binary connectives.

✎

SECTION 16.3

Induction on the natural numbers

Many students come away from the study of induction in math classes with the feeling that it has something special to do with the natural numbers. By now, it should be obvious that this method of proof is far more general than that. We can prove things about many different kinds of sets using induction. In fact, whenever a set is defined inductively, we can prove general claims about its members using an inductive proof. Still, the natural numbers are one of the simplest and most useful examples to which induction applies.

Just how are the natural numbers defined? Intuitively, the definition runs

defining natural numbers

as follows:

1. 0 is a natural number.

2. If n is a natural number, then $n + 1$ is a natural number.

3. Nothing is a natural number except in virtue of repeated applications of (1) and (2).

In set theory, this definition gets codified as follows. The set $\mathbf{N}$ of natural numbers is the smallest set satisfying:

1. $0 \in \mathbf{N}$

2. If $n \in \mathbf{N}$, then $n + 1 \in \mathbf{N}$

Based on this definition, we can prove statements about natural numbers by induction. Suppose we have some set Q of natural numbers and want to prove that the set contains all natural numbers:

$$\forall x\, [x \in \mathbf{N} \to x \in Q]$$

induction on $\mathbf{N}$

If we prove the following two things:

1. $0 \in Q$

2. If $n \in Q$, then $n + 1 \in Q$

then we know that $\mathbf{N} \subseteq Q$, since $\mathbf{N}$ is defined to be the smallest set satisfying these clauses. And this is just another way of stating the universal claim we want to prove.

Let's work through an example that illustrates induction on the natural numbers.

Proposition 4. *For every natural number n, the sum of the first n natural numbers is $n(n-1)/2$.*

Proof: We want to prove the statement $\forall n(n \in \mathbf{N} \to Q(n))$, where $Q(n)$ is the statement: the sum of the first n natural numbers is $n(n-1)/2$. We do this by induction.

Basis: The basis step requires that we prove that the sum of the first 0 natural numbers is 0, which it is. (If you don't like this case, you might check to see that $Q(1)$ holds. You might even go so far as to check $Q(2)$, although it's not necessary.)

Induction: To prove the inductive step, we assume that we have a natural number k for which $Q(k)$ holds, and show that $Q(k+1)$ holds.

That is, our inductive hypothesis is that the sum of the first k natural numbers is $k(k-1)/2$. We must show that the sum of the first $k+1$ natural numbers is $k(k+1)/2$. How do we conclude this? We simply note that the sum of the first $k+1$ natural numbers is k greater than the sum of the first k natural numbers (since the first natural number is zero, the second is one, and so on). We already know by the inductive hypothesis that this latter sum is simply $k(k-1)/2$. Thus the sum of the first $k+1$ numbers is

$$\frac{k(k-1)}{2} + k$$

Getting a common denominator gives us

$$\frac{k(k-1)}{2} + \frac{2k}{2}$$

which we factor to get

$$\frac{k(k+1)}{2}$$

the desired result.

Exercises

16.14 Prove by induction that for all natural numbers n, $n \leq 2n$.

16.15 Prove by induction that for all natural numbers n, $0 + 1 + \ldots + n \leq n^2$. Your proof should not presuppose Proposition 4, which we proved in the text, though it will closely follow the structure of that proof.

16.16 Prove by induction that for all n,

$$1 + 3 + 5 + \ldots + (2n + 1) = (n + 1)^2$$

16.17 Prove that for all natural numbers $n \geq 2$,

$$(1 - \frac{1}{2})(1 - \frac{1}{3}) \ldots (1 - \frac{1}{n}) = \frac{1}{n}$$

16.18 Notice that $1^3 + 2^3 + 3^3 = 36 = 6^2$ and that $1^3 + 2^3 + 3^3 + 4^3 + 5^3 = 225 = 15^2$. Prove that the
✎⋆   sum of the first n perfect cubes is a square. [Hint: This is an instance of the inventor's paradox. You will have to prove something stronger than this.]

16.19 (after Pólya) Examine the following incorrect proof that all logicians have the same shoe size.
✎   Identify and describe clearly why the proof is incorrect.

> *Basis*: Every set containing just one logician, contains logicians all of whom have the same shoe size.
>
> *Induction*: Our induction hypothesis is that any set of n logicians contains logicians with the same shoe size. Now look at any set of $n+1$ logicians. We must show that this set contains logicians with the same shoe size. Number the logicians: $1, 2, 3, \ldots, n, n+1$, and look at the sets $\{1, 2, 3, \ldots, n\}$ and $\{2, 3, 4, \ldots, n+1\}$. Each of these sets contain only n logicians, therefore within each set all of the logicians have the same shoe size. But the two sets overlap, so there must be only one shoe size among all $n+1$ logicians.

SECTION 16.4

Axiomatizing the natural numbers

In giving examples of informal proofs in this book, we have had numerous occasions to use the natural numbers as examples. In proving things about the natural numbers, we have made recourse to any fact about the natural numbers that was obviously true. If we wanted to formalize these proofs, we would have to be much more precise about what we took to be the "obvious" facts about the natural numbers.

Peano Arithmetic (PA)   Over the years, a consensus has arisen that the obviously true claims about the natural numbers can be formalized in what has come to be known as Peano Arithmetic, or PA for short, named after the Italian mathematician Giuseppe Peano. This is a certain first-order theory whose main axiom states a form of induction for natural numbers.

successor function   PA is formulated in a first-order language that has the constant 0 together with the unary successor function s and the identity predicate. It also contains the binary function symbols $+$ and $\times$, but we think of 0 and s as primitive, while the meanings of $+$ and $\times$ will be completely defined by axioms of PA. Intuitively, the successor function applied to any number n gives us the next greatest number (what we would normally write as $n + 1$). So $s(0)$ denotes 1, $s(s(0))$ denotes 2, and so forth.

Here are the first six axioms of Peano Arithmetic. We will get to the important induction axiom in due course.

1. $\forall x \, (s(x) \neq 0)$

2. $\forall x \, \forall y \, (s(x) = s(y) \rightarrow x = y)$

3. $\forall x \, (x + 0 = x)$

4. $\forall x \, \forall y \, [x + s(y) = s(x + y)]$

5. $\forall x \, (x \times 0 = 0)$

6. $\forall x \, \forall y \, [x \times s(y) = (x \times y) + x]$

The first two axioms tell us essential properties of the successor function. Together they ensure that the numbers are arranged in a sequence with no loops. The first axiom tells us that the sequence cannot loop back to zero, since then zero would be the successor of something (contradicting axiom one), and that it can't loop back to some other number, since then that number would be the successor of two different numbers (contradicting axiom two).

Axioms 3 and 4 define the properties of the binary addition operator in terms of the successor function. The definition of + reflects the structure of the natural numbers given by their inductive definition. In the first axiom we state the truth that the result of adding 0 to any number is that number. The second axiom says that the result of adding the successor of n to m is the successor of the result of adding n to m, for any n and m. Together these axioms tell us how to add any pair of numbers, since the first concerns adding 0, and the second the successor of a number, and all numbers are one or the other.

Axioms 5 and 6 follow a similar pattern, and provide a definition for multiplication. The first tells the result of multiplying by 0, and the second the result of multiplying by the successor of a number. Again the second axiom relates the result of multiplying by the successor of a number to the result of multiplying by that number.

We should view the claim that these axioms can serve as *definitions* of addition and multiplication with a little suspicion. The second addition axiom contains + on both sides of the equality making it look like an inductive definition. But we are not giving an inductive definition of a set, but rather defining the function +. The reason the definition works is that the function applies to the natural numbers, a set that is itself inductively defined using the successor function. So while axiom 4 does not allow us to eliminate the + function in one step, we can argue that + can be eliminated after some number of applications of this axiom, followed by an application of axiom 3.

To see how this works, let's see how we could work out the value of $s(s(s(0))) + s(s(0))$ by eliminating the + function. This is our official way of writing what would normally be written as $3 + 2$. Well, this sum is formed by adding $s(s(0))$ to $s(s(s(0)))$. The second clause of the definition can be used since $s(s(0))$ is the successor of something. It tells us that the result is $s(\ s(s(s(0))) + s(0)\ )$.

This expression still contains an occurrence of the addition symbol, so we need to use the definition of $+$ again to simplify further. We can use axiom 4 again, because we are again adding the successor of something, to obtain $s(s(\ s(s(s(0))) + 0\ ))$.

Once again, this expression contains $+$ and so we need to simplify further, but this time the second argument of the $+$ expression is 0. This means that we can use axiom 3 to make the final simplification of the expression.

$$
\begin{aligned}
s(s(s(0))) + s(s(0)) &= s(\ s(s(s(0))) + s(0)\ ) &\text{by axiom 4.} \\
&= s(s(\ s(s(s(0))) + 0\ )) &\text{by axiom 4.} \\
&= s(s(\ s(s(s(0)))\ )) &\text{by axiom 3.}
\end{aligned}
$$

So the final result is $s(s(s(s(s(0)))))$ (better known as 5) as expected.

It should be fairly clear that axioms 3 and 4 allow us to replace any term that contains only $+$, s, and 0 with an expression that contains only s and 0. Once you have convinced yourself of that, see if you understand how axioms 5 and 6 allow us to replace any term containing only $\times$, $+$, s, and 0 with a term that contains only $+$, s, and 0. And we can in turn eliminate the $+$ from *this* term. This is why we say that s and 0 are primitive expressions of the language, while $+$ and $\times$ are defined. Note, however, that this does not mean we could get rid of $+$ and $\times$, and have an equally expressive language. After all, most of the important claims of the language will contain quantified variables, such as $\forall x\, \forall y\ (x + y = y + x)$, and the function symbol $+$ cannot be eliminated from these claims.

induction scheme

In addition to the axioms above, PA has an axiom scheme capturing the principle of mathematical induction on the natural numbers. This can be stated as follows:

$$[Q(0) \wedge \forall x\, (Q(x) \to Q(s(x)))] \to \forall x\, Q(x)$$

What this says is that if $Q(x)$ is satisfied by 0, and if its being satisfied by some number n ensures that it is satisfied by $s(n)$, then $Q(x)$ is satisfied by all natural numbers. (Actually, as with the Axiom of Comprehension in the preceding chapter, the axiom needs to be stated a bit more generally than we have formulated it. The wff $Q(x)$ may have free variables other than x, and these need to be universally quantified.)

There are many other facts about the natural numbers which are obvious. Some of these include the familiar commutative, associative, and distributive laws of addition and multiplication. It turns out, however, that these facts, and all other "obvious" facts, and many not-so-obvious facts, can be proven from the axioms as set out above. Here is one of the simplest; we give an informal proof, using just these axioms, of

$$\forall x \, (s(x) = s(0) + x)$$

Proof: The proof is by the formalized version of mathematical induction. The predicate $Q(x)$ in question is $(s(x) = s(0) + x)$. We need to prove the basis case, $Q(0)$, and the induction step

$$\forall x \, (Q(x) \to Q(s(x)))$$

The basis case requires us to prove that $s(0) = s(0) + 0$, which is obviously true by axiom 3.

We prove the induction step by general conditional proof. Let n be any number such that $Q(n)$. This is our inductive hypothesis. Using it, we need to prove $Q(s(n))$. That is, our induction hypothesis is that $s(n) = s(0) + n$ and our goal is to prove that $s(s(n)) = s(0) + s(n)$. This takes the following steps:

$$
\begin{aligned}
s(\, s(n) \,) &= s(\, s(0) + n \,) & \text{by induction hypothesis.} \\
&= s(0) + s(n) & \text{by axiom 4.}
\end{aligned}
$$

Let's see how we might prove a more complicated result, like the commutativity of $+$:

$$\forall x \, \forall y \, (x + y = y + x)$$

We will not prove this, since it is a valuable exercise that you should complete. But we will sketch the proof, since proving it requires a technique we haven't encountered before, known as *double induction*. It turns out that to prove this claim, we need to use induction on both x and y.

Overall, the proof of this claim will be an induction on x, with the goal of proving the universal claim $\forall x \, Q(x)$, where $Q(x)$ is the formula $\forall y \, (x + y = y + x)$. So the base case of the proof must establish $Q(0)$, that is:

$$\forall y \, (0 + y = y + 0)$$

and the inductive case will assume $Q(n)$:

$$\forall y \, (n + y = y + n)$$

with the goal of showing $Q(s(n))$:

$$\forall y \, (s(n) + y = y + s(n))$$

Now, what's fun about a double inductive proof is that you also need to use induction to prove the claims within the cases. For example, your base case above will be demonstrated by induction on y, with the new predicate $Q'(y)$:

$$0 + y = y + 0$$

This induction on y will have a base case establishing (the trivial) $Q'(0)$:

$$0 + 0 = 0 + 0$$

Its inductive case will assume $Q'(m)$:

$$0 + m = m + 0$$

and try to show $Q'(s(m))$:

$$0 + s(m) = s(m) + 0$$

This all turns out to be easy. But wait, there's more! That was just the base case of our induction on x. So now we assume:

$$\forall y\ (n + y = y + n)$$

and try to prove:

$$\forall y\ (s(n) + y = y + s(n))$$

And this claim we also have to prove by induction on y. That is, we have to prove the base case:

$$s(n) + 0 = 0 + s(n)$$

followed by an inductive case starting from the inductive hypothesis:

$$s(n) + m = m + s(n)$$

and concluding with:

$$s(n) + s(m) = s(m) + s(n)$$

At which point, we're done.

It turns out that none of this is terribly hard. But keeping straight exactly where you are and what you're trying to prove can be terribly confusing. We'll let you try your hand at it in Exercise 16.27, where you'll need to fill out the details of this proof.

Gödel Incompleteness Theorem

Peano Arithmetic is remarkably powerful. Indeed, almost all of the theorems mathematicians have proven about the natural numbers can be proven in PA. There are, however, truths about the natural numbers that cannot be proven from PA. Not only that, but any attempt to set down a list of first-order axioms true of the natural numbers must, in some sense, fail. We will discuss this result, known as Gödel's Incompleteness Theorem, in Chapter 19.

Exercises

Give informal proofs, similar in style to the one in the text, that the following statements are consequences of PA. *Explicitly identify any predicates to which you apply induction. When proving the later theorems, you may assume the results of the earlier problems.*

16.20 $\forall x\,(0 + x = x)$

16.21 $\forall x\,(s(0) \times x = x)$

16.22 $\forall x\,(0 \times x = 0)$

16.23 $\forall x\,(x \times s(0) = s(0) \times x)$

16.24 $\forall x\,\forall y\,(x + s(y) = s(x) + y)$ [Hint: Don't get confused by the two universal quantifiers. Start by assuming that x is an arbitrary number and perform induction on y. This does not require double induction.]

16.25 $\forall x\,\forall y\,\forall z\,(x + z = y + z \to x = y)$ [Hint: this is an easy induction on z.]

16.26 $\forall x\,\forall y\,\forall z\,((x + y) + z = x + (y + z))$ [Hint: This is relatively easy, but you have to perform induction on z. That is, your basis case is to show that $(x + y) + 0 = x + (y + 0)$. You should then assume $(x + y) + n = x + (y + n)$ as your inductive hypothesis and show $(x + y) + s(n) = x + (y + s(n))$.]

16.27 $\forall x\,\forall y\,(x + y = y + x)$ [Hint: this requires double induction.]

16.28 $\forall x\,\forall y\,(x \times y = y \times x)$ [Hint: To prove this, you will first need to prove the lemma $\forall x\,\forall y\,(s(x) \times y = (x \times y) + y)$. Prove this by induction on y.]

SECTION 16.5

Induction in Fitch

Fitch contains an inference rule for induction on the natural numbers which exactly parallels the informal technique that we have just described. In order to complete a formal proof by induction of a universally quantified formula, you must be able to cite the statement of the base case of the induction, and a subproof which represents the step case:

Peano Induction:

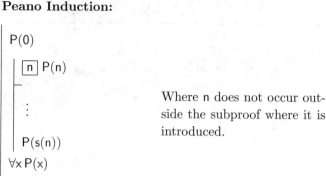

Where n does not occur outside the subproof where it is
introduced.

Like the rules of universal introduction and existential elimination, the Peano Induction rule requires that we use a boxed constant in the subproof where we introduce an arbitrary n for the step case of the induction. This ensures that n is indeed arbitrary, and that all we know about it is that the property P holds of that element of the domain. Our goal is to then prove that the property is inherited by $s(n)$. Once we have that proof, and the proof that 0 has the property P, then we can infer that every natural number has the property.

When we use the Peano induction rule, we must interpret the domain of quantification as being just the set of natural numbers.

Default and generous uses of the Induction rule

Like many of the rules of Fitch, the induction rule has a default use. If no conclusion is given, then Fitch tries to determine the appropriate conclusion on the basis of the citations. Specifically it will try to use the base case formula to decide what you are trying to use the rule to prove.

If you use the **Add Support Steps** command with a step containing a universally quantified formula using this rule, then the necessary base case and step proofs will be inserted into the proof.

Exercises

Use Fitch to construct formal proofs of the following theorems from the six Peano Axioms plus the Peano Induction rule. In the corresponding Exercise file, you will find as premises only the specific axioms needed to prove the goal theorem. In the starred exercises, you may want to use one or more earlier proofs as lemmas. If you have not read Section 10.6 describing the Lemma rule in Fitch, you might want to do so before attempting the starred exercises. (Lemmas are, of course, never necessary — they simply make life a whole lot easier.)

16.29 $\forall x\,(0 + x = x)$
↗

16.30 $\forall x\,(s(0) \times x = x)$
↗

16.31 $\forall x\,(0 \times x = 0)$
↗

16.32 $\forall x\,(x \times s(0) = s(0) \times x)$
↗⋆

16.33 $\forall x\,\forall y\,(x + s(y) = s(x) + y)$ [Hint: This does not require double induction. Your final step will
↗ be universal generalization on x, but within the subproof leading up to that step, you will need
to perform induction on y.]

16.34 $\forall x\,\forall y\,\forall z\,((x + y) + z = x + (y + z))$ [Hint: Prove this by induction on z.]
↗

16.35 $\forall x\,\forall y\,\forall z\,(x + z = y + z \rightarrow x = y)$
↗

16.36 $\forall x\,\forall y\,(x + y = y + x)$ [Hint: this requires double induction.]
↗⋆⋆

16.37 $\forall x\,\forall y\,(s(x) \times y = (x \times y) + y)$ [Hint: This is the key lemma that you will need to prove Exer-
↗⋆⋆ cise 16.38. In order to prove it, you will need the Associativity and Commutativity of Addition
(Exercises 16.34 and 16.36, respectively).]

16.38 $\forall x\,\forall y\,(x \times y = y \times x)$ [Hint: This is pretty easy, once you have Exercises 16.31 and 16.37 to use
↗ as lemmas.]

SECTION 16.6

Ordering the Natural Numbers

We usually think of the natural numbers distributed along a number line with
0 on the left and the numbers increasing toward the right. A number further
to the left is "less than" any number to its right. We can express this idea
formally using the following axiom, which defines the relation $<$.

$$\forall x\,\forall y\,(x < y \leftrightarrow \exists z\,(x + s(z) = y))$$

This says that x is less than y whenever we can obtain y by adding some
non-zero number to x. We will see that the relation $<$ plays an important role
in a second form of induction that we will describe in the next section. First
though, lets investigate the properties of this relation.

Any relation, R, which has the following three properties is called a *total
strict ordering* (TSO):

1. Irreflexive: $\forall x \, \neg x R x$

2. Transitivity: $\forall x \, \forall y \, \forall z \, ((x R y \wedge y R z) \rightarrow x R z)$

3. Trichotomy: $\forall x \, \forall y \, (x R y \vee x = y \vee y R x)$

In addition to $<$ as we have just defined it, other examples of total strict orderings include alphabetical ordering among letters, or among words. Many familiar orderings are not total strict orderings though. For example, the relation *taller than* is not a TSO, since trichotomy fails. Consider for example, two different people who are the same height. Such a relation is called a *partial ordering*.

Anyway, we would like to show that $<$ is a TSO.

Proposition 5. $<$ is irreflexive.

$$\forall x \, \neg x < x$$

Proof: Suppose that there is some number a with $a < a$. Then by the definition of $<$ there is a number k such that $a + s(k) = a$. Axiom 3 of PA tells us that $a + 0 = a$ and so by substitution $a + s(k) = a + 0$. This tells us that $s(k) = 0$ (see Exercise 16.25) contradicting axiom 1 of PA. So there can be no such number a, showing that $<$ is irreflexive.

Proposition 6. $<$ is transitive.

$$\forall x \, \forall y \, \forall z \, ((x < y \wedge y < z) \rightarrow x < z)$$

Proof: Let a, b and c be arbitrary numbers with $a < b$ and $b < c$. This tells us that there are numbers j and k such that $a + s(k) = b$ and $b + s(j) = c$. We can substitute the value of b into the second of these equations to obtain $(a + s(k)) + s(j) = c$. Assuming that $+$ is associative (see exercise 16.26), we have $a + (s(k) + s(j)) = c$, which by axiom 4 gives us that $a + s(s(k) + j) = c$.

Any relation that is transitive and irreflexive is also antisymmetric, that is $\forall x \, \forall y \, (x R y \rightarrow \neg y R x)$. This is easy to show, since if we assume that $a R b$ and $b R a$, then transitivity yields $a R a$ which contradicts irreflexivity. So it follows that if $a < b$ then $\neg b < a$.

The final property required of TSOs is that trichotomy holds. This says that every pair of different objects are related by the ordering relation, one must be less than the other. It is this property that accounts for the appearance of the word "total" in *total strict order*.

Proposition 7. Trichotomy

$$\forall x \, \forall y \, (x < y \lor x = y \lor y < x)$$

Proof: Let a be arbitrary. We will show that either

$$\forall y \, (a < y \lor a = y \lor y < a)$$

by induction on y.

Basis: We must show $(a < 0 \lor a = 0 \lor 0 < a)$. We know that every number, including a, is either 0, or the successor of some number. 0 is less than any successor, so one of the second two disjuncts must hold.

Induction: We assume that $(a < k \lor a = k \lor k < a)$ for some k and show that $(a < s(k) \lor a = s(k) \lor s(k) < a)$. Let's split into cases according to the disjuncts of the induction hypothesis.

$a < k$ In this case $a < s(k)$, and we are done.

$a = k$ In this case too, $a < s(k)$.

$k < a$ This tells us that $\exists y \, k + s(y) = a$. y must be either 0 or the successor of some number. If $y = 0$, then $k + s(0) = a$ or equivalently $s(k) = a$. If on the other hand $y = s(b)$ for some b, then $k + s(s(b)) = a$, which means that $s(k) + s(b) = a$, and so $s(k) < a$ (see Exercise 16.24).

Exercises

Use Fitch to construct formal proofs of the following theorems from the Peano Axioms plus the definition of $<$. In the corresponding Exercise file, you will find as premises only the specific axioms needed to prove the goal theorem.

16.39 $\forall x \, \neg x < 0$

16.40 $\forall x \, x < s(x)$

16.41 $\forall x \, \exists y \, x < y$ [Hint: Notice that the Exercise file contains the definition of $<$ but none of the Peano Axioms! So although this follows from Exercise 16.40, you won't be able to use your proof of that that as a lemma. The proof is actually pretty simple, but requires some thought.]

16.42 $\forall x \, \forall y \, (x < y \to x < s(y))$

For the following exercises you will want to use your proofs from Section 16.5 as lemmas.

16.43 $\forall x \, (x = 0 \lor 0 < x)$

16.44 $\forall x \, \forall y \, (x < y \to s(x) < s(y))$

16.45 $\forall x \, \neg x < x$

16.46 $\forall x \, \forall y \, \forall z \, ((x < y \land y < z) \to x < z)$

16.47 $\forall x \, \forall y \, (x < y \lor x = y \lor y < x)$ [Hint: You'll find some earlier exercises from this section to be handy lemmas for this proof.]

16.48 Define $\leq$ so that $\forall x \, \forall y \, (x \leq y \leftrightarrow \exists z \, x + z = y)$. Give informal proofs that $\leq$ is reflexive and transitive.

16.49 Define $\geq$ so that $\forall x \, \forall y \, (x \geq y \leftrightarrow \neg y < x)$. Give informal proofs that $\geq$ is reflexive and transitive.

SECTION 16.7

Strong Induction

Sometimes the induction principle that we have described does not fit well with the property of natural numbers that we are trying to prove. As an example of this imagine proving that every natural number greater than one is either prime, or can be expressed as the product of primes. Mathematicians call this fact *The Fundamental Theorem of Arithmetic.* Anything that merits such a grand name deserves an investigation. Let's see what happens when we try to prove this using ordinary induction on the natural numbers. We will start with a base case of $n = 2$, since the claim does not apply to 0 or 1.

Proposition 8. For any number n greater than 1, n is either prime or can be expressed as a product of primes, i.e., $n = p_1 \times \ldots \times p_m$, where each of the p's are prime.

Proof (Attempt): We will try to prove this by induction.

Basis (n=2): Since 2 is a prime number, it is either prime or the product of primes, and so the base case is proved.

Inductive: Assume that the claim is true of n. We need to show that the claim is true of $n+1$. There are two cases, either $n+1$ is prime, in which case we are done, or it is not prime, which means that there are two smaller numbers j and k such that $n+1 = j \times k$.

At this point, we're stuck. If we knew that the claim applied to both j and k, then we could complete the inductive step, since we'd take the primes $p_1^j \times \ldots \times p_m^j = j$ and the primes $p_1^k \times \ldots \times p_m^k = k$ and multiply them all together to get $n+1 = p_1^j \times \ldots \times p_m^j \times p_1^k \times \ldots \times p_m^k$, showing that $n+1$ can be expressed as the product of primes. Unfortunately, we don't know that the claim is true of j and k, we only know that it is true of n, the immediate predecessor of $n+1$.

It may have dawned on you that our problem is just another example of the inventor's paradox, discussed on page 459. For the induction to work, we need to prove a somewhat stronger claim than the one we tried using. Suppose we had tried to prove that for any n, n and all numbers less than n are either prime or the product of primes. If we had used this stronger claim, then at the inductive step, we could have assumed not just that the claim applied to n, but also that it applied to the smaller numbers j and k. We could then have finished the proof in the way we indicated.

This is a sufficiently common problem that a separate induction principle has been introduced to handle it. The principle is known as *strong induction* (a.k.a. *complete* or *course of values* induction). The basic idea is that, if you can show that whenever a property holds of the numbers smaller than n, it also holds of n, then you can conclude that it holds of all the natural numbers. The principle can be stated like this:

strong induction

(SI) $\forall n \, [\forall k \, (k < n \rightarrow Q(k)) \rightarrow Q(n)] \rightarrow \forall x \, Q(x)$

How would an informal proof using this principle look? The goal of the proof would be to show that the antecedent of this conditional is true, which would then allow us to conclude $\forall x \, Q(x)$ by $\rightarrow$ **Elim**. So we need to show:

(1) $\forall n \, [\forall k \, (k < n \rightarrow Q(k)) \rightarrow Q(n)]$

This is a universal claim that we prove using general conditional proof. That is, we take an arbitrary number n and assume:

(2) $\forall k \, (k < n \rightarrow Q(k))$

In other words, we assume that every number less than n has the property in question. This is our inductive hypothesis. The goal of the proof is then to show that n has the property, that is, $Q(n)$.

Here, we have to consider two cases. If $n = 0$, then our inductive hypothesis (2) tells us nothing at all, since it holds trivially for any Q. So we need to show $Q(0)$ without the benefit of any other information. But if $n \neq 0$, then our goal is to show $Q(n)$ based on the substantive knowledge that all the predecessors of n have the property Q. So just like ordinary induction, we in effect have a base case, $Q(0)$, and an inductive case, $Q(n)$ for $n > 0$. Only in the latter case will the inductive hypothesis (2) give us any help.

Proofs by strong induction thus take the following form:

Basis (n = 0): Show $Q(0)$.

Inductive (n > 0): Assume $\forall k \ (k < n \rightarrow Q(k))$. Show $Q(n)$.

Conclude: $\forall x \ Q(x)$

Now how can we justify the principle of strong induction? Ordinary induction is justified by virtue of how the set of natural numbers is inductively defined, as we saw in Section 16.3. But strong induction does not follow the clauses of the inductive definition, so it is legitimate to ask how we know it is a valid principle. It turns out that we can justify the principle by proving that it follows from ordinary induction. That is, (SI) can be proven by induction.

The key trick is to remember that strong induction is really just a generalized solution to the inventor's paradox, and so in order to prove it, we will use ordinary induction on a stronger property.

Proposition 9.

$$\forall n \ [\forall k \ (k < n \rightarrow Q(k)) \rightarrow Q(n)] \rightarrow \forall x \ Q(x)$$

Proof: We will prove this by assuming:

(a) $\forall n \ [\forall k \ (k < n \rightarrow Q(k)) \rightarrow Q(n)]$

and showing:

(b) $\forall x \ Q(x)$

But instead of proving (b) directly, we will first prove the stronger:

(c) $\forall x \ \forall y \ ((y = x \lor y < x) \rightarrow Q(y))$

This says for every x, x and all of its predecessors has the property Q. Clearly, (c) $\Rightarrow$ (b), so if we can prove (a) $\Rightarrow$ (c), we will have (a) $\Rightarrow$ (b), as desired.

We prove (c) from (a) by ordinary induction on x.

Basis: We need to show that $\forall y\,((y = 0 \lor y < 0) \to Q(y))$. Let's begin by instantiating our assumption (a) to 0:

$$\forall k\,(k < 0 \to Q(k)) \to Q(0)$$

Notice that the antecedent of this holds trivially, since there are no numbers less than 0. Hence we have $Q(0)$. But then we can also conclude $\forall y\,(y = 0 \to Q(y))$, and so of course $\forall y\,((y = 0 \lor y < 0) \to Q(y))$, since there are no numbers less than 0.

Inductive: We assume as our inductive hypothesis:

$$\forall y\,((y = n \lor y < n) \to Q(y))$$

and our goal is to prove:

$$\forall y\,((y = s(n) \lor y < s(n)) \to Q(y))$$

To prove this we need to consider two cases, $y = s(n)$ and $y < s(n)$. Let's do the second one first.

If $y < s(n)$, then clearly $y = n$ or $y < n$. But then by the inductive hypothesis, we have $Q(y)$, as desired.

Suppose, on the other hand, that $y = s(n)$. Let's instantiate our original assumption (a) with $s(n)$, giving us:

$$\forall k\,(k < s(n) \to Q(k)) \to Q(s(n))$$

Since the only numbers less than $s(n)$ are either equal to n or less than n, we can rewrite this as:

$$\forall k\,((k = n \lor k < n) \to Q(k)) \to Q(s(n))$$

From this and the inductive hypothesis, we can conclude $Q(s(n))$. So whether $y = s(n)$ or $y < s(n)$, we have $Q(y)$. Formally:

$$\forall y\,((y = s(n) \lor y < s(n)) \to Q(y))$$

This concludes our inductive proof of (c). So as desired, (a) $\Rightarrow$ (c). And since (c) $\Rightarrow$ (b), we have our original goal, (a) $\Rightarrow$ (b).

You might think that strong induction is poorly named, since it follows from ordinary "weak" induction. But the point of the name is not that the principle is stronger than ordinary induction. In fact, anything you can prove by one you can also prove by the other. The difference is simply that strong induction allows you to use a stronger inductive hypothesis. You get to assume that all the numbers smaller than n have the property, not just its immediate predecessor.

You should probably be able to guess the form of Fitch's strong induction rule:

Strong Induction:

$\boxed{n}\ \forall x\ (x < n \rightarrow P(x))$

$\vdots$

$P(n)$

$\triangleright$ $\forall x\, P(x)$

Where n does not occur outside the subproof where it is introduced.

Exercises

The following exercises work together to result in a formal proof of Proposition 9. You'll need to do them all to get to the final proof, but it will be worth the work.

16.50 We begin by proving a simple lemma, namely that every number is either 0 or the successor of some other number. If you completed exercise 16.43, then you already proved something similar.

$$\forall x\ (x = 0 \lor \exists y\ x = s(y))$$

16.51 You may have thought that we would use the lemma in the previous exercise in the proof of Proposition 9, but in fact we are going to use it to prove a second lemma. In the informal proof of Proposition 9 we appealed twice to the following fact:

$$\forall x\ \forall y\ (x < s(y) \rightarrow (x = y \lor x < y))$$

Prove this fact, using lemma 16.50 if necessary.

16.52 Open the file Lemma 16.52. This file contains the skeleton of the proof of the inductive step in Proposition 9. Complete the proof by selecting the correct rules, and citations for all of the steps. Submit your file as Proof 16.52.

16.53 Finally open the file Exercise 16.53. This asks you to prove Proposition 9. Formalize the informal proof of this result that we gave in the previous section, using the lemma from Exercise 16.52 where necessary. You may use **Fo Con**, but if you like a challenge you can try to complete the proof without using it.

CHAPTER 17

Advanced Topics in Propositional Logic

This chapter contains some more advanced ideas and results from propositional logic, logic without quantifiers. The most important part of the chapter is the proof of the Completeness Theorem for the propositional proof system $\mathcal{F}_T$ that you learned in Part I. This result was discussed in Section 8.3 and will be used in the final chapter when we prove the Completeness Theorem for the full system $\mathcal{F}$. The final two sections of this chapter treat topics in propositional logic of considerable importance in computer science.

SECTION 17.1
Truth assignments and truth tables

modeling truth tables

In Part I, we kept our discussion of truth tables pretty informal. For example, we did not give a precise definition of truth tables. For some purposes this informality suffices, but if we are going to prove any theorems about FOL, such as the Completeness Theorem for the system $\mathcal{F}_T$, this notion needs to be modeled in a mathematically precise way. As promised, we use set theory to do this modeling.

truth assignments

We can abstract away from the particulars of truth tables and capture what is essential to the notion as follows. Let us define a *truth assignment* for a first-order language to be any function h from the set of all atomic sentences of that language into the set {TRUE, FALSE} provided that h assigns FALSE to the formula $\bot$. That is, for each atomic sentence A of the language, h gives us a truth value, written $h(A)$, either TRUE or FALSE. Intuitively, we can think of each such function h as representing one row of the reference columns of a large truth table.

modeling semantics

Given a truth assignment h, we can define what it means for h to make an arbitrary sentence of the language true or false. There are many equivalent ways to do this. One natural way is to extend h to a function $\hat{h}$ defined on the set of all sentences and taking values in the set {TRUE, FALSE}. Thus if we think of h as giving us a row of the reference column, then $\hat{h}$ fills in the values of the truth tables for all sentences of the language, that is, the values corresponding to h's row. The definition of $\hat{h}$ is what you would expect, given the truth tables:

1. $\hat{h}(\bot) = $ FALSE

2. $\hat{h}(\mathsf{Q}) = h(\mathsf{Q})$ for atomic sentences Q.

3. $\hat{h}(\neg\mathsf{Q}) = $ TRUE if and only if $\hat{h}(\mathsf{Q}) = $ FALSE;

4. $\hat{h}(\mathsf{Q} \wedge \mathsf{R}) = $ TRUE if and only if $\hat{h}(\mathsf{Q}) = $ TRUE and $\hat{h}(\mathsf{R}) = $ TRUE;

5. $\hat{h}(\mathsf{Q} \vee \mathsf{R}) = $ TRUE if and only if $\hat{h}(\mathsf{Q}) = $ TRUE or $\hat{h}(\mathsf{R}) = $ TRUE, or both.

6. $\hat{h}(\mathsf{Q} \rightarrow \mathsf{R}) = $ TRUE if and only if $\hat{h}(\mathsf{Q}) = $ FALSE or $\hat{h}(\mathsf{R}) = $ TRUE, or both.

7. $\hat{h}(\mathsf{Q} \leftrightarrow \mathsf{R}) = $ TRUE if and only if $\hat{h}(\mathsf{Q}) = \hat{h}(\mathsf{R})$.

A truth assignment h assigns values to every atomic sentence in the language. But intuitively, to compute the truth table for a sentence S, we need only fill out the reference rows for the atomic sentences that actually appear in S. In Exercise 17.3, we ask you to prove that the only values of h that matter to $\hat{h}(\mathsf{S})$ are those assigned to the atomic constituents of S.

With this precise model of a truth assignment, we can give a mathematically precise version of our definitions of a tautology and a tt-satisfiable sentence. Namely, we say that S is a *tautology* if every truth assignment h has S coming out true, that is, $\hat{h}(\mathsf{S}) = $ TRUE. More generally, we say that a sentence S is a *tautological consequence* of a set $\mathcal{T}$ of sentences provided every truth assignment that makes all the sentences in $\mathcal{T}$ true also makes S true. Similarly, we say that a sentence S is *tt-satisfiable* provided there is a truth assignment h such that $\hat{h}(\mathsf{S}) = $ TRUE. Similarly, a set $\mathcal{T}$ of sentences is *tt-satisfiable* if there is a single assignment h that makes each of the sentences in $\mathcal{T}$ true.

modeling tautology and consequence

tt-satisfiable

Proposition 1. The sentence S is a tautological consequence of the set $\mathcal{T}$ if and only if the set $\mathcal{T} \cup \{\neg\mathsf{S}\}$ is not tt-satisfiable.

The proof of this result is left as an exercise.

Note that if $\mathcal{T}$ is finite, we can reduce the question of whether S is a tautological consequence of $\mathcal{T}$ to the question of whether a single sentence is not tt-satisfiable, namely the conjunction of the sentences in $\mathcal{T}$ and $\neg\mathsf{S}$.

Remember

A truth assignment is simply a function from the atomic sentences into the set $\{$TRUE, FALSE$\}$. It models a single row of a complete truth table for the language.

Exercises

17.1 Recall the Sheffer stroke symbol from Exercise 7.29, page 197, and the three place symbol ♣ discussed on page 194. Suppose we had included these as basic symbols of our language. Write out the clauses for $\hat{h}(Q \mid R)$ and $\hat{h}(\clubsuit(P, Q, R))$ that would be needed to complete the definition given above.

17.2 Give an informal proof of Proposition 1 on page 485.

17.3 Let h_1 and h_2 be truth assignments that agree on (assign the same value to) all the atomic sentences in S. Show that $\hat{h}_1(\mathsf{S}) = \hat{h}_2(\mathsf{S})$. [Hint: use induction on wffs.]

SECTION 17.2
Completeness for propositional logic

We are now in a position to prove the Completeness Theorem for propositional logic first stated on page 221. Recall that we used the notation $\mathcal{F}_{\mathrm{T}}$ to stand for that part of $\mathcal{F}$ that uses only the introduction and elimination rules for $\wedge, \vee, \neg, \rightarrow, \leftrightarrow$ and $\bot$. Given a set $\mathcal{T}$ of sentences and another sentence S, we write $\mathcal{T} \vdash_{\mathrm{T}} \mathsf{S}$ to mean that there is a formal proof of S in the system $\mathcal{F}_{\mathrm{T}}$ with premises drawn from $\mathcal{T}$. It is not assumed that every sentence in $\mathcal{T}$ is actually used in the proof. For example, it might be that the set $\mathcal{T}$ is an infinite set of sentences while only a finite number can be used in any one proof, of course. Notice that if $\mathcal{T} \vdash_{\mathrm{T}} \mathsf{S}$ and $\mathcal{T}$ is a subset of some other set $\mathcal{T}'$ of sentences, then $\mathcal{T}' \vdash_{\mathrm{T}} \mathsf{S}$. We restate the desired result as follows:

Completeness of $\mathcal{F}_{\mathrm{T}}$

Theorem (Completeness of $\mathcal{F}_{\mathrm{T}}$) If a sentence S is a tautological consequence of a set $\mathcal{T}$ of sentences then $\mathcal{T} \vdash_{\mathrm{T}} \mathsf{S}$.

You might think that the way to prove the Completeness Theorem would be to assume that S is a tautological consequence of $\mathcal{T}$ and then try to construct a proof of S from $\mathcal{T}$. But since we don't know anything about the meaning of S or of the sentences in $\mathcal{T}$, this strategy would get us nowhere. In fact, the way we will prove the theorem is by proving its contrapositive: that if $\mathcal{T} \nvdash_{\mathrm{T}} \mathsf{S}$ (that is, if there is no proof of S from $\mathcal{T}$), then S is not a tautological consequence of $\mathcal{T}$. That is to say, we will show that if $\mathcal{T} \nvdash_{\mathrm{T}} \mathsf{S}$, then there is a truth assignment h that makes all of the sentences in $\mathcal{T}$ true, but S false. In other words, we will show that $\mathcal{T} \cup \{\neg \mathsf{S}\}$ is tt-satisfiable. The following lemma will be helpful in carrying out this proof.

Lemma 2. $\mathcal{T} \cup \{\neg S\} \vdash_{T} \bot$ if and only if $\mathcal{T} \vdash_{T} S$.

Proof: Assume that $\mathcal{T} \cup \{\neg S\} \vdash_{T} \bot$, i.e., that there is a proof of $\bot$ from premises $\neg S$ and certain sentences $P_1, \ldots, P_n$ of $\mathcal{T}$. By rearranging these premises, we can suppose the formal proof has the following form:

We can use this proof to construct a formal proof of S from $\mathcal{T}$. Start a proof with premises $P_1, \ldots, P_n$. Immediately begin a subproof with assumption $\neg S$. In that subproof, repeat the original proof of $\bot$. End the subproof and use ¬ **Intro** to conclude $\neg\neg S$ from $P_1, \ldots, P_n$. Then apply ¬ **Elim** to get S. The resulting proof will look like this:

This formal proof shows that $\mathcal{T} \vdash_{T} S$, as desired. Proving the other direction of this lemma is simple. We leave it as Exercise 17.13.

This lemma shows that our assumption that $\mathcal{T} \nvdash_{T} S$ is tantamount to assuming that $\mathcal{T} \cup \{\neg S\} \nvdash_{T} \bot$. We can state our observations so far in a more positive and memorable way by introducing the following notion. Let us say that a set of sentences $\mathcal{T}$ is *formally consistent* if and only if $\mathcal{T} \nvdash_{T} \bot$, that is, if and only if there is no proof of $\bot$ from $\mathcal{T}$ in $\mathcal{F}_{T}$. With this notion in hand,

formal consistency

we can state the following theorem, which turns out to be equivalent to the Completeness Theorem:

Theorem (Reformulation of Completeness) Every formally consistent set of sentences is tt-satisfiable.

outline of proof

The Completeness Theorem results from applying this to the set $\mathcal{T} \cup \{\neg S\}$. The remainder of the section is devoted to proving this theorem. The proof is quite simple in outline.

formally complete set of sentences

Completeness for formally complete sets: First we will show that this theorem holds of any formally consistent set with an additional property, known as formal completeness. A set $\mathcal{T}$ is *formally complete* if for any sentence S of the language, either $\mathcal{T} \vdash_{\mathrm{T}} S$ or $\mathcal{T} \vdash_{\mathrm{T}} \neg S$. This is really an unusual property of sets of sentences, since it says that the set is so strong that it settles every question that can be expressed in the language, since for any sentence, either it or its negation is provable from $\mathcal{T}$.

Extending to formally complete sets: Once we show that every formally consistent, formally complete set of sentences is tt-satisfiable, we will show that every formally consistent set can be expanded to a set that is both formally consistent and formally complete.

Putting things together: The fact that this expanded set is tt-satisfiable will guarantee that the original set is as well, since a truth value assignment that satisfies the more inclusive set will also satisfy the original set.

The rest of this section is taken up with filling out this outline.

Completeness for formally complete sets of sentences

To prove that every formally consistent, formally complete set of sentences is tt-satisfiable, the following lemma will be crucial.

Lemma 3. Let $\mathcal{T}$ be a formally consistent, formally complete set of sentences, and let R and S be any sentences of the language.

1. $\mathcal{T} \vdash_{\mathrm{T}} (R \wedge S)$ iff $\mathcal{T} \vdash_{\mathrm{T}} R$ and $\mathcal{T} \vdash_{\mathrm{T}} S$

2. $\mathcal{T} \vdash_{\mathrm{T}} (R \vee S)$ iff $\mathcal{T} \vdash_{\mathrm{T}} R$ or $\mathcal{T} \vdash_{\mathrm{T}} S$

3. $\mathcal{T} \vdash_{\mathrm{T}} \neg S$ iff $\mathcal{T} \nvdash_{\mathrm{T}} S$

4. $\mathcal{T} \vdash_{\mathrm{T}} (R \to S)$ iff $\mathcal{T} \nvdash_{\mathrm{T}} R$ or $\mathcal{T} \vdash_{\mathrm{T}} S$

5. $\mathcal{T} \vdash_{\mathrm{T}} (R \leftrightarrow S)$ iff either $\mathcal{T} \vdash_{\mathrm{T}} R$ and $\mathcal{T} \vdash_{\mathrm{T}} S$ or $\mathcal{T} \nvdash_{\mathrm{T}} R$ and $\mathcal{T} \nvdash_{\mathrm{T}} S$

Proof: Let us first prove (1). Since it is an "iff," we need to prove that each side entails the other. Let us first assume that $\mathcal{T} \vdash_{\mathrm{T}} (\mathsf{R} \wedge \mathsf{S})$. We will show that $\mathcal{T} \vdash_{\mathrm{T}} \mathsf{R}$. The proof that $\mathcal{T} \vdash_{\mathrm{T}} \mathsf{S}$ will be exactly the same. Since $\mathcal{T} \vdash_{\mathrm{T}} (\mathsf{R} \wedge \mathsf{S})$, there is a formal proof of $(\mathsf{R} \wedge \mathsf{S})$ from premises in $\mathcal{T}$. Take this proof and add one more step. At this step, write the desired sentence R, using the rule $\wedge$ **Elim**.

Next, let us suppose that $\mathcal{T} \vdash_{\mathrm{T}} \mathsf{R}$ and $\mathcal{T} \vdash_{\mathrm{T}} \mathsf{S}$. Thus, there are proofs of each of R and S from premises in $\mathcal{T}$. What we need to do is "merge" these two proofs into one. Suppose the proof of R uses the premises $\mathsf{P}_1, \ldots, \mathsf{P}_n$ and looks like this:

$$
\begin{array}{|l}
\mathsf{P}_1 \\
\vdots \\
\mathsf{P}_n \\
\hline
\vdots \\
\mathsf{R}
\end{array}
$$

And suppose the proof of S uses the premises $\mathsf{Q}_1, \ldots, \mathsf{Q}_k$ and looks like this:

$$
\begin{array}{|l}
\mathsf{Q}_1 \\
\vdots \\
\mathsf{Q}_k \\
\hline
\vdots \\
\mathsf{S}
\end{array}
$$

To merge these two proofs into a single proof, we simply take the premises of both and put them into a single list above the Fitch bar. Then we follow the Fitch bar with the steps from the proof of R, followed by the steps from the proof of S. The citations in these steps need to be renumbered, but other than that, the result is a legitimate proof in $\mathcal{F}_{\mathrm{T}}$. At the end of this proof, we add a single step containing $\mathsf{R} \wedge \mathsf{S}$ which we justify by $\wedge$ **Intro**. The merged proof looks like this:

$$\begin{array}{|l}
P_1 \\
\vdots \\
P_n \\
Q_1 \\
\vdots \\
Q_k \\[1em]
\vdots \\
R \\
\vdots \\
S \\
R \wedge S
\end{array}$$

We now turn to (2). One half of this, the direction from right to left, is very easy, using the rule of ∨ **Intro**, so let's prove the other direction. Thus, we want to show that if $\mathcal{T} \vdash_{\mathrm{T}} (R \vee S)$ then $\mathcal{T} \vdash_{\mathrm{T}} R$ or $\mathcal{T} \vdash_{\mathrm{T}} S$. (This is not true in general, but it is for formally consistent, formally complete sets.)

Assume that $\mathcal{T} \vdash_{\mathrm{T}} (R \vee S)$, but, toward a proof by contradiction, that $\mathcal{T} \nvdash_{\mathrm{T}} R$ and $\mathcal{T} \nvdash_{\mathrm{T}} S$. Since $\mathcal{T}$ is formally complete, it follows that $\mathcal{T} \vdash_{\mathrm{T}} \neg R$ and $\mathcal{T} \vdash_{\mathrm{T}} \neg S$. This means that we have two formal proofs p_1 and p_2 from premises in $\mathcal{T}$, p_1 having $\neg R$ as a conclusion, p_2 having $\neg S$ as a conclusion. As we have seen, we can merge these two proofs into one long proof p that has both of these as conclusions. Then, by ∧ **Intro**, we can prove $\neg R \wedge \neg S$. But then using the proof of the version of DeMorgan from Exercise 6.25, we can extend this proof to get a proof of $\neg(R \vee S)$. Thus $\mathcal{T} \vdash_{\mathrm{T}} \neg(R \vee S)$. But by assumption we also have $\mathcal{T} \vdash_{\mathrm{T}} (R \vee S)$. By merging the proofs of $\neg(R \vee S)$ and $R \vee S$ we can get a proof of $\bot$ by adding a single step, justified by $\bot$ **Intro**. But this means that $\mathcal{T}$ is formally inconsistent, contradicting our assumption that it is formally consistent.

One direction of part (3) follows immediately from the definition of formal completeness, while the left to right half follows easily from the definition of formal consistency.

Parts (4) and (5) are similar to part (2) and are left as an exercise.

With this lemma in hand, we can now fill in the first step in our outline.

Proposition 4. Every formally consistent, formally complete set of sentences is tt-satisfiable.

> **Proof:** Let $\mathcal{T}$ be the formally consistent, formally complete set of sentences. Define an assignment h on the atomic sentences of the language as follows. If $\mathcal{T} \vdash_T A$ then let $h(A) = \text{TRUE}$; otherwise let $h(A) = \text{FALSE}$. Then the function $\hat{h}$ is defined on all the sentences of our language, atomic or complex. We claim that:
>
> for all wffs S, $\hat{h}(S) = \text{TRUE}$ if and only if $\mathcal{T} \vdash_T S$.
>
> The proof of this is a good example of the importance of proofs by induction on wffs. The claim is true for all atomic wffs from the way that h is defined, and the fact that h and $\hat{h}$ agree on atomic wffs. We now show that if the claim holds of wffs R and S, then it holds of $(R \wedge S)$, $(R \vee S)$, $\neg R$, $(R \rightarrow S)$ and $(R \leftrightarrow S)$. These all follow easily from Lemma 3. Consider the case of disjunction, for example. We need to verify that $\hat{h}(R \vee S) = \text{TRUE}$ if and only if $\mathcal{T} \vdash_T (R \vee S)$. To prove the "only if" half, assume that $\hat{h}(R \vee S) = \text{TRUE}$. Then, by the definition of $\hat{h}$, either $\hat{h}(R) = \text{TRUE}$ or $\hat{h}(S) = \text{TRUE}$ or both. Then, by the induction hypothesis, either $\mathcal{T} \vdash_T R$ or $\mathcal{T} \vdash_T S$ or both. But then by the lemma, $\mathcal{T} \vdash_T (R \vee S)$, which is what we wanted to prove. The other direction is proved in a similar manner.
>
> From the fact that we have just established, it follows that the assignment h makes every sentence provable from $\mathcal{T}$ true. Since every sentence in $\mathcal{T}$ is certainly provable from $\mathcal{T}$, by **Reit** if you like, it follows that h makes every sentence in $\mathcal{T}$ true. Hence $\mathcal{T}$ is tt-satisfiable, which is what we wanted to prove.

Extending to formally complete sets of sentences

The next step in our proof of completeness is to figure out a way to get from formally consistent sets of wffs to sets of wffs that are both formally consistent *and* formally complete. The next lemma shows us that this is not as hard as it may seem at first.

Lemma 5. A set of sentences $\mathcal{T}$ is formally complete if and only if for every atomic sentence A, $\mathcal{T} \vdash_T A$ or $\mathcal{T} \vdash_T \neg A$.

> **Proof:** The direction from left to right is just a consequence of the definition of formal completeness. The direction from right to left

is another example of a proof by induction on wffs. Assume that $\mathcal{T} \vdash_{\mathrm{T}} \mathsf{A}$ or $\mathcal{T} \vdash_{\mathrm{T}} \neg\mathsf{A}$ for every atomic sentence A. We use induction to show that for any sentence S, $\mathcal{T} \vdash_{\mathrm{T}} \mathsf{S}$ or $\mathcal{T} \vdash_{\mathrm{T}} \neg\mathsf{S}$. The basis of the induction is given by our assumption. Let's prove the disjunction case. That is, assume S is of the form $\mathsf{P} \vee \mathsf{Q}$. By our inductive hypothesis, we know that $\mathcal{T}$ settles each of P and Q. If $\mathcal{T}$ proves either one of these, then we know that $\mathcal{T} \vdash_{\mathrm{T}} \mathsf{P} \vee \mathsf{Q}$ by $\vee$ **Intro**. So suppose that $\mathcal{T} \vdash_{\mathrm{T}} \neg\mathsf{P}$ and $\mathcal{T} \vdash_{\mathrm{T}} \neg\mathsf{Q}$. By merging these proofs and adding a step, we get a proof of $\neg\mathsf{P} \wedge \neg\mathsf{Q}$. We can continue this proof to get a proof of $\neg(\mathsf{P} \vee \mathsf{Q})$, showing that $\mathcal{T} \vdash_{\mathrm{T}} \neg\mathsf{S}$, as desired. The other inductive steps are similar.

We can now carry out the second step in our outline of the proof of the Completeness Theorem.

Proposition 6. Every formally consistent set of sentences $\mathcal{T}$ can be expanded to a formally consistent, formally complete set of sentences.

Proof: Let us form a list $\mathsf{A}_1, \mathsf{A}_2, \mathsf{A}_3, \ldots,$ of all the atomic sentences of our language, say in alphabetical order. Then go through these sentences one at a time. Whenever you encounter a sentence A_i such that neither A_i nor $\neg\mathsf{A}_i$ is provable from the set, add A_i to the set. Notice that doing so can't make the set formally inconsistent. If you could prove $\bot$ from the new set, then you could prove $\neg\mathsf{A}_i$ from the previous set, by Lemma 2. But if that were the case, you wouldn't have thrown A_i into the set.

The end result of this process is a set of sentences which, by the preceding lemma, is formally complete. It is also formally consistent; after all, any proof of $\bot$ is a finite object, and so could use at most a finite number of premises. But then it would be a proof of $\bot$ at some stage of this process, when all those premises had been thrown in.

Putting things together

Just for the record, let's put all this together into a proof of the Completeness Theorem for $\mathcal{F}_{\mathrm{T}}$.

Proof: Suppose $\mathcal{T} \nvdash_{\mathrm{T}} \mathsf{S}$. Then by Lemma 2, $\mathcal{T} \cup \{\neg\mathsf{S}\}$ is formally consistent. This set can be expanded to a formally consistent, formally complete set, which by our Proposition 4 is tt-satisfiable. Suppose h is a truth value assignment that satisfies this set. Clearly,

h makes all the members of $\mathcal{T}$ true, but S false, showing that S is not a tautological consequence of $\mathcal{T}$.

There is an interesting and logically important consequence of the Completeness Theorem, known as the *Compactness Theorem*. We state it as follows:

Theorem (Compactness Theorem for Propositional Logic) Let $\mathcal{T}$ be any set of sentences of propositional logic. If every finite subset of $\mathcal{T}$ is tt-satisfiable, then $\mathcal{T}$ itself is tt-satisfiable.

Compactness Theorem

> **Proof:** We prove the contrapositive of the claim. Assume that $\mathcal{T}$ is not tt-satisfiable. Then by the Completeness Theorem, the set $\mathcal{T}$ is not formally consistent. But this means that $\mathcal{T} \vdash_{\mathrm{T}} \bot$. But a proof of $\bot$ from $\mathcal{T}$ can use only finitely many premises from $\mathcal{T}$. Let $P_1, \ldots, P_n$ be these premises. By the Soundness Theorem, $P_1, \ldots, P_n$ are not tt-satisfiable. Consequently, there is a finite subset of $\mathcal{T}$ that is not tt-satisfiable.

Remember

1. The Completeness Theorem is proven by showing that every formally consistent set $\mathcal{T}$ of sentences is tt-satisfiable. This is done in two steps.

2. The first step is to show the result for sets $\mathcal{T}$ which are also formally complete.

3. The second step is to show how to extend any formally consistent set to one that is both formally consistent and formally complete.

Exercises

17.4   Consider the following set $\mathcal{T}$:

$$\{(A \land B) \to \neg A,\ C \lor A,\ \neg A \to A,\ B\}$$

The Fitch files Exercise 17.4A and Exercise 17.4B contain proofs showing that $\mathcal{T} \vdash_{\mathrm{T}} \neg A$ and $\mathcal{T} \vdash_{\mathrm{T}} \neg\neg A$. Take these two proofs and merge them into a third proof showing that $\mathcal{T} \vdash_{\mathrm{T}} \bot$. Submit the merged proof as Proof 17.4.

For the following three exercises, suppose our language contains only two predicates, Cube and Small, two individual constants, a and b, and the sentences that can be formed from these by means of the truth-functional connectives.

17.5 Let $\mathcal{T}$ be the following set of sentences:

$$\{\neg(\mathsf{Cube(a)} \vee \mathsf{Small(a)}),\ \mathsf{Cube(b)} \to \mathsf{Cube(a)},\ \mathsf{Small(a)} \vee \mathsf{Small(b)}\}$$

Show that this set is formally consistent and formally complete. To prove the former, you will have to appeal to the Soundness Theorem. To prove the latter, you will want to refer to Lemma 5.

17.6 Let $\mathcal{T}$ again be the following set of sentences:

$$\{\neg(\mathsf{Cube(a)} \vee \mathsf{Small(a)}),\ \mathsf{Cube(b)} \to \mathsf{Cube(a)},\ \mathsf{Small(a)} \vee \mathsf{Small(b)}\}$$

By Proposition 4, there is a truth assignment h making all these sentences true. What values does h assign to each of the atomic sentences of the language?

17.7 This time let $\mathcal{T}$ be the following set of sentences (note the difference in the first sentence):

$$\{\neg(\mathsf{Cube(a)} \wedge \mathsf{Small(a)}),\ \mathsf{Cube(b)} \to \mathsf{Cube(a)},\ \mathsf{Small(a)} \vee \mathsf{Small(b)}\}$$

This set is not formally complete. Use the procedure described in the proof of Proposition 6 to extend this to a formally consistent, formally complete set. (Use alphabetical ordering of atomic sentences.) What is the resulting set? What is the truth value assignment h that satisfies this set? Submit a world making the sentences in your formally complete set true.

17.8 Suppose our language has an infinite number of atomic sentences $A_1, A_2, A_3, \ldots$. Let $\mathcal{T}$ be the following set of sentences:

$$\{A_1 \to A_2, A_2 \to A_3, A_3 \to A_4, \ldots\}$$

There are infinitely many distinct truth value assignments satisfying this set. Give a general description of these assignments. Which of these assignments would be generated from the procedure we used in our proof of the Completeness Theorem?

Each of the following four exercises contains an argument. Classify each argument as being (A) provable in $\mathcal{F}_T$, (B) provable in $\mathcal{F}$ but not in $\mathcal{F}_T$, or (C) not provable in $\mathcal{F}$. In justifying your answer, make explicit any appeal you make to the Soundness and Completeness Theorems for $\mathcal{F}_T$ and for $\mathcal{F}$. (Of course we have not yet proven the latter.) Recall from Chapter 10 that sentences whose main operator is a quantifier are treated as atomic in the definition of tautological consequence.

17.9
$\forall x \, Dodec(x) \rightarrow \forall x \, Large(x)$
$\forall x \, Dodec(x)$

$\forall x \, Large(x)$

17.10
$\forall x \, (Dodec(x) \rightarrow Large(x))$
$\forall x \, Dodec(x)$

$\forall x \, Large(x)$

17.11
$\forall x \, Dodec(x) \rightarrow \forall x \, Large(x)$
$\exists x \, Dodec(x)$

$\exists x \, Large(x)$

17.12
$\forall x \, (Dodec(x) \rightarrow Large(x))$
$\exists x \, Dodec(x)$

$\exists x \, Large(x)$

17.13 Prove the half of Lemma 2 that we did not prove, the direction from right to left.

17.14 Prove the right-to-left half of Part (4) of Lemma 3.

17.15 Prove the left-to-right half of Part (4) of Lemma 3.

17.16 In the inductive proof of Proposition 4, carry out the step for sentences of the form $R \rightarrow S$.

SECTION 17.3

Horn sentences

In Chapter 4 you learned how to take any sentence built up without quantifiers and transform it into one in conjunctive normal form (CNF), CNF, that is, one which is a conjunction of one or more sentences, each of which is a disjunction of one or more literals. Literals are atomic sentences and their negations. We will call a literal *positive* or *negative* depending on whether it is an atomic sentence or the negation of an atomic sentence, respectively.

positive and negative literals

A particular kind of CNF sentence turns out to be important in computer science. These are the so-called "Horn" sentences, named not after their shape, but after the American logician Alfred Horn, who first isolated them and studied some of their properties. A *Horn Sentence* is a sentence in CNF that has the following additional property: every disjunction of literals in the sentence contains *at most one* positive literal. Later in the section we will find that there is a more intuitive way of writing Horn sentences if we use the connective $\rightarrow$. But for now we restrict attention to sentences involving only $\wedge$, $\vee$, and $\neg$.

Horn sentences

The following sentences are all in CNF but none of them are Horn sentences:

¬Home(claire) ∧ (Home(max) ∨ Happy(carl))
(Home(claire) ∨ Home(max) ∨ ¬Happy(claire)) ∧ ¬Happy(carl)
Home(claire) ∨ Home(max) ∨ ¬Home(carl)

The first sentence fails to be a Horn sentence because the second conjunct contains two positive literals, Home(max) and Happy(carl). The second fails to be a Horn sentence because of the first conjunct. It contains the two positive literals Home(claire) and Home(max). Why does the third fail to be a Horn sentence?

By contrast, the following *are* Horn sentences:

¬Home(claire) ∧ (¬Home(max) ∨ Happy(carl))
Home(claire) ∧ Home(max) ∧ ¬Home(carl)
Home(claire) ∨ ¬Home(max) ∨ ¬Home(carl)
Home(claire) ∧ Home(max) ∧ (¬Home(max) ∨ ¬Home(max))

Examination of each shows that each conjunct contains at most one positive literal as a disjunct. Verify this for yourself to make sure you understand the definition. (Remember that the definition of CNF allows some degenerate cases, as we stressed in Chapter 4.)

conditional form of Horn sentences

The definition of Horn sentences may seem a bit *ad hoc*. Why is this particular type of CNF sentence singled out as special? Using the material conditional, we can put them in a form that is more intuitive. Consider the following sentence:

$$(\text{Home(claire)} \land \text{Home(max)}) \to \text{Happy(carl)}$$

If we replace → by its equivalent in terms of ¬ and ∨, and then use DeMorgan's Law, we obtain the following equivalent form:

$$¬\text{Home(claire)} \lor ¬\text{Home(max)} \lor \text{Happy(carl)}$$

This is a disjunction of literals, with only one positive literal. Horn sentences are just conjunctions of sentences of this sort.

Here are some more examples. Assume that A, B, C, and D are atomic sentences. If we replace → by its definition, and use DeMorgan's laws, we find that each sentence on the left is logically equivalent to the Horn sentence on the right.

$$(A \to B) \land ((B \land C) \to D) \quad \Leftrightarrow \quad (¬A \lor B) \land (¬B \lor ¬C \lor D)$$
$$((B \land C \land D) \to A) \land ¬A \quad \Leftrightarrow \quad (¬B \lor ¬C \lor ¬D \lor A) \land ¬A$$
$$A \land ((B \land C) \to D) \quad \Leftrightarrow \quad A \land (¬B \lor ¬C \lor D)$$

The "typical" Horn sentence consists of a conjunction of sentences, each of which is a disjunction of several negative literals and one positive literal, say,

$$\neg\mathsf{A}_1 \lor \ldots \lor \neg\mathsf{A}_n \lor \mathsf{B}$$

This can be rewritten using $\land$ and $\rightarrow$ as:

$$(\mathsf{A}_1 \land \ldots \land \mathsf{A}_n) \rightarrow \mathsf{B}$$

This is the typical case, but there are the important limiting cases, disjunctions with a positive literal but *no negative literals*, and disjunctions with some negative literals but *no positive literal*. By a logical sleight of hand, though, we can in fact rewrite these in the same conditional form. The sleight of hand is achieved by introducing a couple of rather odd atomic sentences, $\top$ and our old friend $\bot$. The first of these is assumed to be always true. The second, of course, is always false. Using these,

$$\neg\mathsf{A}_1 \lor \ldots \lor \neg\mathsf{A}_n$$

can be rewritten as:

$$(\mathsf{A}_1 \land \ldots \land \mathsf{A}_n) \rightarrow \bot$$

Similarly, we can rewrite the lone atomic sentence B as $\top \rightarrow \mathsf{B}$. We summarize these observations by stating the following result.

Proposition 7. Any Horn sentence of propositional logic is logically equivalent to a conjunction of conditional statements of the following three forms, where the A_i and B stand for ordinary atomic sentences:

1. $(\mathsf{A}_1 \land \ldots \land \mathsf{A}_n) \rightarrow \mathsf{B}$

2. $(\mathsf{A}_1 \land \ldots \land \mathsf{A}_n) \rightarrow \bot$

3. $\top \rightarrow \mathsf{B}$

Using the truth table method, we could program a computer to check to see if a sentence is tt-satisfiable or not since the truth table method is completely mechanical. You can think of our **Taut Con** routine as doing something like this, though actually it is more clever than this brute force method. In general, though, any method of checking arbitrary formulas for tt-satisfiability is quite "expensive." It consumes a lot of resources. For example, a sentence involving 50 atomic sentences has 2^{50} rows in its truth table, a very big number. For Horn sentences, however, we can in effect restrict attention to a single row. It is this fact that accounts for the importance of this class of sentences.

inefficiency of truth tables

This efficient method for checking the satisfiability of Horn sentences, known as the *satisfaction algorithm for Horn sentences*, is really quite simple.

satisfaction algorithm for Horn sentences

We first describe the method, and then apply it to a couple of examples. The idea behind the method is to build a one-row truth table by working back and forth, using the conjuncts of the sentence to figure out which atomic sentences need to have TRUE written beneath them. We will state the algorithm twice, once for the Horn sentences in CNF form, but then also for the conditional form.

satisfaction algorithm for Horn sentences

Satisfaction algorithm for Horn sentences: Suppose we have a Horn sentence S built out of atomic sentences $A_1, \ldots, A_n$. Here is an efficient procedure for determining whether S is tt-satisfiable.

1. Start out as though you were going to build a truth table, by listing all the atomic sentences in a row, followed by S. But do not write TRUE or FALSE beneath any of them yet.

2. Check to see which if any of the atomic sentences are themselves conjuncts of S. If so, write TRUE in the reference column under these atomic sentences.

3. If some of the atomic sentences are now assigned TRUE, then use these to fill in as much as you can of the right hand side of the table. For example, if you have written TRUE under A_5, then you will write FALSE wherever you find $\neg A_5$. This, in turn, may tell you to fill in some more atomic sentences with TRUE. For example, if $\neg A_1 \lor A_3 \lor \neg A_5$ is a conjunct of S, and each of $\neg A_1$ and $\neg A_5$ have been assigned FALSE, then write TRUE under A_3. Proceed in this way until you run out of things to do.

4. One of two things will happen. One possibility is that you will reach a point where you are forced to assign FALSE to one of the conjuncts of S, and hence to S itself. In this case, the sentence is not tt-satisfiable. But if this does not happen, then S is tt-satisfiable. For then you can fill in all the remaining columns of atomic sentences with FALSE. This will give you a truth assignment that makes S come out true, as we will prove below. (There may be other assignments that make S true; our algorithm just generates one of them.)

Let's apply this algorithm to an example.

You try it
. . . .

▶   1. Consider the sentence

Home(claire) $\land$ ¬Home(max) $\land$ (Home(max) $\lor$ ¬Home(claire))

To make this fit on the page, let's abbreviate the two atomic sentences Home(claire) and Home(max) by C and M, respectively. Open Boole and create the following table (it will be easier if you choose **By Row** in the **Edit** menu):

$$\begin{array}{c|c||c} C & M & C \wedge \neg M \wedge (M \vee \neg C) \\ \hline & & \end{array}$$

2. The first step of the above method tells us to put TRUE under any atomic sentence that is a conjunct of S. In this case, this means we should put a TRUE under C. So enter a T under the reference column for C.   ◄

3. We now check to see how much of the right side of the table we can fill in. Using Boole, check which columns on the right hand side call on columns that are already filled in. There is only one, the one under ¬C. Fill it in to obtain the following: ◄

$$\begin{array}{c|c||c} C & M & C \wedge \neg M \wedge (M \vee \neg C) \\ \hline T & & \qquad\qquad F \end{array}$$

4. Looking at the last conjunct, we see that if the whole is to be true, we must also assign TRUE to M. So fill this in to obtain   ◄

$$\begin{array}{c|c||c} C & M & C \wedge \neg M \wedge (M \vee \neg C) \\ \hline T & T & \qquad\qquad F \end{array}$$

5. But this means the second conjunct gets assigned FALSE, so the whole sentence comes out FALSE.   ◄

$$\begin{array}{c|c||c} C & M & C \wedge \neg M \wedge (M \vee \neg C) \\ \hline T & T & \quad F \qquad\quad F \end{array}$$

Thus, the sentence is not tt-satisfiable.

6. Finish this row of your table and save the table as Table Horn 1.   ◄

. . . . *Congratulations*

Let's restate the satisfaction algorithm for Horn sentences in conditional form, since many people find it more intuitive, and then apply it to an example.

Satisfaction algorithm for Horn sentences in conditional form: Suppose we have a Horn sentence S in conditional form, built out of atomic sentences $A_1, \ldots, A_n$, as well as $\top$ and $\bot$.

   1. If there are any conjuncts of the form $\top \to A_i$, write TRUE in the reference column under each such A_i.

algorithm for conditional Horn sentences

2. If one of the conjuncts is of the form $(B_1 \wedge \ldots \wedge B_k) \to A$ where you have assigned TRUE to each of $B_1, \ldots, B_k$, then assign TRUE to A.

3. Repeat step 2 as often as possible.

4. Again, one of two things will happen. You may reach a point where you are forced to assign FALSE to one of a conditional of the form $(B_1 \wedge \ldots \wedge B_k) \to \bot$ because you have assigned TRUE to each of the B_i. In this case you must assign FALSE to S, in which case S is not tt-satisfiable. If this does not happen, then fill in the remaining reference columns of atomic sentences with FALSE. This will give a truth assignment that makes all the conditionals true and hence S true as well.

This time, let's look at the sentence

$$(\neg A \vee \neg B) \wedge (\neg B \vee C) \wedge B$$

Writing this in conditional form, we obtain

$$((A \wedge B) \to \bot) \wedge (B \to C) \wedge (\top \to B)$$

We won't actually write out the table, but instead will just talk through the method. First, we see that if the sentence is to be satisfied, we must assign TRUE to B, since $\top \to B$ is a conjunct. Then, looking at the second conjunct, $B \to C$, we see that assigning TRUE to B forces us to assign TRUE to C. But at this point, we run out of things that we are forced to do. So we can assign A the value FALSE getting get an assignment that makes remaining conditional, and hence the whole sentence, TRUE.

correctness of algorithm   How do we know that this algorithm is correct? Well, we don't, yet. The examples may have convinced you, but they shouldn't have. We really need to give a proof.

Theorem The algorithm for the satisfiability of Horn sentences is correct, in that it classifies as tt-satisfiable exactly the tt-satisfiable Horn sentences.

Proof: There are two things to be proved here. One is that any tt-satisfiable sentence is classified as tt-satisfiable by the algorithm. The other is that anything classified by the algorithm as tt-satisfiable really is tt-satisfiable. We are going to prove this result for the form of the algorithm that deals with conditionals. Before getting down to work, let's rephrase the algorithm with a bit more precision. Define sets $\mathcal{T}_0, \mathcal{T}_1, \ldots$ of atomic sentences, together with $\top$ and $\bot$, as follows. Let $\mathcal{T}_0 = \{\top\}$. Let $\mathcal{T}_1$ be the set consisting of $\top$ together with all atomic sentences A such that $\top \to A$ is a conjunct of S. More generally, given $\mathcal{T}_n$, define $\mathcal{T}_{n+1}$ to be $\mathcal{T}_n$ together with all atomic sentences A such that for some $B_1, \ldots, B_k$ in $\mathcal{T}_n$, $(B_1 \wedge \ldots \wedge B_k) \to A$ is a conjunct of S. Notice that $\mathcal{T}_n \subseteq \mathcal{T}_{n+1}$ for each n. Since there are only finitely many atomic sentences in S, eventually we must have $\mathcal{T}_N = \mathcal{T}_{N+1}$. The algorithm declares S to be tt-satisfiable if and only if $\bot \notin \mathcal{T}_N$. Furthermore, it claims that if $\bot$ is not in $\mathcal{T}_N$, then we can get a truth assignment for S be assigning TRUE to each atomic sentence in $\mathcal{T}_N$ and assigning FALSE to the rest.

To prove the first half of correctness, we will show that if S is tt-satisfiable, then $\bot \notin \mathcal{T}_N$. Let h be any truth assignment that makes S true. An easy proof by induction on n shows that $h(A) = \text{TRUE}$ for each $A \in \mathcal{T}_n$. Hence $\bot \notin \mathcal{T}_N$, since $h(\bot) = \text{FALSE}$.

To prove the other half of correctness, we suppose that $\bot \notin \mathcal{T}_N$ and define an assignment h by letting $h(A) = \text{TRUE}$ for $A \in \mathcal{T}_N$, and letting $h(A) = \text{FALSE}$ for the other atomic sentences of S. We need to show that $\hat{h}(S) = \text{TRUE}$. To do this, it suffices to show that $\hat{h}(C) = \text{TRUE}$ for each conditional C that is a conjunct of S. There are three types of conditionals to consider:

Case 1: The conjunct is of the form $\top \to A$. In this case A is in $\mathcal{T}_1$. But then $\hat{h}$ assigns TRUE to the A and so to the conditional.

Case 2: The conjunct is of the form $(A_1 \wedge \ldots \wedge A_n) \to B$. If each of the A_i gets assigned TRUE, then each is in $\mathcal{T}_N$ and so B is in $\mathcal{T}_{N+1} = \mathcal{T}_N$. But then $\hat{h}$ assigns TRUE to B and so to the conditional. On the other hand, if one of the A_i gets assigned FALSE then the conditional comes out true under $\hat{h}$.

Case 3: The conjunct is of the form $(A_1 \wedge \ldots \wedge A_n) \to \bot$. Since we are assuming $\bot \notin \mathcal{T}_N$, at least one of A_i is not $\mathcal{T}_N$, so it gets assigned FALSE by h. But then the antecedent of conditional comes out false under $\hat{h}$ so the whole conditional come out true.

Remember

1. A Horn sentence is a propositional sentence in CNF such that every disjunction of literals in contains at most one positive literal.

2. The satisfaction algorithm for Horn sentences gives an efficient algorithm to tell whether a Horn sentence is tt-satisfiable.

Exercises

17.17 If you skipped the **You try it** section, go back and do it now. Submit the file **Table Horn 1**.

17.18 A sentence in CNF can be thought of as a list of sentences, each of which is a disjunction of literals. In the case of Horn sentences, each of these disjunctions contains at most one positive literal. Open **Horn's Sentences**. You will see that this is a list of sentences, each of which is a disjunction of literals, at most one of which is positive. Use the algorithm given above to build a world where all the sentences come out true, and save it as **World 17.18**.

17.19 Open **Horn's Other Sentences**. You will see that this is a list of sentences, each of which is a disjunctive Horn sentence. Use the algorithm given above to see if you can build a world where all the sentences come out true. If you can, save the world as **World 17.19**. If you cannot, explain how the algorithm shows this.

17.20 Rewrite the following Horn sentences in conditional form. Here, as usual, A, B, and C are taken to be atomic sentences.
  1. $A \land (\lnot A \lor B \lor \lnot C) \land \lnot C$
  2. $(\lnot A \lor \lnot B \lor C) \land \lnot C$
  3. $(\lnot A \lor B) \land (A \lor \lnot B)$

Use Boole to try out the satisfaction algorithm on the following Horn sentences (two are in conditional form). Give the complete row that results from the application of the algorithm. In other words, the table you submit should have a single row corresponding to the assignment that results from the application of the algorithm. Assume that A, B, C, and D are atomic sentences. (If you use **Verify Table** *to check your table, Boole will tell you that there aren't enough rows. Simply ignore the complaint.)*

17.21 $A \land (\lnot A \lor B) \land (\lnot B \lor C)$

17.22 $A \land (\lnot A \lor B) \land \lnot D$

17.23 $A \land (\lnot A \lor B) \land \lnot B$

17.24 $\lnot A \land (\lnot A \lor B) \land \lnot B$

17.25 $((A \land B) \to C) \land (A \to B) \land A \land ((C \land B) \to D)$

17.26 $((A \land B) \to C) \land (A \to B) \land A \land ((C \land B) \to \bot)$

The programming language Prolog is based on Horn sentences. It uses a slightly different notation, though. The clause

$$(A_1 \land \ldots \land A_n) \to B$$

is frequently written

$$B :- A_1, \ldots, A_n$$

or

$$B \leftarrow A_1, \ldots, A_n$$

and read "B, if A_1 through A_n." The following exercises use this Prolog notation.

17.27 Consider the following Prolog "program."

> AncestorOf(a, b) $\leftarrow$ MotherOf(a, b)
> AncestorOf(b, c) $\leftarrow$ MotherOf(b, c)
> AncestorOf(a, b) $\leftarrow$ Father(a, b)
> AncestorOf(b, c) $\leftarrow$ Father(b, c)
> AncestorOf(a, c) $\leftarrow$ AncestorOf(a, b), AncestorOf(b, c)
> MotherOf(a, b) $\leftarrow$ TRUE
> FatherOf(b, c) $\leftarrow$ TRUE
> FatherOf(b, d) $\leftarrow$ TRUE

The first five clauses state instances of some general facts about the relations *mother of, father of,* and *ancestor of.* (Prolog actually lets you say things with variables, so we would not actually need multiple instances of the same scheme. For example, rather than state both the first two clauses, we could just state AncestorOf(x, y) $\leftarrow$ MotherOf(x, y).) The last three clauses describe some particular facts about a, b, c, and d. Use the Horn satisfaction algorithm to determine whether the above set of Horn sentences (in conditional form) is satisfiable.

17.28 The Prolog program in Exercise 17.27 might be considered as part of a database. To ask whether it entails B, Prolog adds

$$\bot \leftarrow B$$

to the database and runs the Horn algorithm on the enlarged database. If the algorithm fails, then Prolog answers "yes." Otherwise Prolog answers "no." Justify this procedure.

17.29 Use the procedure of the Exercise 17.28 to determine whether the following are consequences of the Prolog program given in Exercise 17.27.

> 1. Ancestor(a, c)
> 2. Ancestor(c, d)
> 3. Mother(a, b)
> 4. Mother(a, d)

17.30 Suppose you have a Horn sentence which can be put into conditional form in a way that does not contain any conjunct of form 3 in Proposition 7. Show that it is satisfiable. Similarly, show that if it can be put into a conditional form that does not contain a conjunct of form 2, then it is satisfiable.

SECTION 17.4

Resolution

People are pretty good at figuring out when one sentence is a tautological consequence of another, and when it isn't. If it is, we can usually come up with a proof, especially when we have been taught the important methods of proof. And when it isn't, we can usually come up with an assignment of truth values that makes the premises true and the conclusion false. But for computer applications, we need a reliable and efficient algorithm for determining when one sentence is a tautological consequence of another sentence or a set of sentences.

Recall that S is a tautological consequence of premises $P_1, \ldots, P_n$ if and only if the set $\{P_1, \ldots, P_n, \neg S\}$ is not tt-satisfiable, that is to say, its conjunction is not tt-satisfiable. Thus, the problem of checking for tautological consequence and the problem of checking to see that a sentence is not tt-satisfiable amount to the same thing. The truth table method provides us with a reliable method for doing this. The trouble is that it can be highly expensive in terms of time and paper (or computer memory). If we had used it in Fitch, there are many problems that would have bogged down your computer intolerably.

In the case of Horn sentences, we have seen a much more efficient method, one that accounts for the importance of Horn sentences in logic programming. In this section, we present a method that applies to arbitrary sentences in CNF. It is not in general as efficient as the Horn sentence algorithm, but it is often much more efficient than brute force checking of truth tables. It also has the advantage that it extends to the full first-order language with quantifiers. It is known as the resolution method, and lies at the heart of many applications of logic in computer science. While it is not the algorithm that we have actually implemented in Fitch, it is closely related to that algorithm.

set of clauses   The basic notion in resolution is that of a *set of clauses*. A *clause* is just any finite set of literals. Thus, for example,

$$C_1 = \{\neg \mathsf{Small(a)}, \mathsf{Cube(a)}, \mathsf{BackOf(b, a)}\}$$

is a clause. So is

$$C_2 = \{\mathsf{Small(a)}, \mathsf{Cube(b)}\}$$

The special notation □ is used for the empty clause. A clause C is said to be satisfied by a truth assignment h provided at least one of the literals in C is assigned TRUE by $\hat{h}$.[1] The empty clause □ clearly is not tt-satisfiable by any assignment, since it does not contain any elements to be made true. If $C \neq$ □ then h satisfies C if and only if the disjunction of the sentences in C is assigned TRUE by $\hat{h}$.

A nonempty set S of clauses is said to be satisfied by the truth assignment h provided each clause C in S is satisfied by h. Again, this is equivalent to saying that the CNF sentence formed by conjoining the disjunctions formed from clauses in S is satisfied by $\hat{h}$.

The goal of work on resolution is to come up with as efficient an algorithm as possible for determining whether a set of clauses is tt-satisfiable. The basic insight of the theory stems from the observation that in trying to show that a particular set S is not tt-satisfiable, it is often easier to show that a larger set S' derived from it is not tt-satisfiable. As long as the method of getting S' from S insures that the two sets are satisfied by the same assignments, we can work with the larger set S'. Indeed, we might apply the same method over and over until it became transparent that the sets in question are not tt-satisfiable. The method of doing this is the so-called *resolution method*.

Resolution Method:

1. Start with a set T of sentences in CNF which you hope to show is not tt-satisfiable. Transform each of these sentences into a set of clauses in the natural way: replace disjunctions of literals by clauses made up of the same literals, and replace conjunctions by sets of clauses. Call the set of all these clauses S. The aim now is to show S is *not* tt-satisfiable.

2. To show S is *not* tt-satisfiable, systematically add clauses to the set in such a way that the resulting set is satisfied by the same assignments as the old set. The new clauses to throw in are called *resolvents* of old clauses. If you can finally get a set of clauses which contains □, and so obviously cannot be satisfied, then you know that our original set S could not be satisfied.

To complete our explanation of the resolution method, we need to explain the idea of a resolvent. To make this more understandable, we first give a couple of examples. Let C_1 and C_2 be the clauses displayed earlier. Notice

---

[1]Notice that we have now given two incompatible definitions of what it means for an assignment to satisfy a set of literals: one where we defined what it means for an assignment to satisfy a set of sentences thought of as a theory, and one where we think of the set as a resolution clause. It would be better if two different words were used. But they aren't, so the reader must rely on context to tell which use is intended.

that in order for an assignment h to satisfy the set $\{C_1, C_2\}$, h will have to assign TRUE to at least one of Cube(a), Cube(b), or BackOf(b,a). So let $C_3 = \{\text{Cube(a)}, \text{Cube(b)}, \text{BackOf(b, a)}\}$ be an additional clause. Then the set of clauses $\{C_1, C_2\}$ and $\{C_1, C_2, C_3\}$ are satisfied by exactly the same assignments. The clause C_3 is a resolvent of the first set of clauses.

For another example, let C_1, C_2, and C_3 be the following three clauses:

$$
\begin{aligned}
C_1 &= \{\text{Home(max)}, \text{Home(claire)}\} \\
C_2 &= \{\neg\text{Home(claire)}\} \\
C_3 &= \{\neg\text{Home(max)}\}
\end{aligned}
$$

Notice that in order for an assignment to satisfy both C_1 and C_2, you will have to satisfy the clause

$$
C_4 = \{\text{Home(max)}\}
$$

Thus we can throw this resolvent C_4 into our set. But when we look at $\{C_1, C_2, C_3, C_4\}$, it is obvious that this new set of clauses cannot be satisfied. C_3 and C_4 are in direct conflict. So the original set is not tt-satisfiable.

With these examples in mind, we now define what it means for one clause, say R, to be a resolvent of two other clauses, say C_1 and C_2.

resolvent defined

Definition A clause R is a *resolvent* of clauses C_1 and C_2 if there is an atomic sentence in one of the clauses whose negation is in the other clause, and if R is the set of all the other literals in either clause.

Here are some more examples. Assume A, B, C, and D are atomic. We use □ as above for the empty clause.

$$\frac{\{A, D\} \quad \{\neg A\}}{\{D\}}$$

$$\frac{\{A, \neg A\} \quad \{A\}}{\{A\}}$$

$$\frac{\{B, C\} \quad \{\neg B, \neg D\}}{\{C, \neg D\}}$$

$$\frac{\{D\} \quad \{\neg D\}}{\square}$$

The key fact about resolution is expressed in the following theorem. The proof will be outlined in Exercise 17.45.

Theorem (Completeness of resolution) If $\mathcal{S}$ is an not tt-satisfiable set of clauses then it is always possible to arrive at □ by successive resolutions.

completeness of
resolution

Here is an example which illustrates the resolution method. Suppose S is the following sentence in CNF:

$$\neg A \wedge (B \vee C \vee B) \wedge (\neg C \vee \neg D) \wedge (A \vee D) \wedge (\neg B \vee \neg D)$$

Applying step 1, we convert the sentence S to a the following clauses:

$$\{\neg A\}, \ \{B, C\}, \ \{\neg C, \neg D\}, \ \{A, D\}, \ \{\neg B, \neg D\}$$

Our new aim is to use resolution to show that this set of clauses (and hence the original sentence S) is not tt-satisfiable.

Successive applications of step 2 is illustrated by the following picture:

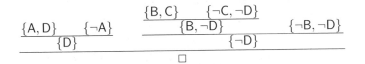

Since we are able to start with clauses in $\mathcal{S}$ and resolve to the empty clause, we know that the original set $\mathcal{T}$ of sentences is not tt-satisfiable. A figure of this sort is sometimes called a *proof by resolution*.

proof by resolution

A proof by resolution shows that a set of sentences, or set of clauses, is not tt-satisfiable. But it can also be used to show that a sentence C is a tautological consequence of premises $P_1, \ldots, P_n$. This depends on the observation, made earlier, that S is a consequence of premises $P_1, \ldots, P_n$ if and only if the set $\{P_1, \ldots, P_n, \neg S\}$ is not tt-satisfiable.

Remember

1. Every set of propositional sentences can be expressed as a set of clauses.

2. The resolution method is an important method for determining whether a set of clauses is tt-satisfiable. The key notion is that of a resolvent for a set of clauses.

3. A resolvent of clauses C_1 and C_2 is a clause R provided there is an atomic sentence in one of the clauses whose negation is in the other clause, and if R is the set of all the remaining literals in either clause.

Exercises

17.31 Open Alan Robinson's Sentences. The sentences in this file are not mutually satisfiable in any world. Indeed, the first six sentences are not mutually satisfiable. Show that the first five sentences are mutually satisfiable by building a world in which they are all true. Submit this as World 17.31. Go on to show that each sentence from 7 on can be obtained from earlier sentences by resolution, if we think of the disjunction in clausal form. The last "sentence," □, is clearly not satisfiable, so this shows that the first six are not mutually satisfiable. Turn in your resolution proof to your instructor.

17.32 Use Fitch to give an ordinary proof that the first six sentences of Alan Robinson's Sentences are not satisfiable.

17.33 Construct a proof by resolution showing that the following CNF sentence is not satisfiable:
$$(A \lor \neg C \lor B) \land \neg A \land (C \lor B \lor A) \land (A \lor \neg B)$$

17.34 Construct a proof by resolution showing that the following sentence is not satisfiable. Since the sentence is not in CNF, you will first have to convert it to CNF.
$$\neg\neg A \land (\neg A \lor ((\neg B \lor C) \land B)) \land \neg C$$

17.35 Resolution can also be used to show that a sentence is logically true. To show that a sentence is logically true, we need only show that its negation is not satisfiable. Use resolution to show that the following sentence is logically true:
$$A \lor (B \land C) \lor (\neg A \land \neg B) \lor (\neg A \land B \land \neg C)$$

Give resolution proofs of the following arguments. Remember, a resolution proof will demonstrate that the premises and the negation of the conclusion form an unsatisfiable set.

17.36

$\neg B$
$\neg A \lor C$
$\neg(C \land \neg B)$

$\neg A$

17.37

$C \lor A$
$\neg C$

$A \lor B$

17.38

$\neg A \lor B$
$C \lor \neg(A \land B)$

$\neg A \lor (B \land C)$

17.39

$A \to B$
A

B

17.40

$B \to C$

$(A \land B) \to C$

17.41

$A \lor B$
$A \to C$
$B \to D$

$C \lor D$

17.42

$A \vee (B \wedge C)$
$\neg E$
$(A \vee B) \rightarrow (D \vee E)$
$\neg A$

$C \wedge D$

17.43

$\neg A \rightarrow B$
$C \rightarrow (D \vee E)$
$D \rightarrow \neg C$
$A \rightarrow \neg E$

$C \rightarrow B$

17.44 (Soundness of resolution) Let $\mathcal{S}$ be a nonempty set of clauses.

1. Let C_1 and C_2 be clauses in $\mathcal{S}$ and let R be a resolvent of C_1 and C_2. Show that $\mathcal{S}$ and $\mathcal{S} \cup \{R\}$ are satisfied by the same assignments.

2. Conclude that if $\mathcal{S}$ is satisfiable, then it is impossible to obtain □ by successive applications of resolution.

17.45 (Completeness of resolution) In this exercise we outline the theorem stated in the section to the effect that the resolution method is complete.

1. Assume that $\mathcal{S}$ is a set of clauses and that the only literals appearing in clauses are A and $\neg A$. Show that if $\mathcal{S}$ is not satisfiable, then □ is a resolvent of clauses in $\mathcal{S}$.

2. Next, assume that $\mathcal{S}$ is a set of clauses and that the only literals appearing in clauses are $A, B, \neg A$, and $\neg B$. Form two new sets of clauses as follows. First, form sets S^B and $S^{\neg B}$ where the first of these consists of all clauses in S that do not contain B and the second consists of all clauses in S that do not contain $\neg B$. Notice that these sets can overlap, since some clauses in S might not contain either. Assume that S is not satisfiable, and that h is any truth assignment. Show that if $h(B) = \text{TRUE}$, then h cannot satisfy S^B. Similarly, show that if $h(B) = \text{FALSE}$, then h fails to satisfy $S^{\neg B}$.

3. With the same setup as above, we now form new sets of clauses S_B and $S_{\neg B}$. The first results from S^B by throwing out $\neg B$ from any clauses that contain it. The second results from $S^{\neg B}$ by throwing out B from its clauses. Show that the observation made above about h still holds for these new sets. Note, however, that neither B nor $\neg B$ appears in any clause in either of these sets. Hence, it follows that no assignment satisfies S_B and that no assignment satisfies $S_{\neg B}$.

4. Still continuing with the same setup, show that if S is not satisfiable, then □ can be obtained as a resolvent of each of S_B and $S_{\neg B}$. Here the result you obtained in part 1 comes into play.

5. Use this result to show that if S is not satisfiable then either □ or $\{\neg B\}$ can be obtained as a resolvent from S^B. Show similarly that either □ or $\{B\}$ can be obtained as a resolvent from $S^{\neg B}$.

6. Use this to show that if S is not satisfiable, then □ can be obtained as an eventual resolvent of S.

7. Now you have shown that any unsatisfiable set $\mathcal{S}$ of clauses built from just two atomic sentences has □ as an eventual resolvent. Can you see how this method generalizes to the case of three atomic sentences? You will need to use your results for one and two atomic sentences.

8. If you have studied the chapter on induction, complete this proof to obtain a general proof of Theorem 17.4. Nothing new is involved except induction.

CHAPTER 18

Advanced Topics in FOL

This chapter presents some more advanced topics in first-order logic. The first three sections deal with a mathematical framework in which the semantics of FOL can be treated rigorously. This framework allows us to make our informal notions of first-order validity and first-order consequence precise, and culminates in a proof of the Soundness Theorem for the full system $\mathcal{F}$. The later sections deal with unification and resolution resolution method, topics of importance in computer science. The Completeness Theorem for $\mathcal{F}$ is taken up in the next chapter, which does not presuppose the sections on unification and resolution.

First-order structures

In our treatment of propositional logic, we introduced the idea of logical consequence in virtue of the meanings of the truth-functional connectives. We developed the rigorous notion of tautological consequence as a precise approximation of the intuitive notion. We achieved this precision thanks to truth table techniques, which we later extended by means of truth assignments. Truth assignments have two advantages over truth tables: First, in assigning truth values to all atomic sentences at once, they thereby determine the truth or falsity of every sentence in the language, which allows us to apply the concept of tautological consequence to infinite sets of sentences. Second, they allow us to do this with complete mathematical rigor.

In Chapter 10, we introduced another approximation of the intuitive notion of consequence, that of first-order consequence, consequence in virtue of the meanings of $\forall, \exists$ and $=$, in addition to the truth-functional connectives. We described a vague technique for determining when a sentence was a first-order consequence of others, but did not have an analog of truth tables that gave us enough precision to prove results about this notion, such as the Soundness Theorem for $\mathcal{F}$.

Now that we have available some tools from set theory, we can solve this problem. In this section, we define the notion of a first-order structure. A first-order structure is analogous to a truth assignment in propositional logic. It represents circumstances that determine the truth values of all of the sentences of a language, but it does so in such a way that identity and the first-order

first-order structures

511

Figure 18.1: Mary Ellen's World.

quantifiers ∀ and ∃ are respected. This will allow us to give a precise definition of first-order consequence and first-order validity.

In our intuitive explanation of the semantics of quantified sentences, we appealed to the notion of a "domain of discourse," defining truth and satisfaction relative to such a domain. We took this notion to be an intuitive one, familiar both from our experience using Tarski's World and from our ordinary experience communicating with others about real-world situations. The notion of a first-order structure results from modeling these domains in a natural way using set theory.

Let's begin with a very simple language, a sublanguage of the blocks language. Assume that we have only three predicates, Cube, Larger, and =, and one name, say c. Even with this simple language there are infinitely many sentences. How should we represent, in a rigorous way, the circumstances that determine the truth values of sentences in this language?

modeling a world

By way of example, consider Mary Ellen's World, shown in Figure 18.1. This world has three cubes, one of each size, and one small tetrahedron. The small cube is named c. Our goal is to construct a mathematical object that represents everything about this world that is relevant to the truth values of sentences in our toy language. Later, we will generalize this to arbitrary first-order languages.

domain of discourse

Since sentences are going to be evaluated in Mary Ellen's World, one thing we obviously need to represent is that the world contains four objects. We do this by using a set $D = \{b_1, b_2, b_3, b_4\}$ of four objects, where b_1 represents the leftmost block, b_2 the next, and so forth. Thus b_4 represents the tetrahedron. This set D is said to be the *domain of discourse* of our first-order structure. To keep first-order structures as clean as possible, we represent only those features of the domain of discourse that are relevant to the truth of sentences in the given first-order language. Given our current sublanguage, there are many features of Mary Ellen's World that are totally irrelevant to the truth of sentences. For example, since we cannot say anything about position, our mathematical structure need not represent any facts about the positions of our blocks. On the other hand, we can say things about size and shape. Namely, we can say that an object is (or is not) a cube and that one object is (or is not) larger than another. So we will need to represent these sorts of facts. We do this by assigning to the predicate Cube a certain subset Cu of the domain of discourse D, namely, the set of cubes. This set is called the *extension* of the predicate Cube in our structure. In modeling the world depicted above, this extension is the set $Cu = \{b_1, b_2, b_3\}$. Similarly, to represent facts about the relative sizes of the objects, we assign to the predicate Larger a set La of ordered pairs $\langle x, y \rangle$, where $x, y \in D$. If $\langle x, y \rangle \in La$, then this represents the fact that x is larger than y. So in our example, we would have

extensions of predicates

$$La = \{\langle b_2, b_1 \rangle, \langle b_3, b_1 \rangle, \langle b_3, b_2 \rangle, \langle b_2, b_4 \rangle, \langle b_3, b_4 \rangle\}$$

La is said to be the *extension* of Larger in the structure.

There is only one more thing we need to make our representation complete, at least as far as the present language is concerned. We have to hook up the individual constant c with the block it names. That is, we need to build into our structure something that tells us that c is a name of block b_1 rather than one of the other blocks. Or, to put it more technically, we need to represent the fact that in our world b_1 is the *referent* of the name c. So we need to pair the name c with the object b_1 that it names. The simplest way to do this in general is to have a function which assigns to each name in the language whatever object in the domain it happens to name. You might call this function the naming function. (The way we actually handle this when we come to the final definition incorporates the naming function in a slightly different way.)

referents of constants

We have neglected to say anything about the identity predicate =. That is because its extension is fixed, once we have the domain D. It is always interpreted as meaning identity, so the extension is just the set of pairs $\langle a, a \rangle$ where $a \in D$. So in this case, it consists of the set $\{\langle b_1, b_1 \rangle, \langle b_2, b_2 \rangle, \langle b_3, b_3 \rangle, \langle b_4, b_4 \rangle\}$.

identity

Let's stick with our little language a bit longer, but consider how we would want to represent other worlds. In general we need: a domain of discourse D, a subset Cu of D to serve as the extension of the predicate Cube, a set La of pairs from D to serve as the extension of the predicate Larger, and a pairing of the name c with its referent, some element of the domain D of discourse. In order to have one single object to represent the whole world, with all its relevant facts, we package the domain of discourse, the extensions of the predicates, and the referents of the names, all into one mathematical object. Just how this packaging is done is not too important, and different textbooks do it in somewhat different ways. The most elegant packaging, and the one we adopt, is to use a single function $\mathfrak{M}$. ("$\mathfrak{M}$" stands for "model," which is another common term for what we are calling a structure.)

The function $\mathfrak{M}$ is defined on the predicates of the language, the names of the language, and the quantifier symbol $\forall$. Such a function is called a *first-order structure* provided the following conditions are satisfied:

definition of first-order structure

1. $\mathfrak{M}(\forall)$ is a nonempty set D, called the *domain of discourse* of $\mathfrak{M}$.

2. If P is an n-ary predicate symbol of the language then $\mathfrak{M}(\mathsf{P})$ is a set of n-tuples $\langle x_1, \ldots, x_n \rangle$ of elements of D. This set is called the *extension* of P in $\mathfrak{M}$. It is required that the extension of the identity symbol consist of all pairs $\langle x, x \rangle$, for $x \in D$.

3. if c is any name of the language, then $\mathfrak{M}(\mathsf{c})$ is an element of D, and is called the *referent* of c in $\mathfrak{M}$.

Instead of writing $\mathfrak{M}(\mathsf{Cube})$, it is more common to write $\mathsf{Cube}^{\mathfrak{M}}$, and similarly for the other predicates and names. And it is common to write just $D^{\mathfrak{M}}$ for the domain of discourse $\mathfrak{M}(\forall)$, or even just D if it is clear from context which structure $\mathfrak{M}$ we are talking about.

Let's think for a minute about the conditions we've imposed on first-order structures. If our goal in defining the notion of a structure were simply to devise set-theoretic models of blocks worlds, then it would be natural to impose much stronger conditions than we have. For example, we might want to require that $D^{\mathfrak{M}}$ be a set of blocks (not just any old objects), that $\mathsf{Cube}^{\mathfrak{M}}$ always be the set of cubes in $D^{\mathfrak{M}}$ and that $\mathsf{Larger}^{\mathfrak{M}}$ always be the ordered pairs $\langle x, y \rangle$ where x is larger than y. These requirements would be analogous to our condition that the extension of the identity symbol always corresponds to the real identity relation.

But remember what we are trying to capture. We are interested in characterizing the first-order consequence relation, and as we have explained, this relation ignores the specific meanings of predicates other than $=$. When we ignore the specific meanings of Cube and Larger, all that we care about is which objects in the domain satisfy the atomic wffs $\mathsf{Cube}(\mathsf{x})$ and $\mathsf{Larger}(\mathsf{x}, \mathsf{y})$. This is why in our definition we allow extensions of these predicates to be arbitrary sets, so long as the arity of the relation is respected.

Exercises

18.1 Write out a complete description of a first-order structure $\mathfrak{M}$ that would represent Mary Ellen's World. This has been done above except for the packaging into a single function.

18.2 (Simon says) Open Mary Ellen's World. The structure $\mathfrak{M}$ that we have used to model this world, with respect to the sublanguage involving only Cube, Larger, and c, is also a good model of many other worlds. What follows is a list of proposed changes to the world. Some of them are allowable changes, in that if you make the change, the model $\mathfrak{M}$ still represents the world with respect to this language. Other changes are not. Make the allowable changes, but not the others.

1. Move everything back one row.
2. Interchange the position of the tetrahedron and the large cube.
3. Make the tetrahedron a dodecahedron.
4. Make the large cube a dodecahedron.
5. Make the tetrahedron (or what was the tetrahedron, if you have changed it) large.
6. Add a cube to the world.
7. Add a dodecahedron to the world.

Now open Mary Ellen's Sentences. Check to see that all these sentences are true in the world you have built. If they are not, you have made some unallowable changes. Submit your modified world.

18.3 In the text we modeled Mary Ellen's World with respect to one sublanguage of Tarski's World. How would our structure have to be modified if we added the following to the language: Tet, Dodec, Between? That is, describe the first-order structure that would represent Mary Ellen's World, in its original state, for this expanded language. [Hint: One of your extensions will be the empty set.]

18.4 Consider a first-order language with one binary predicate Outgrabe. Suppose for some reason we are interested in first-order structures $\mathfrak{M}$ for this language which have the particular domain $\{Alice, Mad\ Hatter\}$. List all the sets of ordered pairs that could serve as the extension of the symbol Outgrabe. How many would there be if the domain had three elements?

18.5 In Section 14.4 (page 396) we promised to show how to make the semantics of generalized quantifiers rigorous. How could we extend the notion of a first-order structure to accommodate the addition of a generalized quantifier Q? Intuitively, as we have seen, a sentence like $\mathsf{Q}\,\mathsf{x}\,(\mathsf{A}(\mathsf{x}), \mathsf{B}(\mathsf{x}))$ asserts that a certain binary relation Q holds between the set A of things that satisfy $\mathsf{A}(\mathsf{x})$ and the set B that satisfies $\mathsf{B}(\mathsf{x})$ in $\mathfrak{M}$. Thus, the natural way to interpret them is by means of a binary relation on $\wp(D^{\mathfrak{M}})$. What quantifier corresponds to the each of the following binary relations on sets?

1. $A \subseteq B$
2. $A \cap B = \emptyset$
3. $A \cap B \neq \emptyset$
4. $|A \cap B| = 1$
5. $|A \cap B| \leq 3$
6. $|A \cap B| > |A - B|$

18.6 While we can't say with precision exactly which binary relation a speaker might have in mind
✎* with the use of some quantifiers, like *many*, we can still use this framework to illustrate the nature of the logical properties like conservativity, monotonicity, and so forth discussed in Section 14.5. Each of the following properties of binary relations Q on subsets of D correspond to a property of quantifiers. Identify them.

1. $Q(A, B)$ if and only if $Q(A, A \cap B)$
2. If $Q(A, B)$ and $A \subseteq A'$ then $Q(A', B)$
3. If $Q(A, B)$ and $A' \subseteq A$ then $Q(A', B)$
4. If $Q(A, B)$ and $B' \subseteq B$ then $Q(A, B')$
5. If $Q(A, B)$ and $B \subseteq B'$ then $Q(A, B')$

Section 18.2

Truth and satisfaction, revisited

In Chapter 9, we characterized the notion of truth in a domain of discourse rather informally. You will recall that in order to define what it means for a quantified sentence (either $\forall x\, S(x)$ or $\exists x\, S(x)$) to be true, we had to have recourse to the notion of satisfaction, what it means for an object b to satisfy a wff $S(x)$ in a domain of discourse. This was defined in terms of what it means for the simpler sentence $S(c)$ to be true, where c was a new name.

modeling satisfaction and truth

Now that we have defined the notion of a first-order structure, we can treat truth and satisfaction more rigorously. The aim here is just to see how our informal treatment looks when you treat it mathematically. There should be nothing surprising in this section, unless it is that these intuitive notions are a bit complicated to define rigorously.

In our earlier discussion, we explained what it meant for an object b in the domain of discourse to satisfy a wff $S(v)$ with one free variable. That was enough to serve our needs in discussing truth and the game. However, for more advanced work, it is important to understand what it means for some objects to satisfy a wff $P(x_1, \ldots, x_n)$ with n-free variables, for any $n \geq 0$. The case of $n = 0$ is the important special case where there are *no* free variables, that is, where P is a sentence.

variable assignments

Let $\mathfrak{M}$ be a first-order structure with domain D. A *variable assignment* in $\mathfrak{M}$ is, by definition, some (possibly partial) function g defined on a set of variables and taking values in the set D. Thus, for example, if $D = \{a, b, c\}$ then the following would all be variable assignments in $\mathfrak{M}$:

1. the function g_1 which assigns b to the variable x

2. the function g_2 which assigns a, b, and c to the variables x, y, and z, respectively

3. the function g_3 which assigns b to all the variables of the language

4. the function g_4 which is the empty function, that is, does not assign values to any variables

The special case of the empty variable assignment g_4 is important, so we denote it by $g_\emptyset$.

empty variable assignment ($g_\emptyset$)

Given a wff P, we say that the variable assignment g is *appropriate* for P if all the free variables of P are in the domain of g, that is, if g assigns objects to each free variable of P. Thus the four variable assignments g_1, g_2, g_3, and g_4 listed above would have been appropriate for the following sorts of wffs, respectively:

appropriate assignments

1. g_1 is appropriate for any wff with the single free variable x, or with no free variables at all;

2. g_2 is appropriate for any wff whose free variables are a subset of $\{x, y, z\}$;

3. g_3 is appropriate for any wff at all; and

4. g_4 (which we just agreed to write as $g_\emptyset$) is appropriate for any wff with no free variables, that is, for sentences, but not for wffs with free variables.

We next come to the definition of truth by way of satisfaction. The definition we gave earlier required us to define this by means of substituting names for variables. The definition we are about to give ends up being equivalent, but it avoids this detour. It works by defining satisfaction more generally. In particular, we will define what it means for an assignment g to satisfy a wff $P(x_1, \ldots, x_n)$ in $\mathfrak{M}$. We will define this by induction, with cases corresponding to the various ways of building up wffs from atomic wffs. This will reduce the problem gradually to the base case of atomic wffs, where we say explicitly what satisfaction means.

In order to handle the two inductive clauses in which P starts with a quantifier, we need a way to modify a variable assignment. For example, if g is defined on x and we want to say what it means for g to satisfy $\forall z\, \mathsf{Likes}(x, z)$, then we need to be able to take any object b in the domain of discourse and consider the variable assignment which is just like g except that it assigns the value b to the variable z. We will say that g satisfies our wff $\forall z\, \mathsf{Likes}(x, z)$ if and only if every such modified assignment g' satisfies $\mathsf{Likes}(x, z)$. To make this a bit easier to say, we introduce the notation "$g[z/b]$" for the modified variable assignment. Thus, in general, $g[v/b]$ is the variable assignment whose

modified variable assignments

domain is that of g plus the variable v and which assigns the same values as g, except that the new assignment assigns b to the variable v.

Here are a couple examples, harking back to our earlier examples of variable assignments given above:

1. g_1 assigns b to the variable x, so $g_1[y/c]$ assigns b to x and c to y. By contrast, $g_1[x/c]$ assigns a value only to x, the value c.

2. g_2 assigns a, b, c to the variables x, y, and z, respectively. Then $g_2[x/b]$ assigns the values b, b, and c to x, y, and z, respectively. The assignment $g_2[u/c]$ assigns the values c, a, b, and c to the variables u, x, y, and z, respectively.

3. g_3 assigns b to all the variables of the language. $g_3[y/b]$ is the same assignment, g_3, but $g_3[y/c]$ is different. It assigns c to y and b to every other variable.

4. g_4, the empty function, does not assign values to any variables. Thus $g_4[x/b]$ is the function which assigns b to x. Notice that this is the same function as g_1.

Notice that what variable assignments do for us is allow us to treat free variables as if they have a temporary denotation, not one assigned by the structure, but one assigned for purposes of the inductive definition of satisfaction. Thus, if a variable assignment g is appropriate for a wff P, then between $\mathfrak{M}$ and g, all the terms (constants and variables) in P have a denotation. For any term t, we write $[\![t]\!]_g^{\mathfrak{M}}$ for the denotation of t. Thus $[\![t]\!]_g^{\mathfrak{M}}$ is $t^{\mathfrak{M}}$ if t is an individual constant and $g(t)$ if t is a variable.

$[\![t]\!]_g^{\mathfrak{M}}$

We are now in a position to define what it means for a variable assignment g to satisfy a wff P in a first-order structure $\mathfrak{M}$. First, it is always required that g be appropriate for P, that is, be defined for all the free variables of P, and maybe other free variables. Second, there is nothing at all surprising in the following definition. There shouldn't be, anyway, since we are just trying to make precise the intuitive idea of satisfaction of a formula by a sequence of objects. We suggest that you work through the example at the end of the definition, referring back to the definition as needed, rather than try to read the definition itself right off.

definition of satisfaction

Definition (Satisfaction) Let P be a wff and let g be an assignment in $\mathfrak{M}$ which is appropriate for P.

1. **The atomic case.** Suppose P is $R(t_1, \ldots, t_n)$, where R is an n-ary predicate. Then g satisfies P in $\mathfrak{M}$ if and only if the n-tuple $\langle [\![t_1]\!]_g^{\mathfrak{M}}, \ldots, [\![t_n]\!]_g^{\mathfrak{M}} \rangle$ is in $R^{\mathfrak{M}}$.

2. **Negation.** Suppose P is ¬Q. Then g satisfies P in $\mathfrak{M}$ if and only if g does not satisfy Q.

3. **Conjunction.** Suppose P is Q ∧ R. Then g satisfies P in $\mathfrak{M}$ if and only if g satisfies both Q and R.

4. **Disjunction.** Suppose P is Q ∨ R. Then g satisfies P in $\mathfrak{M}$ if and only if g satisfies Q or R or both.

5. **Conditional.** Suppose P is Q → R. Then g satisfies P in $\mathfrak{M}$ if and only if g does not satisfy Q or g satisfies R or both.

6. **Biconditional.** Suppose P is Q ↔ R. Then g satisfies P in $\mathfrak{M}$ if and only if g satisfies both Q and R or neither.

7. **Universal quantification.** Suppose P is ∀v Q. Then g satisfies P in $\mathfrak{M}$ if and only if for every $d \in D^{\mathfrak{M}}$, $g[v/d]$ satisfies Q.

8. **Existential quantification.** Suppose P is ∃v Q. Then g satisfies P in $\mathfrak{M}$ if and only if for some $d \in D^{\mathfrak{M}}$, $g[v/d]$ satisfies Q.

It is customary to write

$$\mathfrak{M} \models P\ [g]$$

to indicate that the variable assignment g satisfies wff P in the structure $\mathfrak{M}$.

Let's work through a very simple example. We take a structure $\mathfrak{M}$ with domain $D = \{a, b, c\}$. Let us suppose that our language contains the binary predicate Likes and that the extension of this predicate is the following set of pairs:

$$\text{Likes}^{\mathfrak{M}} = \{\langle a, a\rangle, \langle a, b\rangle, \langle c, a\rangle\}$$

That is, a likes itself and b, c likes a, and b likes no one. Let us consider the wff

$$∃y\,(\text{Likes}(x, y) \land ¬\text{Likes}(y, y))$$

with the single free variable x. If the above definition is doing its stuff, it should turn out that an assignment g satisfies this wff just in case g assigns a to the variable x. After all, a is the only individual who likes someone who does not like himself.

Let's examine the definition of satisfaction to see if this is the way it turns out. First, note that g has to assign *some* value to x, since it has to be appropriate for the formula. Let us call this value e; e is one of a, b, or c. Next, we see from the clause for ∃ that g satisfies our wff just in case there is some object $d \in D$ such that $g[y/d]$ satisfies the wff

$$\text{Likes}(x, y) \land ¬\text{Likes}(y, y)$$

$\mathfrak{M} \models P\ [g]$

But $g[y/d]$ satisfies this wff if and only if it satisfies Likes(x, y) but does not satisfy Likes(y, y), by the clauses for conjunction and negation. Looking at the atomic case, we see that this is true just in case the pair $\langle e, d \rangle$ is in the extension of Likes, while the pair $\langle d, d \rangle$ is not. But this can only happen if $e = a$ and $d = b$. Thus the only way our original g can satisfy our wff is if it assigns a to the variable x, as we anticipated.

Notice in the above example how we started off with a wff with one free variable and an assignment defined on that one variable, but in order to give our analysis, we had to move to consider a wff with two free variables and so to assignments defined on those two free variables. This is typical. After all, what we are really interested in is truth for sentences, that is, wffs with no free variables, but in order to define this, we must define something more general, satisfaction of wffs with free variables by assignments defined on those variables. Indeed, having defined satisfaction, we are now in a position to look at the special case where the wffs have no free variables and use it for our definition of truth.

definition of truth

Definition (Truth) Let $\mathcal{L}$ be some first-order language and let $\mathfrak{M}$ be a structure for $\mathcal{L}$. A sentence P of $\mathcal{L}$ is *true* in $\mathfrak{M}$ if and only if the empty variable assignment $g_\emptyset$ satisfies P in $\mathfrak{M}$. Otherwise P is false in $\mathfrak{M}$.

$\mathfrak{M} \models P$

Just as we write $\mathfrak{M} \models Q\,[g]$ if g satisfies a wff Q in $\mathfrak{M}$, so too we write:

$$\mathfrak{M} \models P$$

if the sentence P is true in $\mathfrak{M}$.

Let's look back at the structure given just above and see if the sentence

$$\exists x \, \exists y \, (\text{Likes}(x, y) \land \neg\text{Likes}(y, y))$$

come out as it should under this definition. First, notice that it is a sentence, that is, has no free variables. Thus, the empty assignment is appropriate for it. Does the empty assignment satisfy it? According to the definition of satisfaction, it does if and only if there is an object that we can assign to the variable x so that the resulting assignment satisfies

$$\exists y \, (\text{Likes}(x, y) \land \neg\text{Likes}(y, y))$$

But we have seen that there is such an object, namely, a. So the sentence is true in $\mathfrak{M}$; in symbols, $\mathfrak{M} \models \exists x \, \exists y \, (\text{Likes}(x, y) \land \neg\text{Likes}(y, y))$.

Consider next the sentence

$$\forall x \, \exists y \, (\text{Likes}(x, y) \land \neg\text{Likes}(y, y))$$

Does the empty assignment satisfy this? It does if and only if for every object e in the domain, if we assign e to x, the resulting assignment g satisfies

$$\exists y \, (\mathsf{Likes}(x, y) \wedge \neg \mathsf{Likes}(y, y))$$

But, as we showed earlier, g satisfies this only if g assigns a to x. If it assigns, say, b to x, then it does not satisfy the wff. Hence, the empty assignment does not satisfy our sentence, i.e., the sentence is not true in $\mathfrak{M}$. So its negation is; in symbols, $\mathfrak{M} \models \neg \forall x \, \exists y \, (\mathsf{Likes}(x, y) \wedge \neg \mathsf{Likes}(y, y))$.

A number of problems are given later to help you understand that this does indeed model the informal, intuitive notion. In the meantime, we will state a proposition that will be important in proving the Soundness Theorem for FOL. Intuitively, whether or not a sentence is true in a structure should depend only on the meanings specified in the structure for the predicates and individual constants that actually occur in the sentence. That this is the case is a consequence of the following, somewhat stronger claim.

Proposition 1. Let $\mathfrak{M}_1$ and $\mathfrak{M}_2$ be structures which have the same domain and assign the same interpretations to the predicates and constant symbols in a wff P. Let g_1 and g_2 be variable assignments that assign the same objects to the free variables in P. Then $\mathfrak{M}_1 \models P[g_1]$ iff $\mathfrak{M}_2 \models P[g_2]$.

The proof of this proposition, which uses induction on wffs, is a good exercise to see if you understand the definition of satisfaction. Consequently, we ask you to prove it in Exercise 18.10.

Once we have truth, we can define the important notions of first-order consequence and first-order validity, our new approximations of the intuitive notions of logical consequence and logical truth. In the following definitions, we assume that we have a fixed first-order language and that all sentences come from that language. By a structure, we mean any first-order structure that interprets all the predicates and individual constants of the language.

Definition[First-order consequence] A sentence Q is a *first-order consequence* of a set $\mathcal{T} = \{P_1, \dots\}$ of sentences if and only if every structure that makes all the sentences in $\mathcal{T}$ true also makes Q true.

definition of FO consequence

You can see that this definition is the exact analogue of our definition of tautological consequence. The only difference is that instead of rows of a truth table (or truth value assignments), we are using first-order structures in the definition. We similarly modify our definition of tautology to get the following definition of first-order validity.

Definition (First-order validity) A sentence P is a *first-order validity* if and only if every structure makes P true.

definition of FO validity

FO-satisfiable

We will also use other notions analogous to those introduced in propositional logic in discussing first-order sentences and sets of sentences. For example, we will call a sentence FO-*satisfiable* if there is a first-order structure that makes it true, and call a set of sentences FO-*satisfiable* if there is a structure that makes all the members of the set true. Sometimes we will leave out the "FO" if the context make it clear what kind of satisfiability we are referring to.

You may have wondered why Tarski's World is so named. It is our way of paying tribute to Alfred Tarski, the logician who played the pivotal role in the development of the semantic conception of logic. It was Tarski who developed the notion of a first-order structure, the notion of satisfaction, and who gave the first analysis of truth, first-order validity, and first-order consequence along the lines we have sketched here.

One final note. If you go on to study logic further, you will discover that our treatment of satisfaction is a bit more general that most. Tarski, and most of those who have followed him, have restricted attention to total variable assignments, that is, to variable assignments that are defined on all variables. Then, to define truth, they pick out one of these total assignments and use it, since they cannot use the empty assignment. The two approaches agree on the resulting notion of truth, and hence on the notion of logical consequence. The approach adopted here using partial assignments is more general, seems to us more natural, and fits in better with our implementation of Tarski's World. It is easy to represent finite partial assignments in the computer's memory, but not so easy to deal with infinite assignments.

Remember

1. First-order structures are mathematical models of the domains about which we make claims using FOL.

2. Variable assignments are functions mapping variables into the domain of some first-order structure.

3. A variable assignment satisfies a wff in a structure if, intuitively, the objects assigned to the variables make the wff true in the structure.

4. Using the notion of satisfaction, we can define what it means for a sentence to be true in a structure.

5. Finally, once we have the notion of truth in a structure, we can model the notions of logical truth, and logical consequence.

Exercises

18.7 (Modifying variable assignments.) Suppose $D = \{a, b, c, d\}$ and let g be the variable assignment
which is defined only on the variable x and takes value b. Describe explicitly each of the
following:

    1. $g[y/c]$
    2. $g[x/c]$
    3. $g[z/b]$
    4. $g[x/b]$
    5. $(g[x/c])[z/d]$
    6. $(g[x/c])[x/d]$

18.8 Consider the language with only one binary predicate symbol P and let $\mathfrak{M}$ be the structure
with domain $D = \{1, 2, 3\}$ and where the extension of P consists of those pairs $\langle n, m \rangle$ such
that $m = n + 1$. For each of the following wffs, first describe which variable assignments are
appropriate for it. Then describe the variable assignments which satisfy it, much the way we
described the variable assignments that satisfy the wff $\forall z\,\mathsf{Likes}(x, z)$ on page 517.

    1. $\mathsf{P(y,z)}$
    2. $\exists y\,\mathsf{P}(y, z)$
    3. $\forall z\,\mathsf{P}(y, z)$
    4. $\mathsf{P(x,x)}$
    5. $\exists x\,\neg\mathsf{P}(x, x)$
    6. $\forall x\,\mathsf{P}(x, x)$
    7. $\mathsf{P}(x, x) \vee \mathsf{P}(y, z)$
    8. $\exists x\,(\mathsf{P}(x, x) \vee \mathsf{P}(y, z))$
    9. $\exists y\,(\mathsf{P}(x, x) \vee \mathsf{P}(y, z))$
    10. $\forall y\,\exists z\,\mathsf{P}(y, z)$
    11. $\forall y\,\exists y\,\mathsf{P}(y, z)$

Now consider the structure $\mathcal{N}$ with the same domain but where the extension of P is the set
of those pairs $\langle n, m \rangle$ such that $n \leq m$. How do your answers change?

18.9 Let g be a variable assignment in $\mathfrak{M}$ which is appropriate for the wff P. Show that the following
three statements are equivalent:

    1. g satisfies P in $\mathfrak{M}$
    2. g' satisfies P in $\mathfrak{M}$ for some extension g' of g
    3. g' satisfies P in $\mathfrak{M}$ for every extension g' of g

Intuitively, this is true because whether a variable assignment satisfies P can depend only on
the free variables of P, but it needs a proof. What does this result say in the case where P is
a sentence? Express your answer using the concept of truth. [Hint: You will need to prove this
by induction on wffs.]

18.10 Give a proof of Proposition 1.
✎⋆

The next two exercises should be attempted before going on to Chapter 19. They contain some key insights that will be important in the proof of the Completeness Theorem for $\mathcal{F}$.

18.11 (From first-order structures to truth assignments.) Recall from Section 10.1 that when dealing
✎ with sentences containing quantifiers, any sentence that starts with a quantifier is treated just like an atomic sentence from the point of view of truth tables and hence truth assignments. Given a first-order structure $\mathfrak{M}$ for a language L, define a truth assignment $h^{\mathfrak{M}}$ as follows: for any sentence S that is atomic or begins with a quantifier,

$$h^{\mathfrak{M}}(\mathsf{S}) = \text{TRUE if and only if } \mathfrak{M} \models \mathsf{S}$$

Show that the same "if and only if" holds for all sentences.

18.12 (From truth assignments to first-order structures.) Let h be any truth assignment for a first-
✎ order language without function symbols. Construct a first-order structure $\mathfrak{M}_h$ as follows. Let the domain of $\mathfrak{M}$ be the set of individual constants of the language. Given a relation symbol R, binary let's say for simplicity of notation, define its extension to be

$$\{\langle \mathsf{c}, \mathsf{d} \rangle \mid h(\mathsf{R}(\mathsf{c}, \mathsf{d})) = \text{TRUE}\}$$

Finally, interpret each individual constant as naming itself.

1. Show that for any sentence S that does not contain quantifiers or the identity symbol:

$$\mathfrak{M}_h \models \mathsf{S} \text{ iff } h(\mathsf{S}) = \text{TRUE}$$

[Hint: use induction on wffs.]
2. Show that the result in (1) does not extend to sentences containing the identity symbol. [Hint: consider an h that assigns FALSE to b = b.]
3. Recall from Section 10.1 that it is possible for a truth assignment h to assign TRUE to Cube(b) but FALSE to ∃x Cube(x). Show that for such an h, the result in (1) does not extend to quantified sentences.

18.13 (An important problem about satisfiability.) Open **Skolem's Sentences**. You will notice that
↗|✎ these sentences come in pairs. Each even-numbered sentence is obtained from the preceding sentence by replacing some names with variables and existentially quantifying the variables. The odd-numbered sentence logically implies the even-numbered sentence which follows it, of course, by existential generalization. The converse does not hold. But something close to it does. To see what, open **Thoralf's First World** and check the truth values of the sentences in the world. The even numbered sentences all come out true, while the odd sentences can't be evaluated because they contain names not in use in the world.

Extend **Thoralf's First World** by assigning the names b, c, d and e in such a way that the odd numbered sentences are also true. Do the same for Thoralf's Second World, saving the resulting worlds as World 18.13.1 and World 18.13.2. Submit these worlds.

Explain under what conditions a world in which $\exists x\, P(x)$ is true can be extended to one in which $P(c)$ is true. Turn in your explanation to your instructor.

Soundness for FOL

Having made the notion of first-order consequence more precise using the notion of first-order structure, we are now in a position to state and prove the Soundness Theorem for FOL. Given a set $\mathcal{T}$ of sentences we write $\mathcal{T} \vdash S$ to mean there is a proof of S from premises in $\mathcal{T}$ in the full system $\mathcal{F}$.[1] As mentioned in Chapter 17, this notation does not mean that all the sentences in $\mathcal{T}$ have to be used in the formal proof of S, only that there is a proof of S whose premises are all elements of $\mathcal{T}$. In particular, the set $\mathcal{T}$ could be infinite (as in the case of proofs from ZFC or PA) whereas only a finite number of premises can be used in any one proof. This notation allows us to state the Soundness Theorem as follows.

Theorem (Soundness of $\mathcal{F}$) If $\mathcal{T} \vdash S$, then S is a first-order consequence of set $\mathcal{T}$.

soundness of $\mathcal{F}$

> **Proof:** The proof is very similar to the proof of the Soundness Theorem for $\mathcal{F}_T$, the propositional part of $\mathcal{F}$, on page 216. We will show that any sentence that occurs at any step in a proof p in $\mathcal{F}$ is a first-order consequence of the assumptions in force at that step (which include the premises of p). This claim applies not just to sentences at the main level of proof p, but also to sentences appearing in sub-proofs, no matter how deeply nested. The theorem follows from this claim because if S appears at the main level of p, then the only assumptions in force are premises drawn from $\mathcal{T}$. So S is a first-order consequence of $\mathcal{T}$.
>
> Call a step of a proof *valid* if the sentence at that step is a first-order consequence of the assumptions in force at that step. Our earlier proof of soundness for $\mathcal{F}_T$ was actually a disguised form of induction on the number of the step in question. Since we had not yet discussed induction, we disguised this by assuming there was an invalid step

---

[1] Recall that the formal proof system $\mathcal{F}$ includes all the introduction and elimination rules, but not the **Con** procedures.

and considering the first of these. When you think about it, you
see that this is really just the inductive step in an inductive proof.
Assuming we have the first invalid step allows us to assume that all
the earlier steps are valid, which is the inductive hypothesis, and
then prove (by contradiction) that the current step is valid after all.
We could proceed in the same way here, but we will instead make
the induction explicit. We thus assume that we are at the nth step,
that all earlier steps are valid, and show that this step is valid as
well.

The proof is by cases, depending on which rule is applied at step
n. The cases for the rules for the truth-functional connectives work
out pretty much as before. We will look at one, to point out the
similarity to our earlier soundness proof.

$\rightarrow$ **Elim**: Suppose the nth step derives the sentence R from an application of $\rightarrow$ **Elim** to sentences Q $\rightarrow$ R and Q appearing earlier in the
proof. Let $A_1, \ldots, A_k$ be a list of all the assumptions in force at step
n. By our induction hypothesis we know that Q $\rightarrow$ R and Q are both
established at valid steps, that is, they are first-order consequences
of the assumptions in force at those steps. Furthermore, since $\mathcal{F}$ only
allows us to cite sentences in the main proof or in subproofs whose
assumptions are still in force, we know that the assumptions in force
at steps Q $\rightarrow$ R and Q are also in force at R. Hence, the assumptions for these steps are among $A_1, \ldots, A_k$. Thus, both Q $\rightarrow$ R and
Q are first-order consequences of $A_1, \ldots, A_k$. We now show that R is
a first-order consequence of $A_1, \ldots, A_k$.

Suppose $\mathfrak{M}$ is a first-order structure in which all of $A_1, \ldots, A_k$ are
true. Then we know that $\mathfrak{M} \models$ Q $\rightarrow$ R and $\mathfrak{M} \models$ Q, since these sentences are first-order consequences of $A_1, \ldots, A_k$. But in that case,
by the definition of truth in a structure we see that $\mathfrak{M} \models$ R as well.
So R is a first-order consequence of $A_1, \ldots, A_k$. Hence, step n is a
valid step.

Notice that the only difference in this case from the corresponding
case in the proof of soundness of $\mathcal{F}_{\mathrm{T}}$ is our appeal to first-order
structures rather than rows of a truth table. The remaining truth-
functional rules are all similar. Let's now consider a quantifier rule.

$\exists$ **Elim**: Suppose the nth step derives the sentence R from an application of $\exists$ **Elim** to the sentence $\exists x P(x)$ and a subproof containing
R at its main level, say at step m. Let c be the new constant intro-

duced in the subproof. In other words, $P(c)$ is the assumption of the subproof containing R:

$$
\begin{array}{l}
\vdots \\
j. \;\; \exists x\, P(x) \\
\vdots \\
\qquad \boxed{c}\;\; P(c) \\
\qquad \vdots \\
\qquad m. \;\; R \\
\qquad \vdots \\
n. \;\; R
\end{array}
$$

Let $A_1, \ldots, A_k$ be the assumptions in force at step n. Our inductive hypothesis assures us that steps j and m are valid steps, hence $\exists x\, P(x)$ is a first-order consequence of the assumptions in force at step j, which are a subset of $A_1, \ldots, A_k$, and R is a first-order consequence of the assumptions in force at step m, which are a subset of $A_1, \ldots, A_k$, *plus the sentence* $P(c)$, the assumption of the subproof in which m occurs.

We need to show that R is a first-order consequence of $A_1, \ldots, A_k$ alone. To this end, assume that $\mathfrak{M}$ is a first-order structure in which each of $A_1, \ldots, A_k$ is true. We need to show that R is true in $\mathfrak{M}$ as well. Since $\exists x\, P(x)$ is a consequence of $A_1, \ldots, A_k$, we know that this sentence is also true in $\mathfrak{M}$. Notice that the constant c cannot occur in any of the sentences $A_1, \ldots, A_k$, $\exists x\, P(x)$, or R, by the restriction on the choice of temporary names imposed by the $\exists$ **Elim** rule. Since $\mathfrak{M} \models \exists x\, P(x)$, we know that there is an object, say b, in the domain of $\mathfrak{M}$ that satisfies $P(x)$. Let $\mathfrak{M}'$ be exactly like $\mathfrak{M}$, except that it assigns the object b to the individual constant c. Clearly, $\mathfrak{M}' \models P(c)$, by our choice of interpretation of c. By Proposition 1 on page 521 $\mathfrak{M}'$ also makes each of the assumptions $A_1, \ldots, A_k$ true. But then $\mathfrak{M}' \models R$, because R is a first-order consequence of these sentences. Since c does not occur in R, R is also true in the original structure $\mathfrak{M}$, again by Proposition 1.

The case of $\forall$ **Intro** is very similar to $\exists$ **Elim**, and the remaining two cases are much simpler. We leave these cases as exercises.

The Soundness Theorem for $\mathcal{F}$ assures us that we will never prove an invalid argument using just the rules of $\mathcal{F}$. It also warns us that we will never be able to prove a valid argument whose validity depends on meanings of predicates other than identity. The Completeness Theorem for $\mathcal{F}$ is significantly harder to prove than the Soundness Theorem for $\mathcal{F}$, or for that matter, than the Completeness Theorem for $\mathcal{F}_{\mathrm{T}}$. In fact, it is the most significant theorem that we prove in this book and forms the main topic of Chapter 19.

Exercises

18.14 Prove the inductive step in the soundness proof corresponding to the rule ∧ **Intro**.

18.15 Prove the inductive step in the soundness proof corresponding to the rule → **Intro**.

18.16 Prove the inductive step in the soundness proof corresponding to the rule ∃ **Intro**.

18.17 Prove the inductive step in the soundness proof corresponding to the rule ∀ **Intro**.

Section 18.4

The completeness of the shape axioms

In Section 12.5 (on page 348), we promised to convince you that the ten axioms we gave for shape are complete, that is, that they completely bridged the gap between first-order consequence and the intuitive notion of logical consequence for the blocks language, as far as shape is concerned. We list the axioms again here for your convenience:

Basic Shape Axioms:

1. $\neg \exists x\, (\mathsf{Cube}(x) \land \mathsf{Tet}(x))$

2. $\neg \exists x\, (\mathsf{Tet}(x) \land \mathsf{Dodec}(x))$

3. $\neg \exists x\, (\mathsf{Dodec}(x) \land \mathsf{Cube}(x))$

4. $\forall x\, (\mathsf{Tet}(x) \lor \mathsf{Dodec}(x) \lor \mathsf{Cube}(x))$

SameShape Introduction Axioms:

5. $\forall x\, \forall y\, ((\mathsf{Cube}(x) \land \mathsf{Cube}(y)) \to \mathsf{SameShape}(x, y))$

6. $\forall x\, \forall y\, ((\mathsf{Dodec}(x) \land \mathsf{Dodec}(y)) \to \mathsf{SameShape}(x, y))$

7. $\forall x \forall y ((\mathsf{Tet}(x) \wedge \mathsf{Tet}(y)) \rightarrow \mathsf{SameShape}(x, y))$

SameShape Elimination Axioms:

8. $\forall x \forall y ((\mathsf{SameShape}(x, y) \wedge \mathsf{Cube}(x)) \rightarrow \mathsf{Cube}(y))$

9. $\forall x \forall y ((\mathsf{SameShape}(x, y) \wedge \mathsf{Dodec}(x)) \rightarrow \mathsf{Dodec}(y))$

10. $\forall x \forall y ((\mathsf{SameShape}(x, y) \wedge \mathsf{Tet}(x)) \rightarrow \mathsf{Tet}(y))$

We need to show that any argument that is logically valid in virtue of the meanings of the shape predicates (and the first-order quantifiers, connectives, and identity) is first-order valid once we add these ten axioms as premises. To show this, it suffices to show that any first-order structure $\mathfrak{M}$ making the axioms true is just like one where the meanings of the four shape predicates is as intended.[2]

The reason this suffices is not hard to see. For suppose we have an argument A that is valid in virtue of the meanings of the shape predicates. We want to show that the result A' of adding the ten axioms gives us an argument that is first-order valid. To do this, it suffices to show that any first-order structure $\mathfrak{M}$ making the original premises and the ten axioms true is just like a structure $\mathfrak{M}'$ where the predicates mean what they should. Hence, by the presumed validity of the argument A in the latter such structures, the conclusion holds in $\mathfrak{M}'$. But since $\mathfrak{M}$ is just like $\mathfrak{M}'$, the conclusion also holds in $\mathfrak{M}$. Hence, since the structure $\mathfrak{M}$ was an arbitrary one making the original premises and the ten axioms true, this will show that A' is first-order valid.

So now let us prove our claim about $\mathfrak{M}$ and $\mathfrak{M}'$. Recall that $\mathfrak{M}$ is any first-order structure making our ten shape axioms true. Let $Cu, Do,$ and Te be the extensions in $\mathfrak{M}$ of Cube, Dodec, and Tet, respectively. Axiom 1 insures that Cu and Te are disjoint. Similarly, by Axioms 2 and 3, all three of the sets are disjoint from the others. Axiom 4 insures us that everything in the domain D of $\mathfrak{M}$ is in one of these three sets.

Recall from Exercise 15.47 that a partition of D is a set $\mathcal{P}$ of non-empty subsets of D with the property that every element of D is in exactly one member of $\mathcal{P}$. As we saw in that exercise, every such partition is the set of equivalence classes of an equivalence relation, the relation of being in the same member of the partition. This applied directly to our setting. Not all of these sets Cu, Do and Te need be non-empty, but if we restrict attention to those that are, the preceding paragraph shows that we have a partition of D.

---

[2]What "just like" means here is that the structures are *isomorphic*, a notion we have not defined. The intuitive notion should be enough to convince you of our claim.

Now let S be the extension of SameShape in $\mathfrak{M}$. The remaining six axioms insure that this relation is the equivalence relation generated by our partition. Replace each object in Cu by a cube, each object in Do by a dodecahedron, and each object in Te by a tetrahedron, making sure to use distinct blocks to replace distinct objects of $\mathfrak{M}$. This is possible because we have a partition of D. Call the resulting structure $\mathfrak{M}'$. The extension of SameShape in $\mathfrak{M}'$ is just like S, except with the new blocks. Hence, the meaning of our predicates is what we want it to be. Since the structures are otherwise unchanged, the structures satisfy the same sentences of the blocks language.

Exercises

18.18 Let $\mathfrak{M}$ be the structure whose domain is the natural numbers, where Cube, Dodec, and Tet have as extensions the sets of natural numbers of the forms $3n$, $3n + 1$, and $3n + 2$. Can we interpret SameShape so as to make the ten shape axioms true? If so, in how many ways can we do this?

18.19 Let $\mathfrak{M}$ be any first-order structure making the first four shape axioms true. Prove that there is a unique way to interpret SameShape so as to make all ten axioms true.

Section 18.5

Skolemization

One important role function symbols play in first-order logic is as a way of simplifying (for certain purposes) sentences that have lots of quantifiers nested inside one another. To see an example of this, consider the sentence

$$\forall x \, \exists y \, \mathsf{Neighbor}(x, y)$$

Given a fixed domain of discourse (represented by a first-order structure $\mathfrak{M}$, say) this sentence asserts that every b in the domain of discourse has at least one neighbor c. Let us write this as

$$\mathfrak{M} \models \mathsf{Neighbor}(x, y)[b, c]$$

rather than the more formal $\mathfrak{M} \models \mathsf{Neighbor}(x, y)[g]$ where g is the variable assignment that assigns b to x and c to y. Now if the original quantified sentence is true, then we can pick out, for each b, one of b's neighbors, say his nearest neighbor or favorite neighbor. Let $f(b)$ be this neighbor, so that we have, for every b

$$\mathfrak{M} \models \mathsf{Neighbor}(\mathsf{x}, \mathsf{y})[b, f(b)]$$

Now, we would like to say the following: if we had a function symbol f expressing our function f.

$$\mathfrak{M} \models \forall \mathsf{x}\, \mathsf{Neighbor}(\mathsf{x}, \mathsf{f}(\mathsf{x}))$$

This would reduce the quantifier string "$\forall\mathsf{x}\, \exists\mathsf{y}$" in the original sentence to the simpler "$\forall\mathsf{x}$." So we need to expand our first-order language and give ourselves such a function symbol f to use as a name of f.

This important trick is known as *Skolemization*, after the Norwegian logician Thoralf Skolem. The function f is called a *Skolem function* for the original quantified sentence. The new sentence, the one containing the function symbol but no existential quantifier, is called the *Skolem normal form* of the original sentence.

Skolemization
Skolem function

Skolem normal form

Notice that we did not say that a sentence is logically equivalent to its Skolemization. The situation is a little more subtle than that. If our language allowed existential quantification to apply to function symbols, we could get a logically equivalent sentence, namely

$$\exists \mathsf{f}\, \forall \mathsf{x}\, \mathsf{P}(\mathsf{x}, \mathsf{f}(\mathsf{x}))$$

This sort of sentence, however, takes us into what is known as second-order logic, which is beyond the scope of this book.

Skolem functions, and Skolem normal form, are very important in advanced parts of logic. We will discuss one application of them later in the chapter, when we sketch how to apply the resolution method to FOL with quantifiers.

One of the reasons that natural language does not get bogged down in lots of embedded quantifiers is that there are plenty of expressions that act like function symbols, so we can usually get by with Skolemizations. Possessives, for example, act as very general Skolem functions. We usually think of the possessive "apostrophe s" as indicating ownership, as in *John's car*. But it really functions much more generally as a kind of Skolem function. For example, if we are trying to decide where the group will eat out, then *Max's restaurant* can refer to the restaurant that Max likes best. Or if we are talking about logic books, we can use *Kleene's book* to refer not to one Kleene owns, but to one he wrote.

Skolemization in
natural language

Remember

(Simplest case of Skolemization) Given a sentence of the form $\forall x \, \exists y \, P(x, y)$ in some first-order language, we Skolemize it by choosing a function symbol f not in the language and writing $\forall x \, P(x, f(x))$. Every world that makes the Skolemization true also makes the original sentence true. Every world that makes the original sentence true can be turned into one that makes the Skolemization true by interpreting the function symbol f by a function f which picks out, for any object b in the domain, some object c such that they satisfy the wff $P(x, y)$.

Exercises

18.20 Discuss the logical relationship between the following two sentences. [Hint: One is a logical consequence of the other, but they are not logically equivalent.]

$$\forall y \, \exists z \, \mathsf{ParentOf}(z, y)$$
$$\forall y \, \mathsf{ParentOf}(\mathsf{bestfriend}(y), y)$$

Explain under what conditions the second would be a Skolemization of the first.

18.21 Skolemize the following sentence using the function symbol f.

$$\forall z \, \exists y \, [(1 + (z \times z)) < y]$$

Which of the following functions on natural numbers could be used as a Skolem function for this sentence?

1. $f(z) = z^2$
2. $f(z) = z^2 + 1$
3. $f(z) = z^2 + 2$
4. $f(z) = z^3$

Section 18.6

Unification of terms

We now turn to a rather different topic, unification, that applies mainly to languages that contain function symbols. Unification is of crucial importance when we come to extend the resolution method to the full first-order language.

The basic idea behind unification can be illustrated by comparing a couple of claims. Suppose first that Nancy tells you that Max's father drives a Honda, and that no one's grandfather drives a Honda. Now this is not true, but there is nothing logically incompatible about the two claims. Note that if Nancy went on to say that Max was a father (so that Max's father was a grandfather) we could then accuse her of contradicting herself. Contrast Nancy's claim with Mary's, that Max's grandfather drives a Honda, and that no one's father drives a Honda. Mary can be accused of contradicting herself. Why? Because grandfathers are, among other things, fathers.

More abstractly, compare the following pairs, where P is a unary predicate symbol, f and g are unary function symbols, and a is an individual constant.

First pair:	P(f(a)),	$\forall x \neg P(f(g(x)))$
Second pair:	P(f(g(a))),	$\forall x \neg P(f(x))$

The first pair is a logical possibility. It is perfectly consistent to suppose that the object $f(a)$ has property P but that no object of the form $f(g(b))$ has property P. This can only happen, though, if a is not of the form $g(b)$. By contrast, the second pair is not a logical possibility. Why? Because if $\forall x \neg P(f(x))$ holds, so does the instance where we substitute g(a) for x: $\neg P(f(g(a)))$. But this contradicts P(f(g(a))).

Unification gives us a useful test to see if sets of claims like the above are contradictory or not. You look at the terms involved and see whether they are "unifiable." The terms f(a) and f(g(x)) in the first pair of sentences are not unifiable, whereas the terms in the second pair, f(g(a)) and f(x), are unifiable. What this means is that in the second case there is a way to substitute a term for x so that the results coincide. This agreement produces a clash between the two original sentences. In the first pair of terms, however, there is no such way to make the terms coincide. No way of substituting a term in for the variable x in f(g(x)) is going to give you the term f(a), since a is an individual constant.

These examples motivate the following definition.

unification

Definition Terms t_1 and t_2 are *unifiable* if there is a substitution of terms for some or all of the variables in t_1 and t_2 such that the terms that result from the substitution are syntactically identical terms.

definition of unifiable terms

Similarly, any set T of terms, of whatever size, is said to be unifiable if there is a single substitution of terms for some or all of the variables that occur in terms of T so that all the resulting terms are identical.

Notice that whether or not terms are unifiable is a purely syntactic notion. It has to do with terms, not with what they denote. The terms father(Max) and father(father(x)) are not unifiable, regardless of whether or not Max is a father. On the other hand, the terms father(father(Max)) and father(y) are unifiable. Just substitute father(Max) in for the variable y. This means that we can decide whether a pair of terms is unifiable without any idea of what the terms happen to stand for.

Let's give a couple more examples of unification. Suppose we have a binary function symbol f and two unary function symbols g and h. Here is an example of three terms that are unifiable. Can you find the substitution that does the trick?

$$f(g(z), x), \quad f(y, x), \quad f(y, h(a))$$

If you said to substitute h(a) for the variable x and g(z) for y you were right. All three terms are transformed into the term $f(g(z), h(a))$. Are there any other substitutions that would work? Yes, there are. We could plug any term in for z and get another substitution. The one we chose was the simplest in that it was the most general. We could get any other from it by means of a substitution.

most general unifiers

Here are some examples of pairs, some of which can, others of which cannot, be unified. See if you can tell which are which before reading on.

$$
\begin{array}{ll}
g(x), & h(y) \\
h(f(x,x)), & h(y) \\
f(x,y), & f(y,x) \\
g(g(x)), & g(h(y)) \\
g(x), & g(h(z)) \\
g(x), & g(h(x))
\end{array}
$$

Half of these go each way. The ones that are unifiable are the second, third, and fifth. The others are not unifiable. The most general unifiers of the three that are unifiable are, in order:

- Substitute $f(x, x)$ for y

- Substitute some other variable z for both x and y

- Substitute h(z) for x

The first pair is not unifiable because no matter what you do, one will always start with g while the other starts with h. Similarly, the fourth pair is not unifiable because the first will always start with a pair of g's, while the second will always start with a g followed by an h. (The reason the last pair cannot be unified is a tad more subtle. Do you see why?)

Unification Algorithm

There is a very general procedure for checking when two (or more) terms are unifiable or not. It is known as the Unification Algorithm. We will not explain it in this book. But once you have done the following exercises, you will basically understand how the algorithm works.

18.22 Which of the following terms are unifiable with father(x) and which are not? If they are, give the substitution. If they are not, then explain why not.
1. Max
2. father(Claire)
3. mother(Max)
4. father(mother(Claire))
5. father(mother(y))
6. father(mother(x))

18.23 Which of the following terms are unifiable with f(x, g(x)) and which are not? If they are, give the most general unifier. If they are not, then explain why not. (Here, as usual, a and b are names, not variables.)
1. f(a, a)
2. f(g(a), g(a))
3. f(g(x), g(g(x)))
4. h(f(a, g(a)))
5. f(f(a, b), g(f(a, b)))

18.24 Find a set of four terms which can simultaneously be unified to obtain the following term:

$$h(f(h(a), g(a)))$$

18.25 Show that there are an infinite number of different substitutions that unify the following pair of terms. Find one that is most general.

$$g(f(x, y)), \quad g(f(h(y), g(z)))$$

18.26 How many substitutions are there that unify the following pair?

$$g(f(x, x)), \quad g(f(h(a), g(b)))$$

Resolution, revisited

In this section we discuss in an informal way how the resolution method for propositional logic can be extended to full first-order logic by combining the tools we have developed above.

The general situation is that you have some first-order premises $P_1, \ldots, P_n$ and a potential conclusion Q. The question is whether Q is a first-order consequence of $P_1, \ldots, P_n$. This, as we have seen, is the same as asking if there is no first-order structure which is a counterexample to the argument that Q follows from $P_1, \ldots, P_n$. This in turn is the same as asking whether the sentence

$$P_1 \wedge \ldots \wedge P_n \wedge \neg Q$$

extending resolution to FOL

is not FO-satisfiable. So the general problem can be reduced to that of determining, of a fixed finite sentence, say S, of FOL, whether it is FO-satisfiable.

The resolution method discussed earlier gives a procedure for testing this when the sentence S contains no quantifiers. But interesting sentences do contain quantifiers. Surprisingly, there is a method for reducing the general case to the case where there are no quantifiers.

An overview of this method goes as follows. First, we know that we can always pull all quantifiers in a sentence S out in front by logically valid transformations, and so we can assume S is in prenex form.

universal sentences

Call a sentence *universal* if it is in prenex form and all of the quantifiers in it are universal quantifiers. That is, a universal sentence S is of the form

$$\forall x_1 \ldots \forall x_n\, P(x_1, \ldots, x_n)$$

For simplicity, let us suppose that there are just two quantifiers:

$$\forall x\, \forall y\, P(x, y)$$

Let's assume that P contains just two names, b and c, and, importantly, that there are no function symbols in P.

We claim that S is FO-satisfiable if and only if the following set $\mathcal{T}$ of quantifier-free sentences is FO-satisfiable:

$$\mathcal{T} = \{P(b, b), P(b, c), P(c, b), P(c, c)\}$$

Note that we are not saying the two are equivalent. S obviously entails $\mathcal{T}$, so if S is FO-satisfiable so is $\mathcal{T}$. $\mathcal{T}$ does not in general entail S, but it is the case that if $\mathcal{T}$ is FO-satisfiable, so is S. The reason is fairly obvious. If you have a structure that makes $\mathcal{T}$ true, look at the substructure that just consists of b and c and the relationships they inherit. This little structure with only two objects makes S true.

reducing to non-quantified sentences

This neat little observation allows us to reduce the question of the unsatisfiability of the universal sentence S to a sentence of FOL containing no quantifiers, something we know how to solve using the resolution method for propositional logic.

caveats

There are a couple of caveats, though. First, since the resolution method for propositional logic gives us truth-assignments, in order for our proof to work must be able to go from a truth-assignment h for the atomic sentences of our language to a first-order structure $\mathfrak{M}_h$ for that language making the same atomic sentences true. This works for sentences that do not contain $=$, as we saw in Exercise 18.12, but not in general. This means that in order to be sure our proof works, the sentence S cannot contain $=$.

Exercise 18.12 also required that the sentence not contain any function symbols. This is a real pity, since Skolemization gives us a method for taking any prenex sentence S and finding another one that is universal and FO-satisfiable if and only if S is: just replace all the $\exists$'s one by one, left to right, by function symbols. So if we could only generalize the above method to the case where function symbols are allowed, we would have a general method. This is where the Unification Algorithm comes to the rescue. The basic strategy of resolution from propositional logic has to be strengthened a bit.

Skolemizing

Resolution method for FOL: Suppose we have sentences S, S′, S″, ... and want to show that they are not simultaneously FO-satisfiable. To do this using resolution, we would carry out the following steps:

resolution method for FOL

1. Put each sentence S into prenex form, say

$$\forall x_1 \exists y_1 \forall x_2 \exists y_2 \ldots P(x_1, y_1, x_2, y_2, \ldots)$$

  We can always make them alternate in this way by introducing some null quantifiers.

2. Skolemize each of the resulting sentences, say

$$\forall x_1 \forall x_2 \ldots P(x_1, f_1(x_1), x_2, f_2(x_1, x_2), \ldots)$$

  using different Skolem function symbols for different sentences.

3. Put each quantifier free matrix P into conjunctive normal form, say

$$P_1 \wedge P_2 \wedge \ldots \wedge P_n$$

  where each P_i is a disjunction of literals.

4. Distribute the universal quantifiers in each sentence across the conjunctions and drop the conjunction signs, ending with a set of sentences of the form

$$\forall x_1 \forall x_2 \ldots P_i$$

5. Change the bound variables in each of the resulting sentences so that no variable appears in two of them.

6. Turn each of the resulting sentences into a set of literals by dropping the universal quantifiers and disjunction signs. In this way we end up with a set of resolution clauses.

7. Use resolution and unification to resolve this set of clauses.

Rather than explain this in great detail, which would take us beyond the scope of this book, let's look at a few examples.

Example. Suppose you want to show that $\forall x\, P(x, b)$ and $\forall y\, \neg P(f(y), b)$ are not jointly FO-satisfiable, that is, that their conjunction is not FO-satisfiable. With this example, we can skip right to step 6, giving us two clauses, each consisting of one literal. Since we can unify x and $f(y)$, we see that these two clauses resolve to □.

Example. Suppose we are told that the following are both true:

$$\forall x\, (P(x, b) \vee Q(x))$$
$$\forall y\, (\neg P(f(y), b) \vee Q(y))$$

and we want to derive the sentence,

$$\forall y\, (Q(y) \vee Q(f(y)))$$

To show that this sentence is a first-order consequence of the first two, we need to show that those sentences together with the negation of this sentence are not simultaneously FO-satisfiable. We begin by putting this negation into prenex form:

$$\exists y\, (\neg Q(y) \wedge \neg Q(f(y)))$$

We now want to Skolemize this sentence. Since the existential quantifier in our sentence is not preceded by any universal quantifiers, to Skolemize this sentence we replace the variable y by a 0-ary function symbol, that is, an individual constant:

$$\neg Q(c) \wedge \neg Q(f(c)))$$

Dropping the conjunction gives us the following two sentences:

$$\neg Q(c)$$
$$\neg Q(f(c))$$

We now have four sentences to which we can apply step 6. This yields the following four clauses:

1. $\{P(x, b), Q(x)\}$
2. $\{\neg P(f(y), b), Q(y)\}$
3. $\{\neg Q(c)\}$
4. $\{\neg Q(f(c))\}$

Applying resolution to these shows that they are not FO-satisfiable. Here is a step-by-step derivation of the empty clause.

	Resolvent	Resolved Clauses	Substitution
5.	$\{Q(y), Q(f(y))\}$	1, 2	$f(y)$ for x
6.	$\{Q(f(c))\}$	3, 5	c for y
7.	□	4, 6	none needed

Example. Let's look at one more example that shows the whole method at work. Consider the two English sentences:

1. *Everyone admires someone who admires them unless they admire Quaid.*
2. *There are people who admire each other, at least one of whom admires Quaid.*

Suppose we want to use resolution to show that under one plausible reading of these sentences, (2) is a first-order consequence of (1). The readings we have in mind are the following, writing $A(x, y)$ for $\mathsf{Admires}(x, y)$, and using q for the name Quaid:

$$(\mathsf{S}_1) \quad \forall x\,[\neg A(x, \mathsf{q}) \to \exists y\,(A(x, y) \wedge A(y, x))]$$
$$(\mathsf{S}_2) \quad \exists x\,\exists y\,[A(x, \mathsf{q}) \wedge A(x, y) \wedge A(y, x)]$$

(When you figure out why S_1 logically entails S_2 in Problem 18.27, you may decide that these are not reasonable translations of the English. But that is beside the point here.)

Our goal is to show that S_1 and $\neg\mathsf{S}_2$ are not jointly FO-satisfiable. The sentence $\neg\mathsf{S}_2$ is equivalent to the following universal sentence, by DeMorgan's Laws:

$$\forall x\,\forall y\,(\neg A(x, \mathsf{q}) \vee \neg A(x, y) \vee \neg A(y, x))$$

The sentence S_1 is not logically equivalent to a universal sentence, so we must Skolemize it. First, note that it is equivalent to the prenex form:

$$\forall x\,\exists y\,[A(x, \mathsf{q}) \vee (A(x, y) \wedge A(y, x))]$$

Skolemizing, we get the universal sentence,

$$\forall x\,[A(x, \mathsf{q}) \vee (A(x, f(x)) \wedge A(f(x), x))]$$

Putting the quantifier-free part of this in conjunctive normal form gives us:

$$\forall x\,[(A(x, \mathsf{q}) \vee A(x, f(x))) \wedge (A(x, \mathsf{q}) \vee A(f(x), x))]$$

This in turn is logically equivalent to the conjunction of the following two sentences:

$$\forall x\,[A(x, \mathsf{q}) \vee A(x, f(x))]$$
$$\forall x\,[A(x, \mathsf{q}) \vee A(f(x), x))]$$

Next, we change variables so that no variable is used in two sentences, drop the universal quantifiers, and form clauses from the results. This leaves us with the following three clauses:

1. $\{A(x, q), A(x, f(x))\}$
2. $\{A(y, q), A(f(y), y)\}$
3. $\{\neg A(z, q), \neg A(z, w), \neg A(w, z)\}$

Finally, we apply resolution to derive the empty clause.

	Resolvent	Resolved Clauses	Substitution
4.	$\{A(q, f(q))\}$	1, 3	q for w, x, z
5.	$\{A(f(q), q)\}$	2, 3	q for w, y, z
6.	$\{\neg A(q, f(q))\}$	3, 5	f(q) for z, q for w
7.	$\square$	4, 6	none needed

The FO Con routine of Fitch

automated deduction

The resolution method provides us with a way to try to demonstrate logical consequence that is really rather different from giving an informal proof, or even giving a formal proof in a system like $\mathcal{F}$ that models normal informal methods. What it really involves is trying to build a counterexample to a claim of first-order consequence. If the method finds an insurmountable obstacle to building such a counterexample, in the form of $\square$, then it declares the conclusion a first-order consequence of the premises.

Methods like this turn out to be much more tractable to implement on computers than trying to find a natural proof in a system like $\mathcal{F}$. The reason is that one can set things up in a systematic way, rather than requiring the kind of semantic insight that is needed in giving proofs. After all, computers cannot, at least not yet, really pay attention to the meanings of sentences, the way we can.

FO Con routine

You have learned that Fitch has a routine, **FO Con**, that checks for first-order consequence. While it does not actually use resolution, it uses a method that is very similar in spirit. It basically tries to build a first-order structure that is a counterexample to the claim of consequence. If it finds an obstruction to building such a structure, then it declares the inference valid. Of course sometimes it is able to build such a counterexample. In these cases, it declares the inference invalid. And then sometimes it simply runs out of steam, room, or time, and, like the best of us, gives up.

Exercises

18.27 Give an informal proof that S_2 is a logical consequence of S_1.

18.28 Give an informal proof that the sentence given as a prenex form of S_1 really is logically equivalent to it.

18.29 There are usually many ways to proceed in resolution. In our derivation of □ in the last two examples, we chose optimal derivations. Work out different derivations for both.

18.30 Use the resolution method to show that the following sentence is a logical truth:

$$\exists x \, (P(x) \rightarrow \forall y \, P(y))$$

CHAPTER 19

Completeness and Incompleteness

This introduction to first-order logic culminates in discussions of two very famous and important results. They are the so-called Completeness Theorem and Incompleteness Theorem of FOL. Both are due to the logician Kurt Gödel, no doubt the greatest logician yet. We present a complete proof of the first, together with a explanation of the second, with just a sketch of its proof.

In this book you have learned the main techniques for giving proofs that one statement is a logical consequence of others. There were simple valid reasoning steps and more intricate methods of proof, like proof by contradiction or the method of general conditional proof. But the definition of logical consequence was fundamentally a semantic one: S is a logical consequence of premises $P_1,\ldots,P_n$ if there is no way for the premises to be true without the conclusion also being true. The question arises as to whether the methods of proof we have given are sufficient to prove everything we would like to prove. Can we be sure that if S is a logical consequence of $P_1,\ldots,P_n$, then we can find a proof of S from $P_1,\ldots,P_n$?

The answer is both yes and no, depending on just how we make the notion of logical consequence precise, and what language we are looking at.

Gödel's Completeness Theorem

The answer to our question is yes if by logical consequence we mean first-order consequence. Gödel's Completeness Theorem for FOL assures us that if S is a first-order consequence of some set $\mathcal{T}$ of first-order sentences then there is a formal proof of S using only premises from $\mathcal{T}$. The main goal of this chapter is to give a full proof of this important result. The first such completeness proof was given by Gödel in his dissertation in 1929. (His proof was actually about a somewhat different formal system, one used by Bertrand Russell and Alfred North Whitehead in their famous work *Pincipia Mathematica*, but the formal systems have the same power.)

Gödel's Incompleteness Theorem

Suppose, though, that we are using some specific first-order language and we are interested in the logical consequence relation *where the meaning of the predicates of the language is taken into account*. Do we need other methods of proof? If so, can these somehow be reduced to those we have studied? Or is it conceivable that there simply is no complete formal system that captures the notion of logical consequence for some languages? We will return to these questions at the end of the chapter, with a discussion of interpreted languages and Gödel's Incompleteness Theorem.

The Completeness Theorem for FOL

The first few sections of this chapter are devoted to giving a complete proof of the Gödel Completeness Theorem just referred to. We use the terms "theory" and "set of sentences" interchangeably. (Some authors reserve the term "theory" for a set of first-order sentences which is "closed under provability," that is, satisfying the condition that if $\mathcal{T} \vdash S$ then $S \in \mathcal{T}$.) In this section we write $\mathcal{T} \vdash S$ to mean there is a proof of S from the theory $\mathcal{T}$ in the full system $\mathcal{F}$.[1] As mentioned in Chapter 17, this notation does not mean that all the sentences in $\mathcal{T}$ have to be used in the formal proof of S, only that there is a proof of S whose premises are all elements of $\mathcal{T}$. In particular, the set $\mathcal{T}$ could be infinite (as in the case of proofs from ZFC or PA) whereas only a finite number of premises can be used in any one proof. This notation allows us to state the Completeness Theorem as follows.

theories

$\mathcal{T} \vdash S$

Theorem (Completeness Theorem for $\mathcal{F}$). Let $\mathcal{T}$ be a set of sentences of a first-order language L and let S be a sentence of the same language. If S is a first-order consequence of $\mathcal{T}$, then $\mathcal{T} \vdash S$.

Completeness Theorem for $\mathcal{F}$

Exactly as in the case of propositional logic, we obtain the following as an immediate consequence of the Completeness Theorem.

Theorem (Compactness Theorem for FOL). Let $\mathcal{T}$ be a set of sentences of a first-order language L. If for each finite subset of $\mathcal{T}$ there is a first-order structure making this subset of $\mathcal{T}$ true, then there is a first-order structure $\mathfrak{M}$ that makes all the sentences of $\mathcal{T}$ true.

Compactness Theorem

The Completeness Theorem for FOL was first established by Kurt Gödel, as we mentioned above. The proof of the Completeness Theorem for first-order consequence is, as we shall see, considerably subtler than for tautological consequence. The proof we give here is simpler than Gödel's original, though, and is based on a proof known as the Henkin method, named after the logician Leon Henkin who discovered it.

Henkin method

Recall from Section 10.1 that the truth table method is too blunt to take account of the meaning of either the quantifiers ∀ and ∃ or the identity symbol

---

[1]Recall that the formal proof system $\mathcal{F}$ includes all the introduction and elimination rules, but not the **Con** procedures.

=. In Exercise 18.12, we illustrated this defect. We noted there, for example, that there are truth assignments that assign true to both the sentences Cube(b) and ¬∃x Cube(x) (since from the point of view of propositional logic, ∃x Cube(x) is an atomic sentence unrelated to Cube(b)), whereas no first-order structure can make both of these sentences true.

Henkin's method finds a clever way to isolate the exact gap between the first-order validities and the tautologies by means of a set of FOL sentences $\mathcal{H}$. In a sense that we will make precise, $\mathcal{H}$ captures exactly what the truth table method misses about the quantifiers and identity.[2] For example, $\mathcal{H}$ will contain the sentence Cube(b) → ∃x Cube(x), thereby ruling out truth assignments like those mentioned above that assign true to both Cube(b) and ¬∃x Cube(x).

Here is an outline of our version of Henkin's proof.

witnessing constants

Adding witnessing constants: Let L be a fixed first-order language. We want to prove that if a sentence S of L is a first-order consequence of a set $\mathcal{T}$ of L sentences, then $\mathcal{T} \vdash$ S. The first step is to enrich L to a language L_H with infinitely many new constant symbols, known as *witnessing constants*, in a particular manner.

Henkin theory ($\mathcal{H}$)

The Henkin theory: We next isolate a particular theory $\mathcal{H}$ in the enriched language L_H. This theory consists of various sentences which are not tautologies but are theorems of first-order logic, plus some additional sentences known as *Henkin witnessing axioms*. The latter take the form ∃x P(x) → P(c) where c is a witnessing constant. The particular constant is chosen carefully so as to make the Henkin Construction Lemma and Elimination Theorem (described next) true.

Elimination Theorem

The Elimination Theorem: The Henkin theory is weak enough, and the formal system $\mathcal{F}$ strong enough, to allow us to prove the following (Theorem 4): Let p be any formal first-order proof whose premises are all either sentences of L or sentences from $\mathcal{H}$, with a conclusion that is also a sentence of L. We can eliminate the premises from $\mathcal{H}$ from this proof in favor of uses of the quantifier rules. More precisely, there exists a formal proof p' whose premises are those premises of p that are sentences of L and with the same conclusion as p.

Henkin construction

The Henkin Construction: On the other hand, the Henkin theory is strong enough, and the notion of first-order structure wide enough, to allow us to prove the following result (Theorem 19.5): for every truth assignment

---

[2]This remark will be further illustrated by Exercises 19.3–19.5, 19.17 and 19.18, which we strongly encourage you to do when you get to them. They will really help you understand the whole proof.

h that assigns TRUE to all wffs in $\mathcal{H}$ there is a first-order structure $\mathfrak{M}_h$ such that $\mathfrak{M}_h \models S$ for all first-order sentences S assigned TRUE by h. This construction of the structure $\mathfrak{M}_h$ from the truth assignment h is sometimes called the *Henkin construction*.

final steps

Let us show how we can use these results to prove the Completeness Theorem. Assume that $\mathcal{T}$ and S are all from the original language L and that S is a first-order consequence of $\mathcal{T}$. We want to prove that $\mathcal{T} \vdash S$. By assumption, there can be no first-order structure in which all of $\mathcal{T} \cup \{\neg S\}$ is true. By the Henkin Construction there can be no truth assignment h which assigns TRUE to all sentences in $\mathcal{T} \cup \mathcal{H} \cup \{\neg S\}$; if there were, then the first-order structure $\mathfrak{M}_h$ would make $\mathcal{T} \cup \{\neg S\}$ true. Hence S is a tautological consequence of $\mathcal{T} \cup \mathcal{H}$.[3] The Completeness Theorem for propositional logic tells us there is a formal proof p of S from $\mathcal{T} \cup \mathcal{H}$. The Elimination Theorem tells us that using the quantifier rules, we can transform p into a formal proof p' of S from premises in $\mathcal{T}$. Hence, $\mathcal{T} \vdash S$, as desired.

The next few sections of this chapter are devoted to filling in the details of this outline. We label the sections to match the names used in our outline.

SECTION 19.2

Adding witnessing constants

Given any first-order language K, we construct a new first-order language K'. The language K' will have the same symbols as K except that it will have a lot of new constant symbols. For example, if K is our blocks language, then in K' will be able to say things like the following:

1. $\exists x\,(\text{Small}(x) \wedge \text{Cube}(x)) \rightarrow \text{Small}(c_1) \wedge \text{Cube}(c_1)$

2. $\exists z\,(z \neq a \wedge z \neq b) \rightarrow (c_2 \neq a \wedge c_2 \neq b)$

3. $\exists y\,\text{Between}(y, a, b) \rightarrow \text{Between}(c_3, a, b)$

4. $\exists x\,\exists y\,\text{Between}(a, x, y) \rightarrow \exists y\,\text{Between}(a, c_4, y)$

More generally, for each wff P of L with exactly one free variable, form a new constant symbol c_P, making sure to form different names for different wffs. This constant is called the *witnessing constant* for P.

witnessing constant for P

---

[3]In this chapter we are using the notions of tautology and tautological consequence defined in Section 10.1, in which every sentence starting with a quantifier is treated as atomic.

You might wonder just how we can form all these new constant symbols. How do we write them down and how do we make sure that distinct wffs get distinct witnessing constants? Good question. There are various ways we could arrange this. One is simply to use a single symbol c not in the language K and have the new symbol be the expression c with the wff as a subscript. Thus, for example, in our above list, the constant symbol c_1 would really be the symbol

$$c_{(\text{Small}(x) \wedge \text{Cube}(x))}$$

This is a pretty awkward symbol to write down, but it at least shows us how we could arrange things in principle.

the language K'

The language K' consists of all the symbols of K plus all these new witnessing constants. Now that we have all these new constant symbols, we can use them in wffs. For example, the language K' allows us to form sentences like

$$\text{Smaller}(a, c_{\text{Between}(x,a,b)})$$

But then we also have sentences like

$$\exists x \, \text{Smaller}(x, c_{\text{Between}(x,a,b)})$$

so we would like to have a witnessing constant symbol subscripted by

$$\text{Smaller}(x, c_{\text{Between}(x,a,b)})$$

Unfortunately, this wff , while in K', is not in our original language K, so we have not added a witnessing constant for it in forming K'.

Bummer. Well, all is not lost. What we have to do is to iterate this construction over and over again. Starting with a language L, we define an infinite sequence of larger and larger languages

$$L_0 \subseteq L_1 \subseteq L_2 \subseteq \ldots$$

the Henkin language L_H

where $L_0 = L$ and $L_{n+1} = L'_n$. That is, the language L_{n+1} results by applying the above construction to the language L_n. Finally, the Henkin language L_H for L consists of all the symbols of L_n for any $n = 0, 1, 2, 3, \ldots$.

date of birth of witnessing constants

Each witnessing constant c_P is introduced at a certain stage $n \geq 1$ of this construction. Let us call that stage the *date of birth* of c_P. When we come to proving the Elimination Theorem it will be crucial to remember the following fact, which is obvious from the construction of L_H.

Lemma 1. (Date of Birth Lemma) Let $n + 1$ be the date of birth of c_P. If Q is any wff of the language L_n, then c_P does not appear in Q.

Exercises

19.1 This exercise and its companion (Exercise 19.2) are intended to give you a better feel for why we have to keep iterating the witnessing constant construction. It deals with the constants that would turn out to be important if our original set $\mathcal{T}$ contained the sentence $\forall x \exists y \, \mathsf{Larger}(x, y)$. Write out the witnessing constants for the following wffs, keeping track of their dates of birth. The constant symbol a is taken from the original language L.

1. $\mathsf{Larger}(\mathsf{a}, x)$
2. $\mathsf{Larger}(c_1, x)$, where c_1 is your constant from 1.
3. $\mathsf{Larger}(c_2, x)$, where c_2 is your constant from 2.
4. $\mathsf{Larger}(c_3, x)$, where c_3 is your constant from 3.

SECTION 19.3

The Henkin theory

We have added witnessing constants for each wff P with exactly one free variable. The free variable of P is going to be important in what follows so we often write the wff in a way that reminds us of the free variable, namely, as $\mathsf{P}(x)$.[4] Consequently, its witnessing constant is now denoted by $c_{\mathsf{P}(x)}$. Notice that by iterating our construction infinitely often, we have managed to arrange things so that for each wff $\mathsf{P}(x)$ of L_H with exactly one free variable, the witnessing constant $c_{\mathsf{P}(x)}$ is also in L_H. This allows us to form the sentence

$$\exists x \, \mathsf{P}(x) \rightarrow \mathsf{P}(c_{\mathsf{P}(x)})$$

in L_H. This sentence is known as the *Henkin witnessing axiom* for $\mathsf{P}(x)$. The intuitive idea is that $\exists x \, \mathsf{P}(x) \rightarrow \mathsf{P}(c_{\mathsf{P}(x)})$ asserts that if there is something that satisfies $\mathsf{P}(x)$, then the object named by $c_{\mathsf{P}(x)}$ provides an example (or "witness") of one such.

witnessing axioms

Lemma 2. (Independence lemma) If c_P and c_Q are two witnessing constants and the date of birth of c_P is less than or equal to that of c_Q, then c_Q does not appear in the witnessing axiom of c_P.

Proof: If the date of birth of c_P is less than that of c_Q, the result follows from the date of birth lemma. If they have the same date

---

[4]Really, we should be writing this as $\mathsf{P}(\nu)$, where ν can be any of the variables of our language, not just the variable x. We are using x here as a representative variable of our language.

of birth, it follows from the fact that different wffs of a language K have distinct new witnessing constants in K'.

Henkin theory $\mathcal{H}$

Definition The *Henkin theory $\mathcal{H}$* consists of all sentences of the following five forms, where c and d are any constants and P(x) is any formula (with exactly one free variable) of the language L_H:

H1: All Henkin witnessing axioms

$$\exists x\, P(x) \rightarrow P(c_{P(x)})$$

H2: All sentences of the form

$$P(c) \rightarrow \exists x\, P(x)$$

H3: All sentences of the form

$$\neg \forall x\, P(x) \leftrightarrow \exists x\, \neg P(x)$$

H4: All sentences of the form

$$c = c$$

H5: All sentences of the form

$$(P(c) \wedge c = d) \rightarrow P(d)$$

connection to quantifier rules

Notice that there is a parallel between these sentences of $\mathcal{H}$ and the quantifier and identity rules of $\mathcal{F}$:

- H1 corresponds roughly to $\exists$ **Elim**, in that both are justified by the same intuition,

- H2 corresponds to $\exists$ **Intro**,

- H3 reduces $\forall$ to $\exists$,

- H4 corresponds to = **Intro**, and

- H5 corresponds to = **Elim**.

Just what this correspondence amounts to is a bit different in the various cases. For example, the axioms of types H2–H5 are all first-order validities, while this is not true of H1, of course. The witnessing axioms make substantive claims about the interpretations of the witnessing constants. The following result, while not needed in the proof of completeness, does explain why the rest of the proof has a chance of working.

Proposition 3. Let $\mathfrak{M}$ be any first-order structure for L. There is a way to interpret all the witnessing constants in the universe of $\mathfrak{M}$ so that, under this interpretation, all the sentences of $\mathcal{H}$ are true.

> **Proof:** (Sketch) The basic idea of the proof is that if $\mathfrak{M} \models \exists x\, P(x)$, then pick any element b of the domain that satisfies $P(x)$ and let the witnessing constant $c_{P(x)}$ name b. If $\mathfrak{M} \models \neg \exists x\, P(x)$, then let $c_{P(x)}$ name any fixed element of the domain. This takes care of the axioms of type H1. As for the other axioms, they are all logical truths, and so will turn out to be true no matter how we interpret the new constants.

The proof of Proposition 3 is illustrated in Exercise 19.6. The proposition shows that our strategy has a chance of working by allowing us to show that even though the theory $\mathcal{H}$ says substantive things about the witnessing constants, none of those claims can be formulated in the original language L. More precisely (as we ask you to show in Exercise 19.9) if a sentence S of L is a first-order consequence of $\mathcal{T} \cup \mathcal{H}$, then it is a first-order consequence of $\mathcal{T}$ alone. This at least makes the Elimination Theorem plausible.

Exercises

19.2 Write out the witnessing axioms associated with the wffs in Exercise 19.1.

The next three exercises are designed to help you understand how the theory $\mathcal{H}$ fills the gap between tautological and first-order consequence. For these exercises we take L to be the blocks language and $\mathcal{T}$ to consist of the following set of sentences:

$$\mathcal{T} = \{\text{Cube}(a),\ \text{Small}(a),\ \exists x\,(\text{Cube}(x) \wedge \text{Small}(x)) \rightarrow \exists y\,\text{Dodec}(y)\}$$

19.3 Give informal proofs that both of the following are first-order consequences of $\mathcal{T}$:
1. $\exists x\,(\text{Cube}(x) \wedge \text{Small}(x))$
2. $\exists y\,\text{Dodec}(y)$

19.4 Give informal proofs that none of the following is a tautological consequence of $\mathcal{T}$:
1. $\exists x\,(\text{Cube}(x) \wedge \text{Small}(x))$
2. $\exists y\,\text{Dodec}(y)$
3. $\text{Dodec}(c_{\text{Dodec}(y)})$

19.5 Give informal proofs that the sentences in Exercise 19.4 are all tautological consequences of $\mathcal{T} \cup \mathcal{H}$.

19.6 Use Tarski's World to open Henkin's Sentences. Take the constants c and d as shorthand for the witnessing constant $c_{Cube(x)}$ and $c_{Dodec(x) \wedge Small(x)}$, respectively.

1. Show that these sentences are all members of $\mathcal{H}$. Identify the form of each axiom from our definition of $\mathcal{H}$.

2. By (1) and Proposition 3, any world in which c and d are not used as names can be turned into a world where all these sentences are true. Open Henkin's World. Name some blocks c and d in such a way that all the sentences are true. Submit this world.

19.7 Show that for every constant symbol c of L_H there is a distinct witnessing constant d such that c = d is a tautological consequence of $\mathcal{H}$. [Hint: consider the wff c = x.]

19.8 Show that for every binary relation symbol R of L and all constants c, c', d, and d' of L_H, the following is a tautological consequence of $\mathcal{H}$:

$$(R(c,d) \wedge c = c' \wedge d = d') \rightarrow R(c',d')$$

[Hint: Notice that $(R(c,d) \wedge c = c') \rightarrow R(c',d)$ is one of the H5 axioms of $\mathcal{H}$. So is $(R(c',d) \wedge d = d') \rightarrow R(c',d')$. Show that the desired sentence is a tautological consequence of these two sentences and hence a tautological consequence of $\mathcal{H}$.]

19.9 Let $\mathcal{T}$ be a theory of L. Use Proposition 3 (but without using the Completeness Theorem or the Elimination Theorem) to show that if a sentence S of L is a first-order consequence of $\mathcal{T} \cup \mathcal{H}$, then it is a first-order consequence of $\mathcal{T}$ alone.

Section 19.4

The Elimination Theorem

It follows from Proposition 3 that if a sentence S of L is a first-order consequence of $\mathcal{T} \cup \mathcal{H}$, then it is a first-order consequence of $\mathcal{T}$ alone. (This is what you established in Exercise 19.9.) This result shows that we've constructed $\mathcal{H}$ in such a way that it doesn't add any substantive new claims, things that give us new first-order consequences (in L) of $\mathcal{T}$. The Elimination Theorem shows us that our deductive system is strong enough to give us a similar result on the formal system side of things.

Elimination Theorem

Proposition 4. (The Elimination Theorem) Let p be any formal first-order proof with a conclusion S that is a sentence of L and whose premises are sentences $P_1, \ldots, P_n$ of L plus sentences from $\mathcal{H}$. There exists a formal proof p' of S with premises $P_1, \ldots, P_n$ alone.

The proof of this result will take up this section. We break the proof down into a number of lemmas.

Proposition 5. (Deduction Theorem). If $\mathcal{T} \cup \{P\} \vdash Q$ then $\mathcal{T} \vdash P \rightarrow Q$ *Deduction Theorem*

The proof of this is very similar to the proof of Lemma 2 from chapter 17 (page 487) and is left as an exercise. It is also illustrated by the following.

You try it
. . . .

1. Open Deduction Thm 1. This contains a formal first-order proof of the ◄ following argument:

$$\forall x\, (\mathsf{Dodec}(x) \rightarrow \exists y\, \mathsf{Adjoins}(x, y))$$
$$\forall x\, \mathsf{Dodec}(x)$$

$$\forall x\, \exists y\, \exists z\, (\mathsf{Adjoins}(x, y) \wedge \mathsf{Adjoins}(y, z))$$

According to the deduction theorem, we should be able to give a proof of

$$\forall x\, (\mathsf{Dodec}(x) \rightarrow \exists y\, \mathsf{Adjoins}(x, y))$$

$$\forall x\, \mathsf{Dodec}(x) \rightarrow \forall x\, \exists y\, \exists z\, (\mathsf{Adjoins}(x, y) \wedge \mathsf{Adjoins}(y, z))$$

We will show you how to do this.

2. Open Proof Deduction Thm 1, which contains only the first premise from ◄ the file Deduction Thm 1. We'll construct the desired proof in this file.

3. For the first step of the new proof, start a subproof with $\forall x\, \mathsf{Dodec}(x)$ as ◄ · premise.

4. Now using Copy and Paste, copy the entire proof (but not the premises) ◄ from Deduction Thm 1 into the subproof you just created in Proof Deduction Thm 1. After you paste the proof, verify it. You'll see that you need to add some support citations to get all the steps to check out, but you can easily do this.

5. End the subproof and use → **Intro** to obtain the desired conclusion. Save ◄ your completed proof as Proof Deduction Thm 1.
. . . . *Congratulations*

The following is proven using the Deduction Theorem and modus ponens repeatedly to get rid of each of the P_i. The details are left as an exercise.
Proposition 6. If $\mathcal{T} \cup \{P_1, \ldots, P_n\} \vdash Q$ and, for each $i = 1, \ldots, n$, $\mathcal{T} \vdash P_i$ then $\mathcal{T} \vdash Q$.

The main step in our proof of the Elimination Theorem is Lemma 9; Lemmas 7 and 8 help in the proof of Lemma 9. Used judiciously, Lemma 9 will allow us to eliminate the Henkin witnessing axioms from proofs.

two simple facts

Lemma 7. Let $\mathcal{T}$ be a set of first-order sentences of some first-order language L, and let P, Q, and R be sentences of L.

1. If $\mathcal{T} \vdash P \rightarrow Q$ and $\mathcal{T} \vdash \neg P \rightarrow Q$ then $\mathcal{T} \vdash Q$.

2. If $\mathcal{T} \vdash (P \rightarrow Q) \rightarrow R$ then $\mathcal{T} \vdash \neg P \rightarrow R$ and $\mathcal{T} \vdash Q \rightarrow R$.

Proof: (1) We have already seen that $P \vee \neg P$ is provable without any premises at all. Hence, $\mathcal{T} \vdash P \vee \neg P$. Thus, our result will follow from Proposition 6 if we can show that the following argument has a proof in $\mathcal{F}$:

$$
\begin{array}{|l}
P \vee \neg P \\
P \rightarrow Q \\
\neg P \rightarrow Q \\
\hline
Q
\end{array}
$$

But this is obvious by $\vee$ **Elim**.

The proof of (2) is similar, using Exercises 19.12 and 19.13.

The following lemma shows how certain constants in proofs can be replaced by quantifiers, using the rule of $\exists$ **Elim**.

replacing constants with quantifiers

Lemma 8. Let $\mathcal{T}$ be a set of first-order sentences of some first-order language L and let Q be a sentence. Let P(x) be a wff of L with one free variable and which does not contain c. If $\mathcal{T} \vdash P(c) \rightarrow Q$ and c does not appear in $\mathcal{T}$ or Q, then $\mathcal{T} \vdash \exists x\, P(x) \rightarrow Q$.

Proof: Assume that $\mathcal{T} \vdash P(c) \rightarrow Q$, where c is a constant that does not appear in $\mathcal{T}$ or Q. It is easy to see that for any other constant d not in $\mathcal{T}$, P(x), or Q, $\mathcal{T} \vdash P(d) \rightarrow Q$. Just take the original proof p and replace c by d throughout; if d happened to appear in the original proof, replace it by some other new constant, c if you like. Let us now give an informal proof, from $\mathcal{T}$, of the desired conclusion $\exists x\, P(x) \rightarrow Q$, being careful to do it in a way that is easily formalizable in $\mathcal{F}$.

Using the method of $\rightarrow$ **Intro**, we take $\exists x\, P(x)$ as a premise and try to prove Q. Toward this end, we use the rule of

∃ **Elim**. Let d be a new constant and assume P(d). But by our first observation, we know we can prove P(d) → Q. By modus ponens (→ **Elim**), we obtain Q as desired.

This informal proof can clearly be formalized within $\mathcal{F}$, establishing our result. (Exercise 19.16 illustrates this method.)

By combining Lemmas 7 and 8, we can prove the following, which is just what we need to eliminate the Henkin witnessing axioms.

Lemma 9. Let $\mathcal{T}$ be a set of first-order sentences of some first-order language L and let Q be a sentence of L. Let P(x) be a wff of L with one free variable which does not contain c. If $\mathcal{T} \cup \{\exists x\, P(x) \to P(c)\} \vdash Q$ and c does not appear in $\mathcal{T}$ or Q, then $\mathcal{T} \vdash Q$.

eliminating witnessing axioms

> **Proof:** Assume $\mathcal{T} \cup \{\exists x\, P(x) \to P(c)\} \vdash Q$, where c is a constant that does not appear in $\mathcal{T}$ or Q. By the Deduction Theorem,

$$\mathcal{T} \vdash (\exists x\, P(x) \to P(c)) \to Q$$

By (2) of Lemma 7,
$$\mathcal{T} \vdash \neg\exists x\, P(x) \to Q$$

and
$$\mathcal{T} \vdash P(c) \to Q$$

From the latter, using (1) of Lemma 8, we obtain $\mathcal{T} \vdash \exists x\, P(x) \to Q$. Then by (1) of Lemma 7, $\mathcal{T} \vdash Q$.

Lemma 9 will allow us to eliminate the Henkin witnessing axioms from formal proofs. But what about the other sentences in $\mathcal{H}$? In conjunction with Lemma 6, the next result will allow us to eliminate these as well.

Lemma 10. Let P(x) be a wff with one free variable, and let c and d be constant symbols. The following are all provable in $\mathcal{F}$:

eliminating other members of $\mathcal{H}$

$$P(c) \to \exists x\, P(x)$$

$$\neg\forall x\, P(x) \leftrightarrow \exists x\, \neg P(x)$$

$$(P(c) \wedge c = d) \to P(d)$$

$$c = c$$

Proof: The only one of these that is not quite obvious from the rules of inference of $\mathcal{F}$ is the DeMorgan biconditional. We essentially proved half of this biconditional on page 364, and gave you the other half as Exercise 13.44.

We have now assembled the tools we need to prove the Elimination Theorem.

proof of Elimination Theorem

Proof of the Elimination Theorem. Let k be any natural number and let p be any formal first-order proof of a conclusion in L, all of whose premises are all either sentences of L or sentences from $\mathcal{H}$, and such that there are at most k from $\mathcal{H}$. We show how to eliminate those premises that are members of $\mathcal{H}$. The proof is by induction on k. The basis case is where $k = 0$. But then there is nothing to eliminate, so we are done. Let us assume the result for k and prove it for $k + 1$. The proof breaks into two cases.

Case 1: At least one of the premises to be eliminated, say P, is of one of the forms mentioned in Lemma 10. But then P can be eliminated by Lemma 6 giving us a proof with at most k premises to be eliminated, which we can do by the induction hypothesis.

Case 2: All of the premises to be eliminated are Henkin witnessing axioms. The basic idea is to eliminate witnessing axioms introducing young witnessing constants before eliminating their elders. Pick the premise of the form $\exists x\, P(x) \to P(c)$ whose witnessing constant c is as young as any of the witnessing constants mentioned in the set of premises to be eliminated. That is, the date of birth n of c is greater than or equal to that of any of witnessing constants mentioned in the premises. This is possible since there are only finitely many such premises. By the independence lemma, c is not mentioned in any of the other premises to be eliminated. Neither is c mentioned in the conclusion. By Lemma 9, $\exists x\, P(x) \to P(c)$ can be eliminated. This gets us to a proof with at most k premises to be eliminated, which we can do by our induction hypothesis.

Exercises

19.10 If you skipped the **You try it** section, go back and do it now. Submit the file Proof Deduction
⤳   Thm 1.

Give formal proofs of the following arguments. Because these results are used in the proof of Completeness, do not use any of the **Con** *rules in your proofs.*

19.11

$P \rightarrow Q$
$\neg P \rightarrow Q$
Q

19.12

$(P \rightarrow Q) \rightarrow R$
$\neg P \rightarrow R$

19.13

$(P \rightarrow Q) \rightarrow R$
$Q \rightarrow R$

19.14 Prove the Deduction Theorem (Proposition 5). [Hint: The proof of this is very similar to the proof of Lemma 2 from chapter 17 (on page 487).]

19.15 Prove Proposition 6. [Hint: Use induction on n and the Deduction Theorem.]

19.16 Use Fitch to open **Exercise 19.16**. Here you will find a first-order proof of the following argument:

$\forall x\,(\text{Cube}(x) \rightarrow \text{Small}(x))$
$\forall x\,\forall y\,(x = y)$
$\text{Cube}(b) \rightarrow \forall x\,\text{Small}(x)$

Using the method of Lemma 8, transform this proof into a proof of

$\forall x\,(\text{Cube}(x) \rightarrow \text{Small}(x))$
$\forall x\,\forall y\,(x = y)$
$\exists y\,\text{Cube}(y) \rightarrow \forall x\,\text{Small}(x)$

Submit your proof as Proof 19.16.

19.17 Open **Exercise 19.17**. This file contains the following argument:

$\forall x\,(\text{Cube}(x) \rightarrow \text{Small}(x))$
$\exists x\,\text{Cube}(x)$
$\exists x\,\text{Cube}(x) \rightarrow \text{Cube}(c)$
$\text{Small}(c) \rightarrow \exists x\,\text{Small}(x)$
$\neg(\text{Cube}(c) \rightarrow \text{Small}(c)) \rightarrow \exists x\,\neg(\text{Cube}(x) \rightarrow \text{Small}(x))$
$\neg\forall x\,(\text{Cube}(x) \rightarrow \text{Small}(x)) \leftrightarrow \exists x\,\neg(\text{Cube}(x) \rightarrow \text{Small}(x))$
$\exists x\,\text{Small}(x)$

First use **Taut Con** to show that the conclusion is a tautological consequence of the premises. Having convinced yourself, delete this step and give a proof of the conclusion that uses only the propositional rules.

19.18 Open Exercise 19.17 again. Take the constant c as shorthand for the witnessing constant $c_{Cube(x)}$. Take $\mathcal{T}$ to be the first two premises of this proof. We saw in Exercise 19.6 that the other sentences are all members of $\mathcal{H}$. The Elimination Theorem thus applies to show that you could transform your proof from the preceding exercise into a proof from the first two premises, one that does not need the remaining premises. Open Exercise 19.18 and give such a proof. [If you were to actually transform your previous proof, using the method we gave, the result would be a very long proof indeed. You'll be far better off giving a new, direct proof.]

Section 19.5

The Henkin Construction

Henkin Construction Lemma

Proposition 3 allows us to take any first-order structure for L and get from it one for L_H that makes all the same sentences true. This, of course, gives rise to a truth assignment h to all the sentences of L_H that respects the truth-functional connectives: just assign TRUE to all the sentences that are true in the structure, FALSE to the others. (You may recall that you were asked to prove this in Exercise 18.11.) The main step in the Henkin proof of the Completeness Theorem is to show that we can reverse this process.

Theorem (Henkin Construction Lemma) Let h be any truth assignment for L_H that assigns TRUE to all the sentences of the Henkin theory $\mathcal{H}$. There is a first-order structure $\mathfrak{M}_h$ such that $\mathfrak{M}_h \models S$ for all sentences S assigned TRUE by the assignment h.

constructing $\mathfrak{M}_h$

In giving the proof of this result, we will assume that our language L contains only relation symbols and constants, no function symbols. We will return at the end to explain how to modify the proof if there are function symbols. The proof of this theorem has two parts. We must first show how to construct $\mathfrak{M}_h$ from h and then show that $\mathfrak{M}_h$ does indeed make true all the sentences to which h assigned TRUE. To construct $\mathfrak{M}_h$, we must do three things. We must define the domain D of $\mathfrak{M}_h$, we must assign to each n-ary relation symbol R some set R of n-tuples from D, and we must assign to each name c of L_H some element of D. We first give the basic idea of the construction. This idea won't quite work, so it will have to be modified, but it's useful to see the flawed idea before digging into the details that correct the flaw.

The basic (flawed) idea in constructing the first-order structure $\mathfrak{M}$ is to use the construction of Exercise 18.12, but to count on the fact that h assigns TRUE to all the sentences in $\mathcal{H}$ to get us past the quantifiers. In more detail, we build $\mathfrak{M}$ as follows:

○ For the domain of $\mathfrak{M}$, use the set of constant symbols of L_H.

○ Let each constant be a name of itself.

○ To specify the interpretation of a (binary, say) relation symbol R, take the set R of pairs $\langle c, d \rangle$ of constant symbols such that h assigns TRUE to the sentence R(c, d).

This almost defines a perfectly good first-order structure. The problem is that the interpretation of = will not be the genuine identity relation, as is required by our definition of first-order structures (page 514). The reason is that inevitably there are distinct constants c and d such that h assigns true to c = d. (See Exercise 19.7.) According to the final clause above, $\langle c, d \rangle$ will be in the extension assigned to = by $\mathfrak{M}$, but c and d are distinct members of the domain of $\mathfrak{M}$, by the first clause.

a problem: identity

This problem is one that is quite familiar in mathematical situations. What we need to do is form what mathematicians call a "quotient" of $\mathfrak{M}$. What this means is that we must "identify" various elements that $\mathfrak{M}$ considers distinct. This is just the sort of thing that equivalence relations and equivalence classes (studied in Chapter 15) are designed to do, and is the reason we included them in our discussion of sets.

equivalence classes to the rescue

Define a binary relation $\equiv$ on the domain of $\mathfrak{M}$ (i.e., on the constants of L_H) as follows:

$$c \equiv d \text{ if and only if } h(c = d) = \text{TRUE}.$$

Lemma 11. The relation $\equiv$ is an equivalence relation.

Proof: This follows immediately from Exercise 19.20.

From this it follows that we can associate with each constant c its equivalence class

$$[c] = \{d \mid c \equiv d\}$$

This allows us to define our desired first-order structure $\mathfrak{M}_h$:

definition of $\mathfrak{M}_h$

○ The domain D of our desired first-order structure $\mathfrak{M}_h$ is the set of all such equivalence classes.

○ We let each constant c of L_H name its own equivalence class [c].

○ We interpret relation symbols R as follows. For this definition let's again assume that R is binary, just to make the notation simpler. Then the interpretation of R is the set

$$\{\langle [c], [d] \rangle \mid h(R(c, d)) = \text{TRUE}\}$$

We now need to prove that $\mathfrak{M}_h$ makes true all and only the sentences to which h assigns TRUE. That is, we need to show that for any sentence S of L_H, $\mathfrak{M}_h \models$ S if and only if $h(\mathsf{S}) =$ TRUE. The natural way to prove this is by induction on the complexity of the sentence S.

For the atomic case, we have basically built $\mathfrak{M}_h$ to guarantee that our claim holds. But there is one important thing we need to check. Suppose we have distinct constants c and c′, and d and d′, where [c] = [c′] and [d] = [d′]. We need to rule out the possibility that $h(\mathsf{R}(\mathsf{c},\mathsf{d})) =$ TRUE and $h(\mathsf{R}(\mathsf{c}',\mathsf{d}')) =$ FALSE. For if this were the case, $\langle[\mathsf{c}'],[\mathsf{d}']\rangle$ would be in the extension of R (since $\langle[\mathsf{c}],[\mathsf{d}]\rangle$ is, and these are the same), and consequently $\mathfrak{M}_h$ would assign the wrong value to the atomic sentence $\mathsf{R}(\mathsf{c}',\mathsf{d}')$! But this situation is impossible, as is shown by the following lemma.

Lemma 12. If $\mathsf{c} \equiv \mathsf{c}'$, $\mathsf{d} \equiv \mathsf{d}'$, and $h(\mathsf{R}(\mathsf{c},\mathsf{d})) =$ TRUE, then $h(\mathsf{R}(\mathsf{c}',\mathsf{d}')) =$ TRUE.

> **Proof:** By Exercise 19.8, the following is a tautological consequence of the Henkin theory $\mathcal{H}$:
>
> $$(\mathsf{R}(\mathsf{c},\mathsf{d}) \wedge \mathsf{c} = \mathsf{c}' \wedge \mathsf{d} = \mathsf{d}') \rightarrow \mathsf{R}(\mathsf{c}',\mathsf{d}')$$
>
> Since h assigns everything in $\mathcal{H}$ TRUE, and it assigns TRUE to each conjunct of $\mathsf{R}(\mathsf{c},\mathsf{d}) \wedge \mathsf{c} = \mathsf{c}' \wedge \mathsf{d} = \mathsf{d}'$, it must also assign TRUE to $\mathsf{R}(\mathsf{c}',\mathsf{d}')$.

This lemma assures us that our construction of $\mathfrak{M}_h$ works for the atomic sentences. That is, $\mathfrak{M}_h$ will make an atomic sentence true if and only if h assigns TRUE to that atomic sentence. The proof of the Henkin Construction Lemma will be completed by proving the full version of this result.

the crucial lemma

Lemma 13. For any sentence S of L_H, $\mathfrak{M}_h \models$ S if and only if $h(\mathsf{S}) =$ TRUE.

> **Proof:** The basic idea of this proof is to use induction. We have explicitly defined the structure $\mathfrak{M}_h$ to make sure the claim is true for atomic sentences. Further, truth assignments work the same way on truth-functional connectives as the definition of truth in a first-order structure. So the only possible problems are the quantifiers and these, we will see, are taken care of by the quantifier axioms in $\mathcal{H}$. The only mild complication to this outline is that the quantifier $\forall$ is not handled directly, but indirectly through the deMorgan sentences in $\mathcal{H}$:
>
> $$\neg\forall\mathsf{x}\, \mathsf{P}(\mathsf{x}) \leftrightarrow \exists\mathsf{x}\, \neg\mathsf{P}(\mathsf{x})$$

What makes this a complication is that the most obvious measures of complexity, say by length or number of logical operations, would count $\forall x\, P(x)$ as simpler than the sentence $\exists x\, \neg P(x)$, whereas any induction proof is going to need to make sure that the latter works out right before it can be sure that the former does. We get around this by defining a different measure of complexity for wffs. Namely, we define the complexity of an atomic wff to be 0, the complexity of $\neg P$ and $\exists x\, P$ to be one greater than the complexity of P, the complexity of $P \wedge Q$, $P \vee Q$, and $P \rightarrow Q$ to be one greater than the maximum of that of P and Q, but the complexity of $\forall x\, P$ to be *three* greater than that of P. Here is a little table showing the complexity of some wffs in the blocks language:

wff	complexity
Small(x)	0
(x = a)	0
¬(x = a)	1
Small(x) → ¬(x = a)	2
¬(Small(x) → ¬(x = a))	3
∃x ¬(Small(x) → ¬(x = a))	4
∀x (Small(x) → ¬(x = a))	5

Notice that the complexity of a wff $P(x)$ is the same as that of $P(c)$. (See, for example, Exercise 19.21.) With this definition of complexity, we can prove the lemma, by induction on the complexity of sentences. As remarked above, the case where the complexity is 0 is true by the way we defined the structure $\mathfrak{M}_h$.

Assume that the lemma holds for all sentences of complexity $\leq k$ and let S have complexity $\leq k + 1$. There are several cases to consider, depending on the main connective or quantifier of S. We treat one of the truth-functional cases, as these are all similar, and then both of the quantifier cases.

Case 1. Suppose S is $P \vee Q$. If $\mathfrak{M}_h \models S$, then at least one of P or Q is true. Assume that P is true. Since the complexity of S is $\leq k+1$, the complexity of P is $\leq k$, so by induction hypothesis, $h(P) = \text{TRUE}$. But then $h(P \vee Q) = \text{TRUE}$, as desired. The proof in the other direction is similar.

Case 2. Suppose that S is $\exists x\, P(x)$. We need to show that $\mathfrak{M}_h \models \exists x\, P(x)$ if and only if h assigns the sentence TRUE. Assume first that the sentence is true in $\mathfrak{M}_h$. Then since every object in the domain is denoted by some constant, there is a constant c such that $\mathfrak{M}_h \models P(c)$. But since the complexity of this sentence is less than that of S, our induction hypothesis shows us that $h(P(c)) = $ TRUE. But now recall that our theory $\mathcal{H}$ contains the sentence

$$P(c) \rightarrow \exists x\, P(x)$$

so h assigns this sentence TRUE. But then, by the truth table for $\rightarrow$, h assigns TRUE to $\exists x\, P(x)$, as desired.

The reverse direction of this case is very similar, but it uses the Henkin witnessing axiom for $P(x)$. Here is how it goes. Assume that h assigns TRUE to $\exists x\, P(x)$. We need to show that $\mathfrak{M}_h \models \exists x\, P(x)$. But recall that h assigns TRUE to the witnessing axiom

$$\exists x\, P(x) \rightarrow P(c_{P(x)})$$

Hence, by the truth table for $\rightarrow$, h assigns TRUE to $P(c_{P(x)})$. Hence, by induction, this sentence is true in $\mathfrak{M}_h$. But then $\exists x\, P(x)$ is true as well.

Case 3. Finally, let us assume that S is $\forall x\, P(x)$. We need to prove that this sentence is true in $\mathfrak{M}_h$ if and only if h assigns the sentence TRUE. Assume first that S is true in $\mathfrak{M}_h$. Then $\exists x\, \neg P(x)$ is false in $\mathfrak{M}_h$. But then, by induction, h assigns FALSE to this sentence. But recall that $\mathcal{H}$ contains the sentence

$$\neg \forall x\, P(x) \leftrightarrow \exists x\, \neg P(x)$$

From this it follows that h assigns FALSE to $\neg \forall x\, P(x)$ and hence TRUE to $\forall x\, P(x)$, as desired. The proof in the other direction is entirely similar.

Function symbols   If there are function symbols in the original language, we have to explain how to interpret them in our structure. Suppose, for example, that our language contains a one-place function symbol f. How should we define its interpretation f? In particular, if d is some constant symbol, what equivalence class should $f([d])$ be? What comes to our rescue here is the witnessing constant for the sentence

$$\exists x\, [f(d) = x]$$

We can define $f([\mathsf{d}])$ to be the equivalence class $[\mathsf{c}_{f(\mathsf{d})=\mathsf{x}}]$ of the witnessing constant $\mathsf{c}_{f(\mathsf{d})=\mathsf{x}}$. Since

$$\exists \mathsf{x}\,[f(\mathsf{d}) = \mathsf{x}] \to f(\mathsf{d}) = \mathsf{c}_{f(\mathsf{d})=\mathsf{x}}$$

is in $\mathcal{H}$, it is not hard to check that all the details of the proof then work out pretty much without change.

This completes our filling in of the outline of the proof of the Completeness Theorem.

Exercises

19.19 Use Tarski's World to open Henkin Construction. This file lists eight sentences. Let's suppose that the predicates used in these sentences (Cube, Dodec, and Small) exhaust the predicates of L. (In particular, we banish = to avoid the complications it caused in the proof of the Henkin Construction Lemma.) Let h be any truth assignment that assigns TRUE to all these sentences. Describe the first-order structure $\mathfrak{M}_h$. (How many objects will it have? What will they be called? What shape and size predicates will hold of them?) Use Tarski's World to build a world that would be represented by this first-order structure. There will be many such. It should, of course, make all the sentences in this list true. Submit your world.

19.20 Show that all sentences of the following forms are tautological consequences of $\mathcal{H}$:

1. $\mathsf{c} = \mathsf{c}$
2. $\mathsf{c} = \mathsf{d} \to \mathsf{d} = \mathsf{c}$
3. $(\mathsf{c} = \mathsf{d} \wedge \mathsf{d} = \mathsf{e}) \to \mathsf{c} = \mathsf{e}$

19.21 What are the complexities of the following wffs, where complexity is measured as in the proof of the Henkin Construction Lemma?

1. $\mathsf{Cube}(\mathsf{y})$
2. $\mathsf{y} = \mathsf{x}$
3. $\mathsf{Cube}(\mathsf{y}) \to \mathsf{y} = \mathsf{x}$
4. $\forall \mathsf{y}\,(\mathsf{Cube}(\mathsf{y}) \to \mathsf{y} = \mathsf{x})$
5. $\exists \mathsf{x}\,\forall \mathsf{y}\,(\mathsf{Cube}(\mathsf{y}) \to \mathsf{y} = \mathsf{x})$
6. $\mathsf{y} = \mathsf{c}_{\forall \mathsf{y}(\mathsf{Cube}(\mathsf{y}) \to \mathsf{y}=\mathsf{x})}$
7. $\forall \mathsf{y}\,(\mathsf{Cube}(\mathsf{y}) \to \mathsf{y} = \mathsf{c}_{\forall \mathsf{y}(\mathsf{Cube}(\mathsf{y}) \to \mathsf{y}=\mathsf{x})})$

19.22 In the inductive proof of Lemma 13, Case 1 considered only one of the truth-functional connectives, namely, $\vee$. Give an analogous proof that covers the case where S is of the form $\mathsf{P} \wedge \mathsf{Q}$.

19.23 In the inductive proof of Lemma 13, Case 3 considered only one direction of the biconditional for $\forall$. Prove the other direction.

Section 19.6

The Löwenheim-Skolem Theorem

the structure $\mathfrak{M}_h$

One of the most striking things about the proof of completeness is the nature of the first-order structure $\mathfrak{M}_h$. Whereas our original language may be talking about physical objects, numbers, sets, what-have-you, the first-order structure $\mathfrak{M}_h$ that we construct by means of the Henkin construction has as elements something quite different: equivalence classes of constant symbols. This observation allows us to exploit the proof to establish something known as the Löwenheim-Skolem Theorem for FOL.

Recall from our discussion of infinite sets in Chapter 15 that there are different sizes of infinite sets. The smallest infinite sets are those that have the same size as the set of natural numbers, those that can be put in one-to-one correspondence with the natural numbers. A set is *countable* if it is finite or is the same size as the set of natural numbers. Digging into the details of the proof of completeness lets us prove the following important theorem, due originally to the logicians Löwenheim and Skolem. They proved it before Gödel proved the Completeness Theorem, using a very different method. Löwenheim proved it for single sentences, Skolem proved it for countably infinite sets of sentences.

Löwenheim-Skolem Theorem

Theorem (Löwenheim-Skolem Theorem) Let $\mathcal{T}$ be a set of sentences in a countable language L. Then if $\mathcal{T}$ is satisfied by a first-order structure, it is satisfied by one whose domain is countable.

Proof: (Sketch) By the Soundness Theorem, if $\mathcal{T}$ is satisfiable, then it is formally consistent. The proof of completeness shows that if $\mathcal{T}$ is formally consistent, it is true in a first-order structure of the form $\mathfrak{M}_h$, for some truth assignment h for L_H. Let's assume that our original language L is countable and ask how big the structure $\mathfrak{M}_h$ is. The answer to this is not hard to determine. There cannot be any more elements of $\mathfrak{M}_h$ than there are constant symbols in the language L_H. Each of these can be written down using the symbol c, with subscripts involving symbols from L, though the subscripts need to be able to be iterated arbitrarily far. Still, if we can list the symbols of L in a list, we can use this list to alphabetize all the witnessing constants of L_H. In this way, it is possible to show that the domain of the structure $\mathfrak{M}_h$ is countable.

The Skolem Paradox

The Löwenheim-Skolem Theorem can seem somewhat puzzling. Consider, for example, our axiomatization ZFC of set theory. In giving these axioms, the intended range of structures was not terribly clear. It was, roughly speaking, the universe, or universes, of small cumulative sets. No such universe is countable, since each will contain among its sets at least one infinite set (by the axiom of infinity) as well as the powerset of this set (by the powerset axiom). But the powerset of an infinite set is uncountable. Indeed, suppose c is an infinite set. We can prove in ZFC that the powerset of c is uncountable. How can it be, then, that the axioms ZFC are true in a countable structure, as the Löwenheim-Skolem Theorem implies?

the Skolem paradox

The resolution to this puzzling state of affairs rests in the nature of the structure we constructed in the proof of the Löwenheim-Skolem Theorem as applied to ZFC. As we noted, it is a structure $\mathfrak{M}_h$ built out of equivalence classes of constant symbols. It is not at all one of the intended structures we had in mind in axiomatizing set theory. If we look at this structure, we will find that it does contain elements which purport to be powersets of infinite sets. Suppose b and c are members of the domain, where b and c satisfy the wff "x is the powerset of y and y is infinite." The "set" b will contain various "elements," each of which can be seen as corresponding to a set of "elements" of c. (Remember that the members of the domain of $\mathfrak{M}_h$ are really equivalence classes of constants. When we speak of the "elements" of b, we do not mean the members of this equivalence class, but rather the members d of the domain such that d and b satisfy the wff $x \in y$ in $\mathfrak{M}_h$. This is why we are using scare quotes around "set," "element," and so forth.) But most sets of "elements" of c will not correspond to anything in b.

resolution of paradox

Another way of understanding what is going on in $\mathfrak{M}_h$ is to think about the definition of a countable set. An infinite set b is countable if there is a one-to-one function from the set of natural numbers onto b. From outside $\mathfrak{M}_h$, we can see that there is such a function enumerating b, but this function does not correspond to anything in the structure $\mathfrak{M}_h$. That is, there is no "function" in $\mathfrak{M}_h$ that enumerates b.

The lesson to be learned from the application of the Löwenheim-Skolem Theorem to ZFC is that the first-order language of set theory is not rich enough to be able to capture various concepts that we implicitly assume when thinking about the intended universe of set theory. In fact, the key concept that we cannot adequately express is the notion of an arbitrary subset of a set, or what comes to the same thing, the notion of the powerset of a set. When we define *powerset* in the first-order language, no first-order axioms can rule

lesson of paradox

out structures, like $\mathfrak{M}_h$, in which "powerset" means something quite different from the intended notion. While more subtle, this is not unlike the fact that our shape axioms do not rule out structures in which Tet, Cube, Dodec, and SameShape mean *small*, *medium*, *large*, and *same size*.

Section 19.7
The Compactness Theorem

As we mentioned before, one of the immediate consequences of the Completeness Theorem is the first-order Compactness Theorem:

Compactness Theorem

Theorem (Compactness Theorem for FOL) Let $\mathcal{T}$ be a set of sentences of a first-order language L. If every finite subset of $\mathcal{T}$ is true in some first-order structure, then there is a first-order structure $\mathfrak{M}$ that makes every sentence of $\mathcal{T}$ true.

This follows from Completeness for the simple reason that proofs in $\mathcal{F}$ are finite and so can only use finitely many premises. If $\mathcal{T}$ is not satisfiable, then (by Completeness) there is a proof of $\bot$ from sentences in $\mathcal{T}$, and this proof can only use finitely many premises from $\mathcal{T}$. So that subset of $\mathcal{T}$ can't be satisfiable (by Soundness).

It turns out that this theorem, like the Löwenheim-Skolem Theorem, shows some important expressive limitations of FOL. In particular, we can show that it is impossible to come up with axioms in the first-order language of arithmetic that characterize the structure of the natural numbers.

nonstandard models

Theorem (Nonstandard models of arithmetic) Let L be the language of Peano arithmetic. There is a first-order structure $\mathfrak{M}$ such that

1. $\mathfrak{M}$ contains all the natural numbers in its domain,

2. $\mathfrak{M}$ also contains elements greater than all the natural numbers, but

3. $\mathfrak{M}$ makes true exactly the same sentences of L as are true about the natural numbers.

Proof: (Sketch) The proof of this result is fairly easy using the Compactness Theorem. The language of Peano arithmetic, as we defined it in Chapter 16, did not contain a symbol for greater than, but we can define $x > y$ by the wff $\exists z\,(z \neq 0 \wedge x = y + z)$. To say that an element n of $\mathfrak{M}$ is greater than all the natural numbers is to say that n satisfies all the wffs:

$$x \quad > \quad 0$$
$$x \quad > \quad 1$$
$$x \quad > \quad 1+1$$
$$x \quad > \quad (1+1)+1$$
$$\vdots$$

Let $\mathcal{T}$ consist of all sentences of L that are true of the natural numbers. Let n be a new constant symbol and let $\mathcal{S}$ be the set consisting of the following sentences:

$$n \quad > \quad 0$$
$$n \quad > \quad 1$$
$$n \quad > \quad 1+1$$
$$n \quad > \quad (1+1)+1$$
$$\vdots$$

Let $\mathcal{T}' = \mathcal{T} \cup \mathcal{S}$. On the intended interpretation of L, the theory $\mathcal{T}'$ is inconsistent. There is no natural number that is greater than all of the numbers 0, 1, 2, ... But when it comes to first-order consequence, $\mathcal{T}'$ is perfectly consistent, as we can see by applying the Compactness Theorem.

To apply compactness, we need to see that any finite subset $\mathcal{T}_0$ of $\mathcal{T}'$ is true in some first-order structure. Such a theory will contain various sentences from $\mathcal{T}$, all true of the natural numbers, plus a finite number of sentences of the form n > k where we use k as shorthand for

$$\underbrace{(((1+1)+1)+\cdots+1)}_{k}$$

We can make all these sentences true in the natural numbers by interpreting the constant symbol n as a name for some number m that is bigger than the largest k for which n > k is in $\mathcal{T}_0$. Hence, the Compactness Theorem tells us that the whole set $\mathcal{T}'$ is true in some first-order structure $\mathfrak{M}$.

The axioms of $\mathcal{T}$ assure us that $\mathfrak{M}$ contains a copy of the natural numbers (and indeed the actual natural numbers, if we replace the interpretation of k by the number k itself). But structure $\mathfrak{M}$ has a "number" that is greater than all of these.

We shouldn't get too upset or puzzled by this admittedly fascinating result. What it shows is that there is no way to uniquely characterize arithmetic's intended domain of discourse using just axioms stated in the first-order language of arithmetic. With first-order axioms, we can't rule out the existence of "natural numbers" (that is, members of the domain) that are infinitely far from zero. The distinction between being finitely far from zero, which holds of all the genuine natural numbers, and being infinitely far from zero, which holds of elements like n from the proof, is not one that we can make in the first-order language.

We can recast this result by considering what would happen if we added to our language a predicate NatNum, with the intended meaning *is a natural number*. If we did nothing to supplement the deductive system $\mathcal{F}$, then it would be impossible to add sufficient meaning postulates to capture the meaning of this predicate or to prove all the consequences expressible using the predicate. For example, if we take the set $\mathcal{T}' = \mathcal{T} \cup \mathcal{S}$ from the proof above, then intuitively the sentence $\neg$NatNum(n) is a consequence of $\mathcal{T}'$. There can, however, be no proof of this in $\mathcal{F}$.

What would happen if we added new rules to $\mathcal{F}$ involving the predicate NatNum? Could we somehow strengthen $\mathcal{F}$ in some way that would allow us to prove $\neg$NatNum(n) from $\mathcal{T}'$? The answer is that if the strengthened proof system allows only finite proofs and is sound with respect to the intended structure, then our attempt is doomed to fail. Any proof of $\neg$NatNum(n) would use only finitely many premises from $\mathcal{T}'$. This finite subset is satisfiable in the natural numbers: just assign n to a large enough number. Consequently, by the soundness of the extended proof system, $\neg$NatNum(n) must not be provable from this finite subset.

While these observations are about the natural numbers, they show something very general about any language that implicitly or explicitly expresses the concept of finiteness. For example, if the language of set theory is supplemented with a predicate with the intended meaning *is a finite set*, the Compactness Theorem can be used to show that there are first-order structures in which this predicate applies to infinite sets, no matter what meaning postulates we specify for the new predicate.

For a more down-to-earth example, we could consider the first-order language for talking about family relations. If this language has a predicate meaning *is an ancestor of*, then however we try to capture its meaning with axioms, we will fail. Implicit in the concept *ancestor* is the requirement that there are only finitely many intermediate relatives. But since there is no fixed, finite limit to how distant an ancestor can be, the Compactness Theorem guarantees that there will be structures allowing infinitely distant ancestors.

These examples are explored in Exercises 19.30 and 19.31.

Exercises

19.24 Consider the following claims:

Smaller than is irreflexive.
Smaller than is asymmetric.
Smaller than is transitive.
Larger than is the inverse of *smaller than*.

1. Express these claims in a Tarski's World sentence file.
2. Verify that (the formal versions of) these claims are all verified as analytical truths by the **Ana Con** routine of Fitch.
3. Give an informal argument to the effect that these form a complete set of meaning postulates for the sublanguage of the blocks language that contains only the predicates Cube, Smaller, and Larger.

Submit your Tarski's World file and turn in your informal argument to your instructor. There is no need to submit a Fitch file.

The next three exercises refer to the following list of sentences. In each exercise, give an informal argument justifying your answer.

1. $\forall x\, \forall y\, \forall z\, [(\mathsf{Larger}(x,y) \land \mathsf{Larger}(y,z)) \rightarrow \mathsf{Larger}(x,z)]$
2. $\forall x\, \forall y\, [\mathsf{Larger}(x,y) \rightarrow \neg\mathsf{Larger}(y,x)]$
3. $\forall x\, \neg\mathsf{Larger}(x,x)$
4. $\forall x\, \forall y\, [\mathsf{Larger}(x,y) \lor \mathsf{Larger}(y,x) \lor x=y]$
5. $\forall y\, \exists^{\leq 12} x\, \mathsf{Larger}(x,y)$
6. $\forall y\, \exists x\, \mathsf{Larger}(x,y)$

19.25 How large is the largest first-order structure making 1–5 true?

19.26 Show that any structure making 1–4 and 6 true is infinite.

19.27 Is there an infinite structure making 1–3 and 5 true?

19.28 Let $\mathcal{T}$ be a set of first-order sentences. Suppose that for any natural number n, there is a structure whose domain is larger than n that satisfies $\mathcal{T}$. Use the Compactness Theorem to show that there is a structure with an infinite domain that satisfies $\mathcal{T}$. [Hint: Consider the sentences that say *there are at least n things*, for each n.]

19.29 Let L' be the language of Peano arithmetic augmented with the predicate NatNum, with the intended interpretation *is a natural number*. Let $\mathcal{T}$ be the set of sentences in this language that are true of the natural numbers. Let $\mathcal{S}$ be as in the proof of the non-standard model theorem. Show that $\neg\mathsf{NatNum}(n)$ is not a first-order consequence of $\mathcal{T} \cup \mathcal{S}$.

19.30 Suppose we add the monadic predicate Finite to the first-order language of set theory, where
✎⋆⋆   this is meant to hold of all and only finite sets. Suppose that $\mathcal{T}$ consists of the axioms of
zFc plus new axioms involving this predicate, insisting only that the axioms are true in the
intended universe of sets. Use the Compactness Theorem to show that there is a first-order
structure satisfying $\mathcal{T}$ and containing an element c which satisfies Finite(x) but has infinitely
many members. [Hint: Add to the language a constant symbol c and infinitely many constants
$b_1, b_2, \ldots$. Form a theory $\mathcal{S}$ that says that the b's are all different and all members of c. Show
that $\mathcal{T} \cup \mathcal{S}$ is satisfiable.]

19.31 Use the Compactness Theorem to show that the first-order language with the binary predicates
✎⋆⋆   Par(x, y) and Anc(x, y), meaning *parent of* and *ancestor of*, respectively, is not axiomatizable.
That is, there is no set of meaning postulates, finite or infinite, which characterize those first-
order structures which represent logically possible circumstances. [Hint: The crucial point is
that a is an ancestor of b if and only if there is some finite chain linking a to b by the *parent
of* relation, but it is logically possible for that chain to be arbitrarily long.]

SECTION 19.8

The Gödel Incompleteness Theorem

The theorem showing the existence of nonstandard models of arithmetic shows
a kind of incompleteness of FOL. There is, however, a far deeper form of incompleteness that was discovered by Kurt Gödel a few years after he proved
the Completeness Theorem. This is the famous result known as Gödel's Incompleteness Theorem.

Students are sometimes puzzled by the fact that Gödel first proved something called the Completeness Theorem, but then turned around and proved
the Incompleteness Theorem. Couldn't he make up his mind? Actually, though,
the senses of "completeness" involved in these two theorems are quite different.
Recall that the Completeness Theorem tells us that our formal rules of proof
adequately capture first-order consequence. The Incompleteness Theorem, by
contrast, involves the notion of formal completeness introduced earlier. Remember that a theory $\mathcal{T}$ is said to be *formally complete* if for any sentence S of
its language, either S or ¬S is provable from $\mathcal{T}$. (We now know, by soundness
and completeness, that this is equivalent to saying that either S or ¬S is a
first-order consequence of $\mathcal{T}$.)

In the early part of the twentieth century, logicians were analyzing mathematics by looking at axiomatic theories like Peano arithmetic and formal proof
systems like $\mathcal{F}$. The aim was to come up with a formally complete axiomatization of arithmetic, one that allowed us to prove all and only the sentences

*completeness vs.
incompleteness*

that were true of the natural numbers. This was part of an ambitious project that came to be known as Hilbert's Program, after its main proponent, David Hilbert. By the early 1930s a great deal of progress had been made in Hilbert's Program. All the known theorems about arithmetic had been shown to follow from relatively simple axiomatizations like Peano arithmetic. Furthermore, the logician Mojżesz Pressburger had shown that any true sentence of the language not mentioning multiplication could be proven from the relevant Peano axioms.

Hilbert's Program

Gödel's Incompleteness Theorem showed the positive progress was misleading, and that in fact the goal of Hilbert's Program could never be accomplished. A special case of Gödel's theorem can be stated as follows:

Theorem (Gödel's Incompleteness Theorem for PA) Peano Arithmetic is not formally complete.

Gödel's Incompleteness Theorem

The proof of this theorem, which we will describe below, shows that the result applies far more broadly than just to Peano's axiomatization, or just to the particular formal system $\mathcal{F}$. In fact, it shows that no reasonable extension of either of these will give you a formally complete theory of arithmetic, in a sense of "reasonable" that can be made precise.

We'll try to give you a general idea how the proof goes. A key insight is that any system of symbols can be represented in a coding scheme like Morse code, where a sequence of dots and dashes, or equivalently, 0's and 1's, is used to represent any individual symbol of the system. With a carefully designed coding system, any string of symbols can be represented by a string of 0's and 1's. But we can think of such a sequence as denoting a number in binary notation. Hence, we can use natural numbers to code strings of our basic symbols. The first thing Gödel established was that all of the important syntactic notions of first-order logic can be represented in the language of Peano arithmetic. For example, the following predicates are representable:

idea of proof

coding system

representability

n *is the code of a wff,*

n *is the code of a sentence,*

n *is the code of an axiom of Peano arithmetic,*

n *and* m *are codes of sentences, the second of which follows from the first by an application of* $\wedge$ **Elim,**

n *is the code of a proof in* $\mathcal{F}$,

n *is the code of a proof of the sentence whose code is* m.

When we say these predicates are representable in Peano arithmetic, we mean something fairly strong: that the axioms and rules of proof allow us to prove all and only the true instances of these predicates. So if p is a proof of S and n and m are their codes, then the formal version of the last sentence on our list would actually be a first-order consequence of the Peano axioms.

A lot of careful work has to be done to show that these notions are representable in Peano arithmetic, work that is very similar to what you have to do to implement a system like $\mathcal{F}$ on a computer. (Perhaps Gödel was the world's first real hacker.) But it is possible, and fairly routine once you get the hang of it.

Diagonal Lemma

Gödel's second key insight was that it is possible to get sentences that express facts about themselves, relative to the coding scheme. This is known as the *Diagonal Lemma*. This lemma states that for any wff P(x) with a single free variable, it is possible to find a number n that codes the sentence P(n) asserting that n satisfies P(x). In other words, P(n) can be thought of as asserting

> *This sentence has the property expressed by* P.

Depending on what property P expresses, some of these will be true and some false. For example, the formal versions of

> *This sentence is a well-formed formula*

and

> *This sentence has no free variables*

are true, while the formal versions of

> *This sentence is a proof*

and

> *This sentence is an axiom of Peano arithmetic*

are false.

Now consider the formal version of the following sentence, whose existence the Diagonal Lemma guarantees:

> *This sentence is not provable from the axioms of Peano arithmetic.*

the Gödel sentence G

This sentence (the one above, not this one) is called G, after Gödel. Let's show that G is true but not provable in PA.

Proof: To show that G is true, we give an indirect proof. Suppose G is not true. Then given what it claims, it must be provable in PA. But since the axioms of PA are true and $\mathcal{F}$ is sound, anything provable from PA must be true. So G is true. This contradicts our assumption, namely that G was not true. So G is indeed true.

Let us now show that G is not provable in PA. We have already shown that G is true. But then, given what it claims, G is not provable.

Gödel's Incompleteness Theorem follows immediately from this. We have found a true sentence in the language of Peano arithmetic that is not provable from the Peano axioms. What's more, the negation of this sentence cannot be provable, again because the provable consequences of the Peano axioms are all true. So Peano arithmetic is not formally complete.

Of course there is nothing particularly sacred about the Peano axioms. Having found a true but unprovable sentence G, we could always add it, or something that would allow us to prove it, as a new axiom. The problem is that these attempts to strengthen our axioms don't escape Gödel's argument, since the argument does not depend on the weakness of Peano arithmetic, but on its strength. As long as the axioms of the extended system $\mathcal{T}$ are true, and the predicate

applying theorem to other theories

> **n** *encodes an axiom of* $\mathcal{T}$

is representable in $\mathcal{T}$, Gödel's whole argument can be repeated, generating yet another true sentence that is unprovable in the extended system.

Gödel's incompleteness result is one of the most important theorems in logic, one whose consequences are still being explored today. We urge the interested reader to explore it further. Detailed proofs of Gödel's theorem can be found in many advanced textbooks in logic, including Smullyan's *Gödel's Incompleteness Theorems*, Enderton's *Mathematical Introduction to Logic*, and Boolos and Jeffrey's *Computability and Logic*.

Exercises

19.32 Gödel's Incompleteness Theorem was inspired by the famous Liar's Paradox, the sentence *This*
✎* *sentence is not true.*

1. Let us assume that this sentence makes an unambiguous claim. Show that the claim is true if and only if it is not true.
2. Conclude that the sentence must not be making an unambiguous claim.
3. One possibility for locating the ambiguity is in a shift in the domain of discourse as the argument proceeds. Discuss this suggestion.

19.33 (Undefinability of Truth) Show that the following predicate cannot be expressed in the language
✎★   of arithmetic:

> *n is the code of a true sentence.*

This is a theorem due to Alfred Tarski. [Hint: Assume it were expressible. Apply the Diagonal
Lemma to obtain a sentence which says of itself that it is not true.]

19.34 (Löb's Paradox) Consider the sentence *If this conditional is true, then logic is the most fasci-*
✎★   *nating subject in the world.* Assume that the sentence makes an unambiguous claim.
  1. Use the method of conditional proof (and modus ponens) to establish the claim.
  2. Use modus ponens to conclude that logic is the most fascinating subject in the world.

Surely a good way to end a logic course.

Summary of Rules

Conjunction Introduction
($\wedge$ Intro)

$$\begin{array}{l} P_1 \\ \Downarrow \\ P_n \\ \vdots \\ \triangleright \quad P_1 \wedge \ldots \wedge P_n \end{array}$$

Conjunction Elimination
($\wedge$ Elim)

$$\begin{array}{l} P_1 \wedge \ldots \wedge P_i \wedge \ldots \wedge P_n \\ \vdots \\ \triangleright \quad P_i \end{array}$$

Disjunction Introduction
($\vee$ Intro)

$$\begin{array}{l} P_i \\ \vdots \\ \triangleright \quad P_1 \vee \ldots \vee P_i \vee \ldots \vee P_n \end{array}$$

Disjunction Elimination
($\vee$ Elim)

$$\begin{array}{l} P_1 \vee \ldots \vee P_n \\ \vdots \\ \quad \begin{array}{|l} P_1 \\ \vdots \\ S \end{array} \\ \Downarrow \\ \quad \begin{array}{|l} P_n \\ \vdots \\ S \end{array} \\ \vdots \\ \triangleright \quad S \end{array}$$

Negation Introduction
(¬ Intro)

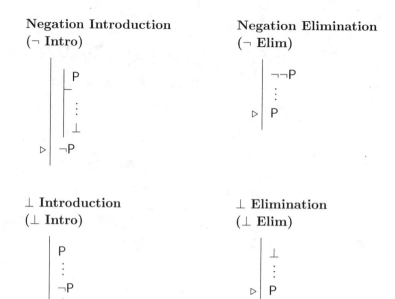

Negation Elimination
(¬ Elim)

⊥ Introduction
(⊥ Intro)

⊥ Elimination
(⊥ Elim)

Conditional Introduction
(→ Intro)

Conditional Elimination
(→ Elim)

Biconditional Introduction
($\leftrightarrow$ Intro)

Biconditional Elimination
($\leftrightarrow$ Elim)

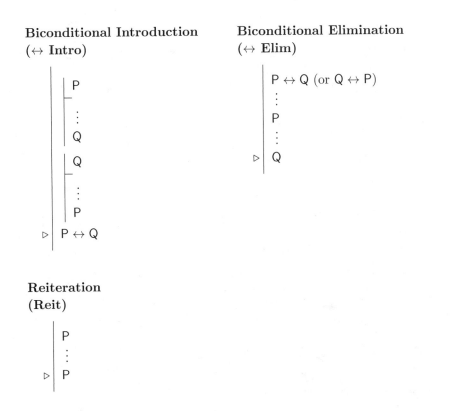

Reiteration
(Reit)

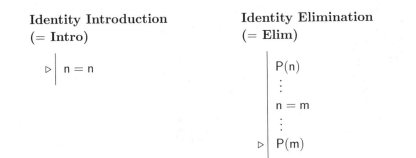

First-order rules ($\mathcal{F}$)

Identity Introduction
(= Intro)

$\triangleright$ | $n = n$

Identity Elimination
(= Elim)

$$P(n)$$
$$\vdots$$
$$n = m$$
$$\vdots$$
$$\triangleright \quad P(m)$$

General Conditional Proof
(∀ Intro)

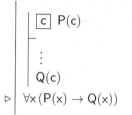

$$\triangleright \quad \forall x\,(P(x) \rightarrow Q(x))$$

Universal Elimination
(∀ Elim)

Universal Introduction
(∀ Intro)

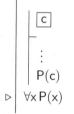

$$\triangleright \quad \forall x\,P(x)$$

where c does not occur outside the subproof where it is introduced.

Existential Introduction
(∃ Intro)

$$\triangleright \quad \exists x\,S(x)$$

Existential Elimination
(∃ Elim)

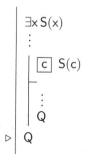

$$\triangleright \quad Q$$

where c does not occur outside the subproof where it is introduced.

Induction rules

Peano Induction:

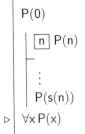

$$\begin{array}{c} P(0) \\ \boxed{n}\ P(n) \\ \vdots \\ P(s(n)) \\ \triangleright\ \forall x\, P(x) \end{array}$$

Where n does not occur outside the subproof where it is introduced.

Strong Induction:

$$\begin{array}{c} \boxed{n}\ \forall x\ (x < n \rightarrow P(x)) \\ \vdots \\ P(n) \\ \triangleright\ \forall x\, P(x) \end{array}$$

Where n does not occur outside the subproof where it is introduced.

Inference Procedures (Con Rules)

Fitch also contains three, increasingly powerful inference procedures. They are not technically inference rules.

**Tautological Consequence
(Taut Con)**

Taut Con allows you to infer any sentence that follows from the cited sentences in virtue of the meanings of the truth-functional connectives alone.

**First-order Consequence
(FO Con)**

FO Con allows you to infer any sentence that follows from the cited sentences in virtue of the meanings of the truth-functional connectives, the quantifiers and the identity predicate.

**Analytic Consequence
(Ana Con)**

In theory, **Ana Con** should allow you to infer any sentence that follows from the cited sentences in virtue of the meanings of the truth-functional connectives, the quantifiers, the identity predicate and the blocks language predicates. The Fitch implementation of **Ana Con**, however, does not take into account the meaning of Adjoins or Between due to the complexity these predicates give rise to.

Glossary

Ambiguity: A feature of natural languages that makes it possible for a single sentence to have two or more meanings. For example, *Max is happy or Claire is happy and Carl is happy*, can be used to claim that either Max is happy or both Claire and Carl are happy, or it can be used to claim that at least one of Max and Claire is happy and that Carl is happy. Ambiguity can also arise from words that have two meanings, as in the case of puns. FOL does not allow for ambiguity.

Analytical consequence: A sentence S is an analytical consequence of some premises if S follows from the premises in virtue of the meanings of the truth-functional connectives, identity, quantifiers, and predicate symbols appearing in S and the premises.

Antecedent: The antecedent of a conditional is its first component clause. In $P \rightarrow Q$, P is the antecedent and Q is the consequent.

Argument: The word "argument" is ambiguous in logic.

1. One kind of argument consists of a sequence of statements in which one (the conclusion) is supposed to follow from or be supported by the others (the premises).

2. Another use of "argument" refers to the term(s) taken by a predicate in an atomic wff. In the atomic wff $LeftOf(x, a)$, x and a are the arguments of the binary predicate LeftOf.

Arity: The arity of a predicate indicates the number of arguments (in the second sense of the word) it takes. A predicate with arity of one is called unary. A predicate with an arity of two is called binary. It's possible for a predicate to have any arity, so we can talk about 6-ary or even 113-ary predicates.

Atomic sentences: Atomic sentences are the most basic sentences of FOL, those formed by a predicate followed by the right number (see arity) of names (or complex terms, if the language contains function symbols). Atomic sentences in FOL correspond to the simplest sentences of English.

Axiom: An axiom is a proposition (or claim) that is accepted as true about some domain and used to establish other truths about that domain.

Boolean connective (Boolean operator): The logical connectives conjunction, disjunction, and negation allow us to form complex claims from simpler claims and are known as the Boolean connectives after the logician George Boole. Conjunction corresponds to the English word *and*, disjunction to *or*, and negation corresponds to the phrase *it is not the case that*. (See also Truth-functional connective.)

Bound variable: A bound variable is an instance of a variable occurring within the scope of a quantifier used with the same variable. For example, in $\forall x\, P(x, y)$ the variable x is bound, but y is "unbound" or "free."

Claim: Claims are made by people using declarative sentences. Sometimes claims are called propositions.

Completeness: "Completeness" is an overworked word in logic.

1. A formal system of deduction is said to be complete if, roughly speaking, every valid argument has a proof in the formal system. This sense is discussed in Section 8.3 and elsewhere in the text. (Compare with Soundness.)

2. A set of sentences of FOL is said to be formally complete if for every sentence of the language, either it or its negation can be proven from the set, using the rules of the given formal system. Completeness, in this sense, is discussed in Section 19.8.

3. A set of truth-functional connectives is said to be truth-functionally complete if every truth-functional connective can be defined using only connectives in the given set. Truth-functional completeness is discussed in Section 7.4.

Conclusion: The conclusion of an argument is the statement that is meant to follow from the other statements, or premises. In most formal systems, the conclusion comes after the premises, but in natural language, things are more subtle.

Conditional: The term "conditional" refers to a wide class of constructions in English including *if... then...*, *...because...*, *...unless....*, and the like, that express some kind of conditional relationship between the two parts. Only some of these constructions are truth functional and can be represented by means of the material conditional of FOL. (See Material conditional.)

Conditional proof: Conditional proof is the method of proof that allows one to prove a conditional statement P → Q by temporarily assuming P and proving Q under this additional assumption.

Conjunct: One of the component sentences in a conjunction. For example, A and B are the conjuncts of A ∧ B.

Conjunction: The Boolean connective corresponding to the English word *and*. A conjunction of sentences is true if and only if each conjunct is true.

Conjunctive normal form (CNF): A sentence is in conjunctive normal form if it is a conjunction of one or more disjunctions of one or more literals.

Connective: An operator for making new statements out of simpler statements. Typical examples are conjunction, negation, and the conditional.

Consequent: The consequent of a conditional is its second component formula. In P → Q, Q is the consequent and P is the antecedent.

Context sensitivity: A predicate, name, or sentence is context sensitive when its interpretation depends on our perspective on the world. For example, in Tarski's World, the predicate Larger is not context sensitive since it is a determinate matter whether one block is larger than another, regardless of our perspective on the world, whereas the predicate LeftOf depends on our perspective on the blocks world. In English many words are context sensitive, including words like *I*, *here*, *now*, *friend*, *home*, and so forth.

Counterexample: A counterexample to an argument is a possible situation in which all the premises of the argument are true but the conclusion is false. Finding even a single counterexample is sufficient to show that an argument is not logically valid.

Contradiction (⊥): Something that cannot possibly be true in any set of circumstances, for example, a statement and its negation. The symbol ⊥ represents contradiction.

Corollary: A corollary is a result which follows with little effort from an earlier theorem. (See Theorem.)

Deductive system: A deductive system is a collection of rules and a specification of the ways they can be use to construct formal proofs. The system

$\mathcal{F}$ defined in the text is an example of a deductive system, though there are many others.

Determinate property: A property is determinate if for any object there is a definite fact of the matter whether or not the object has that property. In first-order logic we assume that we are working with determinate properties.

Determiner: Determiners are words such as *every, some, most,* etc., which combine with common nouns to form quantified noun phrases like *every dog, some horses,* and *most pigs.*

Disjunct: One of the component sentences in a disjunction. For example, A and B are the disjuncts of A ∨ B.

Disjunction: The basic Boolean connective corresponding to the English word *or.* A disjunction is true if at least one of the disjuncts is true. (See also Inclusive disjunction and Exclusive disjunction.)

Disjunctive normal form (DNF): A sentence is in disjunctive normal form if it is a disjunction of one or more conjunctions of one or more literals.

Domain of discourse: When we use a sentence to make a claim, we always implicitly presuppose some domain of discourse. In FOL this becomes important in understanding quantification, since there must be a set of objects under consideration when evaluating claims involving quantifiers. For example, the truth-value of the claim "Every student received a passing grade" depends on our domain of discourse. The truth-values may differ depending on whether our domain of discourse contains all the students in the world, in the university, or just in one particular class.

Domain of quantification: See Domain of discourse.

Empty set: The unique set with no elements, often denoted by $\emptyset$.

Equivalence classes: An equivalence class is the set of all things equivalent to a chosen object with respect to a particular equivalence relation. More specifically, given an equivalence relation R on a set S, we can define an equivalence class for any $x \in D$ as follows:

$$\{y \in D \mid \langle x, y \rangle \in R\}$$

Equivalence relation: An equivalence relation is a binary relation that is reflexive, symmetric, and transitive.

Exclusive disjunction: This is the use of *or* in English that means exactly one of the two disjuncts is true, but not both. For example, when a waiter says "You may have soup or you may have salad," the disjunction is usually meant exclusively. Exclusive disjunctions can be expressed in FOL, but the basic disjunction of FOL is inclusive, not exclusive.

Existential quantifier (∃): In FOL, the existential quantifier is expressed by the symbol ∃ and is used to make claims asserting the existence of some object in the domain of discourse. In English, we express existentially quantified claims with the use of words like *something, at least one thing, a*, etc.

First-order consequence: A sentence S is a first-order consequence of some premises if S follows from the premises simply in virtue of the meanings of the truth-functional connectives, identity, and the quantifiers.

First-order structure: A first-order structure is a mathematical model of the circumstances that determine the truth values of the sentences of a given first-order language. It is analogous to a truth assignment for propositional logic but must also model the domain of quantification and the objects to which the predicates apply.

First-order validity: A sentence S is a first-order validity if S is a logical truth simply in virtue of the meanings of the truth-functional connectives, identity, and the quantifiers. This is the analog, in first-order logic, of the notion of a tautology in propositional logic.

Formal proof: See Proof.

Free variable: A free variable is an instance of a variable that is not bound. (See Bound variable.)

Generalized quantifier: Generalized quantifiers refer to quantified expressions beyond the simple uses of ∀ (everything) and ∃ (something); expressions like *Most students, Few teachers*, and *Exactly three blocks*.

Inclusive disjunction: This is the use of *or* in which the compound sentence is true as long as at least one of the disjuncts is true. It is this sense of *or* that is expressed by FOL's disjunction. Compare Exclusive disjunction.

Indirect proof: See Proof by contradiction.

Individual constant: Individual constants, or names, are those symbols of FOL that stand for objects or individuals. In FOL is it assumed that each individual constant of the language names one and only one object.

Inductive definition: Inductive definitions allow us to define certain types of sets that cannot be defined explicitly in first-order logic. Examples of inductively defined sets include the set of wffs, the set of formal proofs, and the set of natural numbers. Inductive definitions consist of a base clause specifying the basic elements of the defined set, one or more inductive clauses specifying how additional elements are generated from existing elements, and a final clause, which tells us that all the elements are either basic or in the set because of (possibly repeated) application of the inductive clauses.

Inductive proof: Inductive proofs are used to establish claims about inductively defined sets. Given such a set, to prove that some property holds of every element of that set we need a basis step, which shows that the property holds of the basic elements, and an inductive step, which shows that if the property holds of some elements, then it holds of any elements generated from them by the inductive clauses. See Inductive definition.

Infix notation: In infix notation, the predicate or function symbol appears between its two arguments. For example, $a < b$ and $a = b$ use infix notation. Compare with Prefix notation.

Informal proof: See Proof.

Intersection ($\cap$): The operation on sets a and b that returns the set $a \cap b$ whose members are those objects common to both a and b.

Lemma: A lemma is a claim that is proven, like a theorem, but whose primary importance is for proving other claims. Lemmas are of less intrinsic interest than theorems. (See Theorem.)

Literal: A literal is a sentence that is either an atomic sentence or the negation of an atomic sentence.

Logical consequence: A sentence S is a logical consequence of a set of premises if it is impossible for the premises all to be true while the conclusion S is false.

Logical equivalence: Two sentences are logically equivalent if they have the same truth values in all possible circumstances.

Logical necessity: See Logical truth.

Logical possibility: We say that a sentence or claim is logically possible if there is no logical reason it cannot be true, i.e., if there is a possible circumstance in which it is true.

Logical truth: A logical truth is a sentence that is a logical consequence of any set of premises. That is, no matter what the premises may be, it is impossible for the conclusion to be false. This is also called a logical necessity

Logical validity: An argument is logically valid if the conclusion is a logical consequence of the premises.

Material conditional: A truth-functional version of the conditional *if... then...*. The material conditional $P \rightarrow Q$ is false if P is true and Q is false, but otherwise is true. (See Conditional.)

Modus ponens: The Latin name for the rule that allows us to infer Q from P and $P \rightarrow Q$. Also known as $\rightarrow$ Elimination.

Names: See Individual constants.

Necessary condition: A necessary condition for a statement S is a condition that must hold in order for S to obtain. For example, if you must pass the final to pass the course, then your passing the final is a necessary condition for your passing the course. Compare with Sufficient condition.

Negation normal form (NNF): A sentence of FOL is in negation normal form if all occurrences of negation apply directly to atomic sentences. For example, $(\neg A \wedge \neg B)$ is in NNF whereas $\neg(A \vee B)$ is not in NNF.

Numerical quantifier: Numerical quantifiers are those quantifiers used to express numerical claims, for example, *at least two*, *exactly one*, *no more than five*, etc.

Predicate: Predicates are used to express properties of objects or relations between objects. Larger and Cube are examples of predicates in the blocks language.

Prefix notation: In prefix notation, the predicate or relation symbol precedes the terms denoting objects in the relation. Larger(a, b) is in prefix notation. Compare with Infix notation.

Premise: A premise of an argument is one of the statements meant to support (lead us to accept) the conclusion of the argument.

Prenex normal form: A wff of FOL is in prenex normal form if it contains no quantifiers, or all the quantifiers are "out in front."

Proof: A proof is a step-by-step demonstration that one statement (the conclusion) follows logically from some others (the premises). A formal proof is a proof given in a formal system of deduction; an informal proof is generally given in English, without the benefit of a formal system.

Proof by cases: A proof by cases consists in proving some statement S from a disjunction by proving S from each disjunct.

Proof by contradiction: To prove ¬S by contradiction, we assume S and prove a contradiction. In other words, we assume the negation of what we wish to prove and show that this assumption leads to a contradiction.

Proof by induction: See Inductive proof.

Proof of non-consequence: In a proof of non-consequence, we show that an argument is invalid by finding a counterexample. That is, to show that a sentence S is not a consequence of some given premises, we have to show that it is possible for the premises to be true in some circumstance where S is false.

Proposition: Something that is either true or false. Also called a claim.

Quantifier: In English, a quantified expression is a noun phrase using a determiner such as *every*, *some*, *three*, etc. Quantifiers are the elements of FOL that allow us to express quantified expressions like *every cube*. There are only two quantifiers in FOL, the universal quantifier (∀) and the existential quantifier (∃). From these two, we can, however, express more complex quantified expressions.

Reductio ad absurdum: See Proof by contradiction.

Satisfaction: An object named a satisfies an atomic wff S(x) if and only if S(a) is true, where S(a) is the result of replacing all free occurrences of x in S(x) with the name a. Satisfaction for wffs with more than one free variable is defined similarly, using the notion of a variable assignment.

Scope: The scope of a quantifier in a wff is that part of the wff that falls under the "influence" of the quantifier. Parentheses play an important role in determining the scope of quantifiers. For example, in

$$\forall x (P(x) \rightarrow Q(x)) \rightarrow S(x)$$

the scope of the quantifier extends only over $P(x) \rightarrow Q(x)$. If we were to add another set of parentheses, e.g.,

$$\forall x ((P(x) \rightarrow Q(x)) \rightarrow S(x))$$

the scope of the quantifier would extend over the entire sentence.

Sentence: In propositional logic, atomic sentences are formed by combining names and predicates. Compound sentences are formed by combining atomic sentences by means of the truth functional connectives. In FOL, the definition is a bit more complicated. A sentence of FOL is a wff with no free variables.

Soundness: "Sound" is used in two different senses in logic.

1. An argument is sound if it is both valid and all of its premises are true.

2. A formal system is sound if it allows one to construct only proofs of valid arguments, that is, if no invalid arguments are provable within the system. (Compare with Completeness.)

Sufficient condition: A sufficient condition for a statement S is a condition that guarantees that S will obtain. For example, if all you need to do to pass the course is pass the final, then your passing the final is a sufficient condition for your passing the course. Compare with Necessary condition.

Tautological consequence: A sentence S is a tautological consequence of some premises if S follows from the premises simply in virtue of the meanings of the truth-functional connectives. We can check for tautological consequence by means of truth tables, since S is a tautological consequence of the premises if and only if every row of their joint truth table that assigns true to each of premise also assigns true to S. All tautological consequences are logical consequences, but not all logical consequences are tautological consequences.

Tautological equivalence: Two sentences are tautologically equivalent if they are equivalent simply in virtue of the meanings of the truth-functional connectives. We can check for tautological equivalence by means of truth tables since two sentences Q and S are tautologically equivalent if and only if every row of their joint truth table assigns the same value to the main connectives of Q and S.

Tautology: A tautology is a sentence that is logically true in virtue of its truth-functional structure. This can be checked using truth tables since S is a tautology if and only if every row of the truth table for S assigns true to the main connective.

Term: Variables and individual constants are terms of a first-order language, as are the results of combining an n-ary function symbol f with n terms to form a new term.

Theorem: In formal systems, a theorem of is any statement that has been proven from some given set of axioms. Informally, the term "theorem" is usually reserved for conclusions that the author finds particularly interesting or important. (Compare Corollary and Lemma.)

Truth assignment: A function assigning true or false to each atomic sentence of a first-order language. Used to model the informal notion of a world or set of circumstances.

Truth-functional connective: A sentence connective with the property that the truth value of the newly formed sentence is determined solely by the truth value(s) of the constituent sentence(s), nothing more. Examples are the Boolean connectives ($\neg$, $\wedge$, $\vee$) and the material conditional and biconditional ($\rightarrow$, $\leftrightarrow$).

Truth table: Truth tables show the way in which the truth value of a sentence built up using truth-functional connectives depends on the truth values of the sentence's components.

Truth value: The truth value of a statement in some circumstances is true if the statement is true in those circumstances, otherwise its truth value is false. This is an informal notion but also has rigorous counterparts in propositional logic, where circumstances are modeled by truth assignments, and in first-order logic where circumstances are modeled by first-order structures.

Universal quantifier ($\forall$): The universal quantifier is used to express universal claims. Its corresponds, roughly, to English expressions such as *everything, all things, each thing,* etc. (See also Quantifiers.)

Union ($\cup$): The operation on sets a and b that returns the set $a \cup b$ whose members are those objects in either a or b or both.

Validity: "Validity" is used in two ways in logic:

1. Validity as a property of arguments: An argument is valid if the conclusion must be true in any circumstance in which the premises are true. (See also Logical validity and Logical consequence.)

2. Validity as a property of sentences: A first-order sentence is said to be valid if it is logically true simply in virtue of the meanings of its connectives, quantifiers, and identity. (See First-order validity.)

Variable: Variables are expressions of FOL that function somewhat like pronouns in English. They are like individual constants in that they may be the arguments of predicates, but unlike constants, they can be bound by quantifiers. Generally letters from the end of the alphabet, x, y, z, etc., are used for variables.

Variable assignment: A function assigning objects to some or all of the variables of a first-order language. This notion is used in defining truth of sentences in a first-order structure.

Well-formed formula (wff): Wffs are the "grammatical" expressions of FOL. They are defined inductively. First, an atomic wff is any n-ary predicate followed by n terms. Complex wffs are constructed using connectives and quantifiers. The rules for constructing complex wffs are found on page 233. Wffs may have free variables. Sentences of FOL are wffs with no free variables.

File Index

Exercise Index

General Index

and quantification, 259–266

    of FOL, 264

tense, 407

terms, **32**, 231

    complex, 32, 34

    of first-order arithmetic, 39

ternary connective, 197

theorem, 48, 195

theory

    formally complete, 488

    formally consistent, 487

transitivity, 432

    of <, 52

    of ↔, 204, 213

    of identity, 51

translation, 14, 28, 84

    and meaning, 84

    and mixed quantifiers, 298–300, 317

    and paraphrase, 309

    extra exercises, 324–327

    of

      a, 233

      all, 232

      an, 233

      and, 71

      any, 232, 246

      at least n, 375

      at least one, 233

      at most n, 375

      both, 388

      but, 71

      each, 232, 246

      every, 232, 241, 246

      everything, 230

      exactly n, 375

      few, 395

      if, 181

      if and only if, 184

      iff, 184

      just in case, 184

      many, 395

      moreover, 71

      most, 395

      neither, 388

      neither. . . nor, 75

      no, 241, 246

      non-, 68

      not, 68

      only if, 182

      or, 74

      provided, 181

      some, 233, 241

      something, 230

      the, 388

      un-, 68

      unless, 182

    of complex noun phrases, 245–247

    of conditionals, 181

    step-by-step method, 307

    using function symbols, 317

truth, 516

    assignment, 484

    conditions, **84**, 190

    in a structure, **520**, 522

    in all worlds, 372

    logical, 93, 94, 103, 183, 184, 269

    non-logical, 372

    undefinability of, 572

    value, 24, 67

truth table, 67

    disadvantages of, 128

    for ∧, 72

    for ↔, 184

    for ¬, 68

    for →, 180

    for ∨, 75

    joint, **106**, 110